Radiolabelled Molecules for Brain Imaging with PET and SPECT

Radiolabelled Molecules for Brain Imaging with PET and SPECT

Editor

Peter Brust

MDPI • Basel • Beijing • Wuhan • Barcelona • Belgrade • Manchester • Tokyo • Cluj • Tianjin

Editor
Peter Brust
Department of Neuroradiopharmaceuticals,
Institute of Radiopharmaceutical Cancer Research, Helmholtz-Zentrum Dresden-Rossendorf
Germany

Editorial Office
MDPI
St. Alban-Anlage 66
4052 Basel, Switzerland

This is a reprint of articles from the Special Issue published online in the open access journal *Molecules* (ISSN 1420-3049) (available at: https://www.mdpi.com/journal/molecules/special_issues/PET_SPECT).

For citation purposes, cite each article independently as indicated on the article page online and as indicated below:

LastName, A.A.; LastName, B.B.; LastName, C.C. Article Title. *Journal Name* **Year**, *Article Number*, Page Range.

ISBN 978-3-03936-720-7 (Hbk)
ISBN 978-3-03936-721-4 (PDF)

© 2020 by the authors. Articles in this book are Open Access and distributed under the Creative Commons Attribution (CC BY) license, which allows users to download, copy and build upon published articles, as long as the author and publisher are properly credited, which ensures maximum dissemination and a wider impact of our publications.

The book as a whole is distributed by MDPI under the terms and conditions of the Creative Commons license CC BY-NC-ND.

Contents

About the Editor vii

Preface to "Radiolabelled Molecules for Brain Imaging with PET and SPECT" ix

Bright Chukwunwike Uzuegbunam, Damiano Librizzi and Behrooz Hooshyar Yousefi
PET Radiopharmaceuticals for Alzheimer's Disease and Parkinson's Disease Diagnosis, the Current and Future Landscape
Reprinted from: *Molecules* **2020**, *25*, 977, doi:10.3390/molecules25040977 1

Liqun Kuang, Deyu Zhao, Jiacheng Xing, Zhongyu Chen, Fengguang Xiong and Xie Han
Metabolic Brain Network Analysis of FDG-PET in Alzheimer's Disease Using Kernel-Based Persistent Features
Reprinted from: *Molecules* **2019**, *24*, 2301, doi:10.3390/molecules24122301 37

Maria Elisa Serrano, Guillaume Becker, Mohamed Ali Bahri, Alain Seret, Nathalie Mestdagh, Joël Mercier, Frédéric Mievis, Fabrice Giacomelli, Christian Lemaire, Eric Salmon, André Luxen and Alain Plenevaux
Evaluating the In Vivo Specificity of [^{18}F]UCB-H for the SV2A Protein, Compared with SV2B and SV2C in Rats Using microPET
Reprinted from: *Molecules* **2019**, *24*, 1705, doi:10.3390/molecules24091705 51

Cornelius K. Donat, Henrik H. Hansen, Hanne D. Hansen, Ronnie C. Mease, Andrew G. Horti, Martin G. Pomper, Elina T. L'Estrade, Matthias M. Herth, Dan Peters, Gitte M. Knudsen and Jens D. Mikkelsen
In Vitro and In Vivo Characterization of Dibenzothiophene Derivatives [^{125}I]Iodo-ASEM and [^{18}F]ASEM as Radiotracers of Homo- and Heteromeric α7 Nicotinic Acetylcholine Receptors
Reprinted from: *Molecules* **2020**, *25*, 1425, doi:10.3390/molecules25061425 63

Ping Bai, Sha Bai, Michael S. Placzek, Xiaoxia Lu, Stephanie A. Fiedler, Brenda Ntaganda, Hsiao-Ying Wey and Changning Wang
A New Positron Emission Tomography Probe for Orexin Receptors Neuroimaging
Reprinted from: *Molecules* **2020**, *25*, 1018, doi:10.3390/molecules25051018 83

Susann Schröder, Thu Hang Lai, Magali Toussaint, Mathias Kranz, Alexandra Chovsepian, Qi Shang, Sladjana Dukić-Stefanović, Winnie Deuther-Conrad, Rodrigo Teodoro, Barbara Wenzel, Rareș-Petru Moldovan, Francisco Pan-Montojo and Peter Brust
PET Imaging of the Adenosine A$_{2A}$ Receptor in the Rotenone-Based Mouse Model of Parkinson's Disease with [^{18}F]FESCH Synthesized by a Simplified Two-Step One-Pot Radiolabeling Strategy
Reprinted from: *Molecules* **2020**, *25*, 1633, doi:10.3390/molecules25071633 93

Rien Ritawidya, Friedrich-Alexander Ludwig, Detlef Briel, Peter Brust and Matthias Scheunemann
Synthesis and In Vitro Evaluation of 8-Pyridinyl-Substituted Benzo[*e*]imidazo[2,1-*c*][1,2,4]triazines as Phosphodiesterase 2A Inhibitors
Reprinted from: *Molecules* **2019**, *24*, 2791, doi:10.3390/molecules24152791 109

Rien Ritawidya, Barbara Wenzel, Rodrigo Teodoro, Magali Toussaint, Mathias Kranz, Winnie Deuther-Conrad, Sladjana Dukic-Stefanovic, Friedrich-Alexander Ludwig, Matthias Scheunemann and Peter Brust
Radiosynthesis and Biological Investigation of a Novel Fluorine-18 Labeled Benzoimidazotriazine-Based Radioligand for the Imaging of Phosphodiesterase 2A with Positron Emission Tomography
Reprinted from: *Molecules* **2019**, *24*, 4149, doi:10.3390/molecules24224149 . . . . **131**

Paul Cumming, János Marton, Tuomas O. Lilius, Dag Erlend Olberg and Axel Rominger
A Survey of Molecular Imaging of Opioid Receptors
Reprinted from: *Molecules* **2019**, *24*, 4190, doi:10.3390/molecules24224190 . . . . **149**

Jan-Michael Werner, Philipp Lohmann, Gereon R. Fink, Karl-Josef Langen and Norbert Galldiks
Current Landscape and Emerging Fields of PET Imaging in Patients with Brain Tumors
Reprinted from: *Molecules* **2020**, *25*, 1471, doi:10.3390/molecules25061471 . . . . **185**

Lindsey R. Drake, Ansel T. Hillmer and Zhengxin Cai
Approaches to PET Imaging of Glioblastoma
Reprinted from: *Molecules* **2020**, *25*, 568, doi:10.3390/molecules25030568 . . . . **211**

About the Editor

Peter Brust Prof., Dr., is a biologist. He received his M.S. in Immunology in 1981 and his Ph.D. in Neuroscience from Leipzig University in 1986. He worked as a postdoctoral fellow at Montreal Neurological Institute and Johns Hopkins University, Baltimore, from 1990 to 1991. He joined the Research Center Rossendorf (now known as Helmholtz-Zentrum Dresden-Rossendorf, HZDR) in 1992 and headed the Department of Biochemistry. Since 2002, he has been working in Leipzig, first at the Institute of Interdisciplinary Isotope Research and, after an operational transfer in 2010, again at the HZDR, where he leads the Department of Neuroradiopharmaceuticals. His main research interest is in radiotracer development for brain imaging with positron emission tomography, including brain tumor imaging (glioblastoma, brain metastases), imaging of blood–brain barrier transport of radiopharmaceuticals, and neuroimaging of the cholinergic system, second-messenger systems, and neuromodulatory processes. He has authored around 300 peer-reviewed publications and is the owner of numerous patents.

Preface to "Radiolabelled Molecules for Brain Imaging with PET and SPECT"

Positron emission tomography (PET) and single-photon emission computed tomography (SPECT) are in vivo molecular imaging methods which are widely used in nuclear medicine for diagnosis and treatment follow-up of many major diseases. These methods use target-specific molecules as probes, which are labeled with radionuclides of short half-lives that are synthesized prior to the imaging studies. These probes are called radiopharmaceuticals. Their design and development is a rather interdisciplinary process covering many different disciplines of natural sciences and medicine. In addition to their diagnostic and therapeutic applications in the field of nuclear medicine, radiopharmaceuticals are powerful tools for in vivo pharmacology during the process of preclinical drug development to identify new drug targets, investigate pathophysiology, discover potential drug candidates, and evaluate the in vivo pharmacokinetics and pharmacodynamics of drugs.

The use of PET and SPECT for brain imaging is of special significance since the brain controls all the body's functions by processing information from the whole body and the outside world. It is the source of thoughts, intelligence, memory, speech, creativity, emotion, sensory functions, motion control and other important body functions. Protected by the skull and the blood–brain barrier, the brain is somehow a privileged organ with regard to nutrient supply, immune response, and accessibility for diagnostic and therapeutic measures. Invasive procedures are rather limited for the latter purposes. Therefore, noninvasive imaging with PET and SPECT has gained high importance for a great variety of brain diseases, including neurodegenerative diseases, motor dysfunctions, stroke, epilepsy, psychiatric diseases, and brain tumors. This Special Issue focuses on radiolabeled molecules that are used for these purposes, with special emphasis on neurodegenerative diseases and brain tumors.

Molecular imaging of neurodegeneration has become a useful noninvasive clinical tool to early detect pathophysiological changes in the brain and is regarded to be of special importance for prognostic purposes, therapeutic decision making, and therapy follow-up. Alzheimer's disease (AD) and Parkinson's disease (PD) are regarded as the most common and known neurodegenerative disorders, with a growing impact especially in countries with rapidly increased life expectancies during the last decades. Misfolded proteins such as β-amyloid, τ-protein, α-synuclein together with neuronal dystrophy characterize the main pathology of these diseases. Furthermore, multiple neurotransmitter systems are affected and involved in the cellular pathology.

The initial review written by Uzuegbunam, Librizzi, and Yousefi provides an overview of the currently available PET radiopharmaceuticals, examining the timeline and important moments that led to the development of these tracers and offering an outlook that is especially focused on the design of α-synuclein-targeting radiotracers.

This review is followed by a number of articles describing other potential targets for diagnostic and/or therapeutic approaches towards AD and PD. Neuronal dystrophy in AD is accompanied by a reduced glucose metabolism, which can be measured with PET using the radiopharmaceutical 2-deoxy-2-[^{18}F]fluoroglucose ([^{18}F]FDG). However, at the time at which significant reductions of [^{18}F]FDG accumulation in brain regions become evident, AD has usually progressed into the clinical stage. In order to prevent and/or start early treatment of AD, disease diagnosis during the preclinical stage is needed. To address this issue, a novel metabolic brain network analysis of FDG-PET using kernel-based persistent features was proposed by Kuang et al. The FDG imaging data from 140 subjects with AD, 280 subjects with mild cognitive impairment, and 280 healthy normal controls suggest that the approach has the potential of an effective preclinical AD imaging biomarker.

Synaptic loss is well established as the major structural correlate of cognitive impairment in AD. The ability to measure in vivo synaptic density could accelerate the development of disease-modifying treatments for AD. The synaptic vesicle protein 2 (SV2) is involved in synaptic vesicle tracking and regarded as a potential biomarker for the measurement of synaptic density. It consists of the three isoforms, A, B, and C, whereby SV2A, in particular, has been closely related to AD. Therefore, the selectivity of a radiopharmaceutical towards these different isoforms is an important issue. The article of Serrano et al. evaluates the in vivo specificity of [^{18}F]UCB-H, a radiotracer with nanomolar affinity for human SVA2, by comparing the SV2A protein with SV2B and SV2C using microPET in rats.

The potential of nicotinic acetylcholine receptors (nAchRs), as indicators of cholinergic neuronal functions, has previously been reported by a variety of papers, including those of our group, to be reduced in AD and PD. In this Special Issue, the dibenzothiophene derivatives [^{125}I]Iodo-ASEM and [^{18}F]ASEM, isomers of our own ligand [^{18}F]DBT10 (previously published in Molecules 20, 18387-421, 2015), were preclinically characterized in pigs as suitable radiotracers for the imaging of homo- and heteromeric α7 nAchRs with PET and SPECT.

The sleep–wake cycle in patients with AD has been associated with τ pathology and the dysregulation of the neuropeptide orexin, which exerts its action by binding to orexin receptors 1 and 2. There is evidence that the OX2R gene's rs2653349 and rs2292041 polymorphisms may be associated with AD. The FDA has approved orexin as a drug to treat insomnia. For these and other reasons, imaging of the orexin receptor status with PET and/or SPECT appears to be highly impactful. Bai et al. report a new PET radiotracer for orexin receptors neuroimaging which was preclinically used for PET investigations in mice and monkeys.

The following article deals with the adenosine A_{2A} receptor (A2AR), which is regarded as a particularly appropriate target for the non-dopaminergic treatment of PD. Schröder et al. selected the known A2AR-specific radiotracer [^{18}F]FESCH and developed a simplified two-step one-pot radiosynthesis, in order to promote its clinical applicability. The radiotracer was used to investigate the suitability of rotenone-treated mice as an animal model of PD.

In a previous issue (Molecules 21, 650, 2016), the development of ^{18}F-labelled PET ligands for the molecular imaging of the cyclic nucleotide phosphodiesterase 2A, a key enzyme in the cellular metabolism of the second messengers cAMP and cGMP, was reviewed, and PDE2 was proposed as a viable target for future drug development for AD, PD, Huntington's chorea and psychiatric diseases. Two articles by Ritawidya et al. dealing with fluorine-containing benzoimidazotriazine-based PDE2A-selective ligands for potential PET imaging are included in this Special Issue—the first describes the synthesis and in vitro evaluation of 8-pyridinyl-substituted benzo[e]imidazo[2,1-c][1,2,4]triazines as selective PDE2A inhibitors, and the second describes the radiosynthesis and biological evaluation of [^{18}F]BIT1, the best candidate among this series.

Usually, it is broad basic and clinical research on the involvement of potential imaging targets in brain diseases that strongly support the development of related PET/SPECT radiotracers. However, the review of Cummings et al. about the molecular imaging of opioid receptors (ORs) and opioid-receptor-like receptors (ORL) concludes that, in this field, it applies only to μOR, while there is scant documentation of δOR, κOR or ORL1 receptors in healthy human brain or in neurological and psychiatric disorders. Here, clinical PET research must catch up with the recent progress in radiopharmaceutical chemistry.

With the development of radiolabeled amino acids for PET and SPECT imaging, a completely different set of targets for molecular brain imaging was facilitated: brain tumors. The identification of LATI, the sodium- independent L-type amino acid transporter 1, as a light chain of the CD98 heterodimer which is strongly overexpressed in C6 glioma cells, stimulated the radiolabeling of a great variety of amino acids for the purpose of brain tumor imaging. The review by Werner et al. summarizes the clinical value of a variety of tracers that have been used in recent years, for the following indications: the delineation of tumor extent (e.g., for planning of resection or radiotherapy), the assessment of treatment response to systemic treatment options such as alkylating chemotherapy, and the differentiation of treatment-related changes (e.g., pseudoprogression or radiation necrosis) from tumor progression. It also provides an overview of promising newer tracers for the investigation of these questions. The authors conclude that currently, the best-established PET tracers in neuro-oncology are radiolabeled amino acids targeting L-system transporters.

A thematically related review by Drake et al. on brain tumor imaging by PET is focused on glioblastoma. It includes most recent experimental approaches such as sigma receptor imaging, as well as PET imaging of the programmed death ligand 1 (PD-L1), the ADP-ribose polymerase (PARP) and the mutated form of isocitrate dehydrogenase (IDH). The authors conclude that these new PET imaging targets have the potential to enhance diagnosis, staging, and treatment approaches for glioblastoma.

In summary, I regard this to be an interesting collection of papers to get an overview on radiolabeled molecules which are preclinically and clinically used for molecular brain imaging. Future perspectives are also considered, particularly for neurodegenerative diseases and brain cancer.

Peter Brust
Editor

Review

PET Radiopharmaceuticals for Alzheimer's Disease and Parkinson's Disease Diagnosis, the Current and Future Landscape

Bright Chukwunwike Uzuegbunam [1], Damiano Librizzi [2] and Behrooz Hooshyar Yousefi [1,2,*]

1. Nuclear Medicine Department, and Neuroimaging Center, Technical University of Munich, 81675 Munich, Germany; b.uzuegbunam@tum.de
2. Department of Nuclear Medicine, Philipps-University of Marburg, 35043 Marburg, Germany; librizzi@med.uni-marburg.de
* Correspondence: b.yousefi@tum.de or yousefi@med.uni-marburg.de; Tel.: +49-6421-586-5806

Academic Editor: Peter Brust
Received: 5 January 2020; Accepted: 17 February 2020; Published: 21 February 2020

Abstract: Ironically, population aging which is considered a public health success has been accompanied by a myriad of new health challenges, which include neurodegenerative disorders (NDDs), the incidence of which increases proportionally to age. Among them, Alzheimer's disease (AD) and Parkinson's disease (PD) are the most common, with the misfolding and the aggregation of proteins being common and causal in the pathogenesis of both diseases. AD is characterized by the presence of hyperphosphorylated τ protein (tau), which is the main component of neurofibrillary tangles (NFTs), and senile plaques the main component of which is β-amyloid peptide aggregates (Aβ). The neuropathological hallmark of PD is α-synuclein aggregates (α-syn), which are present as insoluble fibrils, the primary structural component of Lewy body (LB) and neurites (LN). An increasing number of non-invasive PET examinations have been used for AD, to monitor the pathological progress (hallmarks) of disease. Notwithstanding, still the need for the development of novel detection tools for other proteinopathies still remains. This review, although not exhaustively, looks at the timeline of the development of existing tracers used in the imaging of Aβ and important moments that led to the development of these tracers.

Keywords: Alzheimer's disease; Parkinson's disease; β-amyloid plaques; neurofibrillary tangles; α-synucleinopathy; positron emission tomography (PET); diagnostic imaging probes

---

1. Introduction

Of all the causes of dementia, AD stands in first place and makes up the largest part—about two-thirds—of all differential diagnoses [1–3], and it is the most common form of dementia in persons older than 65 years [4]. Others have vascular dementia, mixed dementia, PD, Lewy body dementia (LBD) or frontotemporal degeneration (FTD) [2]. Although AD and PD present markedly different clinical and pathological features, many mechanisms involved in AD and PD may be the same, such as mutation in genes, the roles of α-synuclein and tau protein aggregates in oxidative stress and mitochondrial dysfunction, dysregulation in the brain homeostasis of iron [5].

The WHO in 2012 named the prevention and control of neurocognitive disorders (mild cognitive impairment (MCI) or Alzheimer's type dementia) a global public health priority. As of 2012, it was estimated that worldwide 35.6 million people are living with dementia. By 2030 this number will double and by 2050 triple [3]. The World Alzheimer Report also in 2018 estimated that there are 50 million people in the world with dementia. This number by 2050 is likely to rise to about 152 million people [2] a projection not far from that made by the WHO way back in 2012.

In the pathogenesis of AD two proteins are implicated β-amyloid peptide aggregates (Aβ) and tau. Based on several scientific evidences, AD is histopathologically characterized by the progressive deposition of Aβ peptides into the interneuronal space [2,6,7]. The pathogenic pathways leading to AD involve several mechanisms which include the dysfunction of cholinergic neurons and the aggregation of tau, however, it has been shown that the amyloid cascade plays a significant role.

The amyloid cascade assumes that the pathogenesis of AD is as a result of a dysfunction in the synthesis and the secretion of the amyloid precursor protein (APP), usually cleaved by the proteases in the secretase family. Normally, the cleavage of APP by α-secretase within the Aβ domain releases soluble APP-α which is non-pathologic, whereas, in pathology, Aβ is generated from APP via successional cleavages by β-secretase followed by the γ-secretase complex, which cuts the γ-site of the carboxyl-terminal fragment of APP producing two major Aβ isoforms: $Aβ_{1-42}$ and $Aβ_{1-40}$, which subsequently aggregate to form β-amyloid plaques [8,9]. $Aβ_{1-42}$ comprises a major part of amyloid plaques owing to its low solubility and tendency to form aggregates with β-pleated sheet structure [9].

Neurodegeneration and neuronal dysfunction are caused by the binding of extracellular Aβ oligomers to the neuronal surface, leading to functional disruption of a number of receptors, finally culminating in dysfunction and neurodegeneration [2,10]. The accumulation of hyperphosphorylated tau protein in neurons, which normally is a microtubule-associated protein (MAP) abundantly expressed in the central nervous system, is another key player in the pathogenesis of AD. As a result of abnormal hyperphosphorylation the protein self aggregates and forms paired helical filaments (PHF), which leads to the formation of intracellular neurofibrillary tangles, which ultimately block the neuronal transport system [2,11,12].

A definitive diagnosis of AD still requires a histological examination of post-mortem brain sample [13–15]. However, in living patient's cerebrospinal fluid (CSF) biomarkers and positron emission tomography (PET), in combination with several new clinical criteria can assist in the diagnosis [16,17], and for symptomatic patients with familial early-onset AD, it is recommended to undergo clinical genetic testing together with their asymptomatic relatives [18–20].

The European Medicines Agency has presented the measurement of Aβ peptides and total tau protein levels in the CSF as a complementary usable tool in the diagnosis and monitoring of AD [21,22]. Albeit a less expensive method of evaluation, the method is invasive and carries the risks of adverse effects and discomfiture associated with a lumbar puncture [23–25].

Non-invasive modern imaging techniques allow to identify either patients who are at risk of developing AD, and also to monitor disease progression or both [26–28]. Positron emission tomography (PET) imaging especially, which is superior to other imaging techniques in terms of sensitivity, since only picomolar concentrations of the radiotracers are required allows to visualize, characterize and quantify physiological activities at molecular and cellular levels [29,30]. Hence, it may serve as an important diagnostic tool in the field of drug discovery and development, in order to monitor disease progression and the interaction of ligands with their targets.

Aβ is the most studied and first target for the neuroimaging of AD [31], hence it is no surprise that there are already selective PET radiotracers for its imaging. In 2003, Mathis et al. reported the carbon-11 labeled Pittsburgh compound B ([^{11}C]PiB), and the first successful Aβ-selective PET radioligand, which is a derivative of thioflavin (Th-T) an amyloid-binding histological fluorescent dye [32,33].

The discovery of [^{11}C]PiB led to further tracer development of other Aβ tracers. Three of which are already FDA approved and are ^{18}F-labeled [27] (a radioisotope with a relatively longer half-life of 109.7 min [34], in comparison to carbon-11 with a shorter half-life of 20.3 min, a property that logistically limits its use to centers with cyclotron on-site [35]): [^{18}F]florbetaben (Neuraceq) [36]; [^{18}F]florbetapir (Amyvid) [37]; [^{18}F]flutemetamol (Vizamyl) [38].

So far, there are other findings that the density and neocortical spread of NFTs correlate better with neurodegeneration and cognitive decline in AD patients [39–42], in spite of Aβ pathology temporarily preceding tau pathology [27]. Recent evidence further corroborates initial findings of the dominant role of tau in the pathogenesis of AD [39,43,44], backing this protein as a diagnostic as well as a therapeutic target [45].

Moreover, since apart from AD there are other NDD associated with amyloid pathology, amyloid imaging is not enough to differentiate dementia subtypes [27]. Nevertheless, NFTs are also present in other dementias, like FTD, some neurodegenerative movement disorders like corticobasal degeneration (CBD) and progressive supranuclear palsy (PSP) [45]. More recently, Vanhaute et al. reported that the loss of synaptic density in the medial temporal lobe is linked to an increased tau deposition in AD [46,47]. Hence, a radiotracer, that could quantify NFTs would help to understand the pathophysiology and clinical management not only of AD, but these other NDD. Furthermore, when done in conjunction with amyloid diagnosis, PET imaging of NFTs might provide a way to distinguish between AD dementia (when there are NFTs and Aβ present) and non-AD dementia (when NFTs and Aβ are absent). Furthermore, the application of Aβ imaging is just approved for the exclusion of AD in patients with cognitive impairment but amyloid PET-negative [48]. Also, it is being evaluated as a diagnostic tool for the definition of the preclinical stages of AD [49]. Due to the abovementioned reasons, several academic and industrial groups are currently making efforts to develop tau aggregate tracers, which are not only selective, but also with minimal or no off-target binding [50–53].

The α-synucleinopathies: PD, LBD, multiple system atrophy (MSA) have their pathological hallmark as α-syn aggregates included in Lewy body (LB), Lewy neurites (LN), and glial cytoplasmic inclusions (GCI) in MSA [54–57]. α-Synuclein is a small (140 amino acid residues) highly soluble presynaptic protein that normally exists in a native unfolded state. In PD, there is formation of highly ordered insoluble aggregates known as α-syn fibrils, which are stabilized by β-sheet protein structure [58–61].

The identification of point mutations in the SNCA gene in familial cases of PD nearly 23 years ago first linked α-syn to PD [62], and this was corroborated by the additional discovery that increased genetic copies of α-synuclein in the form of duplications and triplications of the SNCA gene are enough to cause PD; the higher gene copy, the earlier the age of disease onset and the more severe the disease [63–65]. More recently, further investigation into the genetic aspects of the disease culminated in genome-wide association studies (GWAS), and candidate gene association studies which have repeatedly validated that statistically relevant signals linked to PD are common variants near the SNCA, LRRK2, MAPT and low-frequency coding variants in GBA (glucocerebrosidase) genes [66]. Moreover, in GWAS so far, not less than 41 risk loci for PD have been identified [67,68]. Even in the sporadic forms of the disease, α-syn as a candidate risk gene has shown significant associations between variation within the SNCA gene and a higher risk of developing PD [69].

It has been known for some time now based on fairly strong evidence that the motor phase of classical PD occurs after a premotor period that could last for a considerable number of years if not decades [70]. Before the appearance of motor symptoms, at least 50% of substantia nigra (stage 3 of the Braak staging) cells have to be lost [71,72] and likely a loss of a higher percentage of dopaminergic nerve endings in the putamen [73]. Based on the findings of Braak et al., there are 6 stages in which the deposition of α-syn in LBs and LNs occurs sequentially and additively [74]. Overall, it is evident that pathophysiological changes in the central nervous system in PD involves the abnormal deposition of α-syn occurs early in PD, hence the earliest definition and most precise detection of premotor PD should be based on the imaging of aggregate α-syn, not dopaminergic alterations.

Despite the high abundance of α-syn in the nervous system, where it constitutes 1% of all cytosolic proteins [75], the amount of α-syn aggregates, however, in LBD and MSA brain is 10-fold or lower than that of Aβ in AD brain, and in advanced cases in the range of 50–200 nM in brainstem and subcortical regions, and moreover, they typically have a small size, which complicates detections [76,77].

Unlike Aβ, but similar to NFTs, LBs are intraneuronal and GCI are intraglial, hence any tracer for the detection α-syn must readily pass through the blood-brain barrier (BBB), and subsequently the cell membrane to access its target [77,78]. Unfortunately, due to the structural similarity of β-pleated sheets amongst different species of amyloid fibrils, and the colocalization of α-syn aggregates with other aggregating amyloid proteins like Aβ plaque and tau fibrils tracer, selectivity for α-syn aggregates over the others is a desired quality. This explains why non-selective ligands are more common than selective tracers [79–82].

Generally, good PET radiotracers for brain amyloid imaging should have the qualities prerequisite for successful central nervous system ligands [83,84]. A good brain penetration via passive diffusion, relatively small molecular weight (< 700 Da), moderate lipophilicity 1–3 at physiological pH (7.4), lack of P-glycoprotein substrate activity, lack of BBB permeable radioactive metabolites or intracerebral radiometabolites, etc. Most importantly, they should with high affinity selectively and reversibly bind to targets in the brain. Target selectivity an important trait depends on factors such as the relative affinities of the tracer to target (specific binding) and non-target (non-specific binding) sites, its brain distribution and the relative concentration of the binding sites. Both target and non-target binding sites should be considered when developing a brain tracer [77,81,82,85,86].

Additionally, a slow and reversible off-rates coupled (k_{off}) with relatively high on-rates (k_{on}), which is reflected by an equilibrium dissociation constant (K_d) in the range of 1 nM. A low K_d value in the nanomolar (nM) range could guarantee that the radioligand-amyloid complex remains intact long enough for a washout of non-specifically bound tracers to occur, hence allowing good signal-to-noise contrast. It is also needed especially when dealing with short-lived PET radioisotopes like ^{11}C with a half-life of 20.3 min and ^{18}F half-life 109.8 min. A standard uptake value (SUV) in the brain > 1.0 within a few min of intravenous injection is also required. Large molecules, antibodies, and nanobodies can cross the BBB, however they are unable to attain an SUV value > 1.0 a few min post-injection (p.i), and this has been a disqualifying criterion for large ligands labeled with short-lived radioisotopes [81,85,86].

2. PET Imaging Agents for the Diagnosis AD and PD

2.1. PET-Tracers for the Imaging of Aβ Plaques

2.1.1. First Generation of Aβ PET Tracers

Benzothiazole (BTA) Derivatives

The development of amyloid-specific imaging compounds is based mostly on conjugated dyes like Th-T (Figure 1) and Congo red, that are used in postmortem AD brain sections for the staining of plaques and tangles [87–90]. The synthesis of the hundreds of the derivatives of the latter by the Pittsburgh group gave rise to a series of pan-amyloid imaging agents that showed nanomolar binding affinities for Aβ, tau, α-syn, and prion aggregates. Notwithstanding, a number of these compounds ionize at physiological pH, and for this reason did not achieve high brain uptake (> 1 SUV) a few min post intravenous injection [32,91].

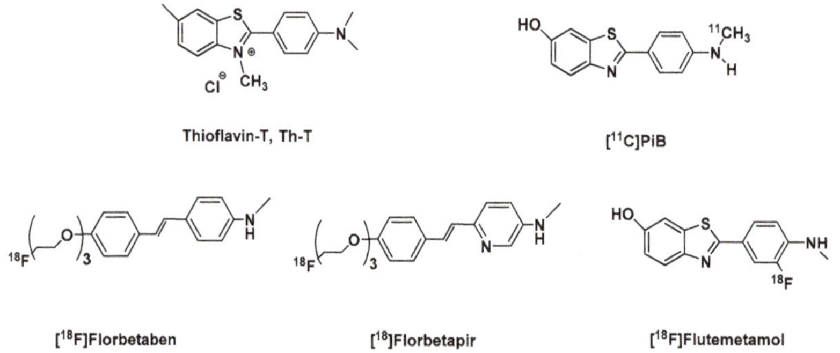

Thioflavin-T, Th-T

[^{11}C]PiB

[^{18}F]Florbetaben

[^{18}F]Florbetapir

[^{18}F]Flutemetamol

Figure 1. Structures of thioflavin-T, [^{11}C]PiB, and the FDA approved Aβ-PET tracers: [^{18}F]florbetaben, [^{18}F]florbetapir, and [^{18}F]flutemetamol.

The examination of the derivatives of Th-T derivatives followed: making the dye neutral by the removal of the methyl group attached to the benzothiazole ring via the nitrogen atom of the ring, hence the positive charge on the benzothiazole ring gave rise to compounds (known as benzothiazole anilines or BTAs) with improved lipophilicity, [^{11}C]6-Me-BTA-1 (Figure 2) being the best in the series. It was 6-fold more lipophilic, and readily crossed the BBB in brains of rodents, and showed 44-fold more affinity for synthetic Aβ fibrils (Table 1) than did Th-T [92,93].

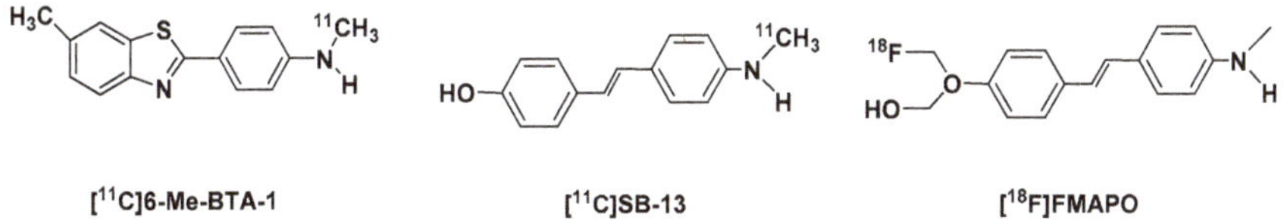

Figure 2. Structures of the predecessors of FDA approved Aβ-PET tracers: [^{11}C]6-Me-BTA-1, [^{11}C]SB-13, [^{18}F]FMAPO.

Further manipulation of the benzothiazole ring by derivatizing the C-6 position and varying the degree of methylation of the aniline nitrogen gave a series of ligands with high affinity for Aβ fibrils. Of these radiotracers, the monomethylated-aniline derivative ([^{11}C]PiB [^{11}C]6-OH-BTA-1 (Figure 1), was selected (which will be referred to as just PiB throughout the paper). It showed a combination of favorable pharmacokinetics as PiB, the highest brain clearance 5 times faster than at 30 min and a high binding affinity to Aβ plaques approximately 207-fold than Th-T [94] (Table 1), with a very low binding affinity to aggregated tau, with a ratio of tau-to-Aβ (($K_{itau}/K_{iAβ}$) greater than 100-fold [33,95–97].

Clinical study with PiB showed that AD patients retained PiB in areas of association cortex known to contain large amounts of amyloid deposits [33]. Further clinical studies to confirm if there is abnormal binding of PiB in clinically healthy individuals showed that PiB-PET not only was able to detect Aβ deposits in AD patients but also in some nondemented patients, hence suggesting that amyloid imaging might be useful in the detection AD in its preclinical stages [98]. Additionally, it was confirmed that there is a direct correlation of the retention of PiB in vivo with region-matched quantitative analyses of Aβ plaques in the same patient, upon post-mortem examination of clinically diagnosed and autopsy-confirmed AD subjects [99]. This too additionally validated PiB-PET as a method for evaluating the amyloid plaque burden in AD subjects [33].

In an experiment carried out by Serdons et al. it was discovered that more than 80% of the tracer remains intact 60 min p.i [100,101]. The radiometabolites of PiB found in animal and human blood, due to their high polarity did not easily pass through the BBB [94,100]. One of the identified radiometabolites 6-sulfato-O-PiB, and others produced in rat brain, built up over time and complicated pharmacokinetic analyses [95,102]. Fortunately, the intracerebral metabolism of PiB is limited only to rats and was not observed in mice, humans, and other nonhuman primates [95].

The success of PiB for in vivo imaging of Aβ plaque deposition led to the development of an ^{18}F analog, which would perform similarly. The development of ^{18}F-labeled radiotracers for the imaging of amyloid deposits in AD was on the basis that, as previously mentioned, carbon-11 with which PiB was labeled has a half-life 20.3 min, and this limits its use to PET centers with cyclotron on-site and with experience in ^{11}C-radiochemistry [33,36].

A variety of structural analogs were developed and evaluated both in vitro [103] and preclinically, out of which flutemetamol also known as [^{18}F]GE067 ([^{18}F]3′F-PiB) (Figure 1) was selected [104]. In vivo studies in rats and mice showed that it has similar pharmacokinetics as PiB. They both readily entered the brain, however, flutemetamol which is more lipophilic was washed out more slowly from the brain approximately 1.4 times slower (Table 1), especially from the white matter [105].

Initial human studies, in which flutemetamol and PiB were compared in AD and control subjects, the former showed similar uptake and specific binding attributes as PiB [104]. A phase-III trial demonstrated that it is safe with high specificity and sensitivity for the in vivo detection of brain Aβ density [106,107]. It was approved by the FDA in 2013 [108].

The Stilbene and Styrylpyridine Derivatives

The discovery of [^{3}H]SB-13, a stilbene derivative which showed a high binding affinity to postmortem AD brain homogenates [109], led to subsequent labeling with carbon-11 to afford [^{11}C]SB-13 (4 methylamino-4′-hydroxystilbene) (Figure 2). The tracer displayed a good brain uptake and brain clearance (Table 1) [110]. In vivo human PET-imaging it displayed properties similar to PiB in discriminating between AD and non-AD patients [111].

The similarities between PiB and SB-13 in addition to their similar biological properties are also in their chemical structures: the presence of a highly conjugated aromatic ring with an electron-donating group (N-methylamine (-NHCH$_3$) or hydroxyl (-OH)) at the end of the molecule and the relative planarity of both ligands [90].

Early attempts at the development of ^{18}F-labeled SB-13 was unsuccessful, due to the high lipophilicity and high nonspecific binding in the brain shown by [^{18}F]SB-13 derivatives with a fluoroalkyl group on either ends of their structures. In order to reduce the lipophilicity of the ligands, the stilbene scaffold was further modified by the introduction of different functional groups. Based on in vitro and in vivo biological assays a NH-CH$_3$ derivative [^{18}F]FMAPO, with a 2-fluoromethyl-1,3-propylenediol group tethered to the phenol end of molecule (Figure 2) was selected for not only exhibiting a selectivity and specific binding to Aβ plaques in AD brain homogenate binding studies but also for showing a higher brain penetration in 2 min, which was nearly three times higher than that of flutemetamol in 5 min (Table 1). Although it displayed a slower washout than the latter, at 60 min p.i. the concentration in the brain was less than 1%ID/g [103,112].

In order to circumvent the complication of in vivo metabolism, which might result due to the presence of a chiral center in the fluorine containing side chain, another series of stilbene derivatives were synthesized with polyethylene glycol (PEG) units of different lengths (n = 2–12) tethered to the 4′-OH group, with ^{18}F attached at the end of PEG side-chain. This also provided a way to maintain a small molecular weight, adjust lipophilicity and facilitate a simple ^{18}F-labeling by nucleophilic substitution. Structure-activity relationship (SAR) studies showed that high binding affinity was maintained when n < 8, and from 8 and above there was a significant reduction in binding affinity. There was a noticeable decrease in brain penetration as shown by in vivo biodistribution studies when n > 5 [113–115], perhaps partly due to increased molecular weight and total polar surface area (tPSA).

Of the four ligands which performed well in in vitro and in vivo assays, florbetaben (Figure 1) also known as AV-1, or BAY94-9172 with n = 3, was selected. Although the tracer did not have the highest affinity for Aβ in comparison with its structural analogs or the fastest washout rate (Table 1) from the brain of healthy mice [114], it, however, showed selectivity for Aβ and non-appreciable binding to NFTs, Pick bodies, LBs and GCIs [112]. Furthermore, binding to postmortem cortex of subjects with FTD or postmortem brain tissue from other NDDs like tauopathies and α-synucleinopathies was not observed [114]. With no observable effects at 100x the expected human dose in preclinical toxicity studies in a different animal species, florbetaben was deemed suitable for human studies [116]. In 2014, it was approved by the FDA [117].

In order to obtain an Aβ tracer with improved in vivo biological properties of targeting Aβ plaques, so that a high signal to noise ratio is quickly and more efficiently achieved, some critical and competing factors were taken into consideration: initial brain uptake, washout from non-afflicted brain regions, in vivo metabolism, and optimal time in the accomplishment of the highest target-to-non-target ratio. For this reason, the stilbene ring was further explored. The fluoropegylation discussed above was extended from stilbene to styrylpyridine series. This was achieved by exchanging one of the stilbene benzene rings for a pyridine ring. This led to the development of florbetapir also known as [^{18}F]AV45 [118]. It displayed 2-fold more binding affinity to Aβ in postmortem AD brain homogenates than florbetaben. Nevertheless, it showed a slightly lower initial uptake and washout rate from the brain of healthy mice than florbetaben (Table 1) [114,119].

An initial clinical trial with a tertiary amine derivative, which was similar to florbetapir but for the dimethylation of the aniline nitrogen suggested lower than expected brain uptake, probably due to a fast in vivo metabolism by N-demethylation. Of all the evaluated ligands, faster brain kinetics was exhibited more by florbetapir, and it also displayed an excellent brain uptake and washout in humans. The signal to noise ratio in the brain approaches an optimal level in 40–60 min post intravenous injection. In vitro metabolic stability assay also demonstrated that it is more stable towards microsomal degradation than florbetaben [115,118].

In AD patients, florbetapir from 30 min p.i. showed a clear separation between cortical and cerebellar activity, hence making it possible to start brain PET scan 30–50 min p.i. [120]. Significant elevations of tracer uptake in several brain regions of AD patients in comparison with controls were observed upon visual evaluation and analysis using semiquantitative methods. Results from phase III clinical trial showed a distinct correlation between the distribution of Aβ and florbetapir PET images at postmortem examination. Furthermore, no serious side effects were recorded in any of the clinical trials of the tracer [121]. It was approved by FDA in 2012 [31,122].

2.1.2. Second Generation of Aβ PET Tracers

Benzofuran, Benzoxazole and Imidazobenzothiazole Derivatives

Other notable Aβ tracers include flutafuranol, also known as [^{18}F]AZD4694 ([^{18}F]NAV4694) (Figure 3) a benzofuran derivative, developed by researchers at AstraZeneca in Sweden [123]. Its development, amongst other second generation of ^{18}F-labeled Aβ imaging agents [124] was spurred by the report that flutemetamol and florbetaben, have high level of non-specific white matter retention [116,125,126], which could be a limitation in situations when insoluble Aβ levels are low, due to a spillover effect of radioactivity to adjacent cortical regions from nonspecific binding in white matter. Hence, they may not be useful for correct mapping of Aβ plaque load in low-density regions and in prodromal phases of AD.

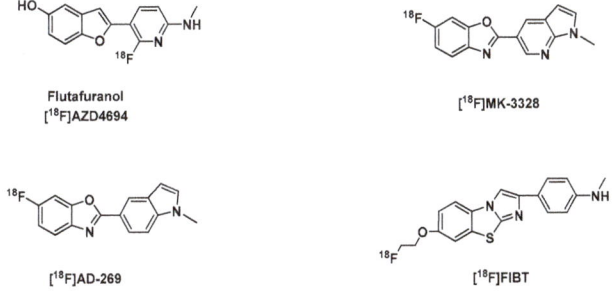

Figure 3. Structures of second generation Aβ-PET tracers: [^{18}F]AZD4694, [^{18}F]MK-3328, [^{18}F]AD-269, [^{18}F]FIBT.

Using the intravenous cassette dosing technique to compare the pharmacokinetics of flutafuranol and flutemetamol, it was seen that they were both readily taken up in the brain tissue and washed out of the brain normal rats between 2 and 30 min, but with less than 10% of concentration of flutafuranol at 2 min remaining at 30 min, a time point at which flutemetamol still had up to 28% of the initial concentration at 2 min (Table 1) [103]. With its fast binding kinetics, it could perform better than other Aβ tracers, like PiB, which display, based on time-activity curves, slower kinetics with a blunt peak of specific binding accompanied by a slower decline [95,127]. Consequently, its rapid binding kinetics makes quantification using data based on short acquisition possible. Furthermore, using the cerebellum as a reference region in approaches like reference Logan, valid estimates of Aβ binding could be easily acquired [128]. It is presently in its phase III of clinical trial for the evaluation of its efficacy and safety for the detection of cerebral Aβ in comparison with postmortem histopathology [129,130].

A benzoxazole derivative [^{18}F]MK-3328 (Figure 3), which was selected amongst four other fluoroazabenzoxazoles owing to its favorable kinetic profile, shown in rhesus monkey PET studies, a relatively low binding potential in white matter and cortical grey matter, which is approximately 2× lower than that of florbetapir, a relatively lower lipophilicity at log D 2.91, in comparison with an analog [^{18}F]AD-269 with similar properties, but more (1.21 fold) lipophilic (Table 1).

In autoradiography studies, it was observed that in an AD patient brain slice that MK-3328 showed punctuate, displaceable binding in the cortical gray matter, with no noticeable binding in the cerebellum [131]. Investigation of the tracer in healthy human volunteers and AD subjects was also being carried out at the time until the premature termination of the clinical trial after the completion of phase 1 of its clinical trial [132].

The best imidazobenzothiazole derivative [^{18}F]FIBT (Figure 3) was reported by the Yousefi et al. group and it was described as the first high-contrast Aβ-imaging agent on par with florbetaben (Figure 4). It also displayed excellent pharmacokinetics, selectivity and high binding affinity to Aβ fibrils in vitro and in vivo comparable to the gold standard PiB [133–135].

Their results also showed that FIBT has a better pharmacokinetic profile and specific binding affinity to Aβ than florbetaben in transgenic mice. This could be expected from a tracer with >300-fold selectivity for Aβ in comparison to the other amyloid protein aggregates a $K_i \gg$ 1000 nM to recombinant tau and $K_i \gg$ 1000 nM to α-syn aggregates [114,136]. Further investigations of the tracer in human subjects are however yet to be carried out [133,135].

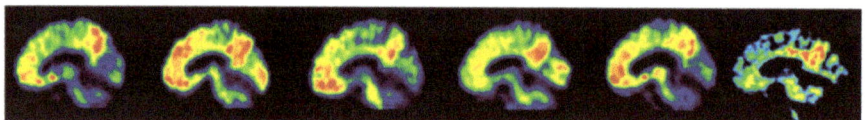

[^{11}C]PiB   [^{18}F]Florbetaben   [^{18}F]Flutemetamol   [^{18}F]Florbetapir   [^{18}F]Flutafuranol   [^{18}F]FIBT

Figure 4. Exemplary sagittal PET images of the FDA approved Aβ PET-tracers of Alzheimer's disease patients with other select featured tracers, [^{11}C]PiB, [^{18}F]Florbetaben, [^{18}F]Flutemetamol, [^{18}F]Florbetapir, [^{18}F]Flutafuranol, and [^{18}F]FIBT (reproduced with permission as agreed by Newlands Press Ltd. [135]).

Table 1. Binding affinities and Pharmacokinetics of featured Aβ PET-tracers.

Tracer	Log P	Aβ(1-40) fibrils, [nM]		Aβ(1-42) Fibrils, [nM]		Aβ plaques in Brain Homogenates, [nM]		Brain Uptake [%ID/g] (2 min p.i.)	Brain Clearance [%ID/g] (30 min p.i.)
		K_i	K_d	K_i	K_d	K_i	K_d		
Th-T [92,96]	0.57	890 580	NA	NA	NA	NA	NA	NA	NA
[^{11}C]PiB [91,94] [105,131]	1.2 2.23	4.3	4.7	NA	NA	IC$_{50}$: 2.3	1.4	0.21%ID-kg/g [1] 1.50 (5 min p.i.)	0.018%ID-kg/g [1] 0.31
[^{18}F]Florbetaben [114,119]	2.41	NA	NA	NA	NA	6.7 2.22	NA	7.77	1.59
[^{18}F]Florbetapir [119]	NA 3.2 [2]	15.3	1.6	NA	NA	2.87	3.72	7.33	1.88 (60 min p.i.)
[^{18}F]Flutemetamol [103,105]	NA	NA	NA	NA	NA	NA	NA	3505 nM 3.67 (5 min p.i.)	980 nM 0.42
[^{11}C][6-Me-BTA-1 [92,94,96]	3.36	20.2 10	NA	NA	NA	NA	NA	7.61 0.223%ID-kg/g [1] 1.15 (cortex) 1.15 (cerebellum)	2.76 0.083%ID-kg/g [1] 0.42 (cortex) 0.41 (cerebellum)
[^{11}C]SB-13 [110,112]	2.36	6.0	NA	NA	NA	1.2	NA	9.75	1.70
[^{18}F]FMAPO [112,114]	2.95	NA	NA	NA	NA	5.0	NA		
[^{18}F]Flutafuranol [^{18}F]AZD4694 [103]	2.8 [2]	18.5	2.3	NA	NA	NA	NA	1550 nM	154 nM
[^{18}F]MK-3328 [131]	2.91	NA	NA	NA	NA	IC$_{50}$: 10.5	NA	NA	NA
[^{18}F]AD-269 [131]	3.42	NA	NA	NA	NA	IC$_{50}$: 8.0	NA	NA	NA
[^{18}F]FIBT [133,134]	1.92	2.1	NA	NA	NA	NA	0.7	~7.3 [3]	~1.25 [3]

The log P values are the partition coefficient (octanol/water) or log D partition coefficient (octanol/PBS) reported in the respective publications. [1] The samples were weighed to determine the percent injected dose per gram tissue (% ID/g), and this value was multiplied by the whole-body weight (in kg) to determine body-weight normalized radioactivity concentration [(% ID-kg/g)] values. [2] ElogD$_{oct}$. [3] The values were estimated from bar-charts presented in the publication. NA: data not available.

2.1.3. The Clinical Utility and Consequences of Clinically Approved PET-Aβ Radiotracers

Since the clinical approval of the abovementioned three FDA approved PET-Aβ tracers as diagnostic tools for the detection of neuritic (Aβ) plaques in live patients, there have been studies to determine their clinical usefulness in the diagnosis AD. These studies have been subsequently and specifically well-reviewed by Kim et al. [137], Barthel et al. [138], Chiotis et al. [139].

In general, the studies have showed that the use of the Aβ-PET tracers led to a moderate to significant change in diagnosis, diagnostic confidence [140–145], and had a substantial impact on change in the treatment and management plan of AD [137,144,146]. It is likely that the new generation of Aβ-PET tracers with improved pharmacokinetics will allow for improved signal-to-noise ratio, hence will be more suitable for the quantification of disease progression and therapeutic monitoring.

2.2. PET-Tracers for the Imaging of Tau Aggregates

As mentioned earlier, the tau protein plays a key role in the pathogenesis of AD [2,6,11,12]. The predominant aggregation of certain MAPT (tau gene) isoforms, either the 4-repeat (4R tau) or the 3-repeat (3R tau) isoforms have been widely described in tauopathies. So, in addition to the already mentioned properties every CNS tracer should possess [65,67,68,73,74], tau tracers must also address 3R and 4R tau deposits. 3R and 4R tau proteins are the classifications of the 6 tau isoforms according to their tubulin-binding domains [147,148]. In a normal brain, there are equal amounts of the 3R and 4R tau proteins, as well as in AD. An imbalance in tau ratio can lead to abnormal tau accumulation and lead to NDD as in tauopathies. For instance, there is an ample amount 4R tau in PSP, CBD, and argyrophilic grain disease, in contrast, there is an abundance of the 3R tau in Pick's disease (PiD) [149]. Furthermore, tau tracers should also be able to bind to different tau folds, all of which will facilitate the detection of tau pathology in both AD and non-AD tauopathies [150,151].

The identification of lead compounds for the imaging of Aβ proteinopathies has been relatively easier since most β-sheet binding ligands have a high affinity for Aβ fibrils, with which NFTs coexist in AD and both are colocalized in the gray matter structures [42,152]. In spite of controversy surrounding the subject, PHFs predominantly found in NFTs in in vitro experiments suggest a β-sheet structured core similar to that characteristic of Aβ and α-syn fibrillar aggregates [42]. There are recent reports that there are α-syn containing aggregates present in AD [79,153]. Therefore, tau PET tracers should be selective for tau aggregates over these aggregates as well.

In AD, there is a distinct difference in the concentration of Aβ relative to tau aggregates. The concentration of Aβ is approximately 5–20 times higher than that of tau aggregates [154]. In spite of this inequality in quantity, there is however a clear-cut regional pattern of Aβ and tau deposition in the neocortex. The frontal cortex has the highest concentration of Aβ aggregates, while the temporoparietal cortices have the highest concentrations of tau aggregates. Different distributions of tau aggregates are also found in the different phenotypes. Although this has its own merits as it will facilitate differential diagnosis of tauopathies, it means that it is unlikely a single tau PET tracer could bind to the whole spectrum of tau polymorphism [42]. In this review, some select selective tau tracers already evaluated in human subjects will be examined, together with other notable tau tracers.

2.2.1. First Generation of Tau-PET Tracers

The Arylquinolines

The THK-compounds (Figure 5) were as a result of the structural modification of the lead compounds BF-158 and BF-170, arylquinoline derivatives. Even though in vitro fluorescence binding affinity assay data and neuropathological suggested that they are good tau ligands, they showed poor selectivity over amyloid plaques, and furthermore were unable to bind to tau present in non-AD tauopathies. However, autoradiographic studies in AD brain section showed an uptake BF-158 in brain regions which were NFT-rich. Biodistribution studies, analyzed using HPLC with a fluorescence detector, showed a good uptake BF-158 (11.3% ID/g at 2 min p.i.) of BF-158 in the brain of normal mice but a slow washout with only 27.4% of the concentration at 2 min washed out at 30 min, which suggested a high unspecific binding (Table 2). In contrast BF-170 performed better in the biodistribution studies with good brain uptake, as well as a faster washout at 30 min than BF-158 [155] (Table 2).

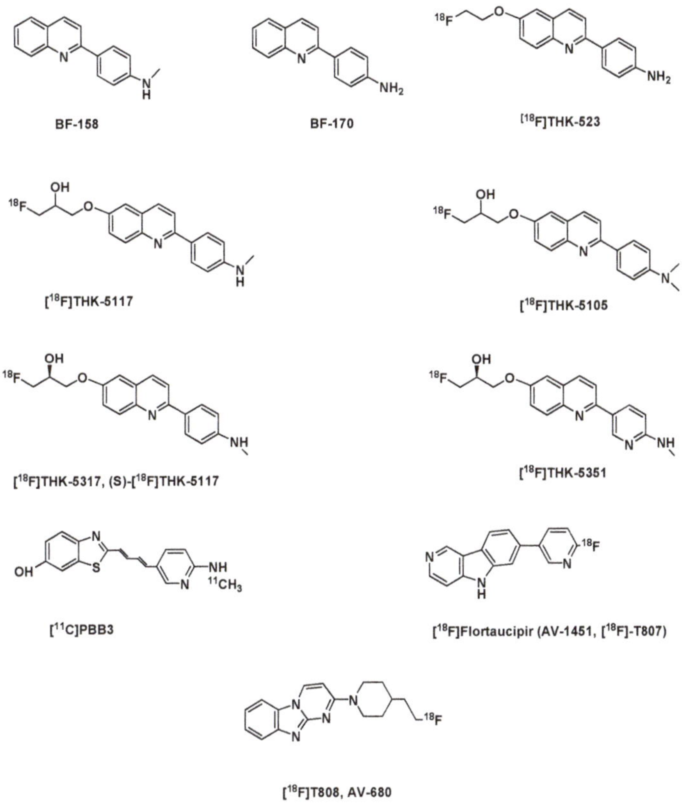

Figure 5. Structures of the first generation tau-PET tracers: BF-158, BF-170, [^{18}F]THK-523, [^{18}F]THK-5105, [^{18}F]THK-5117, [^{18}F]THK-5317(17), (S)-[^{18}F]THK-5117 ([^{18}F]THK-5351), [^{11}C]PBB3, [^{18}F]Flortaucipir (AV-1451, [^{18}F]T807), [^{18}F]T808.

Structural modification of BF-170 led to the development of [^{18}F]THK-523 (Figure 5) a [^{18}F]fluoroethoxy derivative. The introduction of an alkylether in the C6 position of the arylquinoline structure improved its affinity and selectivity for tau aggregates relative to BF-170 [156]. However, competition studies showed that it has a relatively low affinity for recombinant tau fibrils (K_i 59.3 nM) and even lower for PHF in AD brain homogenates (K_d 86.5 nM) than synthetic heparin-induced tau polymers (HITP) K_d 1.67 nM (Table 2), an evidence of the inadequacies of synthetic tau preparations, which fails to completely replicate native tau aggregates in vivo [156,157].

Notwithstanding, it performed better in vivo than it did in vitro. In comparison to healthy controls, it showed higher cortical retention in AD subjects and was distributed in the brain in accordance with reported histopathological brain distribution of PHF in AD. Unfortunately, due to its high white matter retention, a clear visualization of PET scans was not possible, and for this reason, it was not further developed [158].

Introduction of a secondary alcohol in the fluoroethoxy chain in BF-170 and the monomethylation [^{18}F]THK-5117 and dimethylation [^{18}F]THK-5105 (Figure 5) of the aniline moiety gave tracers with higher in vitro affinities (K_d) for both synthetic HITP tau fibrils and for human AD-PHF tau aggregates than [^{18}F]THK-523: they showed a 16-fold and 32-fold increase in affinity for human AD-PHF tau aggregates in comparison to their direct predecessor [^{18}F]THK-523, a higher in vitro selectivity for tau versus Aβ; a coincidence with Gallyas-Braak staining and immunoreactive tau staining in autoradiography staining of human AD brain sections but not with the distribution of PiB and an good initial brain uptake and washout in normal mice than [^{18}F]THK-523 [157] (Table 2).

An improved selectivity could be due to the secondary alcohol present, a polar terminus in the molecules. Likewise the presence of a secondary amine in [^{18}F]THK-5117 and a tertiary amine in [^{18}F]THK-5105 seemed to be behind the enhancement of tau affinity and better pharmacokinetic profile [35].

However, the N-dimethylation of [^{18}F]THK-5105 appeared to be its undoing: due to its relatively higher lipophilicity (1.3x) than [^{18}F]THK-5117 (Table 2), it showed in vivo nonspecific binding in the brainstem, thalamus and subcortical white matter, which hinders interpretation. Notwithstanding, it was still able to differentiate between AD patients and healthy control in its first-in-human PET studies. Its distribution in the mesial and lateral lobes of AD patients is in accordance with the reported NFT distribution in AD brain. Nevertheless, due to its inadequacies in comparison to other known tau-PET tracers, it was not used further [159,160].

On the other hand, first human PET studies with [^{18}F]THK-5117, demonstrated that it has faster kinetics and better signal to noise ratio than seen in [^{18}F]THK-5105, in comparison to which it is less lipophilic. Clinical studies have been conducted with the (S)-enantiomer [^{18}F]THK-5317, owing to its better signal-to-noise ratio and pharmacokinetics than the (R)-enantiomer, a trait observed in the quinoline derivatives [159,161–163]. One of the shortcomings of [^{18}F]THK-5117 and its S-enantiomer [^{18}F]THK-5317 is their significant white matter binding, which might be owing to binding to β-sheet structures of myelin basic protein [164].

In order to reduce white matter binding a feature common among ^{18}F-labeled amyloid tracers, a structural modified (S)-[^{18}F]THK-5117 was developed, the phenyl ring was replaced with a pyridinyl ring, which made [^{18}F]THK-5351 more hydrophilic [164]. It not only displayed a quicker white matter washout (lower white matter retention) and higher specific binding to AD tau-associated regions of interest than [^{18}F]THK-5317, but also its retention correlated with extra-hippocampal sub-regional atrophy rather than hippocampal subfields, proffering hence different underlying mechanisms of atrophy in early AD. Another remarkable advantage it has over other tau tracers was the lack of significant retention in the choroid plexus or venous sinus, which could probably lead to a spill-in of tracer signals into the brain [165,166].

Unfortunately, it has been reported to have high affinity to monoamine oxidase-B (MAO-B) (an isoform of monoamine oxidase whose function is to catalyze the oxidation of monoamines [167,168]) in contrast with [^{18}F]THK-5117, and also showed a greater off-target binding in the midbrain, thalamus and the basal ganglia [169,170].

The Phenylbutadienylbenzothiazoles (PBB)

Following the observation that ligands with a π-electron-conjugated backbone longer than 13Å showed affinities for pathological inclusions in a several tauopathies Maruyama et al. investigated the affinities of a series of compounds with a different structural dimension to tau aggregates and concluded that a core structure with specific distance from 13–19 Å contributes to affinity for non-AD inclusions. Additionally, since ligands with a slender and flat backbone have the ability to transverse and attach to channel-like channels in β-pleated sheets they developed a class of compounds phenyl/pyridinyl-butadienyl-benzothiazoles/benzothiazoliums (PBBs) [171]. They are structural analogs of fluorescent amyloid dye Th-T, with an all-trans butadiene bridge between the aniline and benzothiazolium moieties. Interestingly the resulting tracers were also able to detect tau inclusions in non-AD tauopathies like CBD, Pick's disease and PSP [172].

Amongst a series of analogs, [^{11}C]PBB3 (Figure 5) was selected as the best candidate with an affinity K_d for HITP 2.55 nM and nearly 50 folds selectivity for tau versus Aβ fibrils [35,172] (Table 2). Based on preclinical findings in mice, the tracer was further evaluated in humans. It showed in comparison with control accumulation in the medial temporal region of AD subjects. Its distribution in AD human brains differed from that of PiB, suggesting minimal nonspecific binding to white matter, although in both controls and AD brains it accumulated in dural venous sinuses. The use of the tracer in a CBD patient showed its retention in the basal ganglia, hinting that it could be useful for the imaging non-AD tauopathies additionally [172].

The compound although it seemed to be a likely candidate for the in vivo imaging of tau pathology, regrettably had some in vitro and in vivo instability problems. The in vitro instability was due to its photoisomerization tendencies: the quick interconversion of E/Z isomers in the presence of light. Although, this can be suppressed by shielding it from light during radio- and chemical synthesis it still is an inconvenience. In vivo, it gets quickly metabolized in mice and humans, with 2% remaining unchanged in mice at 1 min p.i. and 8% at 3 min p.i. in humans. Although this radiometabolite is polar, in mice it still made image analysis difficult [173]. It also displayed off-target binding in the basal ganglia, the choroid plexus and the longitudinal sinus [172]. Nevertheless, a new fluorinated PBB compound [^{18}F]PM-PBB3 (Figure 6) has been developed and is being clinically investigated to find out if there would be an improvement in the shortcomings of [^{11}C]PBB3 [51].

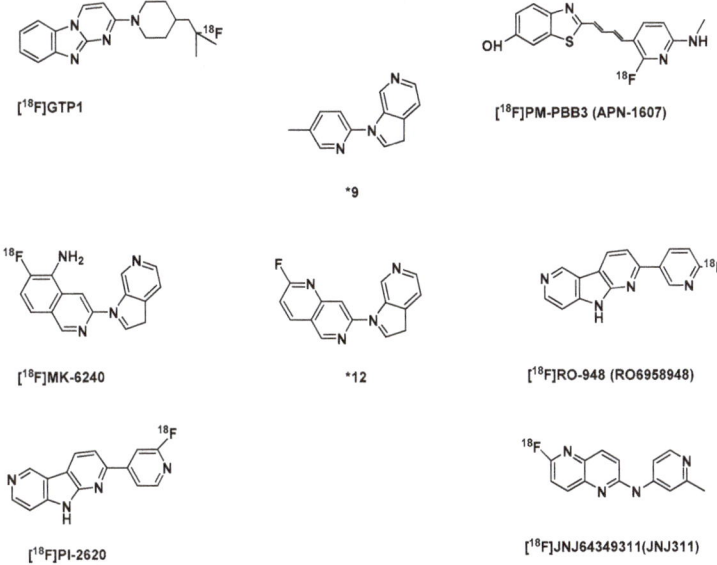

Figure 6. Structures of the selected second-generation tau-PET tracers. [^{18}F]GTP1, [^{18}F]PM-PBB3 (APN-1607), *9, [^{18}F]MK-6240, *12, [^{18}F]RO-948 (RO6958948), [^{18}F]PI-2620, [^{18}F]JNJ64349311(JNJ311).

Interestingly, in in vitro fluorescent study using postmortem DLB and MSA brain sections PBB3 was colocalized on α-syn in LBs, LNs, and GCIs. In contrast, autoradiographic labeling with [^{11}C]PBB3 at 10 nM only showed significant binding in MSA cases in regions with a high density of GCIs in the absence of tau or iron deposits. Since the maximum concentration of [^{11}C]PBB3 in human PET scans is roughly 10 nM as presented by Koga et al., it means that α-syn is only detectable by [^{11}C]PBB3 in MSA patients with a high density of GCIs [174]. A later in vivo human PET study on MSA patients by Perez-Soriano et al. was consistent with the work carried out by Koga et al., that [^{11}C]PBB3 binds to α-syn [175,176].

The Carbazole and Benzimidazole Derivatives

A screening campaign at Siemens MI Biomarker Research led to the discovery of these classes of lead series [177]. Further optimization led to the development of flortaucipir (AV-1451, [^{18}F]T807) and [^{18}F]T808 (AV-680) (Figure 5). They both have sufficient affinity for tau (AD-PHF) 14.6 nM and 22nM based on a Scatchard analysis of autoradiography staining of human PHF-AD brain sections, with $K_{d(A\beta)}/K_{d(tau)}$ 25 and 27 respectively, meaning a higher selectivity of tau aggregates over Aβ fibrils (Table 2). The binding of [^{18}F]T808 to only one type of binding site of the tau aggregates as seen from the degree of linearity in its Scatchard plot further confirmed its selectivity for tau aggregates over Aβ. Most importantly, they have minimal white matter binding and a good pharmacokinetic profile (Table 2) [178,179].

Initial PET scans of flortaucipir in controls and subjects with AD and mild cognitive impairment demonstrated an accumulation of the tracer with a distinct increasing neocortical distribution in tandem with the severity of dementia [180] according to the known mode of spread of PHF in the brain in agreement with Braak's staging [181]. In other tauopathies such as PSP and CBD, it was shown in a head-to-head comparison of [^{11}C]PBB3 and flortaucipir, that the former binds more avidly to neuronal and glial tau lesions relative to the vague binding of flortaucipir [182]. Similar to [^{11}C]PBB3, it is suspected that flortaucipir significantly binds to α-syn in the posterior putamen MSA patients. This, however, is not consistent with in vitro autoradiography results, which so far has proven otherwise [183,184].

First-in-human PET studies with [^{18}F]T808 showed similar results as flortaucipir, but with more rapid kinetics. As early as 30 min p.i. [^{18}F]T808 images stabilized, but flortaucipir SUVR values after 80 min still fluctuated. Notwithstanding, flortaucipir was selected over [^{18}F]T808 for clinical development, because of the metabolic defluorination observed in some cases, and the significant accumulation of fluorine-18 in the skull especially in late time points, that could confound PET images. This prevented further in vivo use of the tracer [185].

Off-target binding has been seen in flortaucipir PET studies in the meninges, striatum, choroid plexus and midbrain. In the analysis of autopsy brain samples, it was found out that flortaucipir also binds to vessels, iron-associated regions, substantia nigra, the leptomeningeal melanin and calcifications in the choroid plexus [186]. Another important off-target of flortaucipir is to both isoforms of the MAO enzyme [167,168,187,188]. Furthermore, there was difficulty in quantification due to the fact that it does not reach a steady-state during a typical imaging duration [189,190].

Table 2. Binding affinities and Pharmacokinetics of featured first-generation tau PET-tracers.

Tracer	Log P	Tau Affinity [nM]		Selectivity tau/Aβ	Aβ Affinity (nM)	Brain Uptake [%ID/g]	Brain Clearance [%ID/g]
		HITP	AD-PHF, K_d [nM]			2 min p.i.	30 min p.i.
BF-158 [155]	1.67	EC_{50}: 399	NA	1.60^1	K_i: > 5000	11.3	3.1
BF-170 [155]	1.85	EC_{50}: 221	NA	3.50^1	K_i: > 5000	9.1	0.25
[^{18}F]THK-523 [156,157]	2.40	K_{d1}:1.67 K_{d2}:21.74 K_i: 59.30	86.50	10^2	K_{d1} (Aβ fibrils): 20.7	2.75	1.47
[^{18}F]THK-5105 [157]	3.03	K_{d1}:1.45 K_{d2}:7.40 K_i: 7.80	2.63	25^2	K_{d1} (Aβ fibrils): 35.9	9.20	3.61
[^{18}F]THK-5117 [157]	2.32	10.50	5.19	30^2	NA	6.06	0.59
[^{18}F]THK-5351 [164]	1.5	NA	2.9	NA	NA	NA	NA
[^{11}C]PBB3 [172,173]	3.3	NA	K_d: 2.55^3	48^2	K_d: 114^3	1.92 (1 min p.i.)	0.11
[^{18}F]Flortaucipir (AV-1451,[^{18}F]T807) [178,191]	1.67	NA	14.6^3	25^2	NA	4.43 (5 min p.i.)7.5	0.620.8
[^{18}F]T808 [179,191]	NA	NA	22	27^2	NA	4.9	0.4

The log P values are the partition coefficient (octanol/water) or log D partition coefficient (octanol/PBS) reported in the respective publications. Selectivity tau vs Aβ: ^{1} $EC_{50(Aβ)}/EC_{50(tau)}$, ^{2} $K_{d(Aβ)}/K_{d(tau)}$. NA: data not available. ^{3} Autoradiographic binding to plaque- and tangle-rich regions in AD brains.

2.2.2. Second Generation of Selective Tau Tracers

Even though several goals were achieved with the first-generation selective tau tracers like improvement in affinity to both 3R and 4R tau deposits, selectivity of the tracers to tau aggregates versus Aβ plaques, and pharmacokinetics, there remains still the problem of lack of selectivity over other protein aggregates, and brain contents: subcortical white matter accumulation, in conjunction with off-target binding especially to MAO-B enzyme in the basal ganglia. Findings in which the THK-radiotracers and flortaucipir have been implicated following in vitro assessments [169,170,184,188]. Clinical validity could be limited in tauopathies where the accumulation of tau is expected in regions with a high concentration of MAO-B, like in PSP and CBS.

There has, furthermore, also been mounting evidence that the binding of certain tracers such as flortaucipir and ^{18}F-labeled THK tracers is not only limited to tau deposits but to other protein deposits, like TDP-43 (transactive response DNA binding protein 43 kDa) predominantly present in patients with semantic dementia. Flortaucipir as well as [^{11}C]PBB3 showed in vivo binding in patients expected to have α-synuclein deposits [174–176,183,184].

Consequently, efforts are being made by various pharmaceutical companies and research institutes to optimize the binding selectivity and enhance the pharmacokinetic profile of tau PET tracers. Hence, more focus will be paid in this section to improvements in the pharmacokinetic profile and specificity of the new tracers in comparison to their predecessors.

Optimized First Generation Tau Tracers

As mentioned earlier [185], [^{18}F]T808 had a propensity to metabolic defluorination, which led to the selection of flortaucipir (AV-1451, [^{18}F]T807) over it, despite its faster kinetics (Table 2). For this reason, it was deuterated to improve its in vivo stability to defluorination, which resulted in the development of [^{18}F]GTP1 (Figure 5). This modification prevented the accumulation of free ^{18}F-flouride in the skull in clinical PET study, in which it also distinctly differentiated AD subjects from healthy controls [189,192,193].

In addition to its low nanomolar affinity to tau aggregates and excellent selectivity to Aβ plaque (Table 3), it was reported to bind to non-AD tau aggregates. It also showed no off-target binding especially to MAO-B. Both preclinical and clinical in vivo kinetic studies showed that the tracer has a good pharmacokinetic profile which allows imaging some minutes earlier than flortaucipir [190,194]. Further investigations however still need to be carried out to properly compare these two tracers [193].

An introduction of fluorine-18 in the structure of the first-generation tracer [^{11}C]PBB3 gave rise to [^{18}F]PM-PBB3 (APN-1607) (Figure 6). In human subjects, it showed in less than 5 min a peak ~2.5 [^{18}F]PM-PBB3 SUV in the brain. It has less off-target signals in the basal ganglia than [^{11}C]PBB3, and a greater signal-to-background ratio. It showed no significant off-target binding in the basal ganglia and thalamus. Furthermore, it did not show radiometabolites in the brain as did its predecessor [^{11}C]PBB3 [192,195]. Phase 0 of its clinical evaluation was completed not so long ago in 2018 [196].

A structurally modified version of flortaucipir whose inadequacies were already discussed [186,187,197], [^{18}F]RO-948 (RO69558948) (Figure 6) was developed and selected from three lead compounds. Of the selected three which also displayed good brain uptake, fast brain clearance, high affinity for NFT (Table 3) and excellent selectivity against Aβ plaques in AD brain tissue, lower affinity for MAO-A and MAO-B in comparison to [^{18}F]T807 and [^{18}F]THK-5351, and based on preclinical binding study RO-948 was selected for further development. This was because in comparison to the other analogs it displayed better pharmacokinetics and metabolic properties both in mice and non-human primates. Moreover, it showed a better signal-to-background ratio than the others in AD patients. Notwithstanding, three of them gave results from their first-in-human study, which were consistent with preclinical data [198–200].

Upon the discovery that affinity for MAO-A is significantly attenuated and high affinity for aggregated tau improved in the presence of pyrrolo[2,3-b:4,5-c']dipyridine core structures in comparison to pyrido[4,3-b]indole core structure a series of fluoropyridine regioisomers were developed from which the 4-pyridine regioisomer [^{18}F]PI-2620 (Figure 6) a regioisomer of RO-948 was selected. In AD brain homogenate competition assays, it demonstrated a high affinity for tau deposits pIC50 8.4 nM (Table 3), and a superior binding to both 3R and 4R tau aggregate folds in self-competition experiments using recombinant K18 fibrils (representing 4R tau pathology) as well as human PSP and PiD brain homogenates.

Besides, it is selective over Aβ and has no off-target binding towards either MAO-A as [^{18}F]RO-948 or MAO-B as flortaucipir, and furthermore showed low off binding in brains of non-demented controls, with rapid and complete washout. It also showed selective binding to pathological tau present in Braak I, III and V human brain sections in autoradiography experiments. In autoradiography studies it also showed to tau aggregates/folds in PSP brain sections, which of course has been controversial, since many tau tracers has been reported not to be bind to tau deposits in PSP in autoradiography experiments. However, off-target binding was observed in the pars compacta portion of the substantia nigra in human brain sections, consistent with the affinity of some tau tracers like flortaucipir, [^{18}F]MK-6240 to melanin-containing cells [52,53,201].

Clinical data are needed to confirm the usefulness of [^{18}F]PI-2620 in non-AD patients. Nonetheless, it is presently being examined in several clinical trials in order to establish its pharmacokinetic profile in humans, and decide its application in in vivo PET-imaging of tau aggregates/folds both in non-AD and AD tauopathies [53].

The Azaindole-Isoquinoline and Naphthyridine Derivatives

Following an SAR study, an azaindole-isoquinoline derivative was developed. The study showed that the azaindole core (*9) (Figure 6) with a 2,4-substituted pyridine shown below was the minimum pharmacophore needed for a high binding affinity to NFTs. The insertion of fluorine in the minimal pharmacophore led to a loss in affinity by >10 fold. This was also observed when either pyridinyl rings were fluorinated. A phenomenon, which hinted at a specific electronic contribution of the basic nitrogen to NFT binding. In [^{18}F]MK-6240 there is a minimum effect of fluorine on the basicity of the heterocyclic nitrogen in the isoquinoline ring and the presence of a primary amine, an additional stronger basic center, must have improved its affinity to NFTs, K_d 0.36, in comparison to the 1,6-naphthyridine derivative (*12), with both basic centers in the ring, with affinity to NFTs, K_d 52.6 [202] (Table 3)

It exhibited favorable pharmacokinetics, with a fast brain uptake and clearance (Table 3). Uptake was higher in AD subjects and was considerably higher in brain regions expected to have NFT like in the hippocampus, but very low uptake in the cerebellar gray matter suggests a potential use of the cerebellar gray matter as a reference region. Based on reliability analysis simplified quantitative approaches could offer informed estimates of NFT load [203]. Furthermore, the spatial patterns of binding of the tracer were in accordance with the neuropathological staging of NFT, as reported from recent clinical studies [204].

Preclinical findings confirmed a lack of binding to MAO-A and MAO-B [203]. Unlike flortaucipir and [^{18}F]THK-5351 off-target binding was not seen in the choroid plexus and basal ganglia [204], but like flortaucipir and various tau PET tracers, off-target binding to neuromelanin- and melanin containing cells like the pigmented neurons in the substantia nigra, and meninges was observed [52,187,201]. To confirm initial observations, there are ongoing clinical trials on non-AD patients. The phase I of its clinical trial was completed in 2016 [205].

A 1,5-napthyridine derivative, [^{18}F]JNJ64349311(JNJ311) (Figure 6) was also reported. It showed moderate initial brain uptake but a fast brain clearance in biodistribution assay using wild-type mice (Table 3). It is quickly metabolized as was seen in NMRI mice and a rhesus monkey 30 min after intravenous injection, where only 22% and 35% respectively of the recovered activity was the intact radioligand. However, all the detected radiometabolites were more polar than the tracer, and none was found in the brain even at 60 min after injection. Furthermore, no bone uptake was detected in a rhesus monkey during a 120 min scan, in the duration of a microPET scan, which also showed a moderate initial brain uptake (SUV of 1.9 at 1 min p.i.) with a rapid wash-out.

Table 3. Binding affinities and Pharmacokinetics of featured tau PET-tracers.

Tracer	Log P	Tau affinity [nM]		Selectivity tau/Aβ	K_i, Aβ fibrils (nM)	Brain Uptake [%ID/g]	Brain Clearance [%ID/g]
		HITP	AD-PHF			2 min p.i.	30 min p.i.
[18F]GTP1 [193]	NA	NA	K_d:10.8	NA	NA	NA	NA
[18F]PM-PBB3 (APN-1607) (*9) [202]	NA	NA	NA	>1136[1]	>10000	NA	NA
[18F]MK-6240 [202]	3.32	NA	K_i: 8.8	>27777[1]	>10000	NA	NA
(*12) [202]	2.90	NA	K_i: 0.36	>190	>10000	NA	NA
			K_i: 52.6				
[18F]RO-948 (RO6958948)[53] [199]	3.22	NA	44%[2] pIC$_{50}$: 8.4[3]	NA	pIC$_{50}$: <6[4]	5.7[5]	10.9[6]
[18F]PI-2620 [53]	NA	NA	pIC$_{50}$: 8.5[7]	NA	pIC$_{50}$: <6[4] >4398[9]	5.9[5]	16.6[6]
[18F]JNJ6349311 (JNJ311) [206]	2.2	NA	K_i: 8[8]	>500	IC$_{50}$: <5[9]	1.9[10]	0.3[10]

The log P values are the partition coefficient (octanol/water) or log D partition coefficient (octanol/PBS) reported in the respective publications. Selectivity tau vs Aβ: [1] $K_{i(Aβ)}/K_{i(tau)}$. [2] % inhibition of 10 nM of [3H]T808 on fresh frozen human brain sections derived from AD cases. [3] Self-competition. [4] In competition with [3H]PiB. [5] Peak uptake (injected dose per gram brain; ID/g); [6] ratio of peak uptake divided by peak at 30 min. [7] In competition with [18F]RO-948. [8] [3H]AV680 as competitor. [9] [3H]Florbetapir as competitor. [10] Data are expressed as SUV mean. NA: data not available. *numbers given to the tracers in the respective publications.

Semi-quantitative autoradiography studies on post-mortem tissue sections of human AD brains displayed highly displaceable binding to NFT-rich regions, but it showed no specific binding to human PSP and CBD brain slices. Based on its in vitro and in vivo preclinical profiling, it was deemed a promising candidate for quantitative tau PET imaging in AD [206]. There is presently no in vivo human data for the tracers of the JNJ series.

2.3. Selective PET-Tracers for the Imaging of α-syn

Most of the PET/SPECT tracers developed and approved so far for the differential differential diagnosis of PD have been geared towards the evaluation of the function of the dopaminergic system [207]. As was mentioned earlier, 50% of substantia nigra cells (stage III of the Braak staging) and a probable loss of a higher percentage of dopaminergic nerve endings in the putamen have to be lost before the appearance of motor symptoms [71–73], in contrast, based on the findings of Braak et al., there is deposition of α-syn in LBs and LNs, which occurs sequentially and additively throughout the VI stages of disease progression [74], therefore the most accurate and earliest detection of premotor PD should be based on imaging α-syn instead of dopaminergic changes [208].

The development of α-syn PET tracers is still an unmet need and in its early stages. Regardless, efforts have been made in the past decades and are still being presently made to develop tracers with a high affinity and selectivity for α-syn over Aβ and tau aggregates.

2.3.1. The Phenothiazine Derivatives

In 2011, to discover selective α-syn tracers Yu et al. synthesized a series of phenothiazine derivatives. Three of the tricyclic compounds (Figure 7) based on in vitro Th-T competition assay to recombinant α-syn fibrils were selected: [^{11}C]SIL5, [^{125}I]SIL23, and [^{18}F]SIL26 based on the fact that they displayed an affinity (K_i) to the α-syn fibrils less than 60 nM [209] (Table 4).

In further tests, [^{125}I]SIL23 with a k_i of 57.9 nM [209] was able to bind to α-syn fibrils in postmortem PD brain homogenates, which indicated that the tracer binding affinity in PD brain samples is comparable to its affinity to recombinant α-syn fibrils. It also displayed 5-fold and 2-fold less affinity for Aβ$_{1-42}$ and tau aggregates respectively in comparison to α-syn fibrils, however this selectivity was insufficient for in vivo imaging. Moreover, a high nonspecific binding in white matter seems to limit autoradiography with the tracer initial experiments. Furthermore, its affinity for α-syn fibrils K_d 148 nM, is also not optimal for the in vivo imaging of α-syn fibrils [210].

[^{11}C]SIL5 and [^{18}F]SIL26 with binding affinities (K_i) 1.8× and 1.2× more than that of [^{125}I]SIL23 respectively, showed a low initial brain uptake in healthy Sprague-Dawley rats 5 min p.i. 0.953%ID/g and 0.758%ID/g respectively, and slow washout with [^{18}F]SIL26 performing poorer of the 2 tracers with not less than 50% of the initial brain concentration at 5 min remaining at 60 min. [^{11}C]SIL5 still had at 60 min 16.57% of its initial brain uptake at 2 min, which suggested a slow washout of the tracer or a lot of unspecific binding (Table 4).

In vivo microPET imaging in a healthy cynomolgus macaque confirmed that [^{11}C]SIL5, with a faster washout kinetics of the two was able to penetrate the BBB into the brain, and also has a homogeneous distribution and fast washout kinetics. The authors believe that both compounds require further structural optimization in order to make them a more suitable α-syn tracer [211].

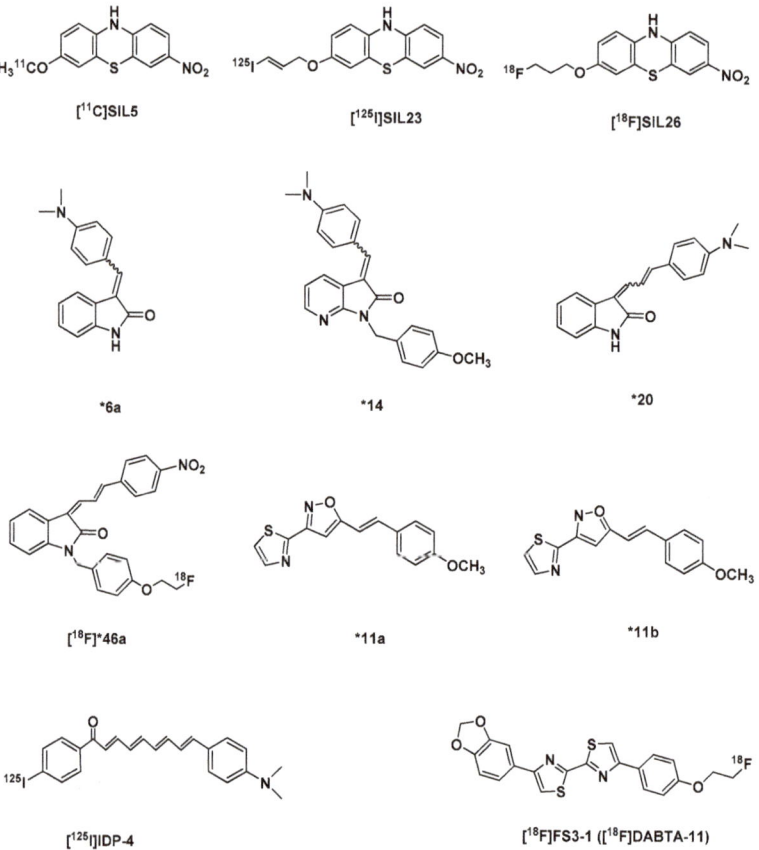

Figure 7. Structures of selective α-syn-PET tracers: [^{11}C]SIL5, [^{125}I]SIL23, [^{18}F]SIL26, *6a, *14, *20, *46a, *11a, *11b, [^{125}I]IDP-4, [^{18}F]FS3-1 ([^{18}F]DABTA-11). *numbers given to the tracers in their respective publications.

2.3.2. The Indolinone and Indolinonediene Derivatives

An SAR study by structural modifications of an indolinone derivative *6a [212] (Figure 7), by the introduction of different alkyl and arylalkyl groups substituted at the indolinone nitrogen led to the identification of an aza-analog-*14 (Figure 7), which was the most successful in this series (Table 3). With a binding affinity K_i 79 nM to recombinant α-syn fibrils less than that of the select 3 phenothiazine tracers [209,213], with 1,4- and 11-fold selectivity over recombinant Aβ and tau aggregates. Another major limitation of the series was the presence of E/Z isomers, which re-equilibrate after separation.

A homologation of the double bond in *6a by the addition of an additional double (to form diene group) bond, in order to increase affinity to α-syn fibrils over the other aggregates gave a series of compounds, which were a mixture of stereoisomers with either an E,E or Z,E configuration, which quickly re-equilibrated after chromatographic purification. *20 (Figure 7) was however selected from the series based on its improved affinity (K_i) for recombinant α-syn fibrils 1.9× more than *14, but unfortunately affinity K_i 27.6 nM for Aβ increased as well, with 1.3-fold selectivity over tau aggregates (Table 4). However, its strong fluorescent attributes allowed for a performance of fluorescent microscopy studies of postmortem AD and PD brain samples. Results showed that it labels both LB and Aβ plaques. Regardless, this showed that indolinonediene derivatives can label α-syn fibrils in LBs.

Introduction of a para nitro group into the pendant benzene ring of the diene moiety made it possible to isolate both the E,E and Z,E stereoisomers, and further explore their in vitro properties. The Z,E regioisomers were generally more active than the corresponding E,E configuration in terms of higher affinity for α-syn over Aβ and tau fibrils. The best in this series was [^{18}F]46a with the highest affinity (K_i) for α-syn fibrils 2 nM, 70-fold and 40-fold less affinity for Aβ and tau fibrils respectively (Table 4).

Unfortunately, due to its high lipophilicity log D 4.18, which prevented obtaining a reliable and reproducible results from binding assays to insoluble α-syn acquired from PD brain and the possible reduction of the nitro group to an amino group in vivo makes it an unsuitable PET probe for imaging LB and LN in PD subjects. In spite of the shortcomings of the compound it showed interesting selectivity for α-syn fibrils over Aβ and tau, and could serve as a good lead for further development of α-syn fibril tracers [213].

2.3.3. Chalcone Derivatives and Structural Cogeners

Hsieh et al. went further to investigate a series of chalcone derivatives, whose enone moiety serves as an isosteric replacement of the diene group in the indolinonediene derivatives while precluding the E,E and Z,E isomerization problem. The indole ring was further replaced with a benzothiazole ring system, based on a previous SAR study, which revealed that an electron-deficient ring like the aza-indole system has a higher affinity for α-syn fibrils relative to the indole ring system and to prevent the Michael-acceptor properties associated with the chalcone system they replaced the enone moiety with an isoxazole and a pyrazole ring system. The results of a competition in vitro binding studies with Th-T led to the identification of a compound *11a,b (37 a,b) (Figure 7), isoxazole derivatives with a modest affinity in comparison to [^{18}F]46a (36) for α-syn at K_i 18.5 nM over Aβ and tau fibrils with 5-fold and over 54-fold less affinity respectively (Table 4). Although the compounds described in their report have modest affinity to serve as a PET radiotracer for in vivo imaging studies, they could, however could be used for further SAR studies [214].

Based on previous research, it was discovered that flavonoids could inhibit not only the formation of Aβ [215–217] but also of α-syn [218–220] aggregates, an indication that they could also bind with α-syn aggregates. With this in mind, Ono et al. developed some α-syn imaging tracers based on the chalcone scaffold: they developed four prospective chalcone derivatives (IDP compounds) with varying molecular lengths made possible by conjugated double bonds. A longer molecular length and a long conjugated π system were believed to lead to increased affinity of probes to Aβ [221], and tau aggregates as was seen in the PBB compounds [172,222].

All the IDP-compounds, unfortunately, displayed almost as much affinity for Aβ as they displayed for α-syn (Table 4), with affinity for α-syn increasing proportionately to molecular length, with not really much change in the selectivity to Aβ, which means that [^{125}I]IDP-4 with a tetraene structure was the best in this regard. It had a high binding affinity to α-syn K_d of 5.4 nM and 3-fold selectivity to recombinant Aβ fibrils, which nevertheless was not as high as [^{18}F]46a, which the authors believe could be attributed to different binding assay conditions.

[^{125}I]IDP-4 (Figure 7) in vivo biodistribution in normal mice performed poorer than the other IDP-compounds, with a brain uptake of 0.45%ID/g at 2 min p.i. and a low brain clearance, 93.3% of the concentration at 2 min still remaining at 60 min (Table 4). The suboptimal pharmacokinetics could be due to the lipophilicity, high molecular weight and chemical structure of the tracer. In any case, this property makes it an unlikely tracer for in vivo imaging of α-syn aggregates. Still, it could be a useful probe for in vitro screening of compounds for affinity to α-syn and Aβ fibrils in vitro [222].

Table 4. Binding affinities and Pharmacokinetics of featured α-syn PET-tracers.

Tracer	Log D	α-syn Affinity [nM]			Aβ Fibrils Affinity [nM]		Tau Fibrils Affinity [nM]		Brain Uptake [%ID/g]	Brain Clearance [%ID/g]
		K_i		K_d	K_i	K_d	K_i	K_d	5 min p.i.	60 min p.i.
		α-syn Fibrils	Human PD Homogenate							
[11C]SIL5 [209–211]	3.79	32.1, 66.2	83.1	NA	110	NA	136	NA	0.953	0.158
[125I]SIL23 [209,210]	5.72	57.9	NA	148	NA	635	NA	230	NA	NA
[18F]SIL26 [209–211]	4.02	49.0, 15.5	33.5	NA	103	NA	125	NA	0.758	0.410
*14 [213]	4.2 [1]	79.5	NA	NA	113.3	NA	853.5	NA	NA	NA
*20 [213]	3.5 [2]	40.7	NA	NA	27.6	NA	53.7	NA	NA	NA
[18F]46a [213]	4.18	2.1	NA	8.9	142.4	NA	80.1	NA	NA	NA
*11a,b [214]	3.54 [3]	18.5	NA	NA	91.5	NA	>1000	NA	NA	NA
[125I]IDP-4 [222]	NA	NA	NA	5.4	NA	16.24	NA	NA	0.45 (2 min p.i.)	0.42

[1,2] and [3] Calculated by ChemDraw Ultra 13.0 and Professional 15.1 respectively. * numbers given to the tracers in their respective publications. NA: data not available.

2.3.4. Diarybisthiazole Compounds

More recently, highly innovative small molecules utilized as diagnostic probes, based on 4,4'-diaryl-2,2'-bithiazole, called DABTAs were developed by Yousefi et al. as sensitive, selective and specific tracers. According to pilot data these compounds suitable for the visualization and quantification of α-syn pathology by nuclear medicine imaging. With the help of these newly developed ligands, the presence, distribution and progression pattern of α-syn can be investigated in the living brain, reflecting the disease entity and stage. Preliminary results with the [^{18}F]FS3-1 ([^{18}F]DABTA-11) (Figure 7) in a rat model overexpressing human E46K mutated α-syn are promising regarding sensitivity [223,224]. In addition, in vivo imaging of the progression and spreading of α-synuclein pathology over time with the tracer greatly motivates the authors to develop tracers for aggregated α-syn in PD, DLB or MSA.

3. Conclusions

Several Aβ PET tracers have been developed and promoted discussion on their clinical values. Among these three tracers have been already approved by the FDA and EMA. Research on tau PET yielded several tau-tracers already entered into clinical investigations, however, still inherent are challenges in selective tau imaging. The lack of any consented and approved tracer for aggregated α-syn greatly motivated scientists in the field to develop tracer for aggregated α-syn in PD, DLB or MSA. Remarkable progress has been made so far in order to fulfill the unmet need for α-syn PET tracers with suitable pharmacokinetics, binding affinity and selectivity. Lately, innovative small molecules have been identified with excellent binding affinity and selectivity for α-syn fibrils relative to Aβ and tau aggregates. Utilizing high-throughput binding assays, in-silico design in the conventional tracer development may accelerate the identification of new leads for a specific α-syn PET imaging.

Funding: The APC was funded by Technical University of Munich, University Library Open Access Publishing Fund. BCU gratefully acknowledge the funding received towards his PhD from the German Academic Exchange Service (DAAD).

Conflicts of Interest: The authors declare no conflict of interest.

References

1. Alzheimer's Disease International. *The Global Impact of Dementia: An Analysis of Prevalence, Incidence, Cost and Trends*; World Alzheimer Report 2015; Alzheimer's Disease International: London, UK, 2015.
2. Alzheimer's Disease International. *World Alzheimer Report 2018—The State of the Art of Dementia Research: New Frontiers*; Alzheimer's Disease International: London, UK, 2018.
3. World Health Organisation-Alzheimer's Disease International. *Dementia: A Public Health Priority*; WHO: Geneva, Switzerland, 2012.
4. Trevisan, K.; Cristina-Pereira, R.; Silva-Amaral, D.; Aversi-Ferreira, T.A. Theories of Aging and the Prevalence of Alzheimer's Disease. *Biomed. Res. Int.* **2019**, 1–9. [CrossRef]
5. Xie, A.; Gao, J.; Xu, L.; Meng, D. Shared mechanisms of neurodegeneration in Alzheimer's disease and Parkinson's disease. *Biomed. Res. Int.* **2014**, *2014*, 648740. [CrossRef]
6. Murphy, M.P.; LeVine, H. Alzheimer's disease and the amyloid-beta peptide. *J. Alzheimers Dis.* **2010**, *19*, 311–323. [CrossRef]
7. Alzheimer's Association 2018 Alzheimer's disease facts and figures. *Alzheimer's Dement.* **2018**, *14*, 367–429. [CrossRef]
8. Zhang, Y.; Thompson, R.; Zhang, H.; Xu, H. APP processing in Alzheimer's disease. *Mol. Brain* **2011**, *4*, 3. [CrossRef] [PubMed]
9. Gu, L.; Guo, Z. Alzheimer's Aβ42 and Aβ40 peptides form interlaced amyloid fibrils. *J. Neurochem.* **2013**, *126*, 305–311. [CrossRef] [PubMed]
10. Kayed, R.; Lasagna-Reeves, C.A. Molecular mechanisms of amyloid oligomers toxicity. *J. Alzheimers Dis.* **2013**, *33* (Suppl. 1), S67–78. [CrossRef]

11. Trojanowski, J.Q.; Clark, C.M.; Schmidt, M.L.; Arnold, S.E.; Lee, V.M. Strategies for Improving the Postmortem Neuropathological Diagnosis of Alzheimer's Disease. *Neurobiol. Aging* **1997**, *18*, S75–S79. [CrossRef]
12. Dubois, B.; Hampel, H.; Feldman, H.H.; Scheltens, P.; Aisen, P.; Andrieu, S.; Bakardjian, H.; Benali, H.; Bertram, L.; Blennow, K.; et al. Preclinical Alzheimer's disease: Definition, natural history, and diagnostic criteria. *Alzheimers Dement.* **2016**, *12*, 292–323. [CrossRef]
13. Delacourte, A. Les diagnostics de la maladie d'Alzheimer. *Ann. Biol. Clin. (Paris)* **1998**, *56*, 133–142.
14. Takizawa, C.; Thompson, P.L.; van Walsem, A.; Faure, C.; Maier, W.C. Epidemiological and economic burden of Alzheimer's disease: A systematic literature review of data across Europe and the United States of America. *J. Alzheimers Dis.* **2015**, *43*, 1271–1284. [CrossRef] [PubMed]
15. Johnson, K.A.; Fox, N.C.; Sperling, R.A.; Klunk, W.E. Brain imaging in Alzheimer disease. *Cold Spring Harb. Perspect. Med.* **2012**, *2*, a006213. [CrossRef] [PubMed]
16. Budson, A.E.; Solomon, P.R. New criteria for Alzheimer disease and mild cognitive impairment: implications for the practicing clinician. *Neurologist* **2012**, *18*, 356–363. [CrossRef] [PubMed]
17. Jack, C.R.; Bennett, D.A.; Blennow, K.; Carrillo, M.C.; Dunn, B.; Haeberlein, S.B.; Holtzman, D.M.; Jagust, W.; Jessen, F.; Karlawish, J.; et al. NIA-AA Research Framework: Toward a biological definition of Alzheimer's disease. *Alzheimers Dement.* **2018**, *14*, 535–562. [CrossRef] [PubMed]
18. Bekris, L.M.; Yu, C.-E.; Bird, T.D.; Tsuang, D.W. Genetics of Alzheimer disease. *J. Geriatr. Psychiatry Neurol.* **2010**, *23*, 213–227. [CrossRef] [PubMed]
19. Freudenberg-Hua, Y.; Li, W.; Davies, P. The Role of Genetics in Advancing Precision Medicine for Alzheimer's Disease-A Narrative Review. *Front. Med. (Lausanne)* **2018**, *5*, 108. [CrossRef]
20. Williamson, J.; Goldman, J.; Marder, K.S. Genetic aspects of Alzheimer disease. *Neurologist* **2009**, *15*, 80–86. [CrossRef]
21. Isaac, M.; Vamvakas, S.; Abadie, E.; Jonsson, B.; Gispen, C.; Pani, L. Qualification opinion of novel methodologies in the predementia stage of Alzheimer's disease: Cerebro-spinal-fluid related biomarkers for drugs affecting amyloid burden—Regulatory considerations by European Medicines Agency focusing in improving benefit/risk in regulatory trials. *Eur. Neuropsychopharmacol.* **2011**, *21*, 781–788. [CrossRef]
22. Blennow, K.; Hampel, H.; Zetterberg, H. Biomarkers in amyloid-β immunotherapy trials in Alzheimer's disease. *Neuropsychopharmacology* **2014**, *39*, 189–201. [CrossRef]
23. Palmqvist, S.; Zetterberg, H.; Mattsson, N.; Johansson, P.; Minthon, L.; Blennow, K.; Olsson, M.; Hansson, O. Detailed comparison of amyloid PET and CSF biomarkers for identifying early Alzheimer disease. *Neurology* **2015**, *85*, 1240–1249. [CrossRef]
24. Hampel, H.; Bürger, K.; Teipel, S.J.; Bokde, A.L.W.; Zetterberg, H.; Blennow, K. Core candidate neurochemical and imaging biomarkers of Alzheimer's disease. *Alzheimers Dement.* **2008**, *4*, 38–48. [CrossRef] [PubMed]
25. Zetterberg, H.; Tullhög, K.; Hansson, O.; Minthon, L.; Londos, E.; Blennow, K. Low incidence of post-lumbar puncture headache in 1,089 consecutive memory clinic patients. *Eur. Neurol.* **2010**, *63*, 326–330. [CrossRef] [PubMed]
26. Reiman, E.M.; Jagust, W.J. Brain imaging in the study of Alzheimer's disease. *Neuroimage* **2012**, *61*, 505–516. [CrossRef] [PubMed]
27. Fawaz, M.V.; Brooks, A.F.; Rodnick, M.E.; Carpenter, G.M.; Shao, X.; Desmond, T.J.; Sherman, P.; Quesada, C.A.; Hockley, B.G.; Kilbourn, M.R.; et al. High Affinity Radiopharmaceuticals Based Upon Lansoprazole for PET Imaging of Aggregated Tau in Alzheimer's Disease and Progressive Supranuclear Palsy: Synthesis, Preclinical Evaluation, and Lead Selection. *ACS Chem. Neurosci.* **2014**, *5*, 718–730. [CrossRef]
28. DeKosky, S.T.; Marek, K. Looking backward to move forward: early detection of neurodegenerative disorders. *Science* **2003**, *302*, 830–834. [CrossRef]
29. Bailey, D.L.; Maisey, M.N.; Townsend, D.W.; Valk, P.E. *Positron Emission Tomography. Basic Sciences*; Springer-Verlag London Limited: London, UK, 2005; ISBN 9781852337988.
30. van der Born, D.; Pees, A.; Poot, A.J.; Orru, R.V.A.; Windhorst, A.D.; Vugts, D.J. Fluorine-18 labelled building blocks for PET tracer synthesis. *Chem. Soc. Rev.* **2017**, *46*, 4709–4773. [CrossRef]
31. Filippi, L.; Chiaravalloti, A.; Bagni, O.; Schillaci, O. 18F-labeled radiopharmaceuticals for the molecular neuroimaging of amyloid plaques in Alzheimer's disease. *Am. J. Nucl. Med. Mol. Imaging* **2018**, *8*, 268–281.
32. Mathis, C.A.; Wang, Y.; Klunk, W.E. Imaging beta-amyloid plaques and neurofibrillary tangles in the aging human brain. *Curr. Pharm. Des.* **2004**, *10*, 1469–1492. [CrossRef]

33. Klunk, W.E.; Engler, H.; Nordberg, A.; Wang, Y.; Blomqvist, G.; Holt, D.P.; Bergström, M.; Savitcheva, I.; Huang, G.-f.; Estrada, S.; et al. Imaging brain amyloid in Alzheimer's disease with Pittsburgh Compound-B. *Ann. Neurol.* **2004**, *55*, 306–319. [CrossRef]
34. Jacobson, O.; Kiesewetter, D.O.; Chen, X. Fluorine-18 radiochemistry, labeling strategies and synthetic routes. *Bioconjug. Chem.* **2015**, *26*, 1–18. [CrossRef]
35. Ariza, M.; Kolb, H.C.; Moechars, D.; Rombouts, F.; Andrés, J.I. Tau Positron Emission Tomography (PET) Imaging: Past, Present, and Future. *J. Med. Chem.* **2015**, *58*, 4365–4382. [CrossRef] [PubMed]
36. Barthel, H.; Sabri, O. Florbetaben to trace amyloid-β in the Alzheimer brain by means of PET. *J. Alzheimers Dis.* **2011**, *26* (Suppl. 3), 117–121. [CrossRef]
37. Johnson, K.A.; Sperling, R.A.; Gidicsin, C.M.; Carmasin, J.S.; Maye, J.E.; Coleman, R.E.; Reiman, E.M.; Sabbagh, M.N.; Sadowsky, C.H.; Fleisher, A.S.; et al. Florbetapir (F18-AV-45) PET to assess amyloid burden in Alzheimer's disease dementia, mild cognitive impairment, and normal aging. *Alzheimers Dement.* **2013**, *9*, S72–S83. [CrossRef] [PubMed]
38. Hatashita, S.; Yamasaki, H.; Suzuki, Y.; Tanaka, K.; Wakebe, D.; Hayakawa, H. 18FFlutemetamol amyloid-beta PET imaging compared with 11CPIB across the spectrum of Alzheimer's disease. *Eur. J. Nucl. Med. Mol. Imaging* **2014**, *41*, 290–300. [CrossRef] [PubMed]
39. Arriagada, P.V.; Growdon, J.H.; Hedley-Whyte, E.T.; Hyman, B.T. Neurofibrillary tangles but not senile plaques parallel duration and severity of Alzheimer's disease. *Neurology* **1992**, *42*, 631–639. [CrossRef] [PubMed]
40. Maccioni, R.B.; Farías, G.; Morales, I.; Navarrete, L. The revitalized tau hypothesis on Alzheimer's disease. *Arch. Med. Res.* **2010**, *41*, 226–231. [CrossRef]
41. Jack, C.R.; Knopman, D.S.; Jagust, W.J.; Petersen, R.C.; Weiner, M.W.; Aisen, P.S.; Shaw, L.M.; Vemuri, P.; Wiste, H.J.; Weigand, S.D.; et al. Tracking pathophysiological processes in Alzheimer's disease: An updated hypothetical model of dynamic biomarkers. *Lancet Neurol.* **2013**, *12*, 207–216. [CrossRef]
42. Villemagne, V.L.; Furumoto, S.; Fodero-Tavoletti, M.; Harada, R.; Mulligan, R.S.; Kudo, Y.; Masters, C.L.; Yanai, K.; Rowe, C.C.; Okamura, N. The challenges of tau imaging. *Future Neurol.* **2012**, *7*, 409–421. [CrossRef]
43. Ke, Y.D.; Suchowerska, A.K.; van der Hoven, J.; de Silva, D.M.; Wu, C.W.; van Eersel, J.; Ittner, A.; Ittner, L.M. Lessons from tau-deficient mice. *Int. J. Alzheimers Dis.* **2012**, *2012*, 873270. [CrossRef]
44. Ludolph, A.C.; Kassubek, J.; Landwehrmeyer, B.G.; Mandelkow, E.; Mandelkow, E.-M.; Burn, D.J.; Caparros-Lefebvre, D.; Frey, K.A.; de Yebenes, J.G.; Gasser, T.; et al. Tauopathies with parkinsonism: clinical spectrum, neuropathologic basis, biological markers, and treatment options. *Eur. J. Neurol.* **2009**, *16*, 297–309. [CrossRef]
45. Medeiros, R.; Baglietto-Vargas, D.; LaFerla, F.M. The role of tau in Alzheimer's disease and related disorders. *CNS Neurosci. Ther.* **2011**, *17*, 514–524. [CrossRef] [PubMed]
46. Winners. Available online: https://www.eanm.org/congresses-events/awards-grants/winners/ (accessed on 22 December 2019).
47. Vanhaute, H.; Ceccarini, J.; Michiels, L.; Koole, M.; Emsell, L.; Lemmens, R.; Vandenbulcke, M.; van Laere, K. *PET-MR Imaging of Tau and Synaptic Density in Prodromal Alzheimer's Disease*; Alzheimer's Disease International: London, UK, 2019.
48. U.S. Food and Drug Administration. Available online: https://www.fda.gov/ (accessed on 20 November 2019).
49. Sperling, R.A.; Aisen, P.S.; Beckett, L.A.; Bennett, D.A.; Craft, S.; Fagan, A.M.; Iwatsubo, T.; Jack, C.R.; Kaye, J.; Montine, T.J.; et al. Toward defining the preclinical stages of Alzheimer's disease: recommendations from the National Institute on Aging-Alzheimer's Association workgroups on diagnostic guidelines for Alzheimer's disease. *Alzheimers Dement.* **2011**, *7*, 280–292. [CrossRef] [PubMed]
50. Villemagne, V.L.; Okamura, N. In vivo tau imaging: obstacles and progress. *Alzheimers Dement.* **2014**, *10*, S254–S264. [CrossRef] [PubMed]
51. Okamura, N.; Harada, R.; Ishiki, A.; Kikuchi, A.; Nakamura, T.; Kudo, Y. The development and validation of tau PET tracers: current status and future directions. *Clin. Transl. Imaging* **2018**, *6*, 305–316. [CrossRef] [PubMed]
52. Aguero, C.; Dhaynaut, M.; Normandin, M.D.; Amaral, A.C.; Guehl, N.J.; Neelamegam, R.; Marquie, M.; Johnson, K.A.; El Fakhri, G.; Frosch, M.P.; et al. Autoradiography validation of novel tau PET tracer F-18-MK-6240 on human postmortem brain tissue. *Acta Neuropathol. Commun.* **2019**, *7*, 37. [CrossRef] [PubMed]

53. Kroth, H.; Oden, F.; Molette, J.; Schieferstein, H.; Capotosti, F.; Mueller, A.; Berndt, M.; Schmitt-Willich, H.; Darmency, V.; Gabellieri, E.; et al. Discovery and preclinical characterization of 18FPI-2620, a next-generation tau PET tracer for the assessment of tau pathology in Alzheimer's disease and other tauopathies. *Eur. J. Nucl. Med. Mol. Imaging* **2019**, *46*, 2178–2189. [CrossRef] [PubMed]
54. Spillantini, M.G.; Goedert, M. The alpha-synucleinopathies: Parkinson's disease, dementia with Lewy bodies, and multiple system atrophy. *Ann. N. Y. Acad. Sci.* **2000**, *920*, 16–27. [CrossRef]
55. Wakabayashi, K.; Yoshimoto, M.; Tsuji, S.; Takahashi, H. α-Synuclein immunoreactivity in glial cytoplasmic inclusions in multiple system atrophy. *Neuroscience Letters* **1998**, *249*, 180–182. [CrossRef]
56. Tu, P.H.; Galvin, J.E.; Baba, M.; Giasson, B.; Tomita, T.; Leight, S.; Nakajo, S.; Iwatsubo, T.; Trojanowski, J.Q.; Lee, V.M. Glial cytoplasmic inclusions in white matter oligodendrocytes of multiple system atrophy brains contain insoluble alpha-synuclein. *Ann. Neurol.* **1998**, *44*, 415–422. [CrossRef]
57. Burré, J.; Sharma, M.; Tsetsenis, T.; Buchman, V.; Etherton, M.R.; Südhof, T.C. Alpha-synuclein promotes SNARE-complex assembly in vivo and in vitro. *Science* **2010**, *329*, 1663–1667. [CrossRef]
58. Vilar, M.; Chou, H.-T.; Lührs, T.; Maji, S.K.; Riek-Loher, D.; Verel, R.; Manning, G.; Stahlberg, H.; Riek, R. The fold of alpha-synuclein fibrils. *Proc. Natl. Acad. Sci. USA* **2008**, *105*, 8637–8642. [CrossRef] [PubMed]
59. Spillantini, M.G.; Crowther, R.A.; Jakes, R.; Hasegawa, M.; Goedert, M. alpha-Synuclein in filamentous inclusions of Lewy bodies from Parkinson's disease and dementia with lewy bodies. *Proc. Natl. Acad. Sci. USA* **1998**, *95*, 6469–6473. [CrossRef] [PubMed]
60. Uéda, K.; Fukushima, H.; Masliah, E.; Xia, Y.; Iwai, A.; Yoshimoto, M.; Otero, D.A.; Kondo, J.; Ihara, Y.; Saitoh, T. Molecular cloning of cDNA encoding an unrecognized component of amyloid in Alzheimer disease. *Proc. Natl. Acad. Sci. USA* **1993**, *90*, 11282–11286. [CrossRef] [PubMed]
61. Clayton, D.F.; George, J.M. Synucleins in synaptic plasticity and neurodegenerative disorders. *J. Neurosci. Res.* **1999**, *58*, 120–129. [CrossRef]
62. Polymeropoulos, M.H.; Lavedan, C.; Leroy, E.; Ide, S.E.; Dehejia, A.; Dutra, A.; Pike, B.; Root, H.; Rubenstein, J.; Boyer, R.; et al. Mutation in the alpha-synuclein gene identified in families with Parkinson's disease. *Science* **1997**, *276*, 2045–2047. [CrossRef] [PubMed]
63. Chartier-Harlin, M.-C.; Kachergus, J.; Roumier, C.; Mouroux, V.; Douay, X.; Lincoln, S.; Levecque, C.; Larvor, L.; Andrieux, J.; Hulihan, M.; et al. α-synuclein locus duplication as a cause of familial Parkinson's disease. *Lancet* **2004**, *364*, 1167–1169. [CrossRef]
64. Singleton, A.B.; Farrer, M.; Johnson, J.; Singleton, A.; Hague, S.; Kachergus, J.; Hulihan, M.; Peuralinna, T.; Dutra, A.; Nussbaum, R.; et al. alpha-Synuclein locus triplication causes Parkinson's disease. *Science* **2003**, *302*, 841. [CrossRef]
65. Ibáñez, P.; Bonnet, A.-M.; Débarges, B.; Lohmann, E.; Tison, F.; Agid, Y.; Dürr, A.; Brice, A.; Pollak, P. Causal relation between α-synuclein locus duplication as a cause of familial Parkinson's disease. *Lancet* **2004**, *364*, 1169–1171. [CrossRef]
66. Billingsley, K.J.; Bandres-Ciga, S.; Saez-Atienzar, S.; Singleton, A.B. Genetic risk factors in Parkinson's disease. *Cell Tissue Res.* **2018**, *373*, 9–20. [CrossRef]
67. Chang, D.; Nalls, M.A.; Hallgrímsdóttir, I.B.; Hunkapiller, J.; van der Brug, M.; Cai, F.; Kerchner, G.A.; Ayalon, G.; Bingol, B.; Sheng, M.; et al. A meta-analysis of genome-wide association studies identifies 17 new Parkinson's disease risk loci. *Nat. Genet.* **2017**, *49*, 1511–1516. [CrossRef]
68. Nalls, M.; Blauwendraat, C.; Vallerga, C.L.; Heilbron, K.; Bandres-Ciga, S.; Chang, D.; Tan, M.; Kia, D.A.; Noyce, A.J.; Xue, A.; et al. Parkinson's disease genetics: Novel risk loci, genomic context, causal insights and heritable risk. *BiorRxiv* 2018. [CrossRef]
69. Vernon, A.C.; Ballard, C.; Modo, M. Neuroimaging for Lewy body disease: is the in vivo molecular imaging of α-synuclein neuropathology required and feasible? *Brain Res. Rev.* **2010**, *65*, 28–55. [CrossRef] [PubMed]
70. Hawkes, C.H. The prodromal phase of sporadic Parkinson's disease: does it exist and if so how long is it? *Mov. Disord.* **2008**, *23*, 1799–1807. [CrossRef] [PubMed]
71. Ross, G.W.; Petrovitch, H.; Abbott, R.D.; Nelson, J.; Markesbery, W.; Davis, D.; Hardman, J.; Launer, L.; Masaki, K.; Tanner, C.M.; et al. Parkinsonian signs and substantia nigra neuron density in decendents elders without PD. *Ann. Neurol.* **2004**, *56*, 532–539. [CrossRef] [PubMed]
72. Fearnley, J.M.; Lees, A.J. Ageing and Parkinson's disease: substantia nigra regional selectivity. *Brain* **1991**, *114*, 2283–2301. [CrossRef] [PubMed]

73. Riederer, P.; Wuketich, S. Time course of nigrostriatal degeneration in parkinson's disease. A detailed study of influential factors in human brain amine analysis. *J. Neural Transm.* **1976**, *38*, 277–301. [CrossRef] [PubMed]
74. Braak, H.; Tredici, K.D.; Rüb, U.; de Vos, R.A.I.; Jansen Steur, E.N.H.; Braak, E. Staging of brain pathology related to sporadic Parkinson's disease. *Neurobiol. Aging* **2003**, *24*, 197–211. [CrossRef]
75. Stefanis, L. α-Synuclein in Parkinson's disease. *Cold Spring Harb. Perspect. Med.* **2012**, *2*, a009399. [CrossRef]
76. Deramecourt, V.; Bombois, S.; Maurage, C.-A.; Ghestem, A.; Drobecq, H.; Vanmechelen, E.; Lebert, F.; Pasquier, F.; Delacourte, A. Biochemical staging of synucleinopathy and amyloid deposition in dementia with Lewy bodies. *J. Neuropathol. Exp. Neurol.* **2006**, *65*, 278–288. [CrossRef]
77. Shah, M.; Seibyl, J.; Cartier, A.; Bhatt, R.; Catafau, A.M. Molecular imaging insights into neurodegeneration: focus on α-synuclein radiotracers. *J. Nucl. Med.* **2014**, *55*, 1397–1400. [CrossRef]
78. Barrett, P.J.; Timothy Greenamyre, J. Post-translational modification of α-synuclein in Parkinson's disease. *Brain Res.* **2015**, *1628*, 247–253. [CrossRef] [PubMed]
79. Irwin, D.J.; Lee, V.M.-Y.; Trojanowski, J.Q. Parkinson's disease dementia: convergence of α-synuclein, tau and amyloid-β pathologies. *Nat. Rev. Neurosci.* **2013**, *14*, 626–636. [CrossRef] [PubMed]
80. Kotzbauer, P.T.; Cairns, N.J.; Campbell, M.C.; Willis, A.W.; Racette, B.A.; Tabbal, S.D.; Perlmutter, J.S. Pathologic accumulation of α-synuclein and Aβ in Parkinson disease patients with dementia. *Arch. Neurol.* **2012**, *69*, 1326–1331. [CrossRef] [PubMed]
81. Mathis, C.A.; Lopresti, B.J.; Ikonomovic, M.D.; Klunk, W.E. Small-molecule PET Tracers for Imaging Proteinopathies. *Semin. Nucl. Med.* **2017**, *47*, 553–575. [CrossRef]
82. Kotzbauer, P.T.; Tu, Z.; Mach, R.H. Current status of the development of PET radiotracers for imaging alpha synuclein aggregates in Lewy bodies and Lewy neurites. *Clin. Transl. Imaging* **2017**, *5*, 3–14. [CrossRef]
83. Pike, V.W. Considerations in the Development of Reversibly Binding PET Radioligands for Brain Imaging. *Curr. Med. Chem.* **2016**, *23*, 1818–1869. [CrossRef]
84. Laruelle, M.; Slifstein, M.; Huang, Y. Relationships between radiotracer properties and image quality in molecular imaging of the brain with positron emission tomography. *Mol. Imaging Biol.* **2003**, *5*, 363–375. [CrossRef]
85. Zhang, L.; Villalobos, A. Strategies to facilitate the discovery of novel CNS PET ligands. *EJNMMI Radiopharm. Chem.* **2017**, *1*, 13. [CrossRef]
86. van de Bittner, G.C.; Ricq, E.L.; Hooker, J.M. A philosophy for CNS radiotracer design. *Acc. Chem. Res.* **2014**, *47*, 3127–3134. [CrossRef]
87. Elghetany, M.T.; Saleem, A. Methods for staining amyloid in tissues: A review. *Stain Technol.* **1988**, *63*, 201–212. [CrossRef]
88. Vassar, P.S.; Culling, C.F. Fluorescent stains, with special reference to amyloid and connective tissues. *Arch. Pathol.* **1959**, *68*, 487–498. [PubMed]
89. Watanabe, H.; Ono, M.; Ariyoshi, T.; Katayanagi, R.; Saji, H. Novel Benzothiazole Derivatives as Fluorescent Probes for Detection of β-Amyloid and α-Synuclein Aggregates. *ACS Chem. Neurosci.* **2017**, *8*, 1656–1662. [CrossRef] [PubMed]
90. Zeng, F.; Goodman, M.M. Fluorine-18 radiolabeled heterocycles as PET tracers for imaging β-amyloid plaques in Alzheimer's disease. *Curr. Top. Med. Chem.* **2013**, *13*, 909–919. [CrossRef] [PubMed]
91. Mathis, C.A.; Mason, N.S.; Lopresti, B.J.; Klunk, W.E. Development of positron emission tomography β-amyloid plaque imaging agents. *Semin. Nucl. Med.* **2012**, *42*, 423–432. [CrossRef]
92. Klunk, W.E.; Wang, Y.; Huang, G.f.; Debnath, M.L.; Holt, D.P.; Mathis, C.A. Uncharged thioflavin-T derivatives bind to amyloid-beta protein with high affinity and readily enter the brain. *Life Sciences* **2001**, *69*, 1471–1484. [CrossRef]
93. Klunk, W.E.; Wang, Y.; Huang, G.-f.; Debnath, M.L.; Holt, D.P.; Shao, L.; Hamilton, R.L.; Ikonomovic, M.D.; DeKosky, S.T.; Mathis, C.A. The Binding of 2-(4′-Methylaminophenyl)Benzothiazole to Postmortem Brain Homogenates Is Dominated by the Amyloid Component. *J. Neurosci.* **2003**, *23*, 2086–2092. [CrossRef]
94. Mathis, C.A.; Wang, Y.; Holt, D.P.; Huang, G.F.; Debnath, M.L.; Klunk, W.E. Synthesis and evaluation of 11C-labeled 6-substituted 2-arylbenzothiazoles as amyloid imaging agents. *J. Med. Chem.* **2003**, *46*, 2740–2754. [CrossRef]

95. Price, J.C.; Klunk, W.E.; Lopresti, B.J.; Lu, X.; Hoge, J.A.; Ziolko, S.K.; Holt, D.P.; Meltzer, C.C.; DeKosky, S.T.; Mathis, C.A. Kinetic modeling of amyloid binding in humans using PET imaging and Pittsburgh Compound-B. *J. Cereb. Blood Flow Metab.* **2005**, *25*, 1528–1547. [CrossRef]
96. Mathis, C.A.; Bacskai, B.J.; Kajdasz, S.T.; McLellan, M.E.; Frosch, M.P.; Hyman, B.T.; Holt, D.P.; Wang, Y.; Huang, G.-f.; Debnath, M.L.; et al. A lipophilic thioflavin-T derivative for positron emission tomography (PET) imaging of amyloid in brain. *Bioorganic & Medicinal Chemistry Letters* **2002**, *12*, 295–298. [CrossRef]
97. Klunk, W.E.; Mathis, C.A. Whatever happened to Pittsburgh Compound-A? *Alzheimer Dis. Assoc. Disord.* **2008**, *22*, 198–203. [CrossRef]
98. Mintun, M.A.; Larossa, G.N.; Sheline, Y.I.; Dence, C.S.; Lee, S.Y.; Mach, R.H.; Klunk, W.E.; Mathis, C.A.; DeKosky, S.T.; Morris, J.C. 11CPIB in a nondemented population: potential antecedent marker of Alzheimer disease. *Neurology* **2006**, *67*, 446–452. [CrossRef] [PubMed]
99. Ikonomovic, M.D.; Klunk, W.E.; Abrahamson, E.E.; Mathis, C.A.; Price, J.C.; Tsopelas, N.D.; Lopresti, B.J.; Ziolko, S.; Bi, W.; Paljug, W.R.; et al. Post-mortem correlates of in vivo PiB-PET amyloid imaging in a typical case of Alzheimer's disease. *Brain* **2008**, *131*, 1630–1645. [CrossRef] [PubMed]
100. Chitneni, S.K.; Serdons, K.; Evens, N.; Fonge, H.; Celen, S.; Deroose, C.M.; Debyser, Z.; Mortelmans, L.; Verbruggen, A.M.; Bormans, G.M. Efficient purification and metabolite analysis of radiotracers using high-performance liquid chromatography and on-line solid-phase extraction. *J. Chromatogr. A* **2008**, *1189*, 323–331. [CrossRef] [PubMed]
101. Serdons, K.; Terwinghe, C.; Vermaelen, P.; van Laere, K.; Kung, H.; Mortelmans, L.; Bormans, G.; Verbruggen, A. Synthesis and evaluation of 18F-labeled 2-phenylbenzothiazoles as positron emission tomography imaging agents for amyloid plaques in Alzheimer's disease. *J. Med. Chem.* **2009**, *52*, 1428–1437. [CrossRef] [PubMed]
102. Mathis, C.A.; Holt, D.; Wang, Y.; Huang, G.-f.; Debnath, M.; Shao, L.; Klunk, W.E. P2-178 Species-dependent formation and identification of the brain metabolites of the amyloid imaging agent [11C]PIB. *Neurobiol. Aging* **2004**, *25*, S277–S278. [CrossRef]
103. Juréus, A.; Swahn, B.-M.; Sandell, J.; Jeppsson, F.; Johnson, A.E.; Johnström, P.; Neelissen, J.A.M.; Sunnemark, D.; Farde, L.; Svensson, S.P.S. Characterization of AZD4694, a novel fluorinated Abeta plaque neuroimaging PET radioligand. *J. Neurochem.* **2010**, *114*, 784–794. [CrossRef]
104. Mason, N.S.; Mathis, C.A.; Klunk, W.E. Positron emission tomography radioligands for in vivo imaging of Aβ plaques. *J. Labelled Comp. Radiopharm.* **2013**, *56*, 89–95. [CrossRef]
105. Snellman, A.; Rokka, J.; Lopez-Picon, F.R.; Eskola, O.; Wilson, I.; Farrar, G.; Scheinin, M.; Solin, O.; Rinne, J.O.; Haaparanta-Solin, M. Pharmacokinetics of ^{18}Fflutemetamol in wild-type rodents and its binding to beta amyloid deposits in a mouse model of Alzheimer's disease. *Eur. J. Nucl. Med. Mol. Imaging* **2012**, *39*, 1784–1795. [CrossRef]
106. Rabinovici, G.D. The translational journey of brain β-amyloid imaging: from positron emission tomography to autopsy to clinic. *JAMA Neurol.* **2015**, *72*, 265–266. [CrossRef]
107. Curtis, C.; Gamez, J.E.; Singh, U.; Sadowsky, C.H.; Villena, T.; Sabbagh, M.N.; Beach, T.G.; Duara, R.; Fleisher, A.S.; Frey, K.A.; et al. Phase 3 trial of flutemetamol labeled with radioactive fluorine 18 imaging and neuritic plaque density. *JAMA Neurol.* **2015**, *72*, 287–294. [CrossRef]
108. GE Healthcare. Gehealthcare Vizamyl Prescribing Information. Available online: http://www3.gehealthcare.com/~{}/media/documents/us-global/products/nuclear-imaging-agents_non-gatekeeper/clinical%20product%20info/vizamyl/gehealthcare-vizamyl-prescribing-information.pdf (accessed on 15 September 2016).
109. Kung, M.-P.; Hou, C.; Zhuang, Z.-P.; Skovronsky, D.; Kung, H.F. Binding of two potential imaging agents targeting amyloid plaques in postmortem brain tissues of patients with Alzheimer's disease. *Brain Res.* **2004**, *1025*, 98–105. [CrossRef] [PubMed]
110. Ono, M.; Wilson, A.; Nobrega, J.; Westaway, D.; Verhoeff, P.; Zhuang, Z.-P.; Kung, M.-P.; Kung, H.F. 11C-labeled stilbene derivatives as Aβ-aggregate-specific PET imaging agents for Alzheimer's disease. *Nucl. Med. Biol.* **2003**, *30*, 565–571. [CrossRef]
111. Verhoeff, N.P.L.G.; Wilson, A.A.; Takeshita, S.; Trop, L.; Hussey, D.; Singh, K.; Kung, H.F.; Kung, M.-P.; Houle, S. In-vivo imaging of Alzheimer disease beta-amyloid with 11CSB-13 PET. *Am. J. Geriatr. Psychiatry* **2004**, *12*, 584–595. [CrossRef] [PubMed]

112. Zhang, W.; Oya, S.; Kung, M.-P.; Hou, C.; Maier, D.L.; Kung, H.F. F-18 stilbenes as PET imaging agents for detecting beta-amyloid plaques in the brain. *J. Med. Chem.* **2005**, *48*, 5980–5988. [CrossRef]
113. Stephenson, K.A.; Chandra, R.; Zhuang, Z.-P.; Hou, C.; Oya, S.; Kung, M.-P.; Kung, H.F. Fluoro-pegylated (FPEG) imaging agents targeting Abeta aggregates. *Bioconjug. Chem.* **2007**, *18*, 238–246. [CrossRef]
114. Zhang, W.; Oya, S.; Kung, M.-P.; Hou, C.; Maier, D.L.; Kung, H.F. F-18 Polyethyleneglycol stilbenes as PET imaging agents targeting Abeta aggregates in the brain. *Nucl. Med. Biol.* **2005**, *32*, 799–809. [CrossRef]
115. Kung, H.F.; Choi, S.R.; Qu, W.; Zhang, W.; Skovronsky, D. 18F stilbenes and styrylpyridines for PET imaging of A beta plaques in Alzheimer's disease: A miniperspective. *J. Med. Chem.* **2010**, *53*, 933–941. [CrossRef]
116. Rowe, C.C.; Ackerman, U.; Browne, W.; Mulligan, R.; Pike, K.L.; O'Keefe, G.; Tochon-Danguy, H.; Chan, G.; Berlangieri, S.U.; Jones, G.; et al. Imaging of amyloid β in Alzheimer's disease with 18F-BAY94-9172, a novel PET tracer: proof of mechanism. *Lancet Neurol.* **2008**, *7*, 129–135. [CrossRef]
117. Frequently Asked Questions about Beta-Amyloid Imaging. Available online: https://www.accessdata.fda.gov/drugsatfda_docs/label/2014/204677s000lbl.pdf (accessed on 20 February 2020).
118. Zhang, W.; Kung, M.-P.; Oya, S.; Hou, C.; Kung, H.F. 18F-labeled styrylpyridines as PET agents for amyloid plaque imaging. *Nucl. Med. Biol.* **2007**, *34*, 89–97. [CrossRef]
119. Choi, S.R.; Golding, G.; Zhuang, Z.; Zhang, W.; Lim, N.; Hefti, F.; Benedum, T.E.; Kilbourn, M.R.; Skovronsky, D.; Kung, H.F. Preclinical properties of 18F-AV-45: A PET agent for Abeta plaques in the brain. *J. Nucl. Med.* **2009**, *50*, 1887–1894. [CrossRef]
120. Wong, D.F.; Rosenberg, P.B.; Zhou, Y.; Kumar, A.; Raymont, V.; Ravert, H.T.; Dannals, R.F.; Nandi, A.; Brasić, J.R.; Ye, W.; et al. In vivo imaging of amyloid deposition in Alzheimer disease using the radioligand 18F-AV-45 (florbetapir corrected F 18). *J. Nucl. Med.* **2010**, *51*, 913–920. [CrossRef] [PubMed]
121. Okamura, N.; Yanai, K. Florbetapir (18F), a PET imaging agent that binds to amyloid plaques for the potential detection of Alzheimer's disease. *IDrugs* **2010**, *13*, 890–899. [PubMed]
122. Kolata, G. Promise Seen for Detection of Alzheimer's. Available online: https://www.nytimes.com/2010/06/24/health/research/24scans.html (accessed on 5 December 2019).
123. Johnson, A.E.; Jeppsson, F.; Sandell, J.; Wensbo, D.; Neelissen, J.A.M.; Juréus, A.; Ström, P.; Norman, H.; Farde, L.; Svensson, S.P.S. AZD2184: A radioligand for sensitive detection of beta-amyloid deposits. *J. Neurochem.* **2009**, *108*, 1177–1186. [CrossRef] [PubMed]
124. Swahn, B.-M.; Sandell, J.; Pyring, D.; Bergh, M.; Jeppsson, F.; Juréus, A.; Neelissen, J.; Johnström, P.; Schou, M.; Svensson, S. Synthesis and evaluation of pyridylbenzofuran, pyridylbenzothiazole and pyridylbenzoxazole derivatives as ^{18}F-PET imaging agents for β-amyloid plaques. *Bioorg. Med. Chem. Lett.* **2012**, *22*, 4332–4337. [CrossRef] [PubMed]
125. Nelissen, N.; van Laere, K.; Thurfjell, L.; Owenius, R.; Vandenbulcke, M.; Koole, M.; Bormans, G.; Brooks, D.J.; Vandenberghe, R. Phase 1 study of the Pittsburgh compound B derivative 18F-flutemetamol in healthy volunteers and patients with probable Alzheimer disease. *J. Nucl. Med.* **2009**, *50*, 1251–1259. [CrossRef] [PubMed]
126. Mathis, C.; Lopresti, B.; Mason, N.; Price, J.; Flatt, N.; Bi, W.; Ziolko, S.; DeKosky, S.; Klunk, W. Comparison of the Amyloid Imaging Agents [F-18]3'-F-PIB and [C-11]PIB in Alzheimer's Disease and Control Subjects. Available online: http://jnm.snmjournals.org/content/48/supplement_2/56P.3.short (accessed on 6 December 2019).
127. Lopresti, B.J.; Klunk, W.E.; Mathis, C.A.; Hoge, J.A.; Ziolko, S.K.; Lu, X.; Meltzer, C.C.; Schimmel, K.; Tsopelas, N.D.; DeKosky, S.T.; et al. Simplified quantification of Pittsburgh Compound B amyloid imaging PET studies: A comparative analysis. *J. Nucl. Med.* **2005**, *46*, 1959–1972.
128. Cselényi, Z.; Jönhagen, M.E.; Forsberg, A.; Halldin, C.; Julin, P.; Schou, M.; Johnström, P.; Varnäs, K.; Svensson, S.; Farde, L. Clinical validation of 18F-AZD4694, an amyloid-β-specific PET radioligand. *J. Nucl. Med.* **2012**, *53*, 415–424. [CrossRef]
129. A Phase 3 Clinical Trial to Evaluate the Efficacy and Safety of [18F]NAV4694 PET for Detection of Cerebral Beta-Amyloid When Compared With Postmortem Histopathology-Full Text View-ClinicalTrials.gov. Available online: https://clinicaltrials.gov/ct2/show/NCT01886820 (accessed on 6 December 2019).
130. Flutafuranol F 18-Navidea Biopharmaceuticals-AdisInsight. Available online: https://adisinsight.springer.com/drugs/800031066 (accessed on 6 December 2019).

131. Hostetler, E.D.; Sanabria-Bohórquez, S.; Fan, H.; Zeng, Z.; Gammage, L.; Miller, P.; O'Malley, S.; Connolly, B.; Mulhearn, J.; Harrison, S.T.; et al. 18FFluoroazabenzoxazoles as potential amyloid plaque PET tracers: synthesis and in vivo evaluation in rhesus monkey. *Nucl. Med. Biol.* **2011**, *38*, 1193–1203. [CrossRef]
132. Merck Clinical Trials. Available online: https://www.merck.com/clinical-trials/study.html?id=3328-002&kw=alzheimer%27s&tab=eligibility (accessed on 6 December 2019).
133. Yousefi, B.H.; Drzezga, A.; von Reutern, B.; Manook, A.; Schwaiger, M.; Wester, H.-J.; Henriksen, G. A Novel (18)F-Labeled Imidazo2,1-bbenzothiazole (IBT) for High-Contrast PET Imaging of β-Amyloid Plaques. *ACS Med. Chem. Lett.* **2011**, *2*, 673–677. [CrossRef]
134. Yousefi, B.H.; von Reutern, B.; Scherübl, D.; Manook, A.; Schwaiger, M.; Grimmer, T.; Henriksen, G.; Förster, S.; Drzezga, A.; Wester, H.-J. FIBT versus florbetaben and PiB: A preclinical comparison study with amyloid-PET in transgenic mice. *EJNMMI Res.* **2015**, *5*, 6519. [CrossRef]
135. Westwell, A.D. *Fluorinated Pharmaceuticals. Development of 18F-labeled compounds for imaging of Aβ plaques by means of PET.*; Future Science Ltd.: London, UK, 2015; ISBN 9781910419007.
136. Yousefi, B.H.; Manook, A.; Grimmer, T.; Arzberger, T.; von Reutern, B.; Henriksen, G.; Drzezga, A.; Förster, S.; Schwaiger, M.; Wester, H.-J. Characterization and first human investigation of FIBT, a novel fluorinated Aβ plaque neuroimaging PET radioligand. *ACS Chem. Neurosci.* **2015**, *6*, 428–437. [CrossRef] [PubMed]
137. Kim, Y.; Rosenberg, P.; Oh, E. A Review of Diagnostic Impact of Amyloid Positron Emission Tomography Imaging in Clinical Practice. *Dement. Geriatr. Cogn. Disord.* **2018**, *46*, 154–167. [CrossRef] [PubMed]
138. Barthel, H.; Sabri, O. Clinical Use and Utility of Amyloid Imaging. *J. Nucl. Med.* **2017**, *58*, 1711–1717. [CrossRef] [PubMed]
139. Chiotis, K.; Saint-Aubert, L.; Boccardi, M.; Gietl, A.; Picco, A.; Varrone, A.; Garibotto, V.; Herholz, K.; Nobili, F.; Nordberg, A. Clinical validity of increased cortical uptake of amyloid ligands on PET as a biomarker for Alzheimer's disease in the context of a structured 5-phase development framework. *Neurobiol. Aging* **2017**, *52*, 214–227. [CrossRef]
140. Ceccaldi, M.; Jonveaux, T.; Verger, A.; Salmon, P.K.; Houzard, C.; Godefroy, O.; Shields, T.; Perrotin, A.; Gismondi, R.; Bullich, S.; et al. Impact of Florbetaben PET Imaging on Diagnosis and Management of Patients with Suspected Alzheimer's Disease Eligible for CSF Analysis in France. Available online: http://jnm.snmjournals.org/content/58/supplement_1/561?related-urls=yes&legid=jnumed;58/supplement_1/561 (accessed on 8 February 2020).
141. Hattori, N.; Ono, S.; UDO, N.; Yamamoto, S.; Ogawa, M.; Sugie, H. Clinical Impact of F-18 Flutemetamol (FMM) PET to Assess Cerebral Aß Pathology in Patients with Various Cognitive Disorders. Available online: http://jnm.snmjournals.org/content/59/supplement_1/480 (accessed on 8 February 2020).
142. Leuzy, A.; Savitcheva, I.; Chiotis, K.; Lilja, J.; Andersen, P.; Bogdanovic, N.; Jelic, V.; Nordberg, A. Clinical impact of 18Fflutemetamol PET among memory clinic patients with an unclear diagnosis. *Eur. J. Nucl. Med. Mol. Imaging* **2019**, *46*, 1276–1286. [CrossRef]
143. Sabri, O.; Seibyl, J.; Rowe, C.; Barthel, H. Beta-amyloid imaging with florbetaben. *Clin. Transl. Imaging* **2015**, *3*, 13–26. [CrossRef]
144. de Wilde, A.; van der Flier, W.M.; Pelkmans, W.; Bouwman, F.; Verwer, J.; Groot, C.; van Buchem, M.M.; Zwan, M.; Ossenkoppele, R.; Yaqub, M.; et al. Association of Amyloid Positron Emission Tomography With Changes in Diagnosis and Patient Treatment in an Unselected Memory Clinic Cohort: The ABIDE Project. *JAMA Neurol.* **2018**, *75*, 1062–1070. [CrossRef]
145. Zannas, A.S.; Doraiswamy, P.M.; Shpanskaya, K.S.; Murphy, K.R.; Petrella, J.R.; Burke, J.R.; Wong, T.Z. Impact of ^{18}F-florbetapir PET imaging of β-amyloid neuritic plaque density on clinical decision-making. *Neurocase* **2014**, *20*, 466–473. [CrossRef]
146. University of Zurich. Investigating the Clinical Consequences of Flutemetamol-PET-Scanning-ICH GCP-Clinical Trials Registry. Available online: https://ichgcp.net/clinical-trials-registry/NCT02353949 (accessed on 8 February 2020).
147. Buée, L.; Bussière, T.; Buée-Scherrer, V.; Delacourte, A.; Hof, P.R. Tau protein isoforms, phosphorylation and role in neurodegenerative disorders11These authors contributed equally to this work. *Brain Res. Rev.* **2000**, *33*, 95–130. [CrossRef]
148. Fichou, Y.; Al-Hilaly, Y.K.; Devred, F.; Smet-Nocca, C.; Tsvetkov, P.O.; Verelst, J.; Winderickx, J.; Geukens, N.; Vanmechelen, E.; Perrotin, A.; et al. The elusive tau molecular structures: can we translate the recent breakthroughs into new targets for intervention? *Acta Neuropathol. Commun.* **2019**, *7*, 31. [CrossRef]

149. Harada, R.; Okamura, N.; Furumoto, S.; Tago, T.; Yanai, K.; Arai, H.; Kudo, Y. Characteristics of Tau and Its Ligands in PET Imaging. *Biomolecules* **2016**, *6*, 7. [CrossRef] [PubMed]
150. Mott, R.T.; Dickson, D.W.; Trojanowski, J.Q.; Zhukareva, V.; Lee, V.M.; Forman, M.; van Deerlin, V.; Ervin, J.F.; Wang, D.-S.; Schmechel, D.E.; et al. Neuropathologic, biochemical, and molecular characterization of the frontotemporal dementias. *J. Neuropathol. Exp. Neurol.* **2005**, *64*, 420–428. [CrossRef] [PubMed]
151. Dickson, D.W.; Kouri, N.; Murray, M.E.; Josephs, K.A. Neuropathology of frontotemporal lobar degeneration-tau (FTLD-tau). *J. Mol. Neurosci.* **2011**, *45*, 384–389. [CrossRef] [PubMed]
152. Harada, R.; Okamura, N.; Furumoto, S.; Yanai, K. Imaging Protein Misfolding in the Brain Using β-Sheet Ligands. *Front. Neurosci.* **2018**, *12*, 585. [CrossRef] [PubMed]
153. Clinton, L.K.; Blurton-Jones, M.; Myczek, K.; Trojanowski, J.Q.; LaFerla, F.M. Synergistic Interactions between Abeta, tau, and alpha-synuclein: Acceleration of neuropathology and cognitive decline. *J. Neurosci.* **2010**, *30*, 7281–7289. [CrossRef] [PubMed]
154. Näslund, J.; Haroutunian, V.; Mohs, R.; Davis, K.L.; Davies, P.; Greengard, P.; Buxbaum, J.D. Correlation between elevated levels of amyloid beta-peptide in the brain and cognitive decline. *JAMA* **2000**, *283*, 1571–1577. [CrossRef]
155. Okamura, N.; Suemoto, T.; Furumoto, S.; Suzuki, M.; Shimadzu, H.; Akatsu, H.; Yamamoto, T.; Fujiwara, H.; Nemoto, M.; Maruyama, M.; et al. Quinoline and benzimidazole derivatives: candidate probes for in vivo imaging of tau pathology in Alzheimer's disease. *J. Neurosci.* **2005**, *25*, 10857–10862. [CrossRef]
156. Fodero-Tavoletti, M.T.; Okamura, N.; Furumoto, S.; Mulligan, R.S.; Connor, A.R.; McLean, C.A.; Cao, D.; Rigopoulos, A.; Cartwright, G.A.; O'Keefe, G.; et al. 18F-THK523: A novel in vivo tau imaging ligand for Alzheimer's disease. *Brain* **2011**, *134*, 1089–1100. [CrossRef]
157. Okamura, N.; Furumoto, S.; Harada, R.; Tago, T.; Yoshikawa, T.; Fodero-Tavoletti, M.; Mulligan, R.S.; Villemagne, V.L.; Akatsu, H.; Yamamoto, T.; et al. Novel 18F-labeled arylquinoline derivatives for noninvasive imaging of tau pathology in Alzheimer disease. *J. Nucl. Med.* **2013**, *54*, 1420–1427. [CrossRef]
158. Villemagne, V.L.; Furumoto, S.; Fodero-Tavoletti, M.T.; Mulligan, R.S.; Hodges, J.; Harada, R.; Yates, P.; Piguet, O.; Pejoska, S.; Doré, V.; et al. In vivo evaluation of a novel tau imaging tracer for Alzheimer's disease. *Eur. J. Nucl. Med. Mol. Imaging* **2014**, *41*, 816–826. [CrossRef]
159. Tago, T.; Furumoto, S.; Okamura, N.; Harada, R.; Adachi, H.; Ishikawa, Y.; Yanai, K.; Iwata, R.; Kudo, Y. Preclinical Evaluation of (18)FTHK-5105 Enantiomers: Effects of Chirality on Its Effectiveness as a Tau Imaging Radiotracer. *Mol. Imaging Biol.* **2016**, *18*, 258–266. [CrossRef] [PubMed]
160. Okamura, N.; Furumoto, S.; Fodero-Tavoletti, M.T.; Mulligan, R.S.; Harada, R.; Yates, P.; Pejoska, S.; Kudo, Y.; Masters, C.L.; Yanai, K.; et al. Non-invasive assessment of Alzheimer's disease neurofibrillary pathology using 18F-THK5105 PET. *Brain* **2014**, *137*, 1762–1771. [CrossRef] [PubMed]
161. Saint-Aubert, L.; Almkvist, O.; Chiotis, K.; Almeida, R.; Wall, A.; Nordberg, A. Regional tau deposition measured by 18FTHK5317 positron emission tomography is associated to cognition via glucose metabolism in Alzheimer's disease. *Alzheimers Res. Ther.* **2016**, *8*, 38. [CrossRef] [PubMed]
162. Chiotis, K.; Saint-Aubert, L.; Savitcheva, I.; Jelic, V.; Andersen, P.; Jonasson, M.; Eriksson, J.; Lubberink, M.; Almkvist, O.; Wall, A.; et al. Imaging in-vivo tau pathology in Alzheimer's disease with THK5317 PET in a multimodal paradigm. *Eur. J. Nucl. Med. Mol. Imaging* **2016**, *43*, 1686–1699. [CrossRef] [PubMed]
163. Tago, T.; Furumoto, S.; Okamura, N.; Harada, R.; Adachi, H.; Ishikawa, Y.; Yanai, K.; Iwata, R.; Kudo, Y. Structure-Activity Relationship of 2-Arylquinolines as PET Imaging Tracers for Tau Pathology in Alzheimer Disease. *J. Nucl. Med.* **2016**, *57*, 608–614. [CrossRef] [PubMed]
164. Harada, R.; Okamura, N.; Furumoto, S.; Furukawa, K.; Ishiki, A.; Tomita, N.; Tago, T.; Hiraoka, K.; Watanuki, S.; Shidahara, M.; et al. 18F-THK5351: A Novel PET Radiotracer for Imaging Neurofibrillary Pathology in Alzheimer Disease. *J. Nucl. Med.* **2016**, *57*, 208–214. [CrossRef]
165. Sone, D.; Imabayashi, E.; Maikusa, N.; Okamura, N.; Furumoto, S.; Kudo, Y.; Ogawa, M.; Takano, H.; Yokoi, Y.; Sakata, M.; et al. Regional tau deposition and subregion atrophy of medial temporal structures in early Alzheimer's disease: A combined positron emission tomography/magnetic resonance imaging study. *Alzheimers Dement.* **2017**, *9*, 35–40. [CrossRef]
166. Betthauser, T.J.; Lao, P.J.; Murali, D.; Barnhart, T.E.; Furumoto, S.; Okamura, N.; Stone, C.K.; Johnson, S.C.; Christian, B.T. In Vivo Comparison of Tau Radioligands 18F-THK-5351 and 18F-THK-5317. *J. Nucl. Med.* **2017**, *58*, 996–1002. [CrossRef]

167. Edmondson, D.E.; Mattevi, A.; Binda, C.; Li, M.; Hubálek, F. Structure and mechanism of monoamine oxidase. *Curr. Med. Chem.* **2004**, *11*, 1983–1993. [CrossRef]
168. Tipton, K.F.; Boyce, S.; O'Sullivan, J.; Davey, G.P.; Healy, J. Monoamine oxidases: certainties and uncertainties. *Curr. Med. Chem.* **2004**, *11*, 1965–1982. [CrossRef]
169. Ishiki, A.; Harada, R.; Kai, H.; Sato, N.; Totsune, T.; Tomita, N.; Watanuki, S.; Hiraoka, K.; Ishikawa, Y.; Funaki, Y.; et al. Neuroimaging-pathological correlations of 18FTHK5351 PET in progressive supranuclear palsy. *Acta Neuropathol. Commun.* **2018**, *6*, 53. [CrossRef] [PubMed]
170. Harada, R.; Ishiki, A.; Kai, H.; Sato, N.; Furukawa, K.; Furumoto, S.; Tago, T.; Tomita, N.; Watanuki, S.; Hiraoka, K.; et al. Correlations of 18F-THK5351 PET with Postmortem Burden of Tau and Astrogliosis in Alzheimer Disease. *J. Nucl. Med.* **2018**, *59*, 671–674. [CrossRef] [PubMed]
171. Krebs, M.R.H.; Bromley, E.H.C.; Donald, A.M. The binding of thioflavin-T to amyloid fibrils: localisation and implications. *J. Struct. Biol.* **2005**, *149*, 30–37. [CrossRef] [PubMed]
172. Maruyama, M.; Shimada, H.; Suhara, T.; Shinotoh, H.; Ji, B.; Maeda, J.; Zhang, M.-R.; Trojanowski, J.Q.; Lee, V.M.-Y.; Ono, M.; et al. Imaging of tau pathology in a tauopathy mouse model and in Alzheimer patients compared to normal controls. *Neuron* **2013**, *79*, 1094–1108. [CrossRef] [PubMed]
173. Hashimoto, H.; Kawamura, K.; Igarashi, N.; Takei, M.; Fujishiro, T.; Aihara, Y.; Shiomi, S.; Muto, M.; Ito, T.; Furutsuka, K.; et al. Radiosynthesis, photoisomerization, biodistribution, and metabolite analysis of 11C-PBB3 as a clinically useful PET probe for imaging of tau pathology. *J. Nucl. Med.* **2014**, *55*, 1532–1538. [CrossRef] [PubMed]
174. Koga, S.; Ono, M.; Sahara, N.; Higuchi, M.; Dickson, D.W. Fluorescence and autoradiographic evaluation of tau PET ligand PBB3 to α-synuclein pathology. *Mov. Disord.* **2017**, *32*, 884–892. [CrossRef]
175. Perez-Soriano, A.; Arena, J.E.; Dinelle, K.; Miao, Q.; McKenzie, J.; Neilson, N.; Puschmann, A.; Schaffer, P.; Shinotoh, H.; Smith-Forrester, J.; et al. PBB3 imaging in Parkinsonian disorders: Evidence for binding to tau and other proteins. *Mov. Disord.* **2017**, *32*, 1016–1024. [CrossRef]
176. Hsieh, C.-J.; Mach, R.H.; Zhude, T.; Kotzbauer, P.T. Imaging of Aggregated Alpha-Synuclein in Parkinson's Disease: A Work in Progress. *The newsletter of the SNMI. Centre for Molecular Imaging Innovation and Translation.* 2018, 12. Available online: http://s3.amazonaws.com/rdcmssnmmi/files/production/public/FileDownloads/MiGateway_2_2018_final.pdf (accessed on 8 December 2019).
177. Cashion, D.K.; Chen, G.; Kasi, D.; Kolb, H.C.; Liu, C.; Sinha, A.; Szardenings, A.K.; Wang, E.; Yu, C.; Zhang, W.; et al. Imaging Agents for Detecting Neurological Disorders. Available online: https://patentscope.wipo.int/search/en/detail.jsf?docId=WO2011119565 (accessed on 8 December 2019).
178. Xia, C.-F.; Arteaga, J.; Chen, G.; Gangadharmath, U.; Gomez, L.F.; Kasi, D.; Lam, C.; Liang, Q.; Liu, C.; Mocharla, V.P.; et al. (18)FT807, a novel tau positron emission tomography imaging agent for Alzheimer's disease. *Alzheimers Dement.* **2013**, *9*, 666–676. [CrossRef]
179. Zhang, W.; Arteaga, J.; Cashion, D.K.; Chen, G.; Gangadharmath, U.; Gomez, L.F.; Kasi, D.; Lam, C.; Liang, Q.; Liu, C.; et al. A highly selective and specific PET tracer for imaging of tau pathologies. *J. Alzheimers Dis.* **2012**, *31*, 601–612. [CrossRef]
180. Chien, D.T.; Bahri, S.; Szardenings, A.K.; Walsh, J.C.; Mu, F.; Su, M.-Y.; Shankle, W.R.; Elizarov, A.; Kolb, H.C. Early clinical PET imaging results with the novel PHF-tau radioligand F-18-T807. *J. Alzheimers Dis.* **2013**, *34*, 457–468. [CrossRef] [PubMed]
181. Mintun, M.; Schwarz, A.; Joshi, A.; Shcherbinin, S.; Chien, D.; Elizarov, A.; Su, M.-Y.; Shankle, W.; Pontecorvo, M.; Tauscher, J.; et al. Exploratory analyses of regional human brain distribution of the PET tau tracer F18-labeled T807 (AV-1541) in subjects with normal cognitive function or cognitive impairment thought to be due to Alzheimer's disease. *Alzheimer's Dement.* **2013**, *9*, P842. [CrossRef]
182. Ono, M.; Sahara, N.; Kumata, K.; Ji, B.; Ni, R.; Koga, S.; Dickson, D.W.; Trojanowski, J.Q.; Lee, V.M.-Y.; Yoshida, M.; et al. Distinct binding of PET ligands PBB3 and AV-1451 to tau fibril strains in neurodegenerative tauopathies. *Brain* **2017**, *140*, 764–780. [CrossRef] [PubMed]
183. Cho, H.; Choi, J.Y.; Lee, S.H.; Ryu, Y.H.; Lee, M.S.; Lyoo, C.H. 18 F-AV-1451 binds to putamen in multiple system atrophy. *Mov. Disord.* **2017**, *32*, 171–173. [CrossRef] [PubMed]
184. Leuzy, A.; Chiotis, K.; Lemoine, L.; Gillberg, P.-G.; Almkvist, O.; Rodriguez-Vieitez, E.; Nordberg, A. Tau PET imaging in neurodegenerative tauopathies-still a challenge. *Mol. Psychiatry* **2019**, *24*, 1112–1134. [CrossRef] [PubMed]

185. Kolb, H.C.; Andrés, J.I. Tau Positron Emission Tomography Imaging. *Cold Spring Harb. Perspect. Biol.* **2017**, *9*. [CrossRef]
186. Lowe, V.J.; Curran, G.; Fang, P.; Liesinger, A.M.; Josephs, K.A.; Parisi, J.E.; Kantarci, K.; Boeve, B.F.; Pandey, M.K.; Bruinsma, T.; et al. An autoradiographic evaluation of AV-1451 Tau PET in dementia. *Acta Neuropathol. Commun.* **2016**, *4*, 58. [CrossRef]
187. Hostetler, E.D.; Walji, A.M.; Zeng, Z.; Miller, P.; Bennacef, I.; Salinas, C.; Connolly, B.; Gantert, L.; Haley, H.; Holahan, M.; et al. Preclinical Characterization of 18F-MK-6240, a Promising PET Tracer for In Vivo Quantification of Human Neurofibrillary Tangles. *J. Nucl. Med.* **2016**, *57*, 1599–1606. [CrossRef]
188. Vermeiren, C.; Motte, P.; Viot, D.; Mairet-Coello, G.; Courade, J.-P.; Citron, M.; Mercier, J.; Hannestad, J.; Gillard, M. The tau positron-emission tomography tracer AV-1451 binds with similar affinities to tau fibrils and monoamine oxidases. *Mov. Disord.* **2018**, *33*, 273–281. [CrossRef]
189. Wooten, D.W.; Guehl, N.J.; Verwer, E.E.; Shoup, T.M.; Yokell, D.L.; Zubcevik, N.; Vasdev, N.; Zafonte, R.D.; Johnson, K.A.; El Fakhri, G.; et al. Pharmacokinetic Evaluation of the Tau PET Radiotracer 18F-T807 (18F-AV-1451) in Human Subjects. *J. Nucl. Med.* **2017**, *58*, 484–491. [CrossRef]
190. Shcherbinin, S.; Schwarz, A.J.; Joshi, A.; Navitsky, M.; Flitter, M.; Shankle, W.R.; Devous, M.D.; Mintun, M.A. Kinetics of the Tau PET Tracer 18F-AV-1451 (T807) in Subjects with Normal Cognitive Function, Mild Cognitive Impairment, and Alzheimer Disease. *J. Nucl. Med.* **2016**, *57*, 1535–1542. [CrossRef] [PubMed]
191. Declercq, L.; Celen, S.; Lecina, J.; Ahamed, M.; Tousseyn, T.; Moechars, D.; Alcazar, J.; Ariza, M.; Fierens, K.; Bottelbergs, A.; et al. Comparison of New Tau PET-Tracer Candidates With 18FT808 and 18FT807. *Mol. Imaging* **2016**, *15*. [CrossRef] [PubMed]
192. Wang, Y.T.; Edison, P. Tau Imaging in Neurodegenerative Diseases Using Positron Emission Tomography. *Curr. Neurol. Neurosci. Rep.* **2019**, *19*, 45. [CrossRef] [PubMed]
193. Sanabria Bohórquez, S.; Marik, J.; Ogasawara, A.; Tinianow, J.N.; Gill, H.S.; Barret, O.; Tamagnan, G.; Alagille, D.; Ayalon, G.; Manser, P.; et al. 18FGTP1 (Genentech Tau Probe 1), a radioligand for detecting neurofibrillary tangle tau pathology in Alzheimer's disease. *Eur. J. Nucl. Med. Mol. Imaging* **2019**, *46*, 2077–2089. [CrossRef]
194. Barret, O.; Alagille, D.; Sanabria, S.; Comley, R.A.; Weimer, R.M.; Borroni, E.; Mintun, M.; Seneca, N.; Papin, C.; Morley, T.; et al. Kinetic Modeling of the Tau PET Tracer 18F-AV-1451 in Human Healthy Volunteers and Alzheimer Disease Subjects. *J. Nucl. Med.* **2017**, *58*, 1124–1131. [CrossRef]
195. Seki, C.; Tagai, K.; Shimada, H.; Takahata, K.; Kubota, M.; Takado, Y.; Shinitoh, H.; Kimura, Y.; Ichise, M.; Okada, M. Establishment of a Simplified Method to Quantify [18F]PM-PBB3 ([18F]APN-1607) Binding in the Brains of Living Human Subjects. 2019. Available online: https://repo.qst.go.jp/?action=pages_view_main&active_action=repository_view_main_item_detail&item_id=78239&item_no=1&page_id=13&block_id=21 (accessed on 18 February 2020).
196. 18F-PM-PBB3 PET Study in Tauopathy Including Alzheimer's Disease, Other Dementias and Normal Controls-Full Text View-ClinicalTrials.gov. Available online: https://clinicaltrials.gov/ct2/show/study/NCT03625128 (accessed on 10 December 2019).
197. Brendel, M.; Yousefi, B.H.; Blume, T.; Herz, M.; Focke, C.; Deussing, M.; Peters, F.; Lindner, S.; von Ungern-Sternberg, B.; Drzezga, A.; et al. Comparison of 18F-T807 and 18F-THK5117 PET in a Mouse Model of Tau Pathology. *Front. Aging Neurosci.* **2018**, *10*, 174. [CrossRef]
198. Wong, D.F.; Comley, R.A.; Kuwabara, H.; Rosenberg, P.B.; Resnick, S.M.; Ostrowitzki, S.; Vozzi, C.; Boess, F.; Oh, E.; Lyketsos, C.G.; et al. Characterization of 3 Novel Tau Radiopharmaceuticals, 11C-RO-963, 11C-RO-643, and 18F-RO-948, in Healthy Controls and in Alzheimer Subjects. *J. Nucl. Med.* **2018**, *59*, 1869–1876. [CrossRef]
199. Gobbi, L.C.; Knust, H.; Körner, M.; Honer, M.; Czech, C.; Belli, S.; Muri, D.; Edelmann, M.R.; Hartung, T.; Erbsmehl, I.; et al. Identification of Three Novel Radiotracers for Imaging Aggregated Tau in Alzheimer's Disease with Positron Emission Tomography. *J. Med. Chem.* **2017**, *60*, 7350–7370. [CrossRef]
200. Honer, M.; Gobbi, L.; Knust, H.; Kuwabara, H.; Muri, D.; Koerner, M.; Valentine, H.; Dannals, R.F.; Wong, D.F.; Borroni, E. Preclinical Evaluation of 18F-RO6958948, 11C-RO6931643, and 11C-RO6924963 as Novel PET Radiotracers for Imaging Tau Aggregates in Alzheimer Disease. *J. Nucl. Med.* **2018**, *59*, 675–681. [CrossRef]
201. Marquié, M.; Normandin, M.D.; Vanderburg, C.R.; Costantino, I.M.; Bien, E.A.; Rycyna, L.G.; Klunk, W.E.; Mathis, C.A.; Ikonomovic, M.D.; Debnath, M.L.; et al. Validating novel tau positron emission tomography tracer F-18-AV-1451 (T807) on postmortem brain tissue. *Ann. Neurol.* **2015**, *78*, 787–800. [CrossRef]

202. Walji, A.M.; Hostetler, E.D.; Selnick, H.; Zeng, Z.; Miller, P.; Bennacef, I.; Salinas, C.; Connolly, B.; Gantert, L.; Holahan, M.; et al. Discovery of 6-(Fluoro-(18)F)-3-(1H-pyrrolo2,3-cpyridin-1-yl)isoquinolin-5-amine ((18)F-MK-6240): A Positron Emission Tomography (PET) Imaging Agent for Quantification of Neurofibrillary Tangles (NFTs). *J. Med. Chem.* **2016**, *59*, 4778–4789. [CrossRef] [PubMed]
203. Pascoal, T.A.; Shin, M.; Kang, M.S.; Chamoun, M.; Chartrand, D.; Mathotaarachchi, S.; Bennacef, I.; Therriault, J.; Ng, K.P.; Hopewell, R.; et al. In vivo quantification of neurofibrillary tangles with 18FMK-6240. *Alzheimers Res. Ther.* **2018**, *10*, 74. [CrossRef] [PubMed]
204. Betthauser, T.J.; Cody, K.A.; Zammit, M.D.; Murali, D.; Converse, A.K.; Barnhart, T.E.; Stone, C.K.; Rowley, H.A.; Johnson, S.C.; Christian, B.T. In Vivo Characterization and Quantification of Neurofibrillary Tau PET Radioligand 18F-MK-6240 in Humans from Alzheimer Disease Dementia to Young Controls. *J. Nucl. Med.* **2019**, *60*, 93–99. [CrossRef] [PubMed]
205. [18F]MK-6240 Positron Emission Tomography (PET) Tracer First-in-Human Validation Study (MK-6240-001)-Full Text View-ClinicalTrials.gov. Available online: https://clinicaltrials.gov/ct2/show/NCT02562989 (accessed on 8 December 2019).
206. Declercq, L.; Rombouts, F.; Koole, M.; Fierens, K.; Mariën, J.; Langlois, X.; Andrés, J.I.; Schmidt, M.; Macdonald, G.; Moechars, D.; et al. Preclinical Evaluation of 18F-JNJ64349311, a Novel PET Tracer for Tau Imaging. *J. Nucl. Med.* **2017**, *58*, 975–981. [CrossRef] [PubMed]
207. Politis, M. Neuroimaging in Parkinson disease: from research setting to clinical practice. *Nat. Rev. Neurol.* **2014**, *10*, 708–722. [CrossRef] [PubMed]
208. Dickson, D.W.; Braak, H.; Duda, J.E.; Duyckaerts, C.; Gasser, T.; Halliday, G.M.; Hardy, J.; Leverenz, J.B.; Del Tredici, K.; Wszolek, Z.K.; et al. Neuropathological assessment of Parkinson's disease: refining the diagnostic criteria. *Lancet Neurol.* **2009**, *8*, 1150–1157. [CrossRef]
209. Yu, L.; Cui, J.; Padakanti, P.K.; Engel, L.; Bagchi, D.P.; Kotzbauer, P.T.; Tu, Z. Synthesis and in vitro evaluation of α-synuclein ligands. *Bioorg. Med. Chem.* **2012**, *20*, 4625–4634. [CrossRef]
210. Bagchi, D.P.; Yu, L.; Perlmutter, J.S.; Xu, J.; Mach, R.H.; Tu, Z.; Kotzbauer, P.T. Binding of the radioligand SIL23 to α-synuclein fibrils in Parkinson disease brain tissue establishes feasibility and screening approaches for developing a Parkinson disease imaging agent. *PLoS ONE* **2013**, *8*, e55031. [CrossRef]
211. Zhang, X.; Jin, H.; Padakanti, P.K.; Li, J.; Yang, H.; Fan, J.; Mach, R.H.; Kotzbauer, P.; Tu, Z. Radiosynthesis and in Vivo Evaluation of Two PET Radioligands for Imaging α-Synuclein. *Appl. Sci.* **2014**, *4*, 66–78. [CrossRef]
212. Honson, N.S.; Johnson, R.L.; Huang, W.; Inglese, J.; Austin, C.P.; Kuret, J. Differentiating Alzheimer disease-associated aggregates with small molecules. *Neurobiol. Dis.* **2007**, *28*, 251–260. [CrossRef]
213. Chu, W.; Zhou, D.; Gaba, V.; Liu, J.; Li, S.; Peng, X.; Xu, J.; Dhavale, D.; Bagchi, D.P.; d'Avignon, A.; et al. Design, Synthesis, and Characterization of 3-(Benzylidene)indolin-2-one Derivatives as Ligands for α-Synuclein Fibrils. *J. Med. Chem.* **2015**, *58*, 6002–6017. [CrossRef] [PubMed]
214. Hsieh, C.-J.; Xu, K.; Lee, I.; Graham, T.J.A.; Tu, Z.; Dhavale, D.; Kotzbauer, P.; Mach, R.H. Chalcones and Five-Membered Heterocyclic Isosteres Bind to Alpha Synuclein Fibrils in Vitro. *ACS Omega* **2018**, *3*, 4486–4493. [CrossRef] [PubMed]
215. Ono, M.; Maya, Y.; Haratake, M.; Ito, K.; Mori, H.; Nakayama, M. Aurones serve as probes of beta-amyloid plaques in Alzheimer's disease. *Biochem. Biophys. Res. Commun.* **2007**, *361*, 116–121. [CrossRef] [PubMed]
216. Ono, M.; Haratake, M.; Mori, H.; Nakayama, M. Novel chalcones as probes for in vivo imaging of beta-amyloid plaques in Alzheimer's brains. *Bioorg. Med. Chem.* **2007**, *15*, 6802–6809. [CrossRef]
217. Ono, M.; Yoshida, N.; Ishibashi, K.; Haratake, M.; Arano, Y.; Mori, H.; Nakayama, M. Radioiodinated flavones for in vivo imaging of beta-amyloid plaques in the brain. *J. Med. Chem.* **2005**, *48*, 7253–7260. [CrossRef]
218. Meng, X.; Munishkina, L.A.; Fink, A.L.; Uversky, V.N. Effects of Various Flavonoids on the α-Synuclein Fibrillation Process. *Parkinsons. Dis.* **2010**, *2010*, 650794. [CrossRef]
219. Zhu, M.; Han, S.; Fink, A.L. Oxidized quercetin inhibits α-synuclein fibrillization. *Biochim. Biophys. Acta* **2013**, *1830*, 2872–2881. [CrossRef]
220. Masuda, M.; Suzuki, N.; Taniguchi, S.; Oikawa, T.; Nonaka, T.; Iwatsubo, T.; Hisanaga, S.-i.; Goedert, M.; Hasegawa, M. Small molecule inhibitors of alpha-synuclein filament assembly. *Biochemistry* **2006**, *45*, 6085–6094. [CrossRef]
221. Cui, M.; Ono, M.; Watanabe, H.; Kimura, H.; Liu, B.; Saji, H. Smart near-infrared fluorescence probes with donor-acceptor structure for in vivo detection of β-amyloid deposits. *J. Am. Chem. Soc.* **2014**, *136*, 3388–3394. [CrossRef]

222. Ono, M.; Doi, Y.; Watanabe, H.; Ihara, M.; Ozaki, A.; Saji, H. Structure–activity relationships of radioiodinated diphenyl derivatives with different conjugated double bonds as ligands for α-synuclein aggregates. *RSC Adv.* **2016**, *6*, 44305–44312. [CrossRef]
223. Fanti, S.; Bonfiglioli, R.; Decristoforo, C. Highlights of the 30th Annual Congress of the EANM, Vienna 2017: "Yes we can-make nuclear medicine great again". *Eur. J. Nucl. Med. Mol. Imaging* **2018**, *45*, 1781–1794. [CrossRef] [PubMed]
224. Wester, H.-J.; Yousefi, B.H. US20170157274A1-Compounds Binding to Neuropathological Aggregates-Google Patents. Available online: https://patents.google.com/patent/US20170157274A1/en (accessed on 24 December 2019).

© 2020 by the authors. Licensee MDPI, Basel, Switzerland. This article is an open access article distributed under the terms and conditions of the Creative Commons Attribution (CC BY) license (http://creativecommons.org/licenses/by/4.0/).

Article

Metabolic Brain Network Analysis of FDG-PET in Alzheimer's Disease Using Kernel-Based Persistent Features

Liqun Kuang [1],*, Deyu Zhao [1], Jiacheng Xing [2], Zhongyu Chen [3], Fengguang Xiong [1] and Xie Han [1],*

1. School of Data Science and Technology, North University of China, Taiyuan 030051, China; 1307084309@st.nuc.edu.cn (D.Z.); hopenxfg@nuc.edu.cn (F.X.)
2. School of Software, Nanchang University, Nanchang 330047, China; xingjiacheng628@163.com
3. School of Software, East China Jiaotong University, Nanchang 330013, China; 15079561913@139.com
* Correspondence: kuang@nuc.edu.cn (L.K.); hanxie@nuc.edu.cn (X.H.)

Academic Editor: Peter Brust
Received: 7 May 2019; Accepted: 20 June 2019; Published: 21 June 2019

Abstract: Recent research of persistent homology in algebraic topology has shown that the altered network organization of human brain provides a promising indicator of many neuropsychiatric disorders and neurodegenerative diseases. However, the current slope-based approach may not accurately characterize changes of persistent features over graph filtration because such curves are not strictly linear. Moreover, our previous integrated persistent feature (IPF) works well on an rs-fMRI cohort while it has not yet been studied on metabolic brain networks. To address these issues, we propose a novel univariate network measurement, kernel-based IPF (KBI), based on the prior IPF, to quantify the difference between IPF curves. In our experiments, we apply the KBI index to study fluorodeoxyglucose positron emission tomography (FDG-PET) imaging data from 140 subjects with Alzheimer's disease (AD), 280 subjects with mild cognitive impairment (MCI), and 280 healthy normal controls (NC). The results show the disruption of network integration in the progress of AD. Compared to previous persistent homology-based measures, as well as other standard graph-based measures that characterize small-world organization and modular structure, our proposed network index KBI possesses more significant group difference and better classification performance, suggesting that it may be used as an effective preclinical AD imaging biomarker.

Keywords: Alzheimer's disease (AD); network measure; graph theory; brain network; positron emission tomography (PET); persistent homology

---

1. Introduction

Alzheimer's disease (AD) is one of the most common neurodegenerative neurological diseases and is the most common form of dementia in the elderly [1]. Its clinical manifestations include long-term memory loss, cognitive decline, language disorders, and other symptoms. AD seriously affects the normal life of the elderly. However, the pathology of AD is not yet clear [1]. Some existing imaging technologies are used to explore the mechanisms of human brain function. Compared to magnetic resonance imaging (MRI), fluorodeoxyglucose positron emission tomography (FDG-PET) has been demonstrated to be a more precise predictor of both AD and mild cognitive impairment (MCI), and is more suitable for monitoring disease progression [2]. It collects and measures changes in glucose metabolism values in brain regions or local brain cells. The signals are then converted into effective three-dimensional images and the connectivity between brain regions are analyzed.

The topological organization of metabolic brain networks have been successfully characterized in many cases using various measures based on graph theory [3–5], such as characteristic path length

(CPL) [6], global efficiency [7], modularity (Mod) [7,8], and network diameter (ND) [9], to name but a few. Specifically, in patients with AD and MCI, several research groups have reported topological alterations in the whole-brain connectome, including a loss of small-worldness [10], a redistribution of hubs [9,11], and a disrupted modular organization [12]. Traditionally, weighted networks usually require defining a set of thresholding values before quantifying network topology [13–15], which may result in inconsistent network features when the thresholding values vary. Generally, the choice of threshold is rather arbitrary and there are no widely accepted criteria [16,17].

Recently, persistent homology [18] in algebraic topology has been studied to detect persistent structures generated over all possible thresholds [19–23] in brain network analysis. There have been significant efforts to model evolution of brain networks and to link network topology to network dynamics. This method constructs multiscale network for all possible thresholds wherever the persistent topological features over the evolution of the network changes are identified. Its ability to handle noisy data and provide homological information has turned it into a successful tool for the analysis of brain network structures [24]. One typical application of persistent homology is in a Betti number plot (BNP) [18], which has been successfully applied to the brain network research on epilepsy [20], autism spectrum disorder, and attention-deficit hyperactivity disorder [19,23]. As BNP ignores the association between persistent features and forthcoming thresholding value changes, we proposed an integrated persistent feature (IPF) by integrating an additional feature of connected component aggregation cost with BNP, and applied it to measure an AD network using resting state functional MRI (rs-fMRI) in our prior study [25]. However, both BNP and IPF applied linear regression analysis for computing the slope of the plot over all thresholds as a univariate network index. Such a slope-based approach may not accurately characterize the changes of persistent features over graph filtration because the curves are not strictly linear. Moreover, our previous IPF works well on an rs-fMRI cohort though it has not been used to study metabolic brain networks yet.

In this paper, we borrow the idea of kernel methods [26,27] on persistent homology and propose a kernel-based IPF (KBI) index based on our prior work on IPF. We hypothesized that our KBI index may help to better reveal the difference between brain networks. With the cross-sectional FDG-PET imaging data of 140 AD, 280 MCI, and 280 normal control (NC) individuals, we set out to test this hypothesis by computing the KBI indices that measure the differences between AD, MCI, and NC groups. We further perform statistical inference and classification to validate the power of KBI.

2. Results

In this section, we use FDG-PET data to evaluate statistical power and classification performance of our proposed KBI index for the analysis of brain metabolic networks related to AD. We further compared it with prior persistent features, BNP [19,21,23] and SIP [25], as well as some other standard graph-based indices.

2.1. Metabolic Brain Networks

After data preprocessing, the summarized point cloud were extracted from PET 3D imaging using predefined automated anatomical labeling atlas with 90 (AAL-90) regions of interests (ROI) [28]. We obtained the SUV matrix for all 700 subjects in all 90 ROIs and plot three histograms, in Figure 1, to show the global distributions of FDG uptake in AD, MCI, and NC. As the number of AD is half of MCI and NC, we have normalized the SUV distribution of AD by doubling its statistics. We observed that the AD cohort has lower glucose metabolism than MCI and NC, but no significant differences were detected in the statistical inference of permutation test. We calculated the Pearson-based correlation distance of FDG uptake between each pair of brain regions using Equation (1) and constructed group-wise brain metabolic networks. The three multiscale networks of AD, MCI, and NC groups are shown in Figure 2, which visualizes the evolution of brain networks over different thresholds.

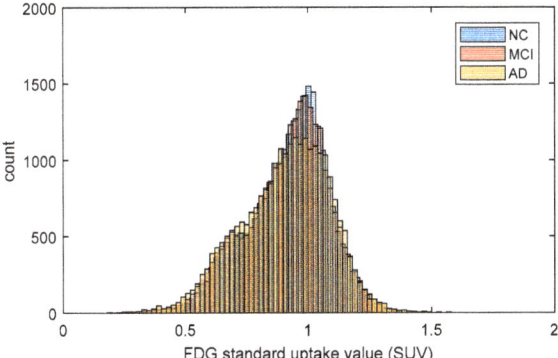

Figure 1. The fluorodeoxyglucose (FDG) uptake distribution of AD, MCI, and NC groups.

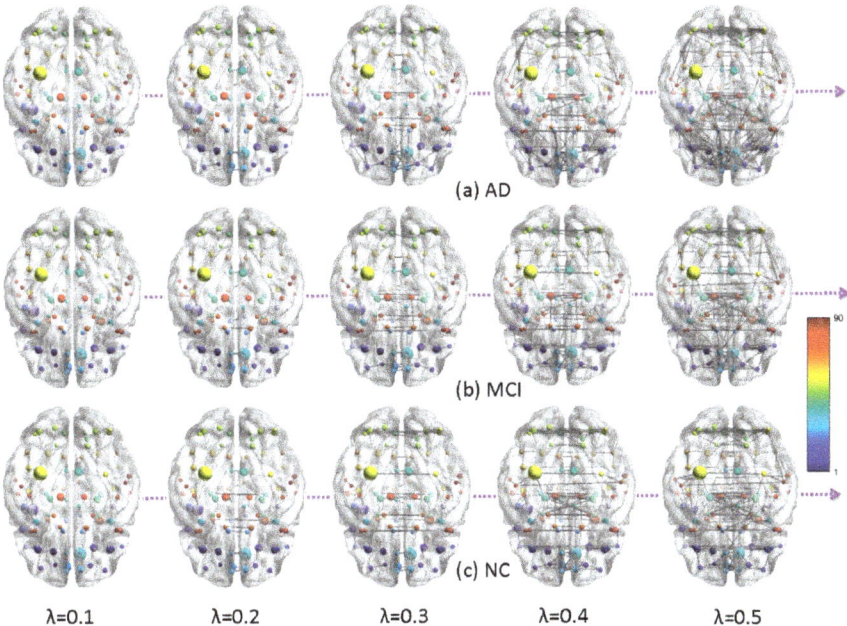

Figure 2. The constructed multiscale networks of AD, MCI, and NC by graph filtration λ, and the node color represents the ROI index predefined in AAL-90 atlas.

2.2. Brain Network Features

We computed the values of graph-based network indices (CPL, ND, and Mod) in three groups based on their weighted networks after filtering their edges, whose corresponding p-values passed a statistical threshold (Bonferroni corrected $p < 0.05$). We then obtained the multiscale network according to graph filtration and computed the BNP and IPF index (i.e., SIP), as well as KBI index. Figure 3 shows three separate IPF plots of AD, MCI, and NC. All brain network index values are shown in Table 1. The differences between groups need to be further verified by statistical inference and classification.

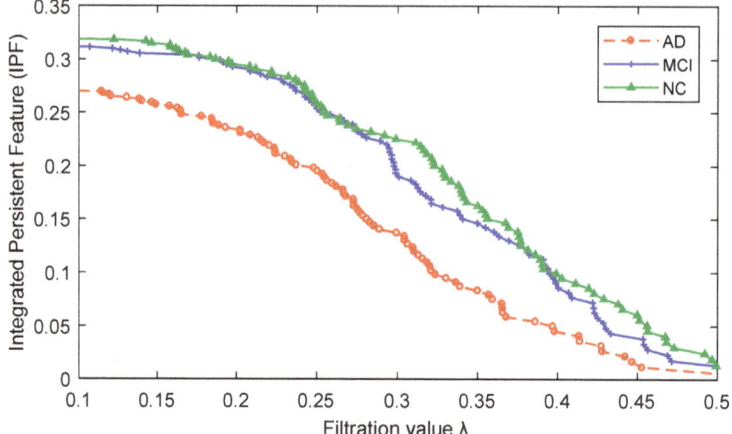

Figure 3. The proposed integrated persistent feature (IPF) plot for three group-wise networks of AD, MCI and NC, respectively.

Table 1. The network index values of three groups.

Cohort	KBI	SIP	BNP	CPL	ND	Mod
AD	0.577	0.790	268.7	1.20	1.601	1.289
MCI	0.826	0.842	265.2	1.17	1.457	0.770
NC	0.988	0.882	245.5	1.15	1.436	1.004

2.3. Statistical Group Difference Performance

In this study, we use the permutation test for 10,000 permutations between any two groups, and show the resulting p-value in Table 2. Only the proposed KBI index obtained a significant difference in any between-group at the significance level of 0.05.

Table 2. Statistical p-values for between-group differences by different graph indices on an AAL-90 atlas.

Cohort	KBI	SIP	BNP	CPL	ND	Mod
AD vs. MCI	**0.017**	0.036	0.048	0.058	0.397	0.074
AD vs. NC	**0.002**	0.042	0.054	0.285	0.266	0.063
MCI vs. NC	**0.036**	0.076	0.339	0.062	**0.031**	0.458

Only the proposed KBI index detected significant difference in any between-group (any p-value < 0.05).

2.4. Classification Performance

Furthermore, we resampled the networks 5000 times for each group with the resampling rate of 0.5, and obtained 5000 values of each network index for each group. We then performed leave-one-out crossvalidation to evaluate the classification powers of two-label (Figures 4 and 5) and three-label (Figure 6) by SVM. Our KBI shows better classification performance than prior persistent features, SIP and BNP, as well as other standard graph-based features, including CPL, ND, and Mod.

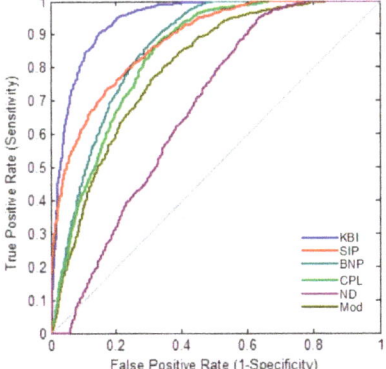

Figure 4. Comparison of ROC curves of different network indices for MCI vs. NC.

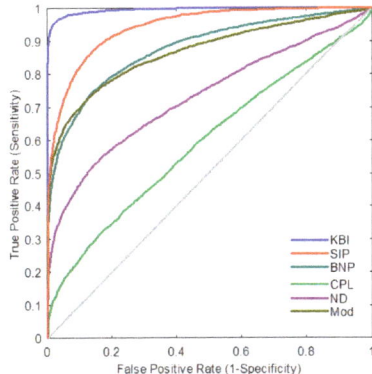

Figure 5. Comparisons of ROC curves of different network indices for AD vs. NC.

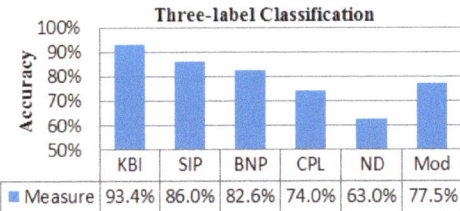

Figure 6. Three-label classification.

3. Discussion

3.1. Present Findings

This study has three main findings.

First, from Figure 1, we found that the AD cohort has lower glucose metabolism than MCI and NC. This may imply cognitive impairment in AD and MCI. Such an inference is further partly confirmed by graph theory analysis because a larger CPL is present in AD and MCI, while the network with smaller CPL is considered to be efficient.

Second, in our previous study [25], we had developed a univariate network index, SIP, based on homology to model graph dynamics over all possible scales and applied it to study the rs-fMRI data of

AD. We found the SIP values of AD were lower than MCI and much lower than NC. In the current PET data, we still find the SIP values show the same pattern AD < MCI < NC, suggesting a slower network integration rate in AD and MCI groups. Thus, the results from both independent cohorts provide consistent empirical evidence for decreased functional integration in AD dementia and MCI.

Finally, we propose a novel univariate network index KBI to enhance our previous study based on persistent homology. With our univariate KBI index, the difference of persistent features between cognitive dysfunction and NC brain network can be measured more accurately. Our preliminary experimental results demonstrate that the proposed KBI may greatly boost prior SIP and BNP power in both statistical inference and classification analyses. The KBI also outperforms other standard graph-based methods, such as CPL, ND, and Mod, suggesting that our method may serve as a valuable preclinical AD imaging biomarker.

3.2. Exploring Other Connectivity Definitions

There are many types of distance functions to construct weighted networks in brain network analysis [29], such as Pearson correlation, partial correlation, psycho–physiological interactions, ReHo, partial least squares, wavelet-based correlation, mutual information, synchronization likelihood, principal component analysis, independent component analysis, cluster analysis, dynamic causal modeling, Granger causality modeling, structural equation modeling, and multivariate autoregressive modeling, among others. At present, it is difficult to put forward the evaluation criteria of these methods, and few studies have compared them comprehensively. Although the Pearson correlation that we used in this study may be the most practical scheme to define the connectome in AD studies, there is still debate about the choice of connectivity definition [29,30]. Therefore, we performed four other connectivity definitions to explore more potentials in defining connectivity network. They were Kendall correlation [31], Spearman correlation [32], partial least squares [33], and Granger causality modeling [34]. The obtained p-values of our proposed KBI with these distance functions are shown in Table 3. There was no significant difference if Granger causality modeling was used, while the other three methods detected at least one significant difference. It should be noted that none of these methods performed significantly better than Pearson correlation (Table 1) in our current dataset. Moreover, when we checked all measures to discriminate AD, MCI, and NC by these methods, we found that the three-label classification accuracy of BNP (88.3%) was improved greatly if the partial least squares method was applied, while the performances in other cases have not been improved significantly. Such an empirical study may justify the connectivity definition adopted in our current work.

Table 3. Statistical p-values for between-group differences of KBI by different connectivity definitions.

Between-Group	Definitions of Distance Function			
	Kendall Correlation	Spearman Correlation	Partial Least Squares	Granger Causality Modeling
AD vs. MCI	0.003	0.007	0.159	0.399
AD vs. NC	0.005	0.003	0.047	0.375
MCI vs. NC	0.109	0.224	0.198	0.238

3.3. Ways of Network Construction

Graph-based brain connectome analyses are sensitive to the choice of parcellation schemes. To assess the effects of different parcellation strategies, we carried out the same set of analyses with another commonly employed atlas, the Harvard–Oxford atlas [35,36] with 110 ROIs (HOA-110). The detailed statistical significances of between-group difference on HOA-110 are presented in Table 4. Again, our proposed KBI achieved better statistical power.

Table 4. Statistical *p*-values for between-group difference of different network indices on HOA-110 atlas.

Between-Group	KBI	SIP	BNP	CPL	ND	Mod
AD vs. MCI	**0.022**	**0.032**	0.063	0.092	0.424	**0.035**
AD vs. NC	**0.013**	0.053	**0.033**	0.458	0.290	0.081
MCI vs. NC	**0.042**	0.051	0.242	0.088	0.076	0.404

Similarly, only the proposed KBI index detected significant difference in any between-group.

In the metabolic network construction, the common practice is building a group-wise brain network for each group as there is only one summarized value (average SUV) in each ROI. However, we notice that some studies [37] constructed subject-wise networks by dividing each ROI of a subject into blocks to obtain the correlation distance between any two ROI. Thus, subject-wise networks were constructed. We did not study this method in as it would require additional discussion that would exceed the scope of this paper.

In addition, we defined the connectivity between two brain regions as 1-Pearson correlation in Equation (1) in our study. Although some existing studies [19,23] also define such kinds of connectivity in analyzing brain network properties, the common practice in brain network analysis based on graph theory is to directly specify the Pearson correlation as the edge weight. To assess the effect of different connectivity definitions on other graph-based measures, we performed statistical inference on the brain networks whose edges were defined as directly based on Pearson correlation, and the statistical *p*-values of graph-based measures are shown in Table 5. We found the results were different from the previous results in Table 2, suggesting that graph-based measures could be affected by way of connectivity definition. We also found that none of the graph-based measures could detect all between-group differences significantly in either connectivity definition.

Table 5. Statistical *p*-values for between-group differences by specifying Pearson correlation as connectivity directly.

Between-Group	CPL	ND	Mod
AD vs. MCI	0.214	0.040	0.081
AD vs. NC	0.063	0.083	0.173
MCI vs. NC	0.004	0.281	0.005

3.4. Limitations and Future Work

Despite the promising results obtained by applying our proposed network index KBI based on persistent homology to PET, there are three important caveats. First, the current study only takes the zeroth persistent homology into account. Higher-order persistent features are also worth studying. In future, we will try to improve the performance of our method by considering higher dimensional persistent homology, such as the first Betti number, which is designed to calculate the number of holes in a graph and may boost the performance, especially in sparse networks that tend to have more holes. Then, although the subject-wise network is more convenient, efficient, and useful than group-wise network for brain network analysis, as we discussed in our prior study [25], we only measured the group-wise metabolic brain network according to regular practice in PET data analysis. In future, we will validate the KBI in a subject-wise metabolic network. Finally, this study was based on cross-sectional PET analysis, and we compared their network indices. With longitudinal PET analysis, we may further study the evolution between longitudinal brain networks by quantifying the difference of their persistent features.

4. Materials and Methods

Figure 7 shows the pipeline of our framework, where the data flow from FDG-PET brain images to some network indices. The details are described in following subsections.

4.1. Participants

Data used in the preparation of this article were obtained from the Alzheimer's Disease Neuroimaging Initiative (ADNI) database (adni.loni.usc.edu) [38,39]. ADNI was launched in 2003 as a public–private partnership led by Principal Investigator Michael W. Weiner, MD. The primary goal of ADNI has been to test whether serial magnetic resonance imaging (MRI), positron emission tomography (PET), other biological markers, and clinical and neuropsychological assessment can be combined to measure the progression of mild cognitive impairment (MCI) and early Alzheimer's disease (AD).

In this study, we chose 700 subjects with FDG-PET data from ADNI2. To match the three research cohorts of AD, MCI, and NC in gender and age, 140 AD, 280 MCI, and 280 NC subjects from 57 sites across North America were selected. The detailed cohort information is described in Table 6.

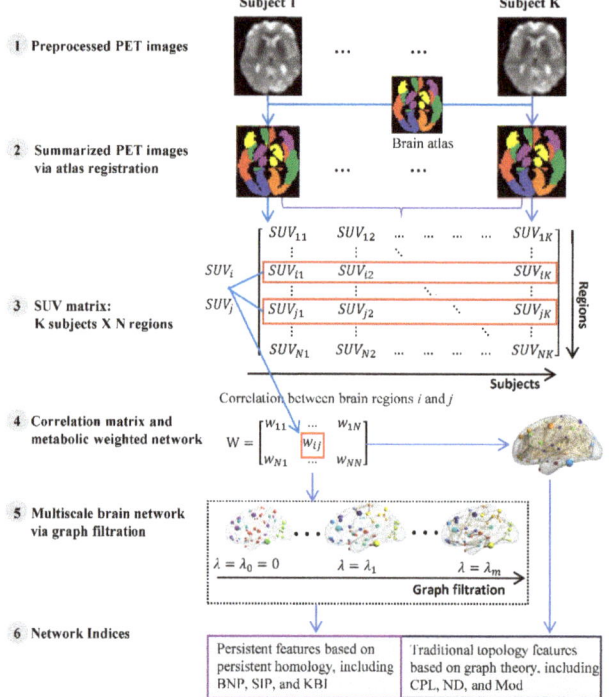

Figure 7. The pipeline of the brain network analysis based on a group of subjects. SUV, support vector machine; BNP, Betti number plot; SIP, slope of IPF plot; KBI, kernel-based IPF; CPL, characteristic path length; ND, network diameter; Mod, modularity.

Table 6. Demographic information of the subjects in this study.

	AD ($n = 140$)	MCI ($n = 280$)	NC ($n = 280$)	p-value [c]
Age [a]	74.2 ± 7.8	73.9 ± 8.0	75.0 ± 6.6	0.20
Gender [b]	70/70	140/140	140/140	1
CDR	≥ 1	0.5	0	–

Key: AD, Alzheimer's disease; MCI, mild cognitive impairment; NC, normal control; [a] mean ± SD; [b] male/female number; [c] statistical group significance using ANOVA test.

4.2. FDG-PET Data Acquisition and Preprocessing

All FDG-PET scans were obtained using Siemens, GE, and Philips PET scanners. Details of the PET data acquisition is described at http://adni.loni.usc.edu/methods/pet-analysis/pre-processing/. All FDG-PET scans used in this research are preprocessed (step 1 of Figure 7) as follows [40]. First, in order to eliminate the individual differences in brain morphology between subjects such that they can completely coincide and be subject to effective statistical analysis, we used the software toolkit Statistical Parametric Mapping (SPM8) [41] in MATLAB (Mathworks Inc, Natick, MA, USA) to linearly align the images into the Montreal Neurological Institute (MNI) space using the TPM.nii template file released with SPM. Second, we borrow a brain mask from SPM, exclude the brain stem and only keep the cerebral cortex (because the cerebral cortex is the object of this study), and then segmented all the images with this cerebral mask. Third, we conducted spatial smoothing with a Gaussian kernel of the full width at half maximum (FWHM) equal to (8,8,8) in three directions (x,y,z) to improve signal-to-noise.

4.3. Network Construction

A weighted graph is a natural and efficient way to represent metabolic brain network because it represents a discretized version of original PET images. In computer graphics, polygon meshes, as a class of graphs with particular connectivity restrictions, are extensively used to represent the topology of an object [42]; however, the mesh representation may not be the most suitable representation for analyzing PET images because of connectivity restrictions [43]. Here, we extend polygon meshes to general graphs by relaxing the connectivity restrictions. Such graphs are easier to construct and are flexible enough to capture metabolic information. We construct a weighted network by encoding the metabolic information through an adjacency matrix $W = \{w_{ij}\}$. The node corresponds to the brain regions, and the edge corresponds to the interregional correlation of brain metabolism. Specifically, the region parcellation in brain imaging is usually defined based on an anatomical atlas. In this study, we applied a predefined atlas, an automated anatomical labeling atlas with 90 (AAL-90) regions of interests (ROI) [28]. Once an ROI is specified, an overall summary measure within it can be calculated to assess the response as a whole, rather than on a voxel-by-voxel basis (step 2 of Figure 7). The most straightforward way to do so is by taking the average standard uptake values (SUV) of all voxels within the ROI. The SUV of a specific ROI is $\frac{1}{M} \sum_{p=1}^{M} v_p$, where M is the total voxel number in a given ROI and v_p is the FDG uptake value of voxel p. Given K subjects and N brain regions, let $SUV_i = \{SUV_{i1}, SUV_{i2}, \ldots, SUV_{iN}\} (1 \leq i \leq N)$ be the vector of average SUV in i-th ROI of all K subjects (step 3 of Figure 7), and the edge weight w_{ij} between two brain regions is defined as 1-Pearson correlation of SUV between them (step 4 of Figure 7), i.e.

$$w_{ij} = 1 - \frac{cov(SUV_i, SUV_j)}{\sigma_{SUV_i}\sigma_{SUV_j}}, \quad (1)$$

where SUV_i, SUV_j are the average SUV in i-th and j-th brain region respectively, cov is the covariance, σ is the standard deviation, and $\frac{cov(SUV_i, SUV_j)}{\sigma_{SUV_i}\sigma_{SUV_j}}$ is coefficient of Pearson correlation.

4.4. Network Indices

In clinical settings, doctors prefer single indices as biomarkers because a single neuroimaging index provides a practical reference for evaluating disease progression and for effective treatments. Generally, there are some available network indices based on graph theory that measure brain global attributes. In addition, we focus on some univariate indices that were developed from persistent homology in algebraic topology, and compare them with the network indices from traditional graph theory in our experiments.

4.4.1. Traditional Graph Theory Indices

Traditionally, graph theoretical analysis has been applied to measure brain network topological features. In this study, three global network indices based on graph theory are investigated, including characteristic path length (CPL) [6], network diameter (ND) [9], and modularity (Mod) [7].

Briefly, CPL can be understood as indicating a network with "easily" transferred information. It is the average shortest path length between all pairs of nodes in the graph, and is calculated as $CPL = \frac{1}{N(N-1)} \sum_{i \in V, j \in V, i \neq j} d_{i,j}$, where $d_{i,j}$ is the shortest path length between nodes i and j. Note that infinitely long paths (i.e., paths between disconnected nodes) are not included in computations. ND is the greatest distance between any pair of nodes, and is defined as $ND = \max_{i \in V} \max_{j \in V} d_{i,j}$. It enables understanding of the size of a network. A graph with a large ND and small CPL would therefore be considered an efficient network. Mod describes the extent to which a network has modules that differ from others, each of which is independent and functionally specialized [7]. Computationally, it is expressed as $Mod = \sum_{i \in M} \left[c_{ii} - \left(\sum_{j \in M} c_{ij} \right)^2 \right]$, where i and j are individual modules in the set of all modules M, and c is the proportion of existing connections between two modules.

In practice, we filtered the weighted network before computing these graph-based indices by only selecting the edges whose corresponding p-values passed through a statistical threshold (Bonferroni corrected $p < 0.05$) and then adopted the Brain Connectivity Toolbox (https://sites.google.com/site/bctnet/) [7] for their implementation (step 6 (right) of Figure 7).

4.4.2. Persistent Features Based on Persistent Homology

Persistent homology is an emerging mathematical concept for characterizing shapes in complex data, and the persistence features based on BNP are widely recognized as a useful feature descriptor. BNPs can distinguish robust and noisy topological properties over a wide range of graph filtrations based on the connectivity of k-dimensional simplicial complexes [18] (step 6 (left) of Figure 7). Graph filtration is an important tool [24] in persistent homology that constructs nesting subnetworks in a coherent manner and avoids thresholding selection (step 5 of Figure 7). BNPs have been successfully applied to measure brain networks based on FDG-PET and structural MRI data [23] in some neurodegenerative diseases. In our previous study [25], we proposed an integrated persistent feature (IPF) by integrating an additional feature of connected component aggregation cost with BNP to achieve holistic descriptions of graph evolutions. The IPF at filtration λ_i is defined as [25]

$$IPF_{\lambda_i} = \begin{cases} \frac{m-i}{m(m-1)} \sum_{k=i+1}^{m-1} \lambda_k, & 0 \leq i \leq m-2 \\ 0, & i = m-1 \end{cases} \quad (2)$$

where the maximal graph filtration is $\lambda_0 = 0 < \lambda_1 < \lambda_2 < \cdots < \lambda_{m-1}$. As the IPF plot over all possible filtration values is a monotonically decreasing convergence function, the absolute value of the slope of IPF plot (SIP) was defined as a univariate network index and was successfully applied to quantify brain network dynamics on rs-fMRI data of AD. Both the BNP and SIP indices indicate the rate of connecting components converging over the filtration value, and can be thought as the information diffusion rate or the convergence speed with said network.

4.4.3. The Kernel-Based IPF (KBI) Index

Although the SIP has been developed as a univariate network index in our previous study, it may not be the most appropriate way to describe IPF plot as it is nonlinear, strictly speaking. Recently, some kernel methods [26,27] have been defined on persistent homology to measure the distance between persistence diagrams, which are not only provably stable but also discriminative. Therefore, we employ the framework of kernel embedding of the IPF plot into reproducing kernel Hilbert spaces [26]. Given a point set of IPF plot $X = \{x_1, x_2, \cdots x_N\}$ and a template $T = \{t_1, t_2, \cdots t_N\}$ that are obtained from an average metabolic network of all NC subjects, the kernel-based IPF (KBI) index is defined as

$$KBI(X) = \frac{1}{N} \sum_{x_i \in X,\ t_i \in T,\ i=1}^{N} \tan^{-1}\left(C(\lambda_i^X)^p\right) \tan^{-1}\left(C(\lambda_i^T)^p\right) e^{-\frac{\|x_i - t_i\|^2}{2\sigma^2}}, \quad (3)$$

where $\tan^{-1}\left(C(\lambda_i^X)^p\right)$ and $\tan^{-1}\left(C(\lambda_i^T)^p\right)$ are both increasing functions with respect to maximal graph filtrations $\lambda^{(X)}$ of X and $\lambda^{(T)}$ of T, and are used for weighting the persistence (λ is a sequence of persistence of zeroth homology in fact). Hence, an essential persistence gives a large weight and a noisy persistence produces a small weight. By adjusting the parameters p, σ, and C, we can control the effect of the persistence. In our practice, we set $p = 5$,

$$\sigma = \underset{\forall X}{median}\{ \underset{x_i, x_j \in X,\ i<j}{median} \|x_i - x_j\| \}, \quad (4)$$

$$C = \left(\underset{\forall \lambda^X}{median} \{ \underset{\lambda_i^X \in \lambda^X}{median}(\lambda_i^X) \} \right)^{-p}, \quad (5)$$

so that they take close values to many $\|x_i - t_i\|$ and λ_i^X, respectively.

4.5. Statistical Analysis

We applied a group-wise statistical analysis of permutation test to all network indices in the last section between AD, MCI, and NC groups. As there are no prior studies on the statistical distribution of any network index, it is difficult to construct a parametric test procedure. Moreover, as there is only one group-wise network for each group, it is necessary to empirically construct the null distribution and determine the p-value. The steps of our permutation test are described as follows. First, the actual network index difference in means between two groups is calculated according to the actual grouping of their subjects. Second, the subjects are randomly assigned to two groups, each of which is assigned the same group size. We then construct two group-wise networks based on such permutated groups and recalculate their indices, whose difference is recorded. This process is repeated 10,000 times and 10,000 permuted differences are obtained. Finally, the total number of permuted differences larger than the actual difference is counted and divided by 10,000. The obtained value is the probability of no difference between the groups, that is, the p-value.

4.6. Classification

We evaluate the power of our method by classification analysis. In this study, our proposed KBI index and other comparison network indices have only one univariate feature to discriminate the global brain network structure. Since the samples are limited, we apply resampling technology beforehand. For a specific group, half subjects are removed randomly at a time and the remaining subjects are used to construct a group-wise network. Resampling all subjects for n times repeatedly in each group until the results are stable, we can obtain n group-wise networks. Each group-wise network is an average of all involved subjects graphs. In our practice, the number of resampling times is set to 5000 ($n = 5000$) and we obtain 5000 resampled networks for each group. Each network can yield a KBI index and other comparable network indices. Then, we compute the values of proposed KBI index, as well as other network measures for all resampled networks, and run support vector machine (SVM) [44] on them. We conduct leave-one-out crossvalidation experiments to evaluate the classification performance. The classification accuracy, sensitivity (i.e., true positive rate), specificity (i.e., true negative rate), and area under the curve (AUC) of receiver operating characteristic (ROC) are severed as criterion of classification performance.

5. Conclusions

This work proposed a novel network index KBI based on our prior work of persistent feature IPF to measure the metabolic brain network of FDG-PET on cognitively impaired cohort. The proposed KBI encoded a great deal of dynamic information over all possible scales that may be inaccessible by standard graph-based measures. Compared to previous slope-based approaches of persistent homology, our kernel-based network index is more accurate regarding the characterization of differences between persistent features. Our current results show that the slope of IPF plot present a pattern AD < MCI < NC in a FDG-PET cohort, consistent with our prior finding in an rs-fMRI cohort, and indicate a slower network integration rate in AD dementia and MCI. Moreover, the enhanced measurement KBI greatly boosted the performance of prior persistent features and outperformed some standard graph-based network indices in both statistical inference and classification experiments, suggesting that our method may serve as a valuable preclinical AD imaging biomarker.

Author Contributions: Conceptualization and methodology, L.K., D.Z. and X.H.; validation: D.Z., J.X. and Z.C.; formal analysis: L.K., F.X., and X.H.; writing: L.K., D.Z., and X.H. visualization, F.X.; software, J.X. and Z.C.

Funding: This work was supported by the National Natural Science Foundation of China (61672473 and 61602426 to L.K., F.X. and X.H.) and Shanxi Province Key R&D Technology Project (201803D121081 to L.K., F.X. and X.H.).

Acknowledgments: Data collection and sharing for this project was funded by the Alzheimer's Disease Neuroimaging Initiative (ADNI) (National Institutes of Health Grant U01 AG024904) and DOD ADNI (Department of Defense award number W81XWH-12-2-0012). ADNI is funded by the National Institute on Aging, the National Institute of Biomedical Imaging and Bioengineering, and through generous contributions from the following: AbbVie, Alzheimer's Association; Alzheimer's Drug Discovery Foundation; Araclon Biotech; BioClinica, Inc.; Biogen; Bristol-Myers Squibb Company; CereSpir, Inc.; Cogstate; Eisai Inc.; Elan Pharmaceuticals, Inc.; Eli Lilly and Company; EuroImmun; F. Hoffmann-La Roche Ltd. and its affiliated company Genentech, Inc.; Fujirebio; GE Healthcare; IXICO Ltd.; Janssen Alzheimer Immunotherapy Research & Development, LLC.; Johnson & Johnson Pharmaceutical Research & Development LLC.; Lumosity; Lundbeck; Merck & Co., Inc.; Meso Scale Diagnostics, LLC.; NeuroRx Research; Neurotrack Technologies; Novartis Pharmaceuticals Corporation; Pfizer Inc.; Piramal Imaging; Servier; Takeda Pharmaceutical Company; and Transition Therapeutics. The Canadian Institutes of Health Research is providing funds to support ADNI clinical sites in Canada. Private sector contributions are facilitated by the Foundation for the National Institutes of Health (www.fnih.org). The grantee organization is the Northern California Institute for Research and Education, and the study is coordinated by the Alzheimer's Therapeutic Research Institute at the University of Southern California. ADNI data are disseminated by the Laboratory for Neuro Imaging at the University of Southern California.

Conflicts of Interest: The authors declare no competing financial interest.

References

1. Lane, C.A.; Hardy, J.; Schott, J.M. Alzheimer's disease. *Eur. J. Neurol.* **2018**, *25*, 59–70. [CrossRef] [PubMed]
2. Ito, K.; Fukuyama, H.; Senda, M.; Ishii, K.; Maeda, K.; Yamamoto, Y.; Ouchi, Y.; Ishii, K.; Okumura, A.; Fujiwara, K. Prediction of outcomes in mild cognitive impairment by using 18F-FDG-PET: A multicenter study. *J. Alzheimers Dis.* **2015**, *45*, 543–552. [CrossRef] [PubMed]
3. Sporns, O. Graph theory methods: Applications in brain networks. *Dialogues Clin. Neuro.* **2018**, *20*, 111–121.
4. Bullmore, E.; Sporns, O. Complex brain networks: Graph theoretical analysis of structural and functional systems. *Nat. Rev. Neurosci.* **2009**, *10*, 186–198. [CrossRef] [PubMed]
5. Sporns, O. The human connectome: A complex network. *Ann. N. Y. Acad. Sci.* **2011**, *1224*, 109–125. [CrossRef] [PubMed]
6. Brier, M.R.; Thomas, J.B.; Fagan, A.M.; Hassenstab, J.; Holtzman, D.M.; Benzinger, T.L.; Morris, J.C.; Ances, B.M. Functional connectivity and graph theory in preclinical alzheimer's disease. *Neurobiol. Aging* **2014**, *35*, 757–768. [CrossRef] [PubMed]
7. Rubinov, M.; Sporns, O. Complex network measures of brain connectivity: Uses and interpretations. *NeuroImage* **2010**, *52*, 1059–1069. [CrossRef]
8. Sporns, O.; Betzel, R.F. Modular brain networks. *Annu. Rev. Psychol.* **2016**, *67*, 613–640. [CrossRef]
9. Fagerholm, E.D.; Hellyer, P.J.; Scott, G.; Leech, R.; Sharp, D.J. Disconnection of network hubs and cognitive impairment after traumatic brain injury. *Brain* **2015**, *138*, 1696–1709. [CrossRef]

10. Stam, C.; Jones, B.; Nolte, G.; Breakspear, M.; Scheltens, P. Small-world networks and functional connectivity in alzheimer's disease. *Cereb. Cortex* **2006**, *17*, 92–99. [CrossRef]
11. Qiu, T.; Luo, X.; Shen, Z.; Huang, P.; Xu, X.; Zhou, J.; Zhang, M. Disrupted brain network in progressive mild cognitive impairment measured by eigenvector centrality mapping is linked to cognition and cerebrospinal fluid biomarkers. *J. Alzheimers Dis.* **2016**, *54*, 1483–1493. [CrossRef] [PubMed]
12. De Haan, W.; Van Der Flier, W.M.; Koene, T.; Smits, L.L.; Scheltens, P.; Stam, C.J. Disrupted modular brain dynamics reflect cognitive dysfunction in alzheimer's disease. *NeuroImage* **2012**, *59*, 3085–3093. [CrossRef] [PubMed]
13. Tong, T.; Aganj, I.; Ge, T.; Polimeni, J.R.; Fischl, B. Functional density and edge maps: Characterizing functional architecture in individuals and improving cross-subject registration. *NeuroImage* **2017**, *158*, 346–355. [CrossRef] [PubMed]
14. Daianu, M.; Jahanshad, N.; Nir, T.M.; Jack, C.R.; Weiner, M.W.; Bernstein, M.A.; Thompson, P.M. Rich club analysis in the alzheimer's disease connectome reveals a relatively undisturbed structural core network. *Hum. Brain Mapp.* **2015**, *36*, 3087–3103. [CrossRef] [PubMed]
15. McKenna, F.; Koo, B.B.; Killiany, R. Comparison of apoe-related brain connectivity differences in early mci and normal aging populations: An fmri study. *Brain Imaging Behav.* **2016**, *10*, 970–983. [CrossRef] [PubMed]
16. Woo, C.-W.; Krishnan, A.; Wager, T.D. Cluster-extent based thresholding in fmri analyses: Pitfalls and recommendations. *NeuroImage* **2014**, *91*, 412–419. [CrossRef]
17. Zalesky, A.; Cocchi, L.; Fornito, A.; Murray, M.M.; Bullmore, E. Connectivity differences in brain networks. *NeuroImage* **2012**, *60*, 1055–1062. [CrossRef] [PubMed]
18. Edelsbrunner, H.; Harer, J. *Computational Topology: An Introduction*; American Mathematical Society: Heidelberg, Germany, 2010.
19. Lee, H.; Kang, H.; Chung, M.K.; Kim, B.-N.; Lee, D.S. Persistent brain network homology from the perspective of dendrogram. *IEEE Trans. Med. Imaging* **2012**, *31*, 2267–2277.
20. Choi, H.; Kim, Y.K.; Kang, H.; Lee, H.; Im, H.-J.; Kim, E.E.; Chung, J.-K.; Lee, D.S. Abnormal metabolic connectivity in the pilocarpine-induced epilepsy rat model: A multiscale network analysis based on persistent homology. *NeuroImage* **2014**, *99*, 226–236. [CrossRef]
21. Chung, M.K.; Hanson, J.L.; Ye, J.; Davidson, R.J.; Pollak, S.D. Persistent homology in sparse regression and its application to brain morphometry. *IEEE Trans. Med. Imaging* **2015**, *34*, 1928–1939. [CrossRef]
22. Yoo, K.; Lee, P.; Chung, M.K.; Sohn, W.S.; Chung, S.J.; Na, D.L.; Ju, D.; Jeong, Y. Degree-based statistic and center persistency for brain connectivity analysis. *Hum. Brain Mapp.* **2017**, *38*, 165–181. [CrossRef] [PubMed]
23. Lee, H.; Kang, H.; Chung, M.K.; Lim, S.; Kim, B.N.; Lee, D.S. Integrated multimodal network approach to pet and mri based on multidimensional persistent homology. *Hum. Brain Mapp.* **2017**, *38*, 1387–1402. [CrossRef] [PubMed]
24. Giusti, C.; Ghrist, R.; Bassett, D.S. Two's company, three (or more) is a simplex: Algebraic-topological tools for understanding higher-order structure in neural data. *J. Comput. Neurosci.* **2016**, *41*, 1–14. [CrossRef] [PubMed]
25. Kuang, L.; Han, X.; Chen, K. A concise and persistent feature to study brain resting-state network dynamics: Findings from the alzheimer's disease neuroimaging initiative. *Hum. Brain Mapp.* **2019**, *40*, 1062–1081. [CrossRef] [PubMed]
26. Kusano, G.; Hiraoka, Y.; Fukumizu, K. *Persistence Weighted Gaussian Kernel for Topological Data Analysis*; International Conference on Machine Learning: New York, NY, USA, 2016; pp. 2004–2013.
27. Carrière, M.; Cuturi, M.; Oudot, S. Sliced wasserstein kernel for persistence diagrams. *Working Papers* **2017**.
28. Tzourio-Mazoyer, N.; Landeau, B.; Papathanassiou, D.; Crivello, F.; Etard, O.; Delcroix, N.; Mazoyer, B.; Joliot, M. Automated anatomical labeling of activations in spm using a macroscopic anatomical parcellation of the mni mri single-subject brain. *NeuroImage* **2002**, *15*, 273–289. [CrossRef] [PubMed]
29. Friston, K.J. Functional and effective connectivity: A review. *Brain Connect* **2011**, *1*, 13–36. [CrossRef]
30. Valdes-Sosa, P.A.; Roebroeck, A.; Daunizeau, J.; Friston, K. Effective connectivity: Influence, causality and biophysical modeling. *NeuroImage* **2011**, *58*, 339–361. [CrossRef]
31. Kendall, M.G. A new measure of rank correlation. *Biometrika* **1938**, *30*, 81–93. [CrossRef]
32. Best, D.; Roberts, D. Algorithm as 89: The upper tail probabilities of spearman's rho. *J. R. Stat. Soc. Ser. C (Appl. Stat.)* **1975**, *24*, 377–379. [CrossRef]

33. McIntosh, A.R.; Chau, W.K.; Protzner, A.B. Spatiotemporal analysis of event-related fmri data using partial least squares. *NeuroImage* **2004**, *23*, 764–775. [CrossRef] [PubMed]
34. Roebroeck, A.; Formisano, E.; Goebel, R. The identification of interacting networks in the brain using fmri: Model selection, causality and deconvolution. *NeuroImage* **2011**, *58*, 296–302. [CrossRef] [PubMed]
35. Kennedy, D.; Lange, N.; Makris, N.; Bates, J.; Meyer, J.; Caviness Jr, V. Gyri of the human neocortex: An mri-based analysis of volume and variance. *Cereb. Cortex* **1998**, *8*, 372–384. [CrossRef] [PubMed]
36. Makris, N.; Meyer, J.W.; Bates, J.F.; Yeterian, E.H.; Kennedy, D.N.; Caviness, V.S. Mri-based topographic parcellation of human cerebral white matter and nuclei: Ii. Rationale and applications with systematics of cerebral connectivity. *NeuroImage* **1999**, *9*, 18–45. [CrossRef] [PubMed]
37. Yu, Z.; Yao, Z.; Zheng, W.; Jing, Y.; Ding, Z.; Mi, L.; Lu, S. *Predicting mci Progression with Individual Metabolic Network Based on Longitudinal FDG-PET*; IEEE International Conference on Bioinformatics & Biomedicine: Kansas City, MO, USA, 2017.
38. Jagust, W.J.; Bandy, D.; Chen, K.; Foster, N.L.; Landau, S.M.; Mathis, C.A.; Price, J.C.; Reiman, E.M.; Skovronsky, D.; Koeppe, R.A. The alzheimer's disease neuroimaging initiative positron emission tomography core. *Alzheimer's Dement.* **2010**, *6*, 221–229. [CrossRef]
39. Jack, C.R.; Bernstein, M.A.; Fox, N.C.; Thompson, P.; Alexander, G.; Harvey, D.; Borowski, B.; Britson, P.J.L.; Whitwell, J.; Ward, C. The alzheimer's disease neuroimaging initiative (adni): Mri methods. *J. Magn. Reson. Imaging* **2008**, *27*, 685–691. [CrossRef] [PubMed]
40. Mi, L.; Zhang, W.; Zhang, J.; Fan, Y.; Goradia, D.; Chen, K.; Reiman, E.M.; Gu, X.; Wang, Y. An optimal transportation based univariate neuroimaging index. In Proceedings of the IEEE International Conference on Computer Vision, Venice, Italy, 2017; pp. 182–191.
41. Penny, W.D.; Friston, K.J.; Ashburner, J.T.; Kiebel, S.J.; Nichols, T.E. *Statistical Parametric Mapping: The Analysis of Functional Brain Images*; Academic press: Pittsburgh, PA, USA, 2011.
42. De Floriani, L.; Magillo, P. Multiresolution mesh representation: Models and data structures. In *Tutorials on Multiresolution in Geometric Modelling*; Springer: Berlin/Heidelberg, Germany, 2002; pp. 363–417.
43. Chen, S.; Tian, D.; Feng, C.; Vetro, A.; Kovačević, J. Fast resampling of 3d point clouds via graphs. *arXiv* **2017**, arXiv:1702.06397. [CrossRef]
44. Cortes, C.; Vapnik, V. Support-vector networks. *Mach. Learn.* **1995**, *20*, 273–297. [CrossRef]

Sample Availability: Samples of the compounds are not available from the authors.

© 2019 by the authors. Licensee MDPI, Basel, Switzerland. This article is an open access article distributed under the terms and conditions of the Creative Commons Attribution (CC BY) license (http://creativecommons.org/licenses/by/4.0/).

Article

Evaluating the In Vivo Specificity of [^{18}F]UCB-H for the SV2A Protein, Compared with SV2B and SV2C in Rats Using microPET

Maria Elisa Serrano [1], Guillaume Becker [1], Mohamed Ali Bahri [1], Alain Seret [1], Nathalie Mestdagh [2], Joël Mercier [2], Frédéric Mievis [3], Fabrice Giacomelli [3], Christian Lemaire [1], Eric Salmon [1], André Luxen [1] and Alain Plenevaux [1,*]

[1] GIGA—CRC In Vivo Imaging, University of Liège, 8 Allée du 6 Août, Building B30, Sart Tilman, 4000 Liège, Belgium; meserrano@uliege.be (M.E.S.); guillaume.becker@sckcen.be (G.B.); M.Bahri@uliege.be (M.A.B.); aseret@uliege.be (A.S.); Christian.Lemaire@uliege.be (C.L.); Eric.Salmon@uliege.be (E.S.); aluxen@uliege.be (A.L.)
[2] UCB Pharma s.a., 1420 Braine-l'Alleud, Belgium; Nathalie.Mestdagh@ucb.com (N.M.); joel.mercier@ucb.com (J.M.)
[3] Nucleis s.a., University of Liège, 8 Allée du 6 Août, Building B30, Sart Tilman, 4000 Liège, Belgium; frederic.mievis@nucleis.eu (F.M.); fabrice.giacomelli@nucleis.eu (F.G.)
* Correspondence: alain.plenevaux@uliege.be; Tel.: +32-4-3662316

Received: 22 March 2019; Accepted: 29 April 2019; Published: 1 May 2019

Abstract: The synaptic vesicle protein 2 (SV2) is involved in synaptic vesicle trafficking. The SV2A isoform is the most studied and its implication in epilepsy therapy led to the development of the first SV2A PET radiotracer [^{18}F]UCB-H. The objective of this study was to evaluate in vivo, using microPET in rats, the specificity of [^{18}F]UCB-H for SV2 isoform A in comparison with the other two isoforms (B and C) through a blocking assay. Twenty Sprague Dawley rats were pre-treated either with the vehicle, or with specific competitors against SV2A (levetiracetam), SV2B (UCB5203) and SV2C (UCB0949). The distribution volume (Vt, Logan plot, t* 15 min) was obtained with a population-based input function. The Vt analysis for the entire brain showed statistically significant differences between the levetiracetam group and the other groups ($p < 0.001$), but also between the vehicle and the SV2B group ($p < 0.05$). An in-depth Vt analysis conducted for eight relevant brain structures confirmed the statistically significant differences between the levetiracetam group and the other groups ($p < 0.001$) and highlighted the superior and the inferior colliculi along with the cortex as regions also displaying statistically significant differences between the vehicle and SV2B groups ($p < 0.05$). These results emphasize the in vivo specificity of [^{18}F]UCB-H for SV2A against SV2B and SV2C, confirming that [^{18}F]UCB-H is a suitable radiotracer for in vivo imaging of the SV2A proteins with PET.

Keywords: SV2A; SV2B; SV2C; microPET; [^{18}F]UCB-H; epilepsy; PBIF; distribution volume; blocking assay; preclinical imaging

1. Introduction

The synaptic vesicle protein 2 (SV2) is an integral membrane protein with twelve transmembrane domains and three N-glycosylation sites in the intravesicular loop. The SV2 protein is ubiquitously present in the nerve terminals of the central and peripheral nervous systems, and in several types of endocrine cells [1]. This protein is critical for the adequate functioning of the central nervous system, acting as a modulator of synaptic transmission [2,3]. Moreover, it has been associated with the pathophysiology of epilepsy [4–6].

Previous studies have identified three SV2 isoforms: SV2A, SV2B and SV2C, characterized by different expression levels during rodent brain development [7] and adulthood [8]. While the SV2A isoform is present across all brain areas, the SV2C isoform can only be found in specific regions, such as the striatum, pallidum, midbrain, brainstem, substantia nigra, and the olfactory bulb [9]. The SV2B isoform is particularly present in the cerebral cortex, and the cornu ammonis sub-region of the hippocampus [10]. The three isoforms present large similarities in their structure: 65% between isoforms A and B, 62% between A and C, and 57% between B and C [8].

Of these three isoforms, SV2A is the most investigated. The antiepileptic drug levetiracetam (Keppra®) binds to SV2A, suggesting a role for SV2A in the pathology underlying certain forms of epilepsy [11–14]. Several studies have shown a correlation between the brain expression of this isoform and the clinical efficacy of this drug [5,13].

To investigate the role of SV2A in vivo, in 2013 [^{18}F]UCB-H was presented as an imaging agent with a nanomolar affinity for human SV2A [3,15–17]. Since then, other PET radiotracers, such as [^{11}C]UCB-J, or [^{11}C]UCB-A, have been synthetized to study this protein [3,18–20] (see Figure 1). These PET radiotracers appear to be more specific than [^{18}F]UCB-H (pIC$_{50}$ = 7.8) [3,16], based on their respective affinity measured in vitro, with pIC$_{50}$ = 8.2 for [^{11}C]UCB-J [18] and pIC$_{50}$ = 7.9 for [^{11}C]UCB-A [20]. The three radiotracers have demonstrated potential for use as synaptic density biomarkers not only in animals, but also in humans [3,21–23]. However, despite the valuable properties of [^{11}C]UCB-J and [^{11}C]UCB-A in assessing brain synaptic density in vivo, their clinical application is limited to facilities with a cyclotron due to the short half-life of ^{11}C (20.3 min) compared to the half-life of ^{18}F (110 min). In addition, the use of a PET radiotracer with a longer half-life (such as ^{18}F) allows the evaluation of a greater number of patients per day with just one production. Therefore, different fluorine-18-labelled derivatives of UCB-J are currently being developed and characterized, such as [^{18}F]SDM-8 [24]. The potential of [^{18}F]UCB-H for detecting variations in SV2A has already been demonstrated in vivo [25,26]. Nevertheless, as the actual specificity of [^{18}F]UCB-H for SV2A against SV2B and SV2C has never been addressed in vivo, we consider that this point deserves more careful evaluation.

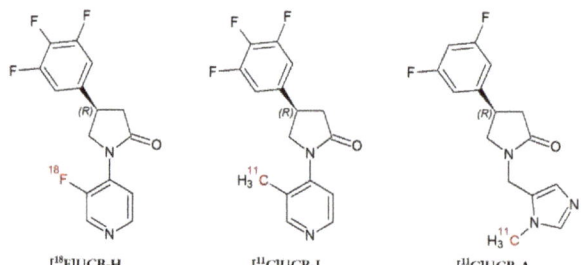

Figure 1. Chemical structures of [^{18}F]UCB-H, [^{11}C]UCB-J and [^{11}C]UCB-A.

This paper, therefore, aims to evaluate for the first time the specificity of [^{18}F]UCB-H for the SV2A isoform against SV2B and SV2C using microPET imaging in rats, by means of a blocking assay between this radiotracer and specific competitors for the three SV2 isoforms. The results will provide highly valuable information about the actual potential of [^{18}F]UCB-H as a radiopharmaceutical candidate to study the SV2A protein with PET in research or clinical practice.

2. Results

Table 1 summarizes the results obtained from the in vitro binding assays. We observe that SV2B$_L$ presents a high affinity for SV2B (pIC$_{50}$ = 7.8), but also has an affinity for SV2A similar to that of SV2A$_L$ (pIC$_{50}$ = 5.6).

Table 1. pIC$_{50}$ of the competitors used for the different SV2 isoforms. Binding affinities measured for human SV2 proteins at 37 °C. Data are presented as mean (n = 3 to 10) from non-linear regression analysis of raw data using a sigmoidal dose-response model. Additional data for SV2B$_L$ (UCB5203) solubility: 0.1 mg/mL, route of administration: ip (suspension in 5% DMSO–1% methyl cellulose in water), CEREP @ 10 µM: all targets < 50% inhibition, mouse brain fraction unbound: 37%, mouse brain exposure (3 mg/kg, 30–60 min): ~1.8 µM total → 0.66 µM free → ~100 fold IC50 SV2B. Additional data for SV2C$_L$ (UCB0949) solubility: 0.055 mg/mL, route of administration: ip (suspension in 5% DMSO—1% methyl cellulose in water), CEREP @ 10 µM: all targets < 50% inhibition, mouse brain fraction unbound: 54%, mouse brain exposure (3 mg/kg, 30–60 min): ~8 µM total → 4.3 µM free → ~270 fold IC$_{50}$ SV2C.

	Synaptic Vesicle Protein Isoforms		
	SV2A	SV2B	SV2C
SV2A$_L$	5.2	−3.1	−3.2
SV2B$_L$	5.6	7.8	5.5
SV2C$_L$	<5	5.9	7.8

In Figure 2, [^{18}F]UCB-H parametric Vt maps are presented for the vehicle group and the three pre-treated groups (SV2A$_L$, SV2B$_L$ and SV2C$_L$). These pictures highlight a clear reduction of the [^{18}F]UCB-H binding throughout the entire brain induced by levetiracetam (SV2A$_L$) pre-treatment at 10 mg/kg (PET image corresponding to the SV2A$_L$ group).

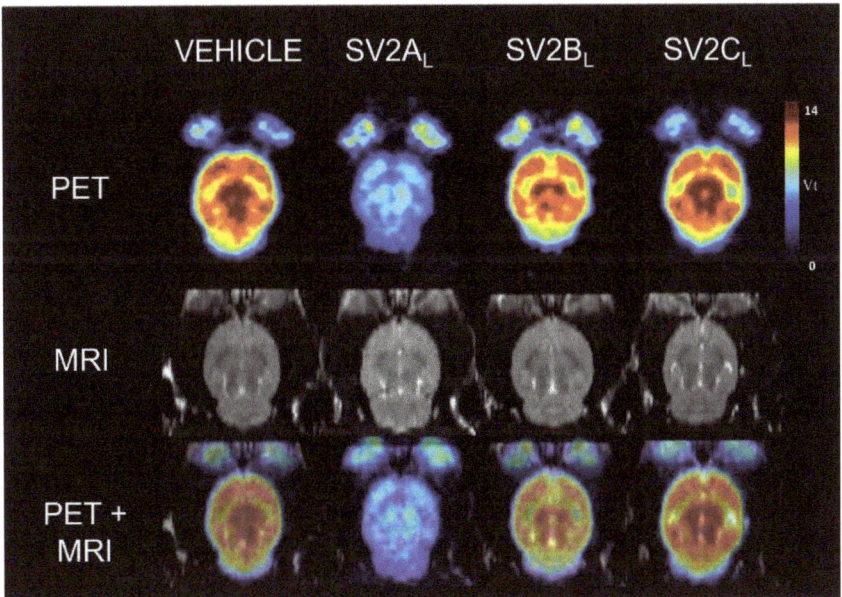

Figure 2. Example of an individual parametric Vt map of [^{18}F]UCB-H binding in rat brain (PET), along with the corresponding individual MRI and the overlay of both images (PET + MRI). Rats were pre-treated 30 min before the 60 min PET acquisition with either vehicle, SV2A competitor (levetiracetam [SV2A$_L$] at 10 mg/kg), SV2B competitor (UCB5203 [SV2B$_L$] at 3 mg/kg) and SV2C competitor (UCB0949 [SV2C$_L$] at 3 mg/kg).

In Figure 3A, we can observe the time activity curves (TACs) corresponding to the four different treatments (vehicle, SV2A$_L$, SV2B$_L$ and SV2C$_L$), for one of the regions of interest (ROIs): the whole brain.

The four TACs reveal a high initial uptake of [^{18}F]UCB-H, which peaks around 5 min post-injection. Subsequently, the radioactivity is quickly washed out of the brain. Some differences can be observed in the kinetic of the TAC after pre-treatment with the respective ligands: The highest peak activity is observed after pre-treatment with the vehicle and with SV2C$_L$. Interestingly, the pre-treatments with SV2A$_L$ and SV2B$_L$ display the same peak of initial uptake. In the case of SV2B$_L$, the kinetics of the TAC from 15 to 60 min are similar to the kinetics of the radiotracer after pre-treatment with either the vehicle or SV2C$_L$. The TACs for all the ROIs are included in Supplementary Figure S1. In addition, the area under each TAC (the AUC) is represented in Figure 3B, where we can observe the differences between the [^{18}F]UCB-H uptake after pre-treatment with SV2A$_L$, and after pre-treatment with the other compounds.

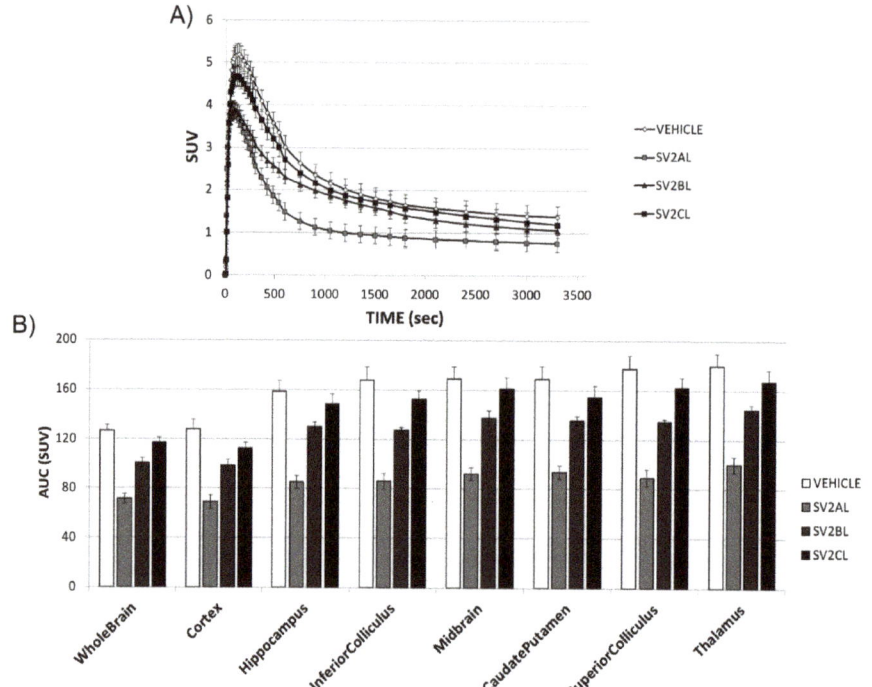

Figure 3. Representative time activity curves (TACs) and AUC (area under each TAC) for the different regions of interest (ROIs). (**A**) TACs extracted from the whole brain as ROI, and normalized by the injected activities and the body weight. Lines represent the [^{18}F]UCB-H uptake over a 60 min acquisition after pre-treatment with the vehicle, SV2A$_L$, SV2B$_L$, or SV2C$_L$. (**B**) The bar plots represent the AUC in the eight ROIs (mean ± SEM, n = 5).

Figure 4 presents the mean Vt values for the eight selected brain structures, calculated from the previous TACs and the population-based input function (PBIF). Comparing Figures 3B and 4, the differences between groups in AUC and in Vt are similar, with the highest value associated to pre-treatment with the vehicle, and the lowest value for pre-treatment with SV2A$_L$.

Table 2 summarizes, for the same regions, the impact on the Vt induced by the blocking experiments, expressed as the relative difference in Vt between the vehicle group and the pre-treated groups. In the whole brain, mean Vt values of 10.4 ± 0.7, 6.0 ± 0.3, 8.3 ± 0.2 and 9.8 ± 0.3 were obtained for the vehicle (control) group, the SV2A$_L$ pre-treated group, the SV2B$_L$ pre-treated group and the SV2C$_L$ pre-treated group, respectively. For the eight ROIs, a statistically significant difference was observed between the SV2A$_L$ pre-treated group and all the other groups ($p < 0.001$). Furthermore, for the whole brain, the cerebral cortex and the inferior and superior colliculus, a statistically significant difference was also detected between the vehicle group and the SV2B$_L$ pre-treated group ($p < 0.05$).

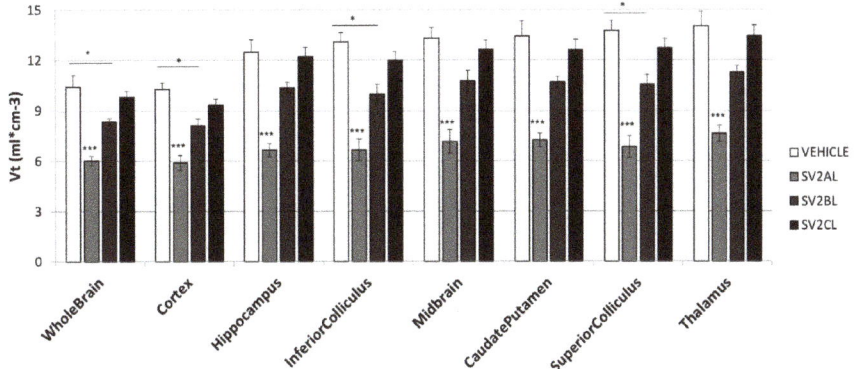

Figure 4. Vt values for the eight selected brain structures. Bars represent the mean ± SEM (n = 5). One-way ANOVA and Scheffe post-hoc tests were performed, with ***$p < 0.001$ and *$p < 0.05$.

Table 2. Illustration of the impact induced by the blocking experiments, expressed as percentage of reduction calculated from the mean Vt values (n = 5) for the eight selected ROIs.

ROIs	Vehicle vs. SV2A$_L$	Vehicle vs. SV2B$_L$	Vehicle vs. SV2C$_L$
Whole brain	42.3	19.9	5.6
Cortex	42.7	21.1	9.1
Hippocampus	46.8	16.9	2.1
Inferior colliculus	49.1	23.7	8.5
Midbrain	46.3	19.1	5.1
Caudate/Putamen	46.2	20.2	6.2
Superior colliculus	50.4	23.4	7.6
Thalamus	45.9	19.7	4.4
Mean	46.2	20.5	6.1
SEM	1.0	0.8	0.8

3. Discussion

The SV2 protein is critical for the adequate functioning of the central nervous system, acting as a modulator of synaptic transmission by priming vesicles in quiescent neurons [4]. The divergent roles of the three isoforms which comprise this family have yet to be clarified, although different pathologies have been associated with them. As previously stated, the SV2A isoform is associated with the physiopathology of epilepsy [5,27]. In contrast, the SV2B isoform is related with prostate small cell carcinoma [28] and the SV2C isoform is generally associated with the correct functioning of basal ganglia nuclei [9,29,30]. Some studies have evaluated the possible relation between SV2C and Parkinson's disease, as SV2C modulates dopamine release [29,31].

This paper's goal was to evaluate, for the first time, in vivo the specificity of the [^{18}F]UCB-H radiotracer in targeting the SV2A isoform compared to SV2B and SV2C. The relevance of such a study stems from the fact that in vivo SV2A quantification can be considered to be an indirect measure of the synaptic density [2,3,18], which is a key parameter for fundamental research and for the clinic.

Before discussing the results obtained during these blocking experiments, we have to address some general considerations. Firstly, the results presented issue from the microPET imaging technique. Like many other microPET cameras, the Focus120 used during this work has a spatial resolution of 1.5 mm, at best hampering the study of brain structures of small size due to the partial volume effect [32,33]. Secondly, the rat brain distributions of the three SV2 protein isoforms [7–9] indicate that almost all major brain structures express at least two SV2 protein isoforms. SV2A, the most extensively studied, is ubiquitously distributed [1,8]. Like SV2A, SV2B can be found in almost all the rat brain structures with few subtle differential expressions in some hippocampal substructures like CA3 and the dentate gyrus, along with the reticular nucleus of the thalamus and some small areas in the brain stem [8,9,34]. Unfortunately, these regions are far too small to be correctly quantified with microPET. Janz and Sudhof showed that unlike SV2A and SV2B, the SV2C protein is characterized by much more restricted localization in brain regions considered to be evolutionarily well preserved in rats: The olfactory bulb, the striatum, the substantia nigra, and some nuclei in the pons and the medulla oblongata [9]. As we can see, it is impossible to find well defined brain structures for in vivo microPET quantification in which one of the three isoforms is uniquely or even mostly expressed. Another important point is that we do not have a clear picture of the respective proportions of each isoform present in the main rat brain structures. All these considerations will have to be taken into account in the following discussion. Accordingly, we have decided to select eight major ROIs to ensure robust in vivo quantification with microPET: Whole brain, cortex, hippocampus, inferior colliculus, superior colliculus, midbrain, caudate putamen and thalamus.

The [^{18}F]UCB-H Vt values obtained during this study for the vehicle pre-treated group, calculated using the PBIF [35] were in good agreement with those previously published for rats [35,36]. This is important in order to establish the consistency of the proposed methodology. The blocking experiments realized with SV2A$_L$ at 10 mg/kg demonstrated a clear significant competition (46.2%) between levetiracetam and [^{18}F]UCB-H in eight selected ROIs. These values are of the same order of magnitude as those previously reported in rats [36]. According to the potency of levetiracetam for SV2A against SV2B and SV2C (Table 2), we can conclude that SV2A is one of the main target of [^{18}F]UCB-H in vivo in rats.

After performing a blocking experiment with SV2C$_L$ at 3 mg/kg, we obtained TACs with similar peaks and kinetics to the TAC corresponding to pre-treatment with the vehicle, in all ROIs. The quantification of the radiotracer uptake, using the Vt, highlighted no statistically significant in vivo competition between SV2C$_L$ and [^{18}F]UCB-H in any of the eight selected ROIs. From this we can infer that SV2C$_L$ pre-treatment has either no impact or an impact of very small size. The population used (n = 5) is not sufficient to demonstrate an effect of small size (f = 0.10), but is optimal to detect medium (f = 0.25) and large effect sizes (f = 0.5). Another important point is that a highly potent SV2C competitor like UCB0949 (pIC$_{50}$ of 7.8) was unable to modify [^{18}F]UCB-H binding in brain structures with a high expression of SV2C, like the midbrain or the caudate/putamen [9]. The reduction measured in these regions was of the same order of magnitude as that found in the other structures. These considerations support the theory that SV2C does not seem to be the main target of [^{18}F]UCB-H in rats.

The pre-treatment with SV2B$_L$ at 3 mg/kg resulted in a TAC with a peak as high as that obtained after pre-treatment with SV2A$_L$. However, it features kinetics which are similar to those obtained after pre-treatment with the vehicle or SV2C$_L$. This lower peak could be attributed to an initial non-specific binding of SV2B$_L$ to the SV2A protein, for which it presents an affinity which is similar to that of SV2A$_L$, with a pIC$_{50}$ = 5.6. After the peak, the SV2B$_L$ TAC follows a similar shape and level to those of pre-treatment with the vehicle and SV2C$_L$, indicating a washing out effect of this fraction of non-specific binding of SV2B$_L$ to the SV2A protein. In order to confirm this hypothesis, a similar experiment with a SV2B$_L$ with a lower affinity for SV2A should be performed. However, the SV2B$_L$ used in this paper is currently the only one available. In addition to the previous analysis, we evaluated the changes in Vt after pre-treatment with SV2B$_L$. In these results, we can observe a consistent mean reduction of 20.5% of the Vt values in the eight selected ROIs. The SV2B$_L$ is characterized by a pIC$_{50}$ of 7.8 for the SV2B isoform. Such a highly potent competitor is expected to effectively impede the binding of any radioligand to the SV2B isoform. If SV2B was the main target for [^{18}F]UCB-H, the blocking induced with this highly efficient SV2B$_L$ would have been very pronounced and much higher than the 20% measured. In order to explain the 20% reduction of [^{18}F]UCB-H Vt values, we have to take into account that SV2B$_L$ also presents some potency for SV2A. SV2B$_L$ has a pIC$_{50}$ of 5.6, which is of the same order of magnitude as that of levetiracetam. Thus, the SV2B$_L$ ligand has some affinity for SV2A, which could lead to partial blocking of SV2A. Hence, the 20% reduction observed is most likely linked to SV2A blocking induced by SV2B$_L$. Accordingly, we can conclude that SV2B does not seem to be the main target of [^{18}F]UCB-H in vivo in rats.

We are aware that the respective affinities of SV2B$_L$ and SV2C$_L$ are a problem for the interpretation of the data, but we have to consider that today UCB5203 and UCB0949 are the only compounds that can be used for this purpose.

4. Materials and Methods

4.1. Animals

Twenty male Sprague Dawley CD rats (five weeks old) were used, bred by Janvier Laboratories (France). The animals were housed in pairs for three weeks under standard 12:12 h light/dark conditions, maintaining room temperature at 22 °C, and humidity at approximately 50%. Standard pellet food and water were provided ad libitum.

The experimental procedures and protocols used in this investigation ("Synap-SV2A project" files 14-1753 and 13-1573) were reviewed and approved by the Institutional Animal Care and Use Committee of the University of Liege, according to the Helsinki declaration, and conducted in accordance with the European guidelines for care of laboratory animals (2010/63/EU). Moreover, the Animal Research Reporting In Vivo Experiments (ARRIVE) guidelines [37] were followed as closely as possible to confer a minimal intrinsic quality to the study.

4.2. Radiopharmaceutical Production and Drugs

[^{18}F]UCB-H was produced through one-step radiolabeling of a pyridyliodonium precursor. This method provides 34% ± 2% of injectable [^{18}F]UCB-H (uncorrected radiochemical yield) from up to 285 GBq (7.7 Ci) of [^{18}F]fluoride (specific activity of 815 ± 185 GBq/µmol and measured purity of 99.8 ± 0.5 wt %); this has previously been reported in Warnier et al. [17].

The ligand for the SV2A isoform (SV2A$_L$) was purchased as an injectable solution (levetiracetam, Keppra®, UCB Pharma S.A. Brussels, Belgium). At the present time, there are no commercially available specific ligands for the other two SV2 isoforms (SV2B and SV2C). The competitors used were obtained from UCB Pharma s.a.: UCB5203 for SV2B (SV2B$_L$, MW: 236.238 g/mol) and UCB0949 for the SV2C (SV2C$_L$, MW: 281.197 g/mol). The information on these compounds was supplied by UCB Pharma s.a. The respective affinities for the different SV2 isoforms are presented in Table 2.

The competitors were prepared daily in a vehicle composed of distilled water containing 1% methyl cellulose (viscosity: 15 cP, Sigma-Aldrich, Overijse, Belgium) and 5% dimethyl sulphoxide (DMSO, Sigma-Aldrich, Belgium). The concentrations differed depending on the product specifications and their respective pharmacokinetics, provided by UCB Pharma s.a. The dosing used was 10 mg/kg for SV2A$_L$, 3 mg/kg for SV2B$_L$, and 3 mg/kg for SV2C$_L$. All solutions were administered through the intraperitoneal (i.p.) route in a total volume of 1 mL per kg of body weight. The animals used as a control group (vehicle) received an equal volume of vehicle through the same route of administration.

4.3. In Vitro Binding Assays

Reagents and reference compounds used were of analytical grade and obtained from various commercial sources. All cell culture reagents were obtained from Invitrogen (Merelbeke, Belgium). Radioligands (3H-UCB30889, 1184 GBq/mmol; 3H-UCB1418435, 925 GBq/mmol; and 3H-UCB101275-1, 1110–1480 GBq/mmol) were obtained from G.E Healthcare, Amersham, UK (now Perkin Elmer, Zaventem, Belgium) and reference compounds (levetiracetam, UCB108649-1 and UCB101275-1) were custom synthesized and stored according to manufacturer's recommendations. Test and reference compounds were dissolved in 100% DMSO or H_2O to give 1 or 10 mM stock solution. The final DMSO concentration in assays was 0.1% unless otherwise stated.

Cell lines generated at UCB Biopharma were human embryonic kidney (HEK) 293 cells expressing human SV2A, SV2B or SV2C proteins. Cells were cultured in Dulbecco's Modified Eagle medium. The culture medium was supplemented with foetal bovine serum (FBS, 10%), 2 mM L-glutamine, 50 to 100 U/mL penicillin, 50 to 100 µg/mL streptomycin, and 200 µg/mL hygromycin B. Cells were grown at 37 °C with 95% air. Confluent cells were detached by 10 min incubation at 37 °C in phosphate buffered saline (PBS) containing 0.02% EDTA. Culture flasks were washed with 15 mL of ice-cold PBS. The cell suspension was centrifuged at 1500× g for 10 min at 4 °C. The pellet was homogenized in 15 mM Tris-HCl buffer (pH 7.5) containing 2 mM $MgCl_2$, 0.3 mM EDTA, and 1 mM EGTA (buffer A) using a glass/teflon homogenizer. The crude homogenate was subjected to a freeze and thaw cycle in liquid nitrogen and DNAse (1 µL/mL) was then added. The homogenate was further incubated for 10 min at 25 °C before being centrifuged at 40,000× g for 25 min at 4 °C. The pellet was re-suspended in buffer A and washed once under the same conditions. The final crude membrane pellet was re-suspended at a protein concentration of 1–3 mg/mL in 7.5 mM Tris-HCl buffer (pH 7.5 at 25 °C) containing 250 mM sucrose and stored in liquid nitrogen until use.

Membranes were incubated in binding buffer (see Table 3) containing test compound or positive control in the presence of the radioligand. The non-specific binding (NSB) was defined as the residual binding observed in the presence of a high concentration (1000 fold its Ki) of a specific unlabeled reference compound. Membrane-bound and free radioligands were separated by rapid filtration through glass fiber filters (GF/C). Samples and filters were rinsed using at least 6 mL of washing buffer. The entire filtration procedure did not exceed 10 s per sample. The radioactivity trapped on the filters was counted by liquid scintillation in a β-counter. To determine the affinity of a compound for a given target, competition curves were performed with at least 10 concentrations of compound spanning at least 5 log units.

Table 3. Details of the in vitro binding assay determination. Percentage of inhibition was calculated as follows: % INHIBITION = 100 − [((BI − NSB)/(B0 − NSB)) × 100], where B0 and BI represent the binding observed in the absence and presence of the test compound, respectively (dpm), NSB is the radioligand non-specific binding (dpm). Raw data were analyzed by non-linear regression using XLfit^{TM} (IDBS, London, Great Britain) according to the following generic equation: B = NSB + [(B0 − NSB)/(1 + (((10^X)/(10^{-pIC50}))^{nH}))], where B is the radioligand bound in the presence of the unlabeled compound (dpm), NSB is the radioligand non-specific binding (dpm), B0 is the radioligand bound in the absence of unlabeled compound (dpm), X is the concentration of unlabeled compound (log M), pIC_{50} is the concentration of unlabeled compound that inhibits the radioligand specific binding by 50% (−log M), and nH is the Hill coefficient.

In Vitro Binding Details	hSV2A Assay	hSV2B	hSV2C
Binding buffer	50 mM Tris-HCl (pH 7.4) containing 2 mM $MgCl_2$		
Filtration buffer	Ice-cold 50 mM Tris-HCl (pH 7.4)		
Incubation time	120 min at 37 °C in 0.5 mL	120 min at 37 °C in 0.5 mL	120 min at 37 °C in 0.2 mL
Radioligand	^{3}H-UCB30889 (4 nM)	^{3}H-UCB1418435 (8 nM)	^{3}H-UCB101275-1 (20 nM)
Proteins	75–125 µg HEK293 membranes	2–5 µg HEK293 membranes	40–60 µg HEK293 membranes
Blocking drug	Levetiracetam (1 mM)	UCB108649-2 (10 µM)	UCB101275-1 (100 µM)

4.4. PET Acquisitions

The animals (n = 5 per group) were anesthetized using 4% isoflurane in air at a flow rate of 1 L/min during induction and 1.5% to 2% isoflurane in air at 0.6 L/min during maintenance. Respiration rate and rectal temperature were continuously measured using a physiological monitoring system (Minerve, France). The temperature was maintained at 37 ± 0.5 °C using an air warming system.

MicroPET scans were performed with a Siemens FOCUS 120 microPET (Siemens, Knoxville, TN, USA). The animals were anesthetized and pre-treated i.p. with vehicle, $SV2A_L$, $SV2B_L$, or $SV2C_L$. Thirty minutes later, they were installed in the microPET scanner and [^{18}F]UCB-H was injected via the lateral tail vein (44.7 ± 3 MBq, 0.55 mL), simultaneously starting a 60 min emission scan, in list mode. Finally, a 10 min transmission scan was performed in a single event acquisition mode, using a ^{57}Co source. The acquired data were then reframed as follows: 6 × 5 s, 6 × 10 s, 3 × 20 s, 5 × 30 s, 5 × 60 s, 8 × 150 s, and 6 × 300 s. For each frame, a total of 95 trans-axial slices were obtained using Fourier rebinning (FORE), followed by 2D ramp filtered backprojection (FBP), in 256 × 256 matrix. The slice thickness was 0.796 mm and the in-slice pixel size was 0.433 mm.

Immediately after the PET acquisition, the anesthetized rats were transferred into a 9.4 Tesla MRI horizontal bore system (Agilent Technologies, Palo Alto, CA, USA), with a 72 mm inner diameter volumetric coil (Rapid Biomedical GmbH, Würzurg, Germany). Anatomical T2-weighted brain images were obtained using a fast spin echo multi-slice sequence with the following parameters: TR = 2000 ms, TE = 40 ms, matrix = 256 × 256, FOV = 45 × 45 mm, 30 contiguous slices of thickness = 0.80 mm and in-plane voxel size = 0.176 × 0.176 mm.

4.5. Imaging Data Processing

PMOD software (Version 3.6, PMOD Technologies, Zurich, Switzerland) was used to process the imaging data. The structural MRI images were firstly co-registered to the corresponding PET images, and subsequently spatially normalized into the PMOD MRI T2 template. Finally, the inverse normalization parameters were calculated and applied to the PMOD rat brain atlas to bring it in the individual PET space. From this atlas, eight relevant regions of interest (ROIs) were chosen according to their differential expression of SV2A, SV2B and SV2C: whole brain, cortex, caudate/putamen, hippocampus, inferior colliculus, superior colliculus, midbrain and thalamus.

Individual time-activity curves (TACs) were extracted for each ROIs and normalized by the body weight and the injected dose of radiotracer to be expressed as standardized uptake value (SUV). A population-based input function (PBIF) published by our laboratory [35] was used to avoid arterial blood sampling during the acquisitions. The distribution volume (Vt), was determined by Logan plot kinetic modelling using the TACs and the PBIF. The equilibration time (t*) was fixed at 15 min (starting point of the range used in the multi-linear regression analysis).

4.6. Statistical Analysis

The results are presented as mean (Vt) ± standard error of the mean (SEM). All the data were tested for normal distribution with Levene's test for homogeneity, and with a Kolmogorov–Smirnov test for normality. Data were analyzed using one-way analysis of variance (ANOVA) followed by Scheffe post-hoc tests.

All statistical analyses were performed with the statistics software Statistica 12 (Statsoft, France) and GraphPad Prism (version 6, GraphPad software, Inc., San Diego, CA, USA). The critical level of statistical significance was always set at $p < 0.05$.

5. Conclusions

For the first time, the specificity of a radiopharmaceutical compound for the three SV2 protein isoforms was assessed in vivo, in rats. The results obtained clearly indicated that SV2A was the main target of [^{18}F]UCB-H, and confirmed that [^{18}F]UCB-H is a suitable radiotracer for in vivo imaging of the SV2A proteins with PET. Consequently, [^{18}F]UCB-H is an interesting candidate to study SV2A-associated pathologies.

Supplementary Materials: The following are available online. Figure S1: TACs extracted from the eight ROIs, and normalized by the injected activities and the body weight. Lines represent the [^{18}F]UCB-H uptake over a 60 min acquisition after pre-treatment with the vehicle, SV2A$_L$, SV2B$_L$, or SV2C$_L$ (mean ± SEM; n = 5).

Author Contributions: Conceptualization, M.E.S., G.B., A.S. and A.P.; methodology, M.A.B., A.S.; formal analysis, M.E.S., G.B., M.A.B., A.S. and A.P.; investigation, M.E.S. and G.B.; resources, N.M., J.M., F.M., F.G., C.L. and A.L.; data curation, M.A.B.; writing—original draft preparation, M.E.S.; writing—review and editing, G.B., M.A.B., A.S., N.M., J.M., F.M., F.G., C.L., E.S., A.L. and A.P.; visualization, M.E.S., M.A.B. and A.P.; supervision, A.L. and A.P.; project administration, E.S. and A.P.; funding acquisition, A.P.

Funding: This research was funded by the University of Liège grant 13/17-07 and UCB Pharma s.a. as partners. The SV2B$_L$ (UCB5203) and SV2C$_L$ (UCB0949) ligands were provided by UCB Pharma s.a., compound structures and specifications being confidential. M.E.S. is supported by the University of Liège grant 13/17-07. A.P. is research director from F.R.S.-FNRS Belgium. J.M. and N.M. are UCB Pharma s.a. employees. F.M. and F.G. are Nucleis s.a. employees.

Conflicts of Interest: The authors declare no conflict of interest. The funders had no role in the design of the study; in the collection, analyses, or interpretation of data; in the writing of the manuscript, or in the decision to publish the results.

References

1. Buckley, K.; Kelly, R.B. Identification of a Transmembrane Glycoprotein Specific for Secretory Vesicles of Neural and Endocrine Cells. *J. Cell Biol.* **1985**, *100*, 1284–1294. [CrossRef] [PubMed]
2. Finnema, S.J.; Nabulsi, N.B.; Eid, T.; Detyniecki, K.; Lin, S.-F.; Chen, M.-K.; Dhaher, R.; Matuskey, D.; Baum, E.; Holden, D.; et al. Imaging synaptic density in the living human brain. *Sci. Transl. Med.* **2016**, *8*, 348ra96. [CrossRef] [PubMed]
3. Mercier, J.; Provins, L.; Valade, A. Discovery and development of SV2A PET tracers: Potential for imaging synaptic density and clinical applications. *Drug Discov. Today Technol.* **2017**, *25*, 45–52. [CrossRef]
4. Custer, K.L.; Austin, N.S.; Sullivan, J.M.; Bajjalieh, S.M. Synaptic Vesicle Protein 2 Enhances Release Probability at Quiescent Synapses. *J. Neurosci.* **2006**, *26*, 1303–1313. [CrossRef] [PubMed]
5. Van Vliet, E.A.; Aronica, E.; Redeker, S.; Boer, K.; Gorter, J.A. Decreased expression of synaptic vesicle protein 2A, the binding site for levetiracetam, during epileptogenesis and chronic epilepsy. *Epilepsia* **2009**, *50*, 422–433. [CrossRef] [PubMed]
6. Bartholome, O.; Van den Ackerveken, P.; Sánchez Gil, J.; de la Brassinne Bonardeaux, O.; Leprince, P.; Franzen, R.; Rogister, B. Puzzling Out Synaptic Vesicle 2 Family Members Functions. *Front. Mol. Neurosci.* **2017**, *10*, 148. [CrossRef]
7. Crèvecœur, J.; Foerch, P.; Doupagne, M.; Thielen, C.; Vandenplas, C.; Moonen, G.; Deprez, M.; Rogister, B. Expression of SV2 isoforms during rodent brain development. *BMC Neurosci.* **2013**, *14*, 87. [CrossRef] [PubMed]
8. Bajjalieh, S.M.; Frantz, G.D.; Weimann, J.M.; McConnell, S.K.; Scheller, R.H. Differential expression of synaptic vesicle protein 2 (SV2) isoforms. *J. Neurosci.* **1994**, *14*, 5223–5235. [CrossRef]
9. Janz, R.; Südhof, T.C. SV2C is a synaptic vesicle protein with an unusually restricted localization: anatomy of a synaptic vesicle protein family. *Neuroscience* **1999**, *94*, 1279–1290. [CrossRef]
10. Bajjalieh, S.M.; Peterson, K.; Linial, M.; Scheller, R.H. Brain contains two forms of synaptic vesicle protein 2. *Proc. Natl. Acad. Sci. USA* **1993**, *90*, 2150–2154. [CrossRef]
11. Stahl, S.M. Psychopharmacology of anticonvulsants: levetiracetam as a synaptic vesicle protein modulator. *J. Clin. Psychiatry* **2004**, *65*, 1162–1163. [CrossRef] [PubMed]
12. Matagne, A.; Margineanu, D.-G.; Kenda, B.; Michel, P.; Klitgaard, H. Anti-convulsive and anti-epileptic properties of brivaracetam (ucb 34714), a high-affinity ligand for the synaptic vesicle protein, SV2A. *Br. J. Pharmacol.* **2008**, *154*, 1662–1671. [CrossRef] [PubMed]
13. Kaminski, R.M.; Gillard, M.; Leclercq, K.; Hanon, E.; Lorent, G.; Dassesse, D.; Matagne, A.; Klitgaard, H. Proepileptic phenotype of SV2A-deficient mice is associated with reduced anticonvulsant efficacy of levetiracetam. *Epilepsia* **2009**, *50*, 1729–1740. [CrossRef] [PubMed]
14. Lynch, B.A.; Lambeng, N.; Nocka, K.; Kensel-Hammes, P.; Bajjalieh, S.M.; Matagne, A.; Fuks, B. The synaptic vesicle protein SV2A is the binding site for the antiepileptic drug levetiracetam. *Proc. Natl. Acad. Sci. USA* **2004**, *101*, 9861–9866. [CrossRef] [PubMed]

15. Bretin, F.; Bahri, M.A.; Bernard, C.; Warnock, G.; Aerts, J.; Mestdagh, N.; Buchanan, T.; Otoul, C.; Koestler, F.; Mievis, F.; et al. Biodistribution and Radiation Dosimetry for the Novel SV2A Radiotracer [18F]UCB-H: First-in-Human Study. *Mol. Imaging Biol.* **2015**, *17*, 557–564. [CrossRef] [PubMed]
16. Bretin, F.; Warnock, G.; Bahri, M.A.; Aerts, J.; Mestdagh, N.; Buchanan, T.; Valade, A.; Mievis, F.; Giacomelli, F.; Lemaire, C.; et al. Preclinical radiation dosimetry for the novel SV2A radiotracer [18F]UCB-H. *EJNMMI Res.* **2013**, *3*, 35. [CrossRef]
17. Warnier, C.; Lemaire, C.; Becker, G.; Zaragoza, G.; Giacomelli, F.; Aerts, J.; Otabashi, M.; Bahri, M.A.; Mercier, J.; Plenevaux, A.; et al. Enabling Efficient Positron Emission Tomography (PET) Imaging of Synaptic Vesicle Glycoprotein 2A (SV2A) with a Robust and One-Step Radiosynthesis of a Highly Potent 18F-Labeled Ligand ([18F]UCB-H). *J. Med. Chem.* **2016**, *59*, 8955–8966. [CrossRef] [PubMed]
18. Nabulsi, N.B.; Mercier, J.; Holden, D.; Carre, S.; Najafzadeh, S.; Vandergeten, M.-C.; Lin, S.-F.; Deo, A.; Price, N.; Wood, M.; et al. Synthesis and Preclinical Evaluation of 11C-UCB-J as a PET Tracer for Imaging the Synaptic Vesicle Glycoprotein 2A in the Brain. *J. Nucl. Med.* **2016**, *57*, 777–784. [CrossRef] [PubMed]
19. Cai, H.; Mangner, T.J.; Muzik, O.; Wang, M.-W.; Chugani, D.C.; Chugani, H.T. Radiosynthesis of 11C-levetiracetam: A potential marker for PET imaging of SV2A expression. *ACS Med. Chem. Lett.* **2014**, *5*, 1152–1155. [CrossRef]
20. Estrada, S.; Lubberink, M.; Thibblin, A.; Sprycha, M.; Buchanan, T.; Mestdagh, N.; Kenda, B.; Mercier, J.; Provins, L.; Gillard, M.; et al. [11C]UCB-A, a novel PET tracer for synaptic vesicle protein 2 A. *Nucl. Med. Biol.* **2016**, *43*, 325–332. [CrossRef] [PubMed]
21. Heurling, K.; Ashton, N.J.; Leuzy, A.; Zimmer, E.R.; Blennow, K.; Zetterberg, H.; Eriksson, J.; Lubberink, M.; Schöll, M. Synaptic vesicle protein 2A as a potential biomarker in synaptopathies. *Mol. Cell. Neurosci.* **2019**. [CrossRef] [PubMed]
22. Rabiner, E.A. Imaging Synaptic Density: A Different Look at Neurologic Diseases. *J. Nucl. Med.* **2018**, *59*, 380–381. [CrossRef] [PubMed]
23. Cai, Z.; Li, S.; Matuskey, D.; Nabulsi, N.; Huang, Y. PET imaging of synaptic density: A new tool for investigation of neuropsychiatric diseases. *Neurosci. Lett.* **2019**, *691*, 44–50. [CrossRef] [PubMed]
24. Li, S.; Cai, Z.; Wu, X.; Holden, D.; Pracitto, R.; Kapinos, M.; Gao, H.; Labaree, D.; Nabulsi, N.; Carson, R.E.; et al. Synthesis and in Vivo Evaluation of a Novel PET Radiotracer for Imaging of Synaptic Vesicle Glycoprotein 2A (SV2A) in Nonhuman Primates. *ACS Chem. Neurosci.* **2019**, *10*, 1544–1554. [CrossRef] [PubMed]
25. Bahri, M.A.; Plenevaux, A.; Aerts, J.; Bastin, C.; Becker, G.; Mercier, J.; Valade, A.; Buchanan, T.; Mestdagh, N.; Ledoux, D.; et al. Measuring brain synaptic vesicle protein 2A with positron emission tomography and [18F]UCB-H. *Alzheimer's Dement.* **2017**, *3*, 481–486.
26. Serrano, M.E.; Bahri, M.A.; Becker, G.; Seret, A.; Mievis, F.; Giacomelli, F.; Lemaire, C.; Salmon, E.; Luxen, A.; Plenevaux, A. Quantification of [18F]UCB-H Binding in the Rat Brain: From Kinetic Modelling to Standardised Uptake Value. *Mol. Imaging Biol.* **2018**. [CrossRef] [PubMed]
27. Hanaya, R.; Hosoyama, H.; Sugata, S.; Tokudome, M.; Hirano, H.; Tokimura, H.; Kurisu, K.; Serikawa, T.; Sasa, M.; Arita, K. Low distribution of synaptic vesicle protein 2A and synaptotagimin-1 in the cerebral cortex and hippocampus of spontaneously epileptic rats exhibiting both tonic convulsion and absence seizure. *Neuroscience* **2012**, *221*, 12–20. [CrossRef] [PubMed]
28. Clegg, N.; Ferguson, C.; True, L.D.; Arnold, H.; Moorman, A.; Quinn, J.E.; Vessella, R.L.; Nelson, P.S. Molecular characterization of prostatic small-cell neuroendocrine carcinoma. *Prostate* **2003**, *55*, 55–64. [CrossRef] [PubMed]
29. Dardou, D.; Monlezun, S.; Foerch, P.; Courade, J.P.; Cuvelier, L.; de Ryck, M.; Schiffmann, S.N. A role for Sv2c in basal ganglia functions. *Brain Res.* **2013**, *1507*, 61–73. [CrossRef]
30. Altmann, V.; Schumacher-Schuh, A.F.; Rieck, M.; Callegari-Jacques, S.M.; Rieder, C.R.; Hutz, M.H. Influence of genetic, biological and pharmacological factors on levodopa dose in Parkinson's disease. *Pharmacogenomics* **2016**, *17*, 481–488. [CrossRef] [PubMed]
31. Dunn, A.R.; Stout, K.A.; Ozawa, M.; Lohr, K.M.; Hoffman, C.A.; Bernstein, A.I.; Li, Y.; Wang, M.; Sgobio, C.; Sastry, N.; et al. Synaptic vesicle glycoprotein 2C (SV2C) modulates dopamine release and is disrupted in Parkinson disease. *Proc. Natl. Acad. Sci. USA* **2017**, *114*, E2253–E2262. [CrossRef] [PubMed]
32. Karp, J.S.; Daube-Witherspoon, M.E.; Muehllehner, G. Factors affecting accuracy and precision in PET volume imaging. *J. Cereb. Blood Flow Metab.* **1991**, *11*, A38–A44. [CrossRef] [PubMed]

33. Aston, J.A.D.; Cunningham, V.J.; Asselin, M.-C.; Hammers, A.; Evans, A.C.; Gunn, R.N. Positron emission tomography partial volume correction: estimation and algorithms. *J. Cereb. Blood Flow Metab.* **2002**, *22*, 1019–1034. [CrossRef] [PubMed]
34. Dardou, D.; Dassesse, D.; Cuvelier, L.; Deprez, T.; De Ryck, M.; Schiffmann, S.N. Distribution of SV2C mRNA and protein expression in the mouse brain with a particular emphasis on the basal ganglia system. *Brain Res.* **2011**, *1367*, 130–145. [CrossRef] [PubMed]
35. Becker, G.; Warnier, C.; Serrano, M.E.; Bahri, M.A.; Mercier, J.; Lemaire, C.; Salmon, E.; Luxen, A.; Plenevaux, A. Pharmacokinetic Characterization of [18F]UCB-H PET Radiopharmaceutical in the Rat Brain. *Mol. Pharm.* **2017**, *14*, 2719–2725. [CrossRef]
36. Warnock, G.I.; Aerts, J.; Bahri, M.A.; Bretin, F.; Lemaire, C.; Giacomelli, F.; Mievis, F.; Mestdagh, N.; Buchanan, T.; Valade, A.; et al. Evaluation of 18F-UCB-H as a novel PET tracer for synaptic vesicle protein 2A in the brain. *J. Nucl. Med.* **2014**, *55*, 1336–1341. [CrossRef] [PubMed]
37. Kilkenny, C.; Browne, W.; Cuthill, I.; Emerson, M.; Altman, D. Improving bioscience research reporting: The ARRIVE guidelines for reporting animal research. *J. Pharmacol. Pharmacother.* **2010**, *1*, 94. [CrossRef] [PubMed]

Sample Availability: Not available.

© 2019 by the authors. Licensee MDPI, Basel, Switzerland. This article is an open access article distributed under the terms and conditions of the Creative Commons Attribution (CC BY) license (http://creativecommons.org/licenses/by/4.0/).

Article

In Vitro and In Vivo Characterization of Dibenzothiophene Derivatives [^{125}I]Iodo-ASEM and [^{18}F]ASEM as Radiotracers of Homo- and Heteromeric α7 Nicotinic Acetylcholine Receptors

Cornelius K. Donat [1,2,*], Henrik H. Hansen [1], Hanne D. Hansen [1], Ronnie C. Mease [3], Andrew G. Horti [3], Martin G. Pomper [3], Elina T. L'Estrade [1,4,5], Matthias M. Herth [4,5], Dan Peters [6], Gitte M. Knudsen [1] and Jens D. Mikkelsen [1,*]

1. Neurobiology Research Unit, Copenhagen University Hospital, Rigshospitalet, DK-2100 Copenhagen, Denmark; hbh@gubra.dk (H.H.H.); Hanne.D.Hansen@nru.dk (H.D.H.); elina.nyberg@sund.ku.dk (E.T.L.); Gitte.Knudsen@nru.dk (G.M.K.)
2. Department of Brain Sciences, Imperial College London, London W12 0 LS, UK
3. Russell H. Morgan Department of Radiology and Radiological Science, The Johns Hopkins University School of Medicine, Baltimore, MD 21287, USA; rmease1@jhmi.edu (R.C.M.); ahorti1@jhmi.edu (A.G.H.); mpomper@jhmi.edu (M.G.P.)
4. Department of Drug Design and Pharmacology, University of Copenhagen, Jagtvej 162, 2100 Copenhagen, Denmark; matthias.herth@nru.dk
5. Department of Clinical Physiology, Nuclear Medicine & PET, Rigshospitalet, Blegdamsvej 9, 2100 Copenhagen, Denmark
6. DanPET AB, 216 19 Malmö, Sweden; info@danpet.eu
* Correspondence: cdonat@imperial.ac.uk (C.K.D.); jens_mikkelsen@dadlnet.dk (J.D.M.); Tel.: +45-40205378 (J.D.M)

Academic Editor: Peter Brust
Received: 24 January 2020; Accepted: 27 February 2020; Published: 20 March 2020

Abstract: The α7 nicotinic acetylcholine receptor (α7 nAChR) is involved in several cognitive and physiologic processes; its expression levels and patterns change in neurologic and psychiatric diseases, such as schizophrenia and Alzheimer's disease, which makes it a relevant drug target. Development of selective radioligands is important for defining binding properties and occupancy of novel molecules targeting the receptor. We tested the in vitro binding properties of [^{125}I]Iodo-ASEM [(3-(1,4-diazabycyclo[3.2.2]nonan-4-yl)-6-(^{125}I-iododibenzo[b,d]thiopentene 5,5-dioxide)] in the mouse, rat and pig brain using autoradiography. The in vivo binding properties of [^{18}F]ASEM were investigated using positron emission tomography (PET) in the pig brain. [^{125}I]Iodo-ASEM showed specific and displaceable high affinity (~1 nM) binding in mouse, rat, and pig brain. Binding pattern overlapped with [^{125}I]α-bungarotoxin, specific binding was absent in α7 nAChR gene-deficient mice and binding was blocked by a range of α7 nAChR orthosteric modulators in an affinity-dependent order in the pig brain. Interestingly, relative to the wild-type, binding in β2 nAChR gene-deficient mice was lower for [^{125}I]Iodo-ASEM (58% ± 2.7%) than [^{125}I]α-bungarotoxin (23% ± 0.2%), potentially indicating different binding properties to heteromeric α7β2 nAChR. [^{18}F]ASEM PET in the pig showed high brain uptake and reversible tracer kinetics with a similar spatial distribution as previously reported for α7 nAChR. Blocking with SSR-180,711 resulted in a significant decrease in [^{18}F]ASEM binding. Our findings indicate that [^{125}I]Iodo-ASEM allows sensitive and selective imaging of α7 nAChR in vitro, with better signal-to-noise ratio than previous tracers. Preliminary data of [^{18}F]ASEM in the pig brain demonstrated principal suitable kinetic properties for in vivo quantification of α7 nAChR, comparable to previously published data.

Keywords: alpha 7; nicotinic acetylcholine receptors; PET; nAChR; autoradiography

1. Introduction

The α7 nicotinic acetylcholine receptor (α7 nAChR) belongs to the superfamily of ligand-gated ion channels and is expressed across all mammalian species [1–4]. The receptor plays an important role in cognition [5], mood [6] and consistent with this, α7 nAChR are particularly abundant in hippocampus and prefrontal cortex [7,8]. Furthermore, α7 nAChR are implied in neuro-immune [9] and immune functions [10] under homeostatic conditions.

Changes in protein and mRNA levels of α7 nAChR have been reported in a number of neuropsychiatric and neurodegenerative diseases [1,11–15]. Notably, certain polymorphisms in the promoter region of the α7 nAChR gene (*CHRNA7*) [16] are probable risk factors for neuropsychiatric diseases, such as major depression [17] and schizophrenia [18] and are associated with developmental disorders and cognitive impairments [19]. Additionally, α7 nAChRs are expressed by several central and peripheral immune cells and activation via agonists and positive allosteric modulators showed neuroprotective and immunomodulatory efficacy in different preclinical disease models [20–24].

Changes of α7 nAChR in the healthy and diseased brain can only be detected in vivo by molecular imaging, such as positron emission tomography (PET) using specific radiotracers. A clinically usable radiotracer requires sufficient selectivity, specificity and suitable affinity, depending on the target [25]. Most of the previously described α7 nAChR PET tracers, among those [^{18}F]NS14490, [^{11}C]NS14492, [^{11}C]CHIBA-1001 and [^{11}C]A-582941 (Table 1), studied in mice, pigs and non-human primates exhibited some shortcomings, such as poor specific and/or high nonspecific binding or radiometabolites crossing the blood–brain barrier [26–32]. Furthermore, the specificity of novel α7 nAChR tracers has not always been tested in respective gene-deficient mice, e.g., using in vitro autoradiography.

While development of novel tracers from different lead structures is still ongoing [33–37], tilorone [38] provided a lead-structure for a number of derivatives subsequently developed into α7 nAChR PET tracers. From those, [^{18}F]ASEM (JHU82132) [39] and the structurally related [^{18}F]DBT-10 (JHU82108) [40] have been most widely investigated. Initial studies have shown that ASEM is a potent antagonist [39] with subnanomolar affinity and high selectivity [39,41], further substantiated by the radiolabelled compounds [^{18}F]ASEM [26] and [^{125}I]Iodo-ASEM [42] as tested in human and rat recombinant α7 nAChR. [^{18}F]ASEM and [^{125}I]Iodo-ASEM readily enter the mouse brain, are displaceable, and accumulate in regions with highest α7 nAChR density [26,39,42,43].

More recent studies using [^{18}F]ASEM and [^{18}F]DBT-10 further supported the suitability of the tracers, showing high and reversible brain uptake with a regional binding pattern consistent with the distribution of α7 nAChR receptors in the non-human primate brain [39,44,45]. Favourable brain pharmacokinetics, excellent test-retest reproducibility and regional uptake [^{18}F]ASEM pattern consistent with post-mortem α7 nAChR distribution have been reported in human PET studies [43,46]. Several recent studies extended the available data in human subjects, showing good agreement with previous distribution volumes (V_T) and test-retest values in nonhuman primates and healthy volunteers [44]. A study in ageing subjects showed a significant positive correlation between age and [^{18}F]ASEM V_T in striatum and several cortical regions [47], however without any correlation between V_T and cognitive measures. A small sample of individuals with schizophrenia on stable antipsychotic medication showed lower [^{18}F]ASEM V_T in cingulate cortex and hippocampus [46] and individuals with recent onset of psychosis were also reported to show lower [^{18}F]ASEM V_T in hippocampus, after controlling for age [48]. Interestingly, patients with mild cognitive impairment showed higher [^{18}F]ASEM V_T when adjusted for age as compared to the control group [49], consistent with post-mortem findings from patients and animal models. Additionally, [^{18}F]ASEM has been employed in a rat 6-OHDA lesion model of Parkinson's disease, showing an initial increase of [^{18}F]ASEM SUVr in the ipsilateral striatum and substantia nigra between 3 and 7 days, which coincided with several histology markers of glia activation [50].

While this data shows the general applicability of [^{18}F]ASEM, binding properties and interpretation of novel α7 nAChR tracers might be complicated by the fact that α7 subunits can form heteromeric receptors together with other subunits, specifically β2 [51]. These receptors can be heterologously expressed in oocytes and are found in the rodent and human basal forebrain and cortex [52,53]. While these heteromeric receptors display different pharmacological properties [52,54], it is not clear how this translates to radiotracer binding. In vitro binding studies of [^{18}F]ASEM or [^{125}I]ASEM in gene-deficient mice could answer the question, and would also reveal the suitability of [^{125}I]ASEM for in vitro autoradiographic studies. The latter would offer a better comparability to in vivo PET data over the current gold-standard tracer [^{125}I]α-bungarotoxin. We therefore investigated the potential of [^{125}I]ASEM for in vitro studies of the α7 nAChR, by comparing binding of [^{125}I]Iodo-ASEM in rat, mouse and pig brain sections. Furthermore, [^{18}F]ASEM was characterized for in vivo brain uptake and target selectivity in a PET study conducted in the pig.

Table 1. Common α7 nAChR ligands and their structure, previously evaluated as radiotracers.

Tracer	Structure
[^{11}C]CHIBA-1001	
[^{11}C]A-582941	
[^{18}F]NS14490	
[^{11}C]NS14492	
[^{18}F]ASEM	
[^{18}F]DBT-10	
[^{125}I]ASEM	

2. Results

2.1. In Vitro Autoradiography

Cerebral binding of [^{125}I]Iodo-ASEM was investigated across several mammalian species, i.e., rat (Figure 1A, upper row), mouse (Figure 2A) and pig brain (Figure 1A, lower row) and compared to [^{125}I]α-bungarotoxin (Figure 1B/2B). Total cortical [^{125}I]Iodo-ASEM binding was highest in the pig, and lower in the rat and mouse (Figures 1C and 2C), as compared to [^{125}I]α-bungarotoxin. [^{125}I]Iodo-ASEM binding was displaceable with (-)-nicotine (1 mMol/L, data not shown) and SSR-180,711 (10 µMol/L, Figures 1A and 2A, nonspecific binding) in all species.

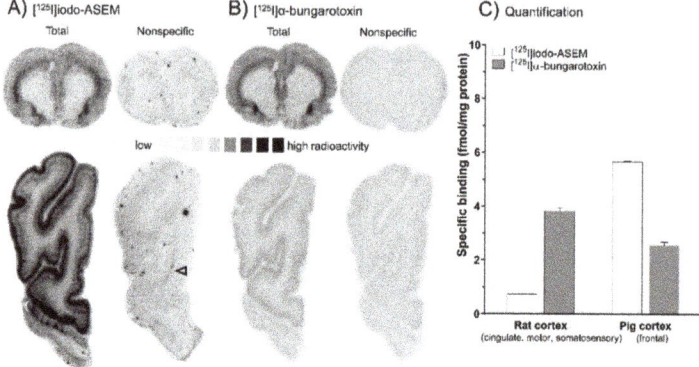

Figure 1. (**A**,**B**) Representative autoradiographs showing [^{125}I]Iodo-ASEM and [^{125}I]α-bungarotoxin total and non-specific binding (determined with 1 mMol/L (-)-nicotine for [^{125}I]α-bungarotoxin and 10 μMol/L SSR-180,711 for [^{125}I]Iodo-ASEM) in 12 μm sections of the rat (upper row) and pig brain (lower row). Arrowheads indicate residual white matter binding. (**C**) Comparative quantitative analysis of specific binding (± S.E.M.) of [^{125}I]Iodo-ASEM and [^{125}I]α-bungarotoxin from autoradiography in the rat ($n = 1$) and pig cortex ($n = 2$). All autoradiographic experiments and quantifications are carried out in 3-4 sections per animal.

A non-specific binding component remained detectable under the described experimental conditions at low levels in white matter structures (arrowheads, Figures 1A and 2A). The distribution pattern of [^{125}I]Iodo-ASEM binding in the rat, mouse and pig brain was comparable to that of [^{125}I]α-bungarotoxin (Figures 1B and 2B). In the pig, [^{125}I]Iodo-ASEM showed a laminar binding pattern in the frontal cortex, with highest density in cortical layers (1–3) (Figure 1A, lower row), while in the rat, binding in motor, cingulate and somatosensory cortex was more prominent in layers 5–6. However, this species difference was also observed for [^{125}I]α-bungarotoxin (Figure 1B).

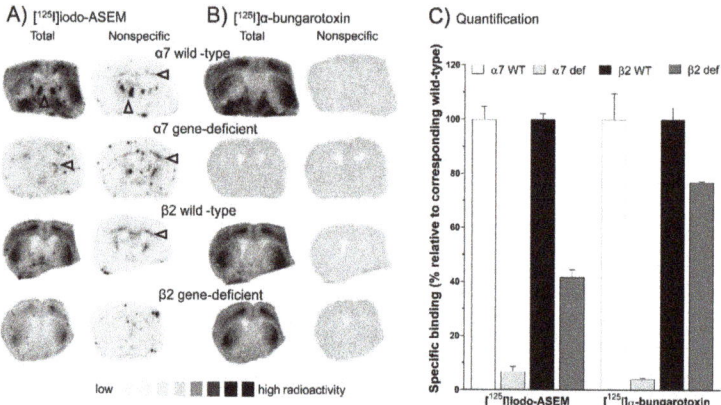

Figure 2. (**A**,**B**) Representative autoradiographs showing total [^{125}I]Iodo-ASEM (**A**) and [^{125}I]a-bungarotoxin (**B**) and non-specific (determined with 1 mMol/L (-)-nicotine for [^{125}I]α-bungarotoxin and 10 μMol/L SSR-180,711 for [^{125}I]Iodo-ASEM) binding in 12 μm brain sections of α7 and β2 nAChR wild-type vs. corresponding gene-deficient (def) mice ($n = 1$ each). Arrowheads indicate residual white matter binding. (**C**) Comparative quantitative analysis of specific binding (± S.E.M.) of [^{125}I]Iodo-ASEM and [^{125}I]a-bungarotoxin in α7 and β2 nAChR wild-type vs. corresponding gene-deficient mice ($n = 1$). All autoradiographic experiments and quantifications are carried out in 3-6 sections per animal.

The specificity of [^{125}I]Iodo-ASEM to α7 nAChR is further substantiated by tracer binding experiments in α7 nAChR gene-deficient mice. Specific [^{125}I]Iodo-ASEM binding was lacking in α7 nAChR gene-deficient mice (Figure 2A), as indicated by the overall reduction in total binding by 93% ± 1.7%, compared to wild-type animals (Figure 2C). Similarly, [^{125}I]α-bungarotoxin total binding (Figure 2B) was 96% ± 0.4% lower in α7 nAChR gene-deficient mice (Figure 2C). In wild-type mice, no difference in [^{125}I]Iodo-ASEM and [^{125}I]α-bungarotoxin binding was observed (Figure 2A,B). However, traces of nonspecific binding were again noted in white matter structures (arrowheads in Figure 2A).

[^{125}I]Iodo-ASEM binding in β2 nAChR gene-deficient mice was different compared to [^{125}I]α-bungarotoxin. An overall 58% ± 2.7% lower specific [^{125}I]Iodo-ASEM binding was observed, as compared to corresponding wild-type controls (Figure 2C). In contrast, [^{125}I]α-bungarotoxin binding was reduced by 23% ± 0.2%, being less affected by β2 nAChR gene-deficiency as compared to the reduction in [^{125}I]Iodo-ASEM binding.

Saturation binding in rat and pig brain sections indicated that [^{125}I]Iodo-ASEM binding was saturable. In the rat, non-linear regression analysis revealed an equilibrium dissociation constant (K_d) of 1.14 nM (cortex, Figure 3B) and 1.17 nM (hippocampus, Figure 3A) with corresponding receptor density (B_{max}) of 0.70 fmol/mg protein (cortex) and 1.44 fmol/mg protein (hippocampus), respectively (Figure 3A,B). In comparison, the pig cortex showed a K_d of 1.21 nM with a B_{max} of 5.47 fmol/mg protein (Figure 3C). The non-specific binding of [^{125}I]Iodo-ASEM at concentrations near the K_d was low (rat hippocampus, 20%; rat cortex, 30%; pig cortex, 10%).

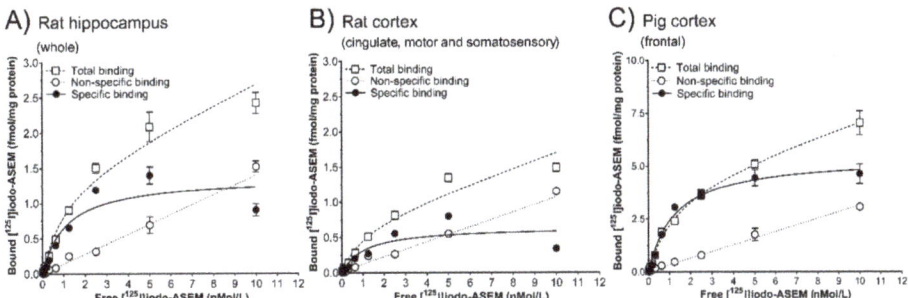

Figure 3. Saturation binding of [^{125}I]Iodo-ASEM (0.02–10 nMol/L) to 12 μm sections from the rat hippocampus and cortex (**A**, **B**, $n = 1$) and pig frontal cortex (**C**, $n = 1$) brain. Non-specific binding was determined in the presence of 10 μMol/L SSR-180,711. Optical density of the autoradiograms was converted into ligand binding (fmol/mg protein ± S.E.M.) from a representative experiment. Data from saturation binding experiments were analysed by non-linear regression. Individual K_d and B_{max} values are indicated in Section 2.1. All autoradiographic experiments and quantifications are carried out in 2–4 sections per animal using 10 radioligand concentrations.

A range of selective α7 nAChR ligands (10 μMol/L each), including the α7 nAChR preferring antagonist methyllycaconitine (MLA), were used to test whether in vitro [^{125}I]Iodo-ASEM binding (0.5 nMol/L) could be blocked in the pig cortex (Table 2). The partial agonists, NS14492, TC-5619, EVP-6124, A-582941, and SSR-180,711, showed almost complete (>90%) blocking of [^{125}I]Iodo-ASEM binding in receptor dense areas of the cortex, e.g. layers 1–3. In contrast, GTS-21 (weak α7 nAChR agonist, ~70% reduction) and MLA (α7 nAChR preferring antagonist, ~80% reduction) exhibited less efficacious blockade of [^{125}I]Iodo-ASEM binding in the pig cortex.

Table 2. Blocking of [^{125}I]Iodo-ASEM binding in the pig cortex by a series of α7 nAChR ligands.

Ligand (10 µMol/L)	[^{125}I]Iodo-ASEM Binding in the Pig Cortex, Layers 1–3 (%, mean ± S.E.M.)	[^{125}I]Iodo-ASEM Binding in the Pig Cortex, Layers 4–6 (%, mean ± S.E.M.)
NS14492	4.04 ± 0.55	9.25 ± 0.83
TC-5619	7.88 ± 1.65	7.60 ± 0.30
EVP-6124	2.50 ± 0.20	5.50 ± 0.07
A-582941	3.22 ± 0.28	4.44 ± 0.93
SSR-180,711	2.92 ± 0.35	3.53 ± 0.50
GTS-21	30.92 ± 2.55	31.39 ± 2.15
MLA	20.46 ± 2.18	18.16 ± 2.41

Results are given in % remaining binding of total binding (mean ± S.E.M.).

2.2. In Vivo PET Imaging in the Pig Using [^{18}F]ASEM

[^{18}F]ASEM readily entered the pig brain and highest tracer accumulation was found in the thalamus followed by cortex, striatum and cerebellum (Figure 4A,C). [^{18}F]ASEM uptake in the white matter was initially lower than in the grey matter regions, however the tracer kinetics were also slower, resulting in lower grey to white matter ratio at the end of the scans. The metabolism of [^{18}F]ASEM in pigs was relatively slow, with 60% of the radioactivity at 120 min still being parent radioligand (data not shown). Kinetic modelling was performed to quantify the tracer uptake. Baseline V_T values varied between animals but after correcting for free fraction in plasma(f_P), there was only a 5% difference in V_T/f_P values between the two baseline animals (Table 3). This also suggest that V_T/f_P values are unaffected by relatively large differences in injected mass (0.35 µg and 1.78 µg).

Table 3. Kinetic modelling of [^{18}F]ASEM with the Logan Graphical Analysis model in different pig brain regions.

	Comparison of Baseline V_T Values.			
Kinetic Modelling 0–90 min	Animal 1		Animal 2	
	V_T	V_T/f_P	V_T	V_T/f_P
Frontal cortex	7.87	43.70	3.75	41.66
Somatosensory cortex	8.33	46.27	4.15	46.14
Occipital cortex	8.03	44.63	3.77	41.86
Remaining cortex	7.63	42.37	3.78	41.94
Thalamus	8.83	49.06	4.12	45.73
Striatum	7.41	41.17	3.77	41.94
Hippocampus	7.53	41.83	3.59	39.93
Cerebellum	6.66	36.99	3.16	35.07
	Comparison of V_T Values at Baseline and After Pre-treatment with SSR-180,711			
Kinetic Modelling 0–150 min	Animal 1		Animal 3	
	V_T	V_T/f_P	V_T	V_T/f_P
Frontal cortex	6.73	37.38	3.61	22.57
Somatosensory cortex	7.26	40.35	3.94	24.60
Occipital cortex	6.96	38.65	3.90	24.40
Remaining cortex	6.68	37.08	3.60	22.50
Thalamus	7.42	41.24	4.34	27.15
Striatum	6.51	36.19	3.83	23.92
Hippocampus	6.55	36.41	3.37	21.09
Cerebellum	5.55	30.82	3.39	21.17

In a third animal, we evaluated the specificity of [^{18}F]ASEM binding in vivo, by administering SSR-180,711 (1 mg/kg) prior to injection of [^{18}F]ASEM. Compared to the baseline studies, we found an increase in [^{18}F]ASEM uptake in all brain areas investigated (Figure 4A). Quantification of uptake and subsequent correction for f_P revealed that SSR-180,711 administration decreased the V_T/f_P compared to baseline (Table 3). Occupancy was computed with the Lassen plot using V_T/f_P values comparing baseline data from animal 1 and blocking data from animal 3 (0-150 min scan data). We found that the 1 mg/kg SSR-180,711 dose resulted in a 49% occupancy (Figure 4B).

From the Lassen plot, the volume of non-displaceable binding (V_{ND}/f_P) was found to be 9.2 mL/cm^3. When comparing the V_{ND}/f_P to the V_T/f_P in the thalamus, we found that 78% of the signal observed in the thalamus is specific binding, leaving 22% as non-displaceable binding.

In one animal, [^{18}F]ASEM acquisition time was 240 min, which allowed subsequent analysis of the time-stability of the parameters estimated with kinetic modelling. Again, the LGA model was used to determine V_T with different scan length and V_T values were found to decrease with more time included in the kinetic modelling. Using all data (0–240 min), V_T values were 5.4 mL/cm^3 (thalamus), 5.0 mL/cm^3 (frontal cortex) and 4.0 mL/cm^3 (cerebellum).

The upper half of the table shows the baseline distribution volumes (V_T) values with and without correction for free fraction in plasma (f_P) in two different animals. Bottom part of the table describes V_T values with and without correction for (f_P) at baseline (animal 1) and after pre-treatment with SSR-180,711 (animal 3). See Table 4 (Material and Methods) for f_P values in the individual animals. Because animal 2 was only scanned for 90 min, the acquisition time of animal 1 was truncated to 90 min to allow for comparison.

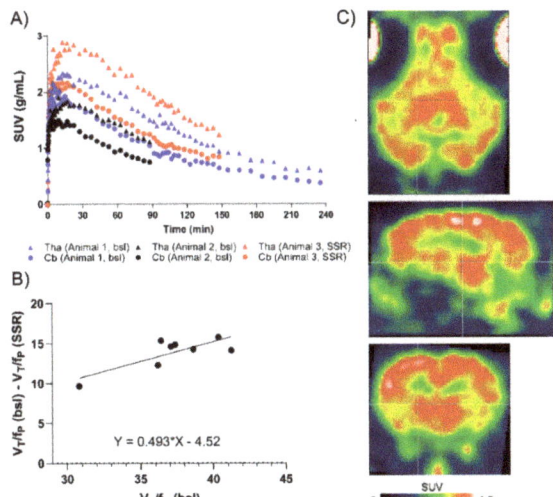

Figure 4. [^{18}F]ASEM binding in the pig brain. (**A**) Time-activity curves of [^{18}F]ASEM in three different animals: Animal 1, bsl, blue; Animal 2, bsl, black; Animal 3, SSR-180,711 pre-treated, red. The regions shown are: Thalamus (tha, triangles) and cerebellum (cb, circles). (**B**) Lassen plot with total distribution volumes (V_T) corrected for free fraction in plasma (f_P) using values from animal 1 and animal 3. Each point represents one region of interest (ROI), please refer to the method section for the complete list of ROIs. (**C**) Summed PET image (0–240 min) from animal 1 showing the distribution of [^{18}F]ASEM in the pig brain. SUV: standard uptake value. Bsl: baseline. SSR: SSR-180,711 (1 mg/kg).

3. Discussion

In this study, we investigated binding properties of radiolabelled ASEM in vitro ([^{125}I]Iodo-ASEM] and in vivo ([^{18}F]ASEM). Autoradiography was used to determine the applicability of [^{125}I]Iodo-ASEM for in vitro assessment of α7 nAChR receptor distribution and occupancy in the mammalian brain. [^{125}I]Iodo-ASEM showed high-affinity and specific binding to α7 nAChR in the rat, mouse and pig brain. Specific binding was absent in α7 gene-deficient mice, indicating high specificity and selectivity. Saturation binding experiments in rat and pig brain sections revealed low nanomolar K_d values (approximately 1 nM) in both species. B_{max} in the pig cortex was considerably higher as compared to the binding found in the mouse and rat brain cortex. Such species differences are well documented in the literature, e.g., for metabotropic glutamate 5 receptors and the 18 kDa translocator protein between monkey and humans [55,56]. As affinity and selectivity are major criteria for radiotracers, our data further substantiates the suitability of ASEM derivatives as favourable α7 nAChR tracers [57,58]

We found that the affinity of [^{125}I]Iodo-ASEM in the pig brain as determined with autoradiography was in a similar range as [^{3}H]NS14492 [59]. In contrast, higher affinities and receptor densities are reported for radioligand binding assay in brain homogenates for a number of different α7 nAChR ligands, such as [^{3}H]NS14492 and [^{3}H]A-585539 [60], including other dibenzothiophenes [26,38].

Across the brain and specifically regions with high α7 nAChR expression, such as hippocampus and superficial cortical layers, non-specific binding of [^{125}I]Iodo-ASEM at concentrations near the K_d was low (10%–30% of total binding) and produced a robust specific signal. However, a consistent nonspecific binding component in white matter was observed in all investigated species, in particular the corpus callosum and subcortical tracts. This is supported by previously reported in vivo findings in human and non-human primate subjects [39,43], where tracer uptake was lowest in white matter structures, such as the corpus callosum. Furthermore, we also observed slower in vivo kinetics in white matter structures in the pig brain. This could be caused by lower perfusion or kinetics may be different when the tracer interacts with lipid membranes, compared to interaction with the receptor. As in vitro binding conditions are distinctly different due to absent metabolism and blood flow, it is possible that these effects may limit pronounced non-specific white matter binding in vivo or that the nonspecific binding exhibits much slower kinetics. Under the employed incubation conditions, [^{125}I]α-bungarotoxin shows no white matter residual binding. However, under the same conditions, the overall non-specific binding in grey matter for [^{125}I]α-bungarotoxin is approximately 45% in human (data not shown) and 55% in pig brain tissue, where it is much lower for [^{125}I]Iodo-ASEM (~10%–30%).

[^{125}I]Iodo-ASEM binding enables an important distinction between grey and white matter structures, e.g., the distinct cortical laminar binding pattern observable in the pig. While [^{125}I]Iodo-ASEM binding was prominent in the deeper cortical layers in the mouse and rat, superficial cortical layers were intensely labelled in the pig. Using in vitro autoradiography, similar laminar cortical binding pattern in the pig brain has also recently been reported for a structurally different α7 nAChR radioligand, [^{3}H]NS14492 [61]. Binding of both tracers was matching the pattern of [^{125}I]α-bungarotoxin, the in vitro gold-standard radioligand for α7 nAChR. However, the spatial binding pattern in the rodent brain was only similar between [^{125}I]Iodo-ASEM and [^{125}I]α-bungarotoxin, but not for [^{3}H]NS14492, suggesting different binding profiles of antagonists and agonists, or species differences in affinity.

[^{125}I]Iodo-ASEM proved specific to the α7 nAChR, as evidenced by the lack of specific binding in α7 nAChR gene-deficient mice and a virtually complete block of cortical [^{125}I]Iodo-ASEM binding by a wide range of structurally different α7 nAChR selective ligands and MLA, with the rank order being NS14492=TC-5619=EVP-6124=A-582941=SSR-180,711>MLA>GTS-21. While this corresponds well with the individual high affinities in the nanomolar range (NS14492, TC5619, EVP-6124, A-582941, SSR-180,711) [31,62–65], as compared to the lower affinity of a partial agonist (GTS-21) [66], it could also reflect the general differences between antagonists and agonist in terms of binding sites and kinetics.

When comparing to [^{125}I]α-bungarotoxin, the specific binding of [^{125}I]Iodo-ASEM was lower in the rat and mouse brain, but higher in the pig cortex. While species differences in receptor structure may account for the discrepancies, it should also be considered that different incubation protocols were used for the determination of optimal [^{125}I]Iodo-ASEM and [^{125}I]α-bungarotoxin binding. Hence, a relatively high detergent concentration was required in the experiments to obtain optimal total tissue binding of [^{125}I]Iodo-ASEM, which may potentially affect binding of the radioligand in the mammalian species tested, e.g., through differences in lipid content and myelination. For example, [^{125}I]Iodo-ASEM showed some degree of non-displaceable binding to white matter structures, which could be caused by the ligands' lipophilicity and/or different kinetics in white matter structures.

Interestingly, radioligand binding in β2 nAChR gene-deficient mice was more strongly reduced for [^{125}I]Iodo-ASEM than [^{125}I]α-bungarotoxin. This observation suggests different binding properties and subtype selectivity to heteromeric α7β2 nAChR, compared to the homomeric receptors. In the CNS, heteromeric α7β2 nAChR are identified in the mouse forebrain and hippocampal neurons, rat basal forebrain cholinergic neurons, as well as in the human basal forebrain and cerebral cortex. Importantly, α7β2 nAChR display distinct functional properties as compared to homomeric α7 nAChR [51,67], owing to their slower whole cell decay kinetics and current amplitudes in both transfected cell systems and native rodent neurons [53,54,68–70]. Accordingly, co-expression of α7 and β2 nAChR subunits in *Xenopus* oocytes also results in lower maximal responses (evoked current amplitudes) of selective α7 nAChR agonists but does not shift pharmacology to a more β2-like profile [52–54,71]. These in vitro studies in transfected cell systems therefore suggest that α7 nAChR agonists bind to the α7-α7 subunit interface, and β2 subunits likely do not contribute to the ligand binding site on heteromeric α7β2 nAChR [67,71]. When using selective α7 nAChR antagonists, including MLA and α-bungarotoxin, to alter the response to some nicotinic agonists in either homomeric α7 and heteromeric α7β2 nAChR, results have been less consistent, as they show unaltered [54] or reduced potency [52] and efficacy [53] in comparison to homomeric α7 nAChR expressed in *Xenopus* oocytes. The functional significance of heteromeric α7 nAChR expression is not well understood, with recent work suggesting that this subtype combination might be more sensitive to inhibition by oligomeric amyloid β$_{1-42}$ [68,69] and isoflurane [72], as compared to homomeric α7 nAChR. Our finding that binding of [^{125}I]Iodo-ASEM, an antagonist, was markedly reduced in the forebrain of β2 gene-deficient mice therefore may suggest that [^{125}I]Iodo-ASEM binds to heteromeric α7β2 nAChR in the brain, as opposed to [^{125}I]α-bungarotoxin. Whether this is due to different affinity for homomeric α7 and heteromeric α7β2 nAChR requires further in vitro studies. Although speculative, this may offer a chance to probe the binding of amyloid β$_{1-42}$ to heteromeric α7β2 nAChR in vivo using PET.

In vivo uptake of [^{18}F]ASEM into the pig brain occurred rapidly within the first 10–20 min and a reversible but slower washout was found, as observed in human and non-human primate subjects [39,43]. The in vivo distribution of [^{18}F]ASEM found here is very similar to that of [^{11}C]NS14492 and importantly, also in accordance with the distribution of α7 nAChR in the pig brain [31,73]. Furthermore, our data with [^{18}F]ASEM matches previous reports with the structurally similar analogue [^{18}F]DBT-10 in piglets [40].

We found variations in brain uptake and f$_P$ in the two baseline animals, and this result is consistent with the interpretation that lower f$_P$ will lead to lower brain uptake [74]. Due to the limited number of animals in this study, this observation merits further investigations. Our finding is however supported by PET studies in non-human primates with [^{18}F]ASEM and [^{18}F]DBT-10, where V$_T$/f$_P$ was shown to be a more stable outcome measure than V$_T$ [44,45]. This has also been shown for radiotracers binding to other neurotransmitter receptors [75].

V$_T$ was found to increase slightly when the scan time was prolonged. This phenomenon was most pronounced in the thalamus and least pronounced in the white matter and is evident from the TACs (Figure 4A), where the ratio between e.g., thalamus and cerebellum was lower at 240 min than at 90 min. This finding is in contrast with the non-human primate and human data, where V$_T$ was underestimated when reducing the PET data from 180 to 60 min [44]. Given that ASEM is an antagonist, it is unlikely that internalization of the receptor-ligand complex is an explanation for the decrease in V_T. We cannot exclude that other receptor adaptations, such as (de)sensitization, could be responsible for this observation. Desensitization could occur if experiments were not conducted at tracer dose, i.e., but we did not attempt to identify the mass dose limit of unlabelled ASEM. Although the injected doses of ASEM varied in the two baseline animals, we only found a 5% difference in the calculated V_T/f_P, which suggest that the studies were conducted at tracer doses. The injected doses in this study (0.007–0.085 ug/kg) are higher than the doses used in the non-human primate evaluation of [^{18}F]ASEM, where injected doses ranged from 0.009 to 0.056 µg/kg [44]. Further studies are needed to identify the mass dose limit of unlabelled ASEM.

While pre-treatment with 1 mg/kg SSR-180,711 resulted in an increased uptake of [^{18}F]ASEM, kinetic modelling for quantification of tracer uptake showed that SSR-180,711 at this dose resulted in 49% occupancy. A similar phenomenon has also been reported in piglets when [^{18}F]DBT-10 was blocked by the weak agonist NS6740, which was ascribed to a potential blood flow-driven effect of NS6740 leading to greater central uptake of [^{18}F]DBT-10 [40]. The increased tracer uptake could also be a result of peripheral α7 nAChR binding sites having been blocked by SSR-180,771. The occupancy found by us is in line with previous work of Horti et al., reporting 39% and 81% occupancy for doses of 0.5 and 5 mg/kg SSR-180,711, respectively [39]. The occupancy computed in this study should be interpreted with care, as the baseline and blocking study is conducted in two different pigs. Due to the half-life of [^{18}F]ASEM, it was not possible to conduct the study in the same animal on the same day. A further limitation to this in vivo study is the low number of PET scans and animals and thus we can only provide a descriptive presentation of the data, without statistical evaluations.

From our results, [^{125}I]Iodo-ASEM therefore offers several advantages over [^{125}I]α-bungarotoxin: 1) low nonspecific binding, 2) similar high affinity and selectivity and 3) in vivo applicability and direct comparison of PET data with autoradiographic data. The lower nonspecific binding of [^{125}I]Iodo-ASEM is advantageous primarily in vitro, as it allows for a better signal-to-noise ratio over [^{125}I]α-bungarotoxin at very similar affinities. While a low nonspecific binding would also be favourable under in vivo conditions (e.g., PET), having two nearly identical molecules as tracers offers interesting avenues, especially for preclinical studies. In vivo PET/SPECT imaging data can be acquired through [^{18}F]ASEM, [^{18}F]DBT10 or [^{123}I]Iodo-ASEM and results can be validated or extended by using the advantages of in vitro autoradiography (e.g. resolution) with [^{125}I]Iodo-ASEM.

In conclusion, [^{125}I]-Iodo-ASEM is applicable for visualizing α7 nAChR binding in vitro, its binding is different between species, and may potentially bind to heteromeric α7β2 nAChR. In addition, [^{18}F]ASEM is demonstrated to have suitable kinetic properties for in vivo quantification of α7 nAChR in the pig.

4. Materials and Methods

4.1. Compounds and Radioligands

[^{125}I]Iodo-ASEM [(3-(1,4-diazabycyclo[3.2.2]nonan-4-yl)-6-(^{125}I-iododibenzo[b,d]thiopentene 5,5-dioxide)] was labelled according to previously published procedures [42]. Mean molar activity was 59.94 ± 6.25 TBq/mmol. [^{125}I]Tyr-54-mono-Iodo-α-bungarotoxin (81.4 TBq/mmol) was purchased from Perkin-Elmer (Skovlunde, Denmark). (-)-nicotine tartrate was purchased from Sigma-Aldrich (St. Louis, MO). Unlabelled ASEM and precursor for radiosynthesis was provided by DanPET (Malmoe, Sweden). The α7-selective ligands were purchased from Sigma-Aldrich (MLA) or provided by DanPET (NS11492) or NeuroSearch A/S (Copenhagen, Denmark) (SSR-180,711, TC-5619, EVP-6124, A-58294, and GTS-21.

4.2. Tissue Origin and Sectioning for In Vitro Autoradiography

All animal procedures were approved by the Danish Animal Experimentation Inspectorate (J. No. 2012-15-2034-00156) and treated in concordance with the European Communities Council Directive of 24th November 1986 (86/609ECC).

One female Sprague-Dawley rat (250 g, obtained from Charles River, Sulzfeld, Germany) was euthanized with an intraperitoneal overdose of pentobarbital, the brain was quickly removed and snap-frozen in −50 °C 2-methylbutane, then stored at −80 °C until further processing.

Mice deficient for the α7 subunits (The Jackson Laboratory) and β2 (Institut Pasteur, Paris, France) and their corresponding wild-type littermates were bred (C57BL/6J background) in an animal care facility at Virginia Commonwealth University. Brains from α7 and β2 gene-deficient mice and corresponding wild-type littermates were kindly provided by Dr. M. Imad Damaj (Dept. of Pharmacology and Toxicology, Virginia Commonwealth University, Richmond, VA, USA).

One two-month old female Danish domestic pig (Landrace x Yorkshire x Duroc, 22 kg) was euthanized with an intravenous injection of pentobarbital, the brain was quickly excised, separated in two hemispheres and frozen on dry ice, before being stored at −80 °C.

All brain specimens were cut in 12 μm serial sections on a cryostat (Microm HM 500 OM, Walldorf, Germany), thaw-mounted onto Super Frost slides (Thermo Scientific, Hvidovre, Denmark), briefly air dried and stored at −80 °C until further processing. Protein concentration was determined from single or three sections with the Bio-Rad Protein Assay (Bio-Rad, Hercules, CA, USA) based on the method of Bradford [76].

4.3. In Vitro Autoradiography with [^{125}I]Iodo-ASEM

Initial optimization of assay conditions was performed to maximize total binding while keeping non-specific binding low. Adjustments included buffer composition and pH, detergent concentration, wash and incubation time and temperature. An assay buffer with 50 mMol/L Tris-HCl pH 7.4, 21 °C (termed Tris-HCl buffer) provided best preserved tissue integrity and lowest non-specific binding, as compared to physiologic Tris, Tris-EDTA-EGTA or HEPES-KRH buffer (data not shown). For all further experiments, tissue from 1–2 animals was used, with experiments and quantifications carried out using 3–4 sections for pig and rat tissue and 3–6 sections for mouse tissue. Adjacent sections were used for autoradiography for all similar experiments (e.g. saturation binding). Sections were brought to room temperature and pre-incubated for 20 min in Tris-HCl buffer (pH 7.4, 21 °C), then incubated for 60 min in the same buffer (21 °C) containing 1.5% Triton X-100 (v/v) and 0.5-1.0 nMol/L [^{125}I]Iodo-ASEM in a humidified chamber. Non-specific binding of [^{125}I]Iodo-ASEM was assessed in the presence of 10 μMol/L SSR-180,711, added to the buffer. Blocking of [^{125}I]Iodo-ASEM binding was investigated with a series of individual α7 nAChR selective compounds (10 μMol/L) added to the incubation buffer, i.e., methyllycaconitine (MLA) [77], SSR-180,711 [63], NS14492 [31], TC-5619 [64], EVP-6124 [65], A-582941 [62], and GTS-21 [66]. Following incubation, slides were rinsed in Tris-HCl buffer (pH 7.4, 21 °C), washed 2 × 5 min in Tris-HCl buffer (pH 7.4, 4 °C) and rinsed (5 sec) in ice-cold distilled water. Slides were gently dried under an air stream and exposed to 4% paraformaldehyde vapour overnight at 4 °C, followed by another drying step in a desiccator for 1 h. Adjacent sections were used for autoradiography.

4.4. In Vitro Autoradiography with [^{125}I]α-bungarotoxin

Slides were thawed at room temperature (21 °C) for 30 min, followed by 30 min of rehydration in 50 mMol/L Tris buffer with 0.1% BSA (w/v), pH 7.3 (binding buffer). For assessment of α-bungarotoxin binding, the binding buffer contained 0.5 mMol/L [^{125}I]α-bungarotoxin and 4.5 nMol/L unlabelled α-bungarotoxin (Tocris, Denmark) yielding a total of 5 nMol/L α-bungarotoxin (incubation buffer). Total binding was determined using one set of slides incubated with the radioligand for 2 h at room temperature in a humidified chamber. Non-specific binding was determined in the presence of 1 mMol/L (-)-nicotine added to the incubation buffer. Afterwards, slides were briefly rinsed in binding buffer, followed by 2 × 30 min of washing in ice-cold binding buffer (4 °C). Finally, slides were briefly rinsed (5 sec) in ice-cold distilled water, dried under a gentle air stream and exposed to 4% paraformaldehyde vapour overnight at 4 °C. On the next day, the slides were dried for 1 h in a desiccator.

4.5. Saturation Binding and Kinetic Analysis Using In Vitro Autoradiography

Saturation binding was carried out in rat and pig brain sections as described above (Section 4.3). Sections were incubated with ten serial dilutions of [^{125}I]Iodo-ASEM ranging from 0.02 to 10 nMol/L, with concentrations measured by gamma-counting. Non-specific binding was determined in adjacent sections in the presence of 10 µMol/L SSR-180,711 for each radioligand concentration. Binding was terminated by washing the sections in ice-cold binding buffer. The equilibrium dissociation constant (K_d) and maximum number of binding sites (B_{max}) were determined by non-linear regression analysis of a one-site saturation binding model using GraphPad Prism 6.0 (GraphPad Software, Inc., San Diego, CA, USA).

4.6. Autoradiographic Image Acquisition and Analysis

BAS SR2040 phosphor imaging plates (Fujifilm, Toyko, Japan) were exposed to the samples along with [^{125}I] standards (ARI 0133A; American Radiolabeled Chemicals, St. Louis, USA) for 24-72 hours. Imaging plates were scanned using a Phosphor Imager BAS-2500 (Fujifilm Europe GmbH, Düsseldorf, Germany). Images were converted to TIF-files using the manufacturer's software and analysed in QuantityOne (BioRad, Waltham, MA, USA). Regions of interest (ROIs) were drawn over grey and white matter structures, depending on the investigated species. High intensity circular spots were occasionally observed in [^{125}I]Iodo-ASEM autoradiographs and were excluded from the analysis. For α7 and β2 gene-deficient mice, only one ROI was drawn over the whole brain, again excluding spots and irregular white matter binding. In the rat brain, the ROIs were drawn over cortex and hippocampus. From the pig brain, only frontal cortex sections were cut, the ROIs therefore contained the frontal cortex and white matter tracts. The mean values of optical density per mm^2 (averaged from the replicates) were converted to radioactive concentration using a linear regression derived from the [^{125}I] radioactive standards. A global background of the imaging plate and individual non-specific binding were subtracted. Final values were expressed as fmol/mg protein, based on the protein measurements from individual sections.

4.7. Radiosynthesis of [^{18}F]ASEM

The radiosynthesis of [^{18}F]ASEM was performed as previously published [26]. No-carrier-added aqueous ^{18}F-fluoride from the target was collected at a non-conditioned activated (10 mL ethanol, 20 mL water and dried with air) anion-exchange cartridge (QMA). A solution of 20 mg of 1,10-diaza-4,7,13,16,21,24-hexaoxabicyclo[8.8.8]hexacosane (Kryptofix-222) and 3.3 mg of K_2CO_3 dissolved in a 0.65 mL methanol-water mixture (97/3 v/v) was used to elute the ^{18}F-fluoride off the cartridge. The elute was thereafter dried by evaporation at 90 °C under nitrogen and then further dried twice with 1 mL dry acetonitrile. To the dried Kryptofix®222/[^{18}F]fluoride complex, 2.4 mg (0.006 mmol)/L of 3-(1,4-Diazabicyclo[3.2.2]nonan-4-yl)-6-nitrodibenzo[b,d]- thiophene 5,5-Dioxide dissolved in 0.8 mL DMSO was added. The reaction was performed at 160 °C for 15 min and afterwards the crude was quenched with 3.5 mL H$_2$O. Reactants and by-products were separated from [^{18}F]ASEM by semi-preparative HPLC [Luna column, Phenomenex Ltd. Aschaffenburg, Germany; 10 µm C18(2) 10×250 mm column, flow rate 6 mL/min, eluent: Ethanol/0.1% H$_3$PO$_4$ in water (25:75) with 6 mM ascorbic acid to prevent radiolysis]. The retention time for [^{18}F]ASEM was 400-450 s and the product was collected into a vial containing 9 mL of PBS (phosphate-buffered saline). The product was visually inspected for clarity, absence of colour and visible particles. Chemical and radiochemical purities were assessed by analytical HPLC [Kinetex column, Phenomenex Ltd. Aschaffenburg, Germany; 2.6µ C18

4.60 × 50 mm, eluent: ACN/0.1% H_3PO_4 in water (25:75) RT: [^{18}F]ASEM = 1.3 min; nitro precursor = 1 min; flow rate 1.5 mL/min]. Molar activity (A_m) of the radiotracer was determined as follows: the area of the UV absorbance peak corresponding to the radiolabelled product was measured (integrated) on the HPLC chromatogram. This value was then converted into a molar mass by comparison with an average of integrated areas (triplet) of a known standard of the reference compound.

4.8. In Vivo Imaging in the Pig

Three female pigs (21, 22 and 23 kg) were used for in vivo PET imaging on a HRRT PET scanner (Siemens Healthcare, Erlangen, Germany). All animal procedures were approved by the Danish Council for Animal Ethics (journal no. 2012-15-2934-00156).

4.8.1. Animal Procedures

Before scanning, anaesthesia was induced with i.m. injection of 0.13 mL/kg Zoletil veterinary mixture (Virbac, Kolding, Denmark; 10.87 mg/kg xylazine + 10.87 mg/kg ketamine + 1.74 mg/kg methadone + 1.74 mg/kg butorphanol + 10.87 mg/kg tiletamine + 10.87 mg/kg zolezepam). Hereafter, anaesthesia was maintained with constant propofol infusion (1.5 mg/kg/h intravenous (i.v.); B. Braun, Melsungen, Germany). An arterial i.v. catheter was employed for blood sampling from the right femoral artery and two venous i.v. catheters for injections were placed in the left and right mammary veins. During anaesthesia, animals were endotracheally intubated and ventilated. Vital parameters (heart rate, body temperature, blood pressure, oxygen saturation and end tidal CO_2) were continuously monitored during the scan.

4.8.2. PET Scanning

[^{18}F]ASEM was given as intravenous i.v. bolus, with experimental details described in Table 4.

Table 4. Experimental details of [^{18}F]ASEM PET scans in pigs.

Details	Animal 1	Animal 2	Animal 3
Type of experiment	Baseline	Baseline	SSR-180,711; 1 mg/kg
Scan length	240 min	90 min	150 min
Molar activity	20 GBq/µmol	345 GBq/µmol	388 GBq/µmol
Injected activity	99 MBq	335 MBq	189 MBq
Injected mass	1.78 µg	0.35 µg	0.18 µg
Free plasma fraction	18%	16%	9%

4.8.3. Blood Sampling

During the first 30 min of the scans, radioactivity in the whole blood was continuously measured using an ABSS autosampler (Allogg Technology, Mariefred, Sweden) counting coincidences in a lead-shielded detector. Concurrently, arterial whole blood was sampled manually at times 2.5, 5, 10, 20, 30, 40, 50, 70, 89, 91, 120 and 150 min after injection of [^{18}F]ASEM. Total radioactivity in plasma (500 µL) and whole blood (500 µL) was measured in a well counter (Cobra 5003; Packard Instruments, Meriden, CT, USA), which was cross-calibrated to the HRRT scanner and autosampler. All measurements of radioactivity were decay corrected to the time of radioligand injection.

4.8.4. Metabolite Analysis

Radiolabelled parent compound and metabolites were determined by direct injection of plasma into a radio-HPLC system (Dionex Ultimate 3000; Thermo Fisher Scientific, Hvidovre, Denmark) configured for column switching. Manually drawn arterial whole blood samples were centrifuged (1500 g, 7 min, 4 °C), and plasma was filtered through a syringe filter (Whatman GD/X 13 mm or 25 mm, PVDF membrane, 0.45 µm pore size; Frisenette ApS, Knebel, Denmark) prior to the analysis by HPLC as previously described [78]. To increase sensitivity on gamma counts from samples with low levels of radioactivity, eluent from the HPLC was collected into fractions (10 mL) using a fraction collector (Foxy Jr FC144; Teledyne, Lincoln, NE, USA) and counted offline in a well counter (2480 Wizard^2 Automatic Gamma Counter, Wallac Oy, Turku, Finland).

4.8.5. Determination of Free Fraction

The free, non-protein bound fraction of [^{18}F]ASEM in pig plasma, f_p, was estimated using an equilibrium dialysis chamber method as previously described [79].

4.8.6. Reconstruction and Pre-Processing of PET Data

150-minute list-mode PET data were reconstructed in 58 dynamic frames (6 × 10, 6 × 20, 6 × 30, 6 × 60, 4 × 120, 14 × 300, 8 × 150, 8 × 300 s). One animal was scanned for 240 min using the mentioned framing protocol but adding 9 frames of 600 s). Images consisted of 207 planes of 256 × 256 voxels of 1.22 × 1.22 × 1.22 mm. A summed picture of all counts in the 150-min scan was reconstructed for each pig and used for co-registration to a standardized MRI-based atlas of the domestic pig brain, similar to that previously published [80]. The time activity curveds (TACs) were calculated for the following volumes of interest (VOIs): thalamus, striatum, hippocampus, cerebellum, white matter, frontal cortex, somatosensory cortex, occipital cortex, rest of the cortex. Radioactivity in all VOIs was calculated as the average of radioactive concentration (Bq/mL) in the left and right sides. Outcome measure in the TACs was calculated as radioactive concentration in VOI (in kBq/mL) normalized to the injected dose corrected for animal weight (in kBq/kg), yielding standardized uptake values (g/mL).

4.8.7. Kinetic Modelling of PET Data

The PET imaging data were analysed with the Logan graphical analysis (LGA) model, using the metabolite corrected arterial plasma concentration to calculate the primary outcome measure: total distribution volume (V_T). The secondary outcome measure was V_T values corrected for free fraction in plasma (V_T/f_P). The parent fraction curve for [^{18}F]ASEM was fitted with a Watabe fit. Both curves were constrained to 1.0 at time = 0. Kinetic modeling was performed in PMOD version 3.0 (PMOD Technologies).

Author Contributions: Conceptualization: C.K.D., J.D.M., A.G.H., M.G.P., D.P; Methodology: C.K.D., H.D.H., G.M.K., M.M.H.; Formal analysis: C.K.D.; H.D.H., H.H.H.; Investigation: C.K.D.; H.D.H., E.T.L.; Resources: D.P., R.C.M., M.M.H., A.G.H.; Writing—original draft preparation: C.K.D.; H.H.H., H.D.H.; Writing—review and editing: C.K.D., H.D.H., J.D.M., A.G.H., M.G.P., G.M.K., M.M.H.,; Visualization: C.K.D., H.D.H.; Supervision: J.D.M., G.M.K.; Funding acquisition: J.D.M., G.M.K., M.G.P.; All authors have read and agreed to the published version of the manuscript.

Funding: This study was supported by the Danish Strategic Research Council (project COGNITO), EB024495, the Augustinus, Elsass and The Lundbeck Foundations.

Acknowledgments: The authors would like to thank the animal facilities at the Department of Experimental Medicine, University of Copenhagen.

Conflicts of Interest: The authors declare no conflict of interest.

References

1. Marutle, A.; Zhang, X.; Court, J.; Piggott, M.; Johnson, M.; Perry, R.; Perry, E.; Nordberg, A. Laminar distribution of nicotinic receptor subtypes in cortical regions in schizophrenia. *J. Chem. Neuroanat.* **2001**, *22*, 115–126. [CrossRef]
2. Kulak, J.M.; Carroll, F.I.; Schneider, J.S. [125I]Iodomethyllycaconitine binds to alpha7 nicotinic acetylcholine receptors in monkey brain. *Eur. J. Neurosci.* **2006**, *23*, 2604–2610. [CrossRef] [PubMed]

3. Whiteaker, P.; Davies, A.R.; Marks, M.J.; Blagbrough, I.S.; Potter, B.V.; Wolstenholme, A.J.; Collins, A.C.; Wonnacott, S. An autoradiographic study of the distribution of binding sites for the novel alpha7-selective nicotinic radioligand [3H]-methyllycaconitine in the mouse brain. *Eur. J. Neurosci.* **1999**, *11*, 2689–2696. [CrossRef] [PubMed]
4. Quik, M.; Choremis, J.; Komourian, J.; Lukas, R.J.; Puchacz, E. Similarity between rat brain nicotinic alpha-bungarotoxin receptors and stably expressed alpha-bungarotoxin binding sites. *J. Neurochem.* **1996**, *67*, 145–154. [CrossRef] [PubMed]
5. Lendvai, B.; Kassai, F.; Szajli, A.; Nemethy, Z. alpha7 nicotinic acetylcholine receptors and their role in cognition. *Brain Res. Bull.* **2013**, *93*, 86–96. [CrossRef] [PubMed]
6. Picciotto, M.R.; Lewis, A.S.; van Schalkwyk, G.I.; Mineur, Y.S. Mood and anxiety regulation by nicotinic acetylcholine receptors: A potential pathway to modulate aggression and related behavioral states. *Neuropharmacology* **2015**, *96*, 235–243. [CrossRef] [PubMed]
7. Albuquerque, E.X.; Pereira, E.F.; Alkondon, M.; Rogers, S.W. Mammalian nicotinic acetylcholine receptors: From structure to function. *Physiol. Rev.* **2009**, *89*, 73–120. [CrossRef]
8. Dani, J.A.; Bertrand, D. Nicotinic acetylcholine receptors and nicotinic cholinergic mechanisms of the central nervous system. *Annu Rev Pharm. Toxicol* **2007**, *47*, 699–729. [CrossRef]
9. Maurer, S.V.; Williams, C.L. The Cholinergic System Modulates Memory and Hippocampal Plasticity via Its Interactions with Non-Neuronal Cells. *Front Immunol.* **2017**, *8*, 1489. [CrossRef]
10. Bosmans, G.; Shimizu Bassi, G.; Florens, M.; Gonzalez-Dominguez, E.; Matteoli, G.; Boeckxstaens, G.E. Cholinergic Modulation of Type 2 Immune Responses. *Front Immunol.* **2017**, *8*, 1873. [CrossRef]
11. Freedman, R.; Hall, M.; Adler, L.E.; Leonard, S. Evidence in postmortem brain tissue for decreased numbers of hippocampal nicotinic receptors in schizophrenia. *Biol. Psychiatry* **1995**, *38*, 22–33. [CrossRef]
12. Guillozet-Bongaarts, A.L.; Hyde, T.M.; Dalley, R.A.; Hawrylycz, M.J.; Henry, A.; Hof, P.R.; Hohmann, J.; Jones, A.R.; Kuan, C.L.; Royall, J.; et al. Altered gene expression in the dorsolateral prefrontal cortex of individuals with schizophrenia. *Mol. Psychiatry* **2014**, *19*, 478–485. [CrossRef]
13. Kunii, Y.; Zhang, W.; Xu, Q.; Hyde, T.M.; McFadden, W.; Shin, J.H.; Deep-Soboslay, A.; Ye, T.; Li, C.; Kleinman, J.E.; et al. CHRNA7 and CHRFAM7A mRNAs: Co-Localized and Their Expression Levels Altered in the Postmortem Dorsolateral Prefrontal Cortex in Major Psychiatric Disorders. *Am. J. Psychiatry* **2015**. [CrossRef] [PubMed]
14. Thomsen, M.S.; Weyn, A.; Mikkelsen, J.D. Hippocampal alpha7 nicotinic acetylcholine receptor levels in patients with schizophrenia, bipolar disorder, or major depressive disorder. *Bipolar Disord.* **2011**, *13*, 701–707. [CrossRef] [PubMed]
15. Sugaya, K.; Giacobini, E.; Chiappinelli, V.A. Nicotinic acetylcholine receptor subtypes in human frontal cortex: Changes in Alzheimer's disease. *J. Neurosci. Res.* **1990**, *27*, 349–359. [CrossRef]
16. Araud, T.; Graw, S.; Berger, R.; Lee, M.; Neveu, E.; Bertrand, D.; Leonard, S. The chimeric gene CHRFAM7A, a partial duplication of the CHRNA7 gene, is a dominant negative regulator of alpha7*nAChR function. *Biochem. Pharmacol.* **2011**, *82*, 904–914. [CrossRef]
17. Gillentine, M.A.; Lozoya, R.; Yin, J.; Grochowski, C.M.; White, J.J.; Schaaf, C.P.; Calarge, C.A. CHRNA7 copy number gains are enriched in adolescents with major depressive and anxiety disorders. *J. Affect. Disord.* **2018**, *239*, 247–252. [CrossRef]
18. Sinkus, M.L.; Graw, S.; Freedman, R.; Ross, R.G.; Lester, H.A.; Leonard, S. The human CHRNA7 and CHRFAM7A genes: A review of the genetics, regulation, and function. *Neuropharmacology* **2015**. [CrossRef]
19. Gillentine, M.A.; Berry, L.N.; Goin-Kochel, R.P.; Ali, M.A.; Ge, J.; Guffey, D.; Rosenfeld, J.A.; Hannig, V.; Bader, P.; Proud, M.; et al. The Cognitive and Behavioral Phenotypes of Individuals with CHRNA7 Duplications. *J. Autism Dev. Disord.* **2017**, *47*, 549–562. [CrossRef]
20. Hua, S.; Ek, C.J.; Mallard, C.; Johansson, M.E. Perinatal hypoxia-ischemia reduces alpha 7 nicotinic receptor expression and selective alpha 7 nicotinic receptor stimulation suppresses inflammation and promotes microglial Mox phenotype. *Biomed Res. Int.* **2014**, *2014*, 718769. [CrossRef]
21. Han, Z.; Li, L.; Wang, L.; Degos, V.; Maze, M.; Su, H. Alpha-7 nicotinic acetylcholine receptor agonist treatment reduces neuroinflammation, oxidative stress, and brain injury in mice with ischemic stroke and bone fracture. *J. Neurochem.* **2014**, *131*, 498–508. [CrossRef] [PubMed]

22. Dash, P.K.; Zhao, J.; Kobori, N.; Redell, J.B.; Hylin, M.J.; Hood, K.N.; Moore, A.N. Activation of Alpha 7 Cholinergic Nicotinic Receptors Reduce Blood-Brain Barrier Permeability following Experimental Traumatic Brain Injury. *J. Neurosci.* **2016**, *36*, 2809–2818. [CrossRef] [PubMed]
23. Mavropoulos, S.A.; Khan, N.S.; Levy, A.C.J.; Faliks, B.T.; Sison, C.P.; Pavlov, V.A.; Zhang, Y.; Ojamaa, K. Nicotinic acetylcholine receptor-mediated protection of the rat heart exposed to ischemia reperfusion. *Mol. Med.* **2017**, *23*. [CrossRef] [PubMed]
24. Gatson, J.W.; Simpkins, J.W.; Uteshev, V.V. High therapeutic potential of positive allosteric modulation of alpha7 nAChRs in a rat model of traumatic brain injury: Proof-of-concept. *Brain Res. Bull.* **2015**, *112*, 35–41. [CrossRef] [PubMed]
25. Pike, V.W. Considerations in the Development of Reversibly Binding PET Radioligands for Brain Imaging. *Nicotin4* **2016**, *23*, 1818–1869.
26. Gao, Y.; Kellar, K.J.; Yasuda, R.P.; Tran, T.; Xiao, Y.; Dannals, R.F.; Horti, A.G. Derivatives of dibenzothiophene for positron emission tomography imaging of alpha7-nicotinic acetylcholine receptors. *J. Med. Chem.* **2013**, *56*, 7574–7589. [CrossRef]
27. Toyohara, J.; Hashimoto, K. α7 Nicotinic Receptor Agonists: Potential Therapeutic Drugs for Treatment of Cognitive Impairments in Schizophrenia and Alzheimer's Disease. *Open Med. Chem. J.* **2010**, *4*, 37–56. [CrossRef]
28. Hashimoto, K.; Nishiyama, S.; Ohba, H.; Matsuo, M.; Kobashi, T.; Takahagi, M.; Iyo, M.; Kitashoji, T.; Tsukada, H. [^{11}C]CHIBA-1001 as a novel PET ligand for α7 nicotinic receptors in the brain: A PET study in conscious monkeys. *PLoS ONE* **2008**, *3*, e3231. [CrossRef]
29. Rötering, S.; Scheunemann, M.; Fischer, S.; Hiller, A.; Peters, D.; Deuther-Conrad, W.; Brust, P. Radiosynthesis and first evaluation in mice of [(18)F]NS14490 for molecular imaging of alpha7 nicotinic acetylcholine receptors. *Bioorganic Med. Chem.* **2013**, *21*, 2635–2642.
30. Kim, S.W.; Ding, Y.S.; Alexoff, D.; Patel, V.; Logan, J.; Lin, K.S.; Shea, C.; Muench, L.; Xu, Y.; Carter, P.; et al. Synthesis and positron emission tomography studies of C-11-labeled isotopomers and metabolites of GTS-21, a partial alpha7 nicotinic cholinergic agonist drug. *Nucl. Med. Biol.* **2007**, *34*, 541–551. [CrossRef]
31. Ettrup, A.; Mikkelsen, J.D.; Lehel, S.; Madsen, J.; Nielsen, E.O.; Palner, M.; Timmermann, D.B.; Peters, D.; Knudsen, G.M. ^{11}C-NS14492 as a novel PET radioligand for imaging cerebral alpha7 nicotinic acetylcholine receptors: In vivo evaluation and drug occupancy measurements. *J. Nucl. Med.* **2011**, *52*, 1449–1456. [CrossRef] [PubMed]
32. Deuther-Conrad, W.; Fischer, S.; Hiller, A.; Becker, G.; Cumming, P.; Xiong, G.; Funke, U.; Sabri, O.; Peters, D.; Brust, P. Assessment of alpha7 nicotinic acetylcholine receptor availability in juvenile pig brain with [(18)F]NS10743. *Eur. J. Nucl. Med. Mol. Imaging* **2011**. [CrossRef] [PubMed]
33. Ouach, A.; Vercouillie, J.; Bertrand, E.; Rodrigues, N.; Pin, F.; Serriere, S.; Boiaryna, L.; Chartier, A.; Percina, N.; Tangpong, P.; et al. Bis(het)aryl-1,2,3-triazole quinuclidines as alpha7 nicotinic acetylcholine receptor ligands: Synthesis, structure affinity relationships, agonism activity, [(18)F]-radiolabeling and PET study in rats. *Eur. J. Med. Chem.* **2019**, *179*, 449–469. [CrossRef] [PubMed]
34. Huan, W.; Aiqin, W.; Jianping, L.; Qianqian, X.; Xia, L.; Lei, Y.; Yu, F.; Huabei, Z. Radiosynthesis and in-vivo evaluation of [125I]IBT: A single-photon emission computed tomography radiotracer for alpha7-nicotinic acetylcholine receptor imaging. *Nucl. Med. Commun.* **2017**, *38*, 683–693. [CrossRef] [PubMed]
35. Wang, S.; Fang, Y.; Wang, H.; Gao, H.; Jiang, G.; Liu, J.; Xue, Q.; Qi, Y.; Cao, M.; Qiang, B.; et al. Design, synthesis and biological evaluation of 1,4-Diazobicylco[3.2.2]nonane derivatives as alpha7-Nicotinic acetylcholine receptor PET/CT imaging agents and agonists for Alzheimer's disease. *Eur. J. Med. Chem.* **2018**, *159*, 255–266. [CrossRef] [PubMed]
36. Teodoro, R.; Scheunemann, M.; Wenzel, B.; Peters, D.; Deuther-Conrad, W.; Brust, P. Synthesis and radiofluorination of novel fluoren-9-one based derivatives for the imaging of alpha7 nicotinic acetylcholine receptor with PET. *Bioorg. Med. Chem. Lett.* **2018**, *28*, 1471–1475. [CrossRef]
37. Sarasamkan, J.; Scheunemann, M.; Apaijai, N.; Palee, S.; Parichatikanond, W.; Arunrungvichian, K.; Fischer, S.; Chattipakorn, S.; Deuther-Conrad, W.; Schuurmann, G.; et al. Varying Chirality Across Nicotinic Acetylcholine Receptor Subtypes: Selective Binding of Quinuclidine Triazole Compounds. *Acs Med. Chem. Lett.* **2016**, *7*, 890–895. [CrossRef]

38. Schrimpf, M.R.; Sippy, K.B.; Briggs, C.A.; Anderson, D.J.; Li, T.; Ji, J.; Frost, J.M.; Surowy, C.S.; Bunnelle, W.H.; Gopalakrishnan, M.; et al. SAR of alpha7 nicotinic receptor agonists derived from tilorone: Exploration of a novel nicotinic pharmacophore. *Bioorg. Med. Chem. Lett.* **2012**, *22*, 1633–1638. [CrossRef]
39. Horti, A.G.; Gao, Y.; Kuwabara, H.; Wang, Y.; Abazyan, S.; Yasuda, R.P.; Tran, T.; Xiao, Y.; Sahibzada, N.; Holt, D.P.; et al. 18F-ASEM, a radiolabeled antagonist for imaging the alpha7-nicotinic acetylcholine receptor with PET. *J. Nucl. Med.* **2014**, *55*, 672–677. [CrossRef]
40. Teodoro, R.; Scheunemann, M.; Deuther-Conrad, W.; Wenzel, B.; Fasoli, F.M.; Gotti, C.; Kranz, M.; Donat, C.K.; Patt, M.; Hillmer, A.; et al. A Promising PET Tracer for Imaging of alpha(7) Nicotinic Acetylcholine Receptors in the Brain: Design, Synthesis, and In Vivo Evaluation of a Dibenzothiophene-Based Radioligand. *Molecules* **2015**, *20*, 18387–18421. [CrossRef]
41. Horti, A.G. Development of [(18)F]ASEM, a specific radiotracer for quantification of the alpha7-nAChR with positron-emission tomography. *Biochem. Pharmacol.* **2015**, *97*, 566–575. [CrossRef] [PubMed]
42. Gao, Y.; Mease, R.C.; Olson, T.T.; Kellar, K.J.; Dannals, R.F.; Pomper, M.G.; Horti, A.G. [(^{125})I]Iodo-ASEM, a specific in vivo radioligand for alpha7-nAChR. *Nucl. Med. Biol.* **2015**, *42*, 488–493. [CrossRef] [PubMed]
43. Wong, D.F.; Kuwabara, H.; Pomper, M.; Holt, D.P.; Brasic, J.R.; George, N.; Frolov, B.; Willis, W.; Gao, Y.; Valentine, H.; et al. Human brain imaging of alpha7 nAChR with [(18)F]ASEM: A new PET radiotracer for neuropsychiatry and determination of drug occupancy. *WinnieBbb13* **2014**, *16*, 730–738.
44. Hillmer, A.T.; Li, S.; Zheng, M.Q.; Scheunemann, M.; Lin, S.F.; Nabulsi, N.; Holden, D.; Pracitto, R.; Labaree, D.; Ropchan, J.; et al. PET imaging of alpha7 nicotinic acetylcholine receptors: A comparative study of [18F]ASEM and [18F]DBT-10 in nonhuman primates, and further evaluation of [18F]ASEM in humans. *Eur. J. Nucl. Med. Mol. Imaging* **2017**, *44*, 1042–1050. [CrossRef]
45. Hillmer, A.T.; Zheng, M.Q.; Li, S.; Scheunemann, M.; Lin, S.F.; Holden, D.; Labaree, D.; Ropchan, J.; Teodoro, R.; Deuther-Conrad, W.; et al. PET imaging evaluation of [(18)F]DBT-10, a novel radioligand specific to alpha7 nicotinic acetylcholine receptors, in nonhuman primates. *Eur. J. Nucl. Med. Mol. Imaging* **2016**, *43*, 537–547. [CrossRef]
46. Wong, D.F.; Kuwabara, H.; Horti, A.G.; Roberts, J.M.; Nandi, A.; Cascella, N.; Brasic, J.; Weerts, E.M.; Kitzmiller, K.; Phan, J.A.; et al. Brain PET Imaging of alpha7-nAChR with [18F]ASEM: Reproducibility, Occupancy, Receptor Density, and Changes in Schizophrenia. *Int. J. Neuropsychopharmacol. /Off. Sci. J. Coll. Int. Neuropsychopharmacol.* **2018**, *21*, 656–667.
47. Coughlin, J.M.; Du, Y.; Rosenthal, H.B.; Slania, S.; Min Koo, S.; Park, A.; Solomon, G.; Vranesic, M.; Antonsdottir, I.; Speck, C.L.; et al. The distribution of the alpha7 nicotinic acetylcholine receptor in healthy aging: An in vivo positron emission tomography study with [(18)F]ASEM. *NeuroImage* **2017**, *165*, 118–124. [CrossRef]
48. Coughlin, J.; Du, Y.; Crawford, J.L.; Rubin, L.H.; Behnam Azad, B.; Lesniak, W.G.; Horti, A.G.; Schretlen, D.J.; Sawa, A.; Pomper, M.G. The availability of the alpha7 nicotinic acetylcholine receptor in recent-onset psychosis: A study using (18)F-ASEM PET. *J. Nucl. Med.* **2018**. [CrossRef]
49. Coughlin, J.M.; Rubin, L.H.; Du, Y.; Rowe, S.P.; Crawford, J.L.; Rosenthal, H.B.; Frey, S.M.; Marshall, E.S.; Shinehouse, L.K.; Chen, A.; et al. High availability of the alpha7 nicotinic acetylcholine receptor in brains of individuals with mild cognitive impairment: A pilot study using (18)F-ASEM PET. *J. Nucl. Med.* **2019**. [CrossRef]
50. Vetel, S.; Vercouillie, J.; Buron, F.; Vergote, J.; Tauber, C.; Busson, J.; Chicheri, G.; Routier, S.; Serriere, S.; Chalon, S. Longitudinal PET Imaging of alpha7 Nicotinic Acetylcholine Receptors with [(18)F]ASEM in a Rat Model of Parkinson's Disease. *WinnieBbb13* **2019**. [CrossRef]
51. Wu, J.; Liu, Q.; Tang, P.; Mikkelsen, J.D.; Shen, J.; Whiteaker, P.; Yakel, J.L. Heteromeric alpha7beta2 Nicotinic Acetylcholine Receptors in the Brain. *Trends Pharm. Sci.* **2016**, *37*, 562–574. [CrossRef] [PubMed]
52. Thomsen, M.S.; Zwart, R.; Ursu, D.; Jensen, M.M.; Pinborg, L.H.; Gilmour, G.; Wu, J.; Sher, E.; Mikkelsen, J.D. alpha7 and beta2 Nicotinic Acetylcholine Receptor Subunits Form Heteromeric Receptor Complexes that Are Expressed in the Human Cortex and Display Distinct Pharmacological Properties. *PLoS ONE* **2015**, *10*, e0130572. [CrossRef] [PubMed]
53. Moretti, M.; Zoli, M.; George, A.A.; Lukas, R.J.; Pistillo, F.; Maskos, U.; Whiteaker, P.; Gotti, C. The novel alpha7beta2-nicotinic acetylcholine receptor subtype is expressed in mouse and human basal forebrain: Biochemical and pharmacological characterization. *Mol. Pharmacol.* **2014**, *86*, 306–317. [CrossRef] [PubMed]

54. Zwart, R.; Strotton, M.; Ching, J.; Astles, P.C.; Sher, E. Unique pharmacology of heteromeric alpha7beta2 nicotinic acetylcholine receptors expressed in Xenopus laevis oocytes. *Eur. J. Pharmacol.* **2014**, *726C*, 77–86. [CrossRef]
55. Patel, S.; Hamill, T.G.; Connolly, B.; Jagoda, E.; Li, W.; Gibson, R.E. Species differences in mGluR5 binding sites in mammalian central nervous system determined using in vitro binding with [18F]F-PEB. *Pet§Cholin2* **2007**, *34*, 1009–1017. [CrossRef]
56. Fujita, M.; Imaizumi, M.; Zoghbi, S.S.; Fujimura, Y.; Farris, A.G.; Suhara, T.; Hong, J.; Pike, V.W.; Innis, R.B. Kinetic analysis in healthy humans of a novel positron emission tomography radioligand to image the peripheral benzodiazepine receptor, a potential biomarker for inflammation. *NeuroImage* **2008**, *40*, 43–52. [CrossRef]
57. Van de Bittner, G.C.; Ricq, E.L.; Hooker, J.M. A philosophy for CNS radiotracer design. *Acc Chem. Res.* **2014**, *47*, 3127–3134. [CrossRef]
58. Laruelle, M.; Slifstein, M.; Huang, Y. Relationships between radiotracer properties and image quality in molecular imaging of the brain with positron emission tomography. *Nicotin4* **2003**, *5*, 363–375. [CrossRef]
59. Magnussen, J.H.; Ettrup, A.; Donat, C.K.; Peters, D.; Pedersen, M.H.F.; Knudsen, G.M.; Mikkelsen, J.D. Radiosynthesis and in vitro validation of 3H-NS14492 as a novel high affinity α7 nicotinic acetylcholine receptor radioligand *Mol. Cell. Neurosci.* **2014**. [CrossRef]
60. Anderson, D.J.; Bunnelle, W.; Surber, B.; Du, J.; Surowy, C.; Tribollet, E.; Marguerat, A.; Bertrand, D.; Gopalakrishnan, M. [^{3}H]A-585539 [(1S,4S)-2,2-dimethyl-5-(6-phenylpyridazin-3-yl)-5-aza-2-azoniabicyclo[2.2.1]hept ane], a novel high-affinity alpha7 neuronal nicotinic receptor agonist: Radioligand binding characterization to rat and human brain. *J. Pharmacol. Exp. Ther.* **2008**, *324*, 179–187. [CrossRef]
61. Magnussen, J.H.; Ettrup, A.; Donat, C.K.; Peters, D.; Pedersen, M.H.; Knudsen, G.M.; Mikkelsen, J.D. Radiosynthesis and in vitro validation of (3)H-NS14492 as a novel high affinity alpha7 nicotinic receptor radioligand. *Eur. J. Pharmacol.* **2015**, *762*, 35–41. [CrossRef] [PubMed]
62. Bitner, R.S.; Bunnelle, W.H.; Anderson, D.J.; Briggs, C.A.; Buccafusco, J.; Curzon, P.; Decker, M.W.; Frost, J.M.; Gronlien, J.H.; Gubbins, E.; et al. Broad-spectrum efficacy across cognitive domains by alpha7 nicotinic acetylcholine receptor agonism correlates with activation of ERK1/2 and CREB phosphorylation pathways. *J. Neurosci.* **2007**, *27*, 10578–10587. [CrossRef] [PubMed]
63. Biton, B.; Bergis, O.E.; Galli, F.; Nedelec, A.; Lochead, A.W.; Jegham, S.; Godet, D.; Lanneau, C.; Santamaria, R.; Chesney, F.; et al. SSR180711, a novel selective alpha7 nicotinic receptor partial agonist: (1) binding and functional profile. *Neuropsychopharmacology* **2007**, *32*, 1–16. [CrossRef] [PubMed]
64. Hauser, T.A.; Kucinski, A.; Jordan, K.G.; Gatto, G.J.; Wersinger, S.R.; Hesse, R.A.; Stachowiak, E.K.; Stachowiak, M.K.; Papke, R.L.; Lippiello, P.M.; et al. TC-5619: An alpha7 neuronal nicotinic receptor-selective agonist that demonstrates efficacy in animal models of the positive and negative symptoms and cognitive dysfunction of schizophrenia. *Biochem. Pharmacol.* **2009**, *78*, 803–812. [CrossRef] [PubMed]
65. Prickaerts, J.; van Goethem, N.P.; Chesworth, R.; Shapiro, G.; Boess, F.G.; Methfessel, C.; Reneerkens, O.A.; Flood, D.G.; Hilt, D.; Gawryl, M.; et al. EVP-6124, a novel and selective alpha7 nicotinic acetylcholine receptor partial agonist, improves memory performance by potentiating the acetylcholine response of alpha7 nicotinic acetylcholine receptors. *Neuropharmacology* **2012**, *62*, 1099–1110. [CrossRef] [PubMed]
66. Meyer, E.M.; Tay, E.T.; Papke, R.L.; Meyers, C.; Huang, G.L.; de Fiebre, C.M. 3-[2,4-Dimethoxybenzylidene]anabaseine (DMXB) selectively activates rat alpha7 receptors and improves memory-related behaviors in a mecamylamine-sensitive manner. *Brain Res.* **1997**, *768*, 49–56. [CrossRef]
67. Nielsen, B.E.; Minguez, T.; Bermudez, I.; Bouzat, C. Molecular function of the novel α7β2 nicotinic receptor. *Cell. Mol. Life Sci. Cmls* **2018**, *75*, 2457–2471. [CrossRef]
68. Liu, Q.; Huang, Y.; Shen, J.; Steffensen, S.; Wu, J. Functional alpha7beta2 nicotinic acetylcholine receptors expressed in hippocampal interneurons exhibit high sensitivity to pathological level of amyloid beta peptides. *Bmc Neurosci.* **2012**, *13*, 155. [CrossRef]
69. Liu, Q.; Huang, Y.; Xue, F.; Simard, A.; DeChon, J.; Li, G.; Zhang, J.; Lucero, L.; Wang, M.; Sierks, M.; et al. A novel nicotinic acetylcholine receptor subtype in basal forebrain cholinergic neurons with high sensitivity to amyloid peptides. *J. Neurosci.* **2009**, *29*, 918–929. [CrossRef]

70. Khiroug, S.S.; Harkness, P.C.; Lamb, P.W.; Sudweeks, S.N.; Khiroug, L.; Millar, N.S.; Yakel, J.L. Rat nicotinic ACh receptor alpha7 and beta2 subunits co-assemble to form functional heteromeric nicotinic receptor channels. *J Physiol* **2002**, *540*, 425–434. [CrossRef]
71. Murray, T.A.; Bertrand, D.; Papke, R.L.; George, A.A.; Pantoja, R.; Srinivasan, R.; Liu, Q.; Wu, J.; Whiteaker, P.; Lester, H.A.; et al. alpha7beta2 nicotinic acetylcholine receptors assemble, function, and are activated primarily via their alpha7-alpha7 interfaces. *Mol. Pharmacol.* **2012**, *81*, 175–188. [CrossRef] [PubMed]
72. Mowrey, D.D.; Liu, Q.; Bondarenko, V.; Chen, Q.; Seyoum, E.; Xu, Y.; Wu, J.; Tang, P. Insights into distinct modulation of α7 and α7β2 nicotinic acetylcholine receptors by the volatile anesthetic isoflurane. *J. Biol. Chem.* **2013**, *288*, 35793–35800. [CrossRef] [PubMed]
73. Teodoro, R.; Moldovan, R.P.; Lueg, C.; Gunther, R.; Donat, C.K.; Ludwig, F.A.; Fischer, S.; Deuther-Conrad, W.; Wunsch, B.; Brust, P. Radiofluorination and biological evaluation of N-aryl-oxadiazolyl-propionamides as potential radioligands for PET imaging of cannabinoid CB2 receptors. *Org. Med. Chem. Lett.* **2013**, *3*, 11. [CrossRef] [PubMed]
74. Innis, R.B.; Cunningham, V.J.; Delforge, J.; Fujita, M.; Gjedde, A.; Gunn, R.N.; Holden, J.; Houle, S.; Huang, S.-C.; Ichise, M.; et al. Consensus nomenclature for in vivo imaging of reversibly binding radioligands. *J. Cereb. Blood Flow Metab: Off. J. Int. Soc. Cereb. Blood Flow Metab.* **2007**, *27*, 1533–1539. [CrossRef] [PubMed]
75. Gallezot, J.-D.; Weinzimmer, D.; Nabulsi, N.; Lin, S.-F.; Fowles, K.; Sandiego, C.; McCarthy, T.J.; Maguire, R.P.; Carson, R.E.; Ding, Y.-S. Evaluation of [(11)C]MRB for assessment of occupancy of norepinephrine transporters: Studies with atomoxetine in non-human primates. *NeuroImage* **2011**, *56*, 268–279. [CrossRef]
76. Bradford, M.M. A rapid and sensitive method for the quantitation of microgram quantities of protein utilizing the principle of protein-dye binding. *Anal. Biochem.* **1976**, *72*, 248–254. [CrossRef]
77. Ward, J.M.; Cockcroft, V.B.; Lunt, G.G.; Smillie, F.S.; Wonnacott, S. Methyllycaconitine: A selective probe for neuronal alpha-bungarotoxin binding sites. *Febs Lett.* **1990**, *270*, 45–48. [CrossRef]
78. Gillings, N. A restricted access material for rapid analysis of [(11)C]-labeled radiopharmaceuticals and their metabolites in plasma. *Nucl. Med. Biol.* **2009**, *36*, 961–965. [CrossRef]
79. Kornum, B.R.; Lind, N.M.; Gillings, N.; Marner, L.; Andersen, F.; Knudsen, G.M. Evaluation of the novel 5-HT4 receptor PET ligand [11C]SB207145 in the Gottingen minipig. *J. Cereb. Blood Flow Metab.* **2009**, *29*, 186–196. [CrossRef]
80. Villadsen, J.; Hansen, H.D.; Jorgensen, L.M.; Keller, S.H.; Andersen, F.L.; Petersen, I.N.; Knudsen, G.M.; Svarer, C. Automatic delineation of brain regions on MRI and PET images from the pig. *J. Neurosci. Methods* **2018**, *294*, 51–58. [CrossRef]

Sample Availability: Samples of the compounds are not available.

© 2020 by the authors. Licensee MDPI, Basel, Switzerland. This article is an open access article distributed under the terms and conditions of the Creative Commons Attribution (CC BY) license (http://creativecommons.org/licenses/by/4.0/).

Brief Report

A New Positron Emission Tomography Probe for Orexin Receptors Neuroimaging

Ping Bai [1,2,3], Sha Bai [2], Michael S. Placzek [2], Xiaoxia Lu [1], Stephanie A. Fiedler [2], Brenda Ntaganda [2], Hsiao-Ying Wey [2] and Changning Wang [2,*]

[1] Chengdu Institute of Biology, Chinese Academy of Sciences, Chengdu 610041, China; pbai@mgh.harvard.edu (P.B.); luxx@cib.ac.cn (X.L.)
[2] Athinoula A. Martinos Center for Biomedical Imaging, Department of Radiology, Massachusetts General Hospital, Harvard Medical School, Charlestown, MA 02129, USA; sbai@mgh.harvard.edu (S.B.); michael.placzek@mgh.harvard.edu (M.S.P.); sfiedler@mgh.harvard.edu (S.A.F.); bNtaganda@mgh.harvard.edu (B.N.); hsiaoying.wey@mgh.harvard.edu (H.-Y.W.)
[3] University of Chinese Academy of Sciences, Beijing 100049, China
* Correspondence: cwang15@mgh.harvard.edu

Received: 17 December 2019; Accepted: 20 February 2020; Published: 25 February 2020

Abstract: The orexin receptor (OX) is critically involved in motivation and sleep–wake regulation and holds promising therapeutic potential in various mood disorders. To further investigate the role of orexin receptors (OXRs) in the living human brain and to evaluate the treatment potential of orexin-targeting therapeutics, we herein report a novel PET probe ([^{11}C]CW24) for OXRs in the brain. CW24 has moderate binding affinity for OXRs (IC$_{50}$ = 0.253 µM and 1.406 µM for OX$_1$R and OX$_2$R, respectively) and shows good selectivity to OXRs over 40 other central nervous system (CNS) targets. [^{11}C]CW24 has high brain uptake in rodents and nonhuman primates, suitable metabolic stability, and appropriate distribution and pharmacokinetics for brain positron emission tomography (PET) imaging. [^{11}C]CW24 warrants further evaluation as a PET imaging probe of OXRs in the brain.

Keywords: orexin receptors; PET; radiotracer; imaging

1. Introduction

Orexin (OX) is a hypothalamic hypocretin peptide that mediates multiple functions such as arousal, attention, neuroendocrine, water balance, and pain modulation [1–6]. The orexin system has two peptide members, orexin-A and orexin-B with 33 amino acids and 28 amino acids, respectively [7]. Orexin receptors (OXRs) also have two subtypes (OX$_1$R and OX$_2$R). Orexin-A has similar binding affinities to OX$_1$R and OX$_2$R, while orexin-B has a 10-fold higher affinity for OX$_2$R compared with OX$_1$R. OXRs are differently expressed throughout the brain. OX$_1$R is mainly expressed in the prefrontal and infralimbic cortex, hippocampus, paraventricular thalamic nucleus, and locus coeruleus [8]. OX$_2$R is mainly distributed in the cerebral cortex, septal nuclei, lateral hypothalamus, hippocampus, and hypothalamic nuclei [9]. The differential distribution of OXRs in the brain may be responsible for the various physiological and psychiatric functions mediated by the OXR system [10–12].

Therapeutics targeting the orexin system have been investigated for various brain disorders such as insomnia [1,3], cluster headache [13], substance abuse [4,14], and maladaptation to stress [15,16]. This has led to the design and evaluation of orexin antagonists, including dual-OX$_1$R and OX$_2$R antagonists (DORA) and selective OX$_1$R and OX$_2$R antagonists (SORA). [7] Almorexant [5] (ACT-426606) [17] from Actelion, SB-649868 [18] from GSK, and Suvorexant [6] from Merck are a few examples demonstrating the efforts from pharmaceutical companies. Among these compounds, many have entered into clinical trials, however, Suvorexant is the only U.S. Food and Drug Administration (FDA) approved orexin drug for the insomnia market. The discovery of ORX antagonists provided powerful tool compounds to investigate the roles of the orexin system in a variety of biological processes. However, the roles of the orexin system in the brain are not well understood and further investigations with translational methods, such as in vivo imaging, are warranted.

A non-invasive molecular imaging with positron emission tomography (PET) is a unique tool to study the living brain, which provides a distinct advantage to quantify receptor expression and/or drug occupancy of novel ligands in vivo. Over the last decade, though there have been major advances in the development of orexin antagonists [19–25], no PET radiotracer is available for OXR imaging in the human brain. This limits our ability to investigate the orexin system in the human brain and hinders the potential of orexin-targeting neurotherapeutics for brain disorders.

To bridge the gap, several orexin antagonists have been radiolabeled by our group, as well as others. Most of the early discovery attempts resulted in compounds with low brain uptake or strong P-glycoprotein (P-gp) binding, including [^{11}C]MK-1064 [26], [^{11}C]EMPA [27], [^{11}C]BBAC [28], and [^{11}C]CW4 [29] (Figure 1). Towards the development of a successful OXR brain-PET imaging probe, we report the discovery and evaluation of [^{11}C]CW24 with good brain uptake, suitable metabolic stability, and pharmacokinetics based on animal imaging studies in rodents and non-human primates.

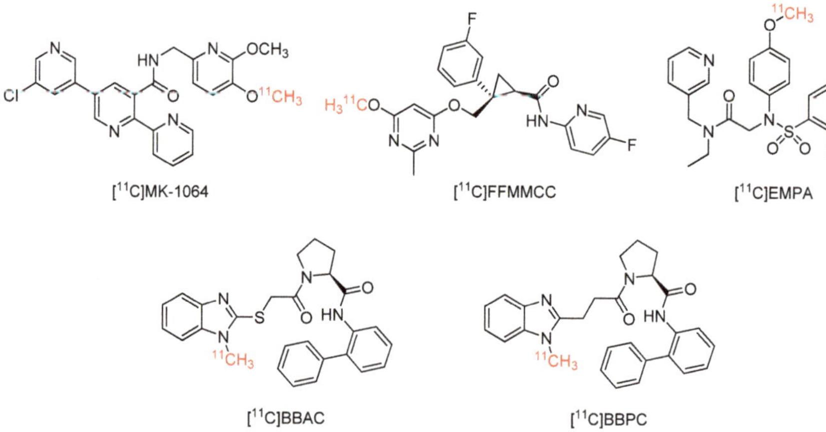

Figure 1. Structures of orexin receptor radioligands.

2. Results and Discussion

2.1. Selection of Scaffold for Orexin Imaging Probe Development

Several small molecules have been reported as DORA or SORA [20,30,31]. Previously, our group has radiolabeled several DORAs based on a 1,4-diazepane scaffold [27,29]. However, the low brain uptake limited their further evaluation as a brain-PET radiotracer. Recently, IPSU was reported as an OX$_2$R selective antagonist with good brain uptake [32,33]. It has been studied in rodents, including blood and brain pharmacokinetics, as well as EEG (electrocorticogram/electroencephalogram) and EMG (electromyogram) recordings following treatment with these OXR antagonists in mice. Based on the structure-activity relationship (SAR) study of ISPU analogs, methylation of the indole imine does not disrupt OXR antagonism [32]. Hence, we chose the indole imine of IPSU as the radiolabeling site and synthesized CW24. The affinity of CW24 for orexin receptors (OX$_1$R and OX$_2$R) in cells was measured in a competition binding assay against [^{125}I] orexin. CW24 displayed potent activity at OXRs (IC$_{50}$ 0.253 µM OX$_1$R, 1.406 µM OX$_2$R). Compared to IPSU binding, this was ~5 fold (OX$_1$R) and ~2 fold (OX$_2$R) increase in potency, warranting further evaluation.

2.2. Physicochemical Properties of CW24

Lipophilicity (logP) and molecular weight need to be optimized for sufficient blood-brain barrier (BBB) penetration while also avoiding high non-specific binding. Studies have shown that it is preferred for clogP ≤ 4, clogD ≤ 3, total polar surface area (tPSA) between 30–75 and molecular weight (M.W.) < 500 for successful brain-PET imaging probes [34,35]. IPSU is reported as a non-selective OXR antagonist with good brain uptake [33]. CW24, the methylated form of IPSU, had favorable properties for CNS penetration (M.W. = 419.5, Log D$_{(7.4)}$ = 2.3, tPSA = 60.74, IC$_{50}$ = 0.253 µM and 1.406 µM for OX$_1$R and OX$_2$R, respectively) (Figure 2). Furthermore, we tested the off-target binding of CW24 in vitro, [36] and were pleased to see that CW24 was selective for OXRs compared to 40 other CNS targets (National Institute of Mental Health Psychoactive Drug Screening Program, NIMH PDSP) (Table S1).

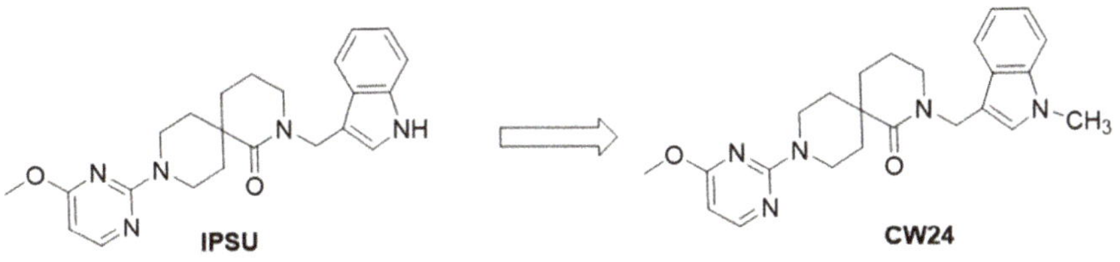

	Mw.	Log D	tPSA	IC$_{50}$ (nM)	
				OXR1	OXR2
IPSU	405.5	n.t.	69.53	1090	3135
CW24	419.5	2.3	60.74	253	1406
Suvorexant	450.93	n.t.	73.1	27	16

Figure 2. Structures and physicochemical properties of IPSU and CW24 (n.t. = not tested). IC50 of IPSU, CW24 and suvorexant were measured by a radioligand competition binding assay.

2.3. Chemical Synthesis for CW24 and [^{11}C]CW24

CW24 was prepared from IPSU via direct methylation in the presence of a base (Figures S1 and S2). [^{11}C]CW24 was radiosynthesized from IPSU (0.5 mg) in DMF (1 mL) with KOH (10 mg) and allowed to react with [^{11}C]CH$_3$I for 3 min at 100 °C (Scheme 1). The average synthesis time was 30–35 minutes from end of cyclotron bombardment (EOB) to end of synthesis (EOS), the radiochemical yield (RCY) was 10–21% (non-decay corrected, from trapped [^{11}C]CH$_3$I), the specific activity (A$_s$) was 1.28 ± 0.2 mCi/nmol (EOS), the chemical and radiochemical purities of [^{11}C]CW24 were ≥ 97%.

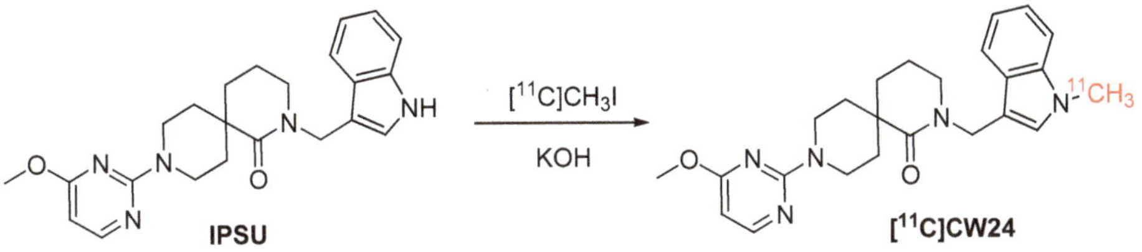

Scheme 1. Radiolabeling condition: IPSU (precursor, 0.5 mg), [^{11}C]CH$_3$I, KOH (10 mg), in 1.0 mL DMF, 3 min, 100 °C. Radiochemical yield (RCY): 10–21% (non-decay corrected from trapped [^{11}C]CH$_3$I).

2.4. Mouse Imaging with [^{11}C]CW24

We performed PET-computed tomography (CT) imaging on mice to test [^{11}C]CW24 as a neuroimaging probe for OXR imaging in vivo. [^{11}C]CW24 exhibited high brain uptake (%ID/cc = 3.5% at C$_{max}$, 1.5 min post-injection). In general, %ID/cc above 0.1% in rat or 0.01% in non-human primate within 5 min post-injection is above the threshold for suitable BBB penetration for CNS PET imaging studies [37]. Next, to evaluate the specificity of [^{11}C]CW24 for OXRs, PET-CT imaging studies were performed in mice after 5 min pretreatment with IPSU (0.5 and 2.0 mg/kg; i.v.) or Suvorexant (5.0 mg/kg; i.v.). Time-activity curves (TACs) were normalized for peak uptake (~20 seconds post-injection in these studies) (Figure 3). Approximately 10% reduction in [^{11}C]CW24 uptake with 0.5 mg/kg IPSU administration in 30–60 mins post-injection of [^{11}C]CW24. In addition, approximately 50% reduction was found when the blocking dose of IPSU was increased to 2.0 mg/kg. Suvorxant (5.0 mg/kg) administration at 30–60 mins post-injection of [^{11}C]CW24 also showed ~50% reduction in the mouse brain. Following IPSU pretreatment, [^{11}C]CW24 binding was reduced in a dose-dependent manner in the brain for the relative radioactivity changes after administration. These results strongly support that [^{11}C]CW24 binds selectively to the OXRs in vivo.

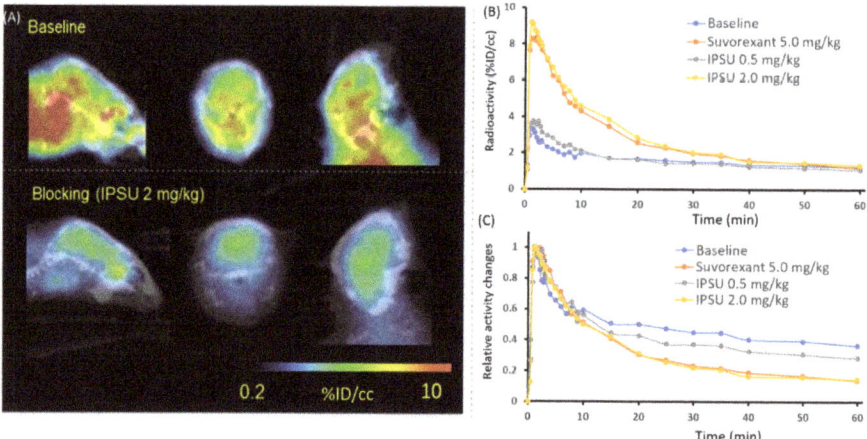

Figure 3. (**A**) The PET/CT imaging of [^{11}C]CW24 focused on mice brain (20-60 min after intravenous administration (i.v.)); (**B**) baseline and blocking time-activity curve (TAC) of [^{11}C]CW24 and (**C**) TAC after normalized the brain uptake curves with the highest uptake time point (20 seconds post-injection).

2.5. Non-human Primate (NHP) Imaging with [^{11}C]CW24

Encouraged by the promising imaging results of [^{11}C]CW24 in mice, we performed PET- Magnetic resonance imaging (MR) imaging in a NHP (Figures 4 and 5). In macaque, [^{11}C]CW24 demonstrated high brain uptake. We observed a peak standard uptake value (SUV) (SUV=C(T)/(injected dose/body weight) between 1.5–3.0 for all brain regions examined (Figure 4A). Furthermore, relatively higher uptake was observed in the midbrain and the thalamus and lower uptake in the hippocampus which suggests heterogenous expressions of OXR in the macaque brain. To further investigate the specific binding of [^{11}C]CW24, a blocking study was performed in which 2.0 mg/kg IPSU was administered 5 min prior the PET-MR acquisition. Similar to the blocking results shown in mice, a reduction of radioactivity uptake was found in several brain regions including hippocampus, nucleus accumbens, thalamus, hypothalamus, putamen and cerebellum (Figure 4B). Using the arterial blood data as input function, Logan graphical analysis was applied to quantify volume of distribution (V_T, mL/cm^3) of [^{11}C]CW24 (Figure 5). The metabolite-corrected arterial plasma after [^{11}C]CW24 bolus injection is showed in Figure 5C.

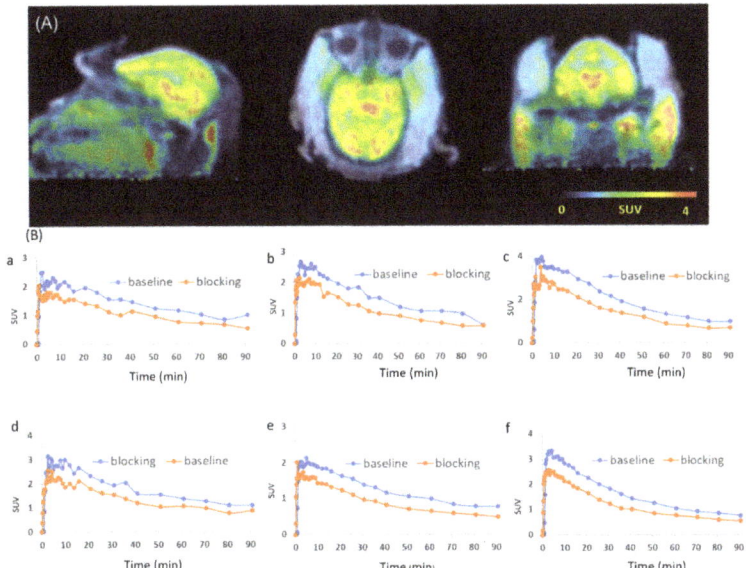

Figure 4. (**A**) Positron emission tomography (PET)-Magnetic resonance (MR) Imaging (macaque brain). Summed PET images (30–90 min after [^{11}C]CW24 injection) superimposed structural MRI from the same macaque. (**B**) Time−activity curves for brain regions of interest from baseline and blocking scans are shown (2.0 mg/kg IPSU) (a) hippocampus, (b) nucleus accumbens, (c) thalamus, (d) hypothalamus, (e) putamen and (f) cerebellum.

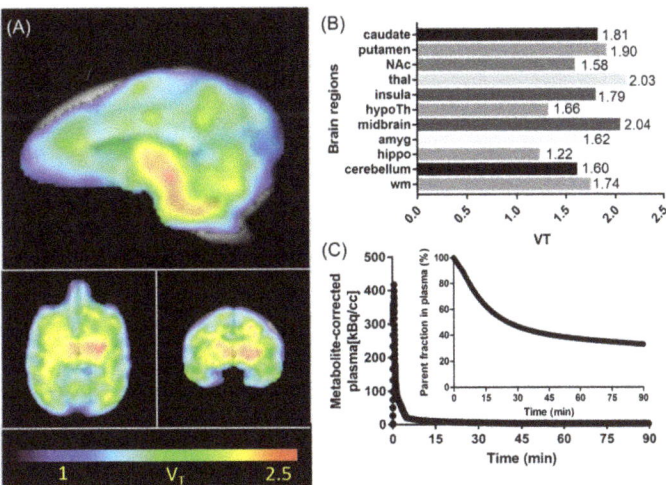

Figure 5. Kinetic modeling results with [^{11}C]CW24 in the macaque brain. (**A**) The total volume of distribution (V_T) images derived from Logan plot with arterial blood data as input function (color bar indicates V_T values from 1–2.5 mL/cm^3 (blue to red); (**B**) Regional V_T values showed that the expression difference in brain regions; (**C**) Arterial plasma analysis showed that [^{11}C]CW24 radioactivity was rapidly cleared from blood and [^{11}C]CW24 stability evaluated in plasma over time showed lasting presence of ~47% of parent compound at 30 min.

Based on our evaluation both in rodents and NHPs in vivo, [^{11}C]CW24 is a promising OXR PET radioligand candidate. It has been reported that OXR was typically present in the cerebral cortex, septal nuclei, hippocampus, medial thalamic groups, raphe nuclei, and abundantly expressed in hypothalamic nuclei that regulate the homeostasis in rodents [12]. In addition, OX_2R was found to have expression in the hypothalamus that enables its unique role of sleep-wake regulation [11]. Compare to the previous OXR probes we developed [27,29] [^{11}C]CW24 had higher brain uptake, increased isoform selectivity over IPSU, appropriate kinetics and distribution, which warrants further development of OXR PET radioligands based on [^{11}C]CW24.

3. Materials and Methods

3.1. Synthesis of Compound CW24

A mixture of IPSU (5 mg, 0.012 mmol, purchased form MedChemExpress), KOH (4 mg, 0.06 mmol), and CH_3I (5 mg, 0.036 mmol) in DMF (0.1 mL) were reacted at room temperature overnight. The reaction mixture was purified by reverse phase chromatography (H_2O: CH_3CN = 1:1) to give the desired product (2 mg, 0.005 mmol, 39% yield) as a white solid. ^{1}H-NMR (500 MHz, Chloroform-d) δ 8.04 (d, J = 5.6 Hz, 1H), 7.64 (d, J = 7.9 Hz, 1H), 7.29 (d, J = 8.2 Hz, 1H), 7.22 (t, J = 7.7 Hz, 1H), 7.10 (t, J = 7.4 Hz, 1H), 7.02 (s, 1H), 5.93 (s, 1H), 4.72 (s, 2H), 4.40 (dt, J = 13.7, 4.7 Hz, 2H), 3.87 (s, 3H), 3.75 (s, 3H), 3.38 (m, 2H), 3.24 (t, J = 6.1 Hz, 2H), 2.24 (m, 2H), 1.83–1.77 (m, 2H), 1.73 (m, 2H), 1.50 (m, 2H). ^{13}C-NMR (125 MHz, Chloroform-d) δ 174.50, 169.94, 161.78, 157.98, 137.12, 128.79, 127.61, 121.89, 119.68, 119.41, 110.84, 109.29, 95.96, 53.00, 46.79, 41.77, 40.27, 39.80(2), 33.84(2), 32.82, 30.61, 18.91. LC-MS [M]$^+$ 419.85.

3.2. Radiosynthesis of [^{11}C]CW24

[^{11}C]CH_3I was trapped in a reactor (TRACERlab FX-M synthesizer, General Electric) preloaded with the precursor (0.5 mg), KOH (5.0 mg) in 1.0 mL dry DMF. The mixture was stirred for 3 min at 100 °C and followed by adding water (1.2 mL). The product was separated by reverse phase semi-preparative HPLC (Phenomenex Luna 5u C8(2), 250 mm × 10 mm, 5 μm; 5.0 mL/min; 60% H_2O + ammonium formate (0.1 M)/ 40% CH_3CN; isocratic). The collected final product was loaded onto a C-18 sep-pak cartridge, and rinsed with water (5 mL), eluted with EtOH (0.3 mL), and saline (0.9%, 2.7 mL). The average time required for the synthesis from EOB to EOS was 30–35 min. The average radiochemical yield (RCY) was 10%–21% (non-decay corrected to trapped [^{11}C]CH_3I). Chemical and radiochemical purities were ≥ 95 % (measured with HPLC equipped with a UV detector and a gamma detector) with a specific activity (A_s) of 1.28 ± 0.2 mCi/nmol (EOS).

3.3. Assessment of Lipophilicity (Log D; pH 7.4)

Log D was determined according to methods identical to those we previously reported [38].

3.4. Human Orexin GPCR Binding (Agonist Radioligand) Assay

Radioligand competition binding assays were performed by Panlabs, Eurofins Pharma Discovery Services. Briefly, human recombinant orexin OX1 receptors expressed in CHO-S cells were used in modified HEPES buffer pH 7.4. An aliquot was incubated with 0.1 nM [^{125}I] Orexin A for 60 minutes at 25 °C. Receptors were filtered and then counted to determine [^{125}I] Orexin A specifically binding. Human recombinant OX2 receptors were used in modified HEPES buffer pH 7.4. An aliquot was incubated with 0.04 nM [^{125}I] Orexin A for 180 minutes at 25 °C. Membranes were filtered and then counted to determine [^{125}I] Orexin A specifically bound.

3.5. Rodent PET/CT Acquisition

The Subcommittee on Research Animal Care (SRAC) serves as the Institutional Animal Care and Use Committee (IACUC) for the Massachusetts General Hospital (MGH). SRAC reviewed and approved all procedures detailed in this paper. B6C3F1/J mice (male, 18-month old; n = 4 total) were utilized in this study. IPSU and Suvorexant were purchased form MedChemExpress and dissolved in 1.0 % DMSO + 1.0 % Tween 80 + 98.0 % saline to make a solution of 1.0 mg/mL.

Animals were anesthetized with 1–1.5% isoflurane and maintained with isoflurane during the imaging scan. In a single imaging session, mice were arranged in a Triumph PET/CT scanner (Gamma Medica, Northridge, CA). The mice were administered [^{11}C]CW24 (3700–7400 KBq per animal) after 5-min pre-treatment of IPSU (0.5 and 2.0 mg/kg; $n = 1$ for each dose; i.v.), Suvorexant (5.0 mg/kg; $n = 1$; i.v.), or vehicle ($n = 1$) via a lateral tail vein catheter. Animals underwent a 60 min dynamic PET scan followed by computed tomography (CT).

3.6. Rodent PET/CT Image Analysis

PET data were reconstructed using a 3D-MLEM method resulting in a full width at half-maximum resolution of 1 mm. PET and CT images in DICOM format were imported to PMOD (PMOD Technologies, Ltd. Zürich, Switzerland) and co-registered to the brain atlas. Volumes of interest (VOIs) were drawn as spheres in brain regions guided by CT structural images and summed PET data. Time-activity curves (TACs) were exported as activity per unit volume (%ID/cc) for analysis.

3.7. Macaque PET-MR Acquisition

A male rhesus macaque (14.4 kg) was included in this study. After endotracheal intubation, the macaque was catheterized antecubitally for radiotracer injection and a radial arterial line was placed for arterial blood sampling and radiometabolite analysis. The animal was anesthetized with 1.2–1.3% Isoflurane in medical oxygen throughout the imaging session. PET-MR data was acquired on a 3T Siemens MRI scanner (Munich, Germany) with a BrainPET insert. A baseline and a blocking scan were performed on the same animal. Dynamic PET image acquisition was initiated followed by a bolus administration of [^{11}C]CW24 (179 MBq for the baseline scan and 192 MBq for the blocking scan). To characterize the specific binding of [^{11}C]CW24, a blocking study was carried out in which 2.0 mg/kg IPSU (1.0 % DMSO + 1.0 % Tween 80 + 98.0 % saline to make a solution of 1.0 mg/mL) was administered 5 min prior the acquisition. A T1-weighting multi-echo magnetization-prepared rapid gradient-echo (MEMPRAGE) sequence began as the PET scan for anatomic co-registration. Dynamic PET data were corrected for decay, scatter, and attenuation and reconstructed using methods similar to our previous studies [39].

3.8. PET-MR Imaging Data Analysis for the Macaque Study

Motion-corrected PET data were registered to the INIA19 macaque MRI Template. VOIs were selected according to the INIA19 Template and NeuroMaps Atlas [40]. TACs from the whole *Macaque* brain as well as a few brain regions were exported. With the availability of a metabolite-corrected arterial plasma data, the regional volume of distribution (V_T) was calculated using Logan Plot analysis (with a fixed t* of 40 min).

3.9. Plasma and Metabolite Analysis

Briefly, arterial blood samples collected (every 10 s at first 3 min post-administration of [^{11}C]CW24, then 5 min, 10 min, 20 min, 30 min, 45 min, 60 min, and 90 min) during macaque imaging were centrifuged to separate plasma, which was then removed and placed in an automated gamma counter [41]. Metabolite analysis of blood samples collected from 5 to 90 min was conducted on a custom automated robot fitted with Phenomenex SPE Strata-X 500 mg solid phase extraction cartridges that were primed with ethanol (2 mL) and deionized water (20 mL). Protein precipitation was achieved by the addition of plasma (300 µL) to acetonitrile (300 µL), which was centrifuged for 1 min to obtain protein-free plasma (PFP). Three hundred microliters of PFP/acetonitrile solution was diluted into deionized water (3 mL), loaded onto the C18 cartridge, and removed of polar metabolites with 100% water. Next, a series of extractions was performed using water and acetonitrile.

Each sample was counted in a WIZARD2 Automatic Gamma Counter to determine the presence of radiolabeled metabolites.

4. Conclusions

In summary, our developed PET imaging radio-ligands, [^{11}C]CW24, provides a non-invasive quantitative imaging tool for evaluating OXR expression in the brain. Based on the in vitro and in vivo evaluations, [^{11}C]CW24 had high brain uptake and good target-selectivity, but suffers from high non-specific binding. Therefore, [^{11}C]CW24 could be used as a lead compound to develop brain-PET probes for OXRs to investigate not only the roles of OXR in a variety of disease applications, but also the development of OXR neurotherapeutics. In future studies, it is important that we characterize and optimize [^{11}C]CW24 to improve specific binding for developing potential PET imaging probes for human imaging.

Supplementary Materials: The following are available online, Figure S1: Representative 1H, 13C-NMR and HPLC data for CW24; Figure S2: The HPLC profile of CW24; Table S1: Off-target binding data of CW24.

Author Contributions: Investigation, P.B., S.B.; writing—original draft preparation, P.B.; writing—review and editing, M.S.P., B.N., S.A.F., H.-Y.W.; supervision, X.L.; project administration, C.W.; funding acquisition, C.W. All authors have read and agreed to the published version of the manuscript.

Funding: This research was funded by National Institutes of Health (NIH), grant number DA048123 and the APC was funded by NIH-DA048123.

Acknowledgments: This work was supported by NIH funding (DA048123) and pilot funding from the Athinoula A. Martinos Center for Biomedical Imaging at the Massachusetts General Hospital (C.W. and H.-Y.W.). The imaging studies were carried out at the Martinos Center, using resources provided by the *Center for Functional Neuroimaging Technologies, P41EB015896*, a P41 Regional Resource supported by the National Institute of Biomedical Imaging and Bioengineering (NIBIB), National Institutes of Health. This work also involved the use of instrumentation supported by the NIH Shared Instrumentation Grant Program and/or High-End Instrumentation Grant Program; specifically, grant numbers: S10RR017208, S10RR026666, S10RR022976, S10RR019933, S10RR023401. The author Ping Bai gratefully acknowledges financial support by China Scholarship Council (CSC) for this training at the Martinos Center. The authors are grateful to (Judit Sore, the Martinos Center radioligand lab, PET/MR imaging staff (Grae Arabasz, Shirley Hsu and Regan Butterfield), and Helen Deng for assistance with non-human primate imaging.

Conflicts of Interest: The authors declare no conflict of interest.

References

1. Ammoun, S.; Holmqvist, T.; Shariatmadari, R.; Oonk, H.B.; Detheux, M.; Parmentier, M.; Akerman, K.E.; Kukkonen, J.P. Distinct recognition of OX1 and OX2 receptors by orexin peptides. *J. Pharmacol. Exp. Ther.* **2003**, *305*, 507–514. [CrossRef] [PubMed]
2. Bettica, P.; Squassante, L.; Groeger, J.A.; Gennery, B.; Winsky-Sommerer, R.; Dijk, D.J. Differential effects of a dual orexin receptor antagonist (SB-649868) and zolpidem on sleep initiation and consolidation, SWS, REM sleep, and EEG power spectra in a model of situational insomnia. *Neuropsychopharmacology* **2012**, *37*, 1224–1233. [CrossRef]
3. Bingham, M.J.; Cai, J.; Deehan, M.R. Eating, sleeping and rewarding: Orexin receptors and their antagonists. *Curr. Opin Drug Discov. Devel.* **2006**, *9*, 551–559.
4. Boutrel, B.; Kenny, P.J.; Specio, S.E.; Martin-Fardon, R.; Markou, A.; Koob, G.F.; de Lecea, L. Role for hypocretin in mediating stress-induced reinstatement of cocaine-seeking behavior. *Proc. Natl. Acad. Sci. USA* **2005**, *102*, 19168–19173. [CrossRef]
5. Brisbare-Roch, C.; Dingemanse, J.; Koberstein, R.; Hoever, P.; Aissaoui, H.; Flores, S.; Mueller, C.; Nayler, O.; van Gerven, J.; de Haas, S.L.; et al. Promotion of sleep by targeting the orexin system in rats, dogs and humans. *Nat. Med.* **2007**, *13*, 150–155. [CrossRef]
6. Cada, D.J.; Levien, T.L.; Baker, D.E. Suvorexant. *Hosp Pharm.* **2015**, *50*, 59–71. [CrossRef]
7. Roecker, A.J.; Cox, C.D.; Coleman, P.J. Orexin receptor antagonists: New therapeutic agents for the treatment of insomnia. *J. Med. Chem.* **2016**, *59*, 504–530. [CrossRef]
8. Bingham, S.; Davey, P.T.; Babbs, A.J.; Irving, E.A.; Sammons, M.J.; Wyles, M.; Jeffrey, P.; Cutler, L.; Riba, I.; Johns, A.; et al. Orexin-A, an hypothalamic peptide with analgesic properties. *Pain* **2001**, *92*, 81–90. [CrossRef]

9. Tabaeizadeh, M.; Motiei-Langroudi, R.; Mirbaha, H.; Esmaeili, B.; Tahsili-Fahadan, P.; Javadi-Paydar, M.; Ghaffarpour, M.; Dehpour, A.R. The differential effects of OX_1R and OX_2R selective antagonists on morphine conditioned place preference in naive versus morphine-dependent mice. *Behav. Brain Res.* **2013**, *237*, 41–48. [CrossRef] [PubMed]

10. De Lecea, L.; Kilduff, T.S.; Peyron, C.; Gao, X.; Foye, P.E.; Danielson, P.E.; Fukuhara, C.; Battenberg, E.L.; Gautvik, V.T.; Frankel, W.N.; et al. The hypocretins: Hypothalamus-specific peptides with neuroexcitatory activity. *Proc. Natl. Acad Sci. USA* **1998**, *95*, 322–327. [CrossRef] [PubMed]

11. Sakurai, T.; Amemiya, A.; Ishii, M.; Matsuzaki, I.; Chemelli, R.M.; Tanaka, H.; Williams, S.C.; Richardson, J.A.; Kozlowski, G.P.; Wilson, S.; et al. Orexins and orexin receptors: A family of hypothalamic neuropeptides and G protein-coupled receptors that regulate feeding behavior. *Cell* **1998**, *92*, 573–585. [CrossRef]

12. Marcus, J.N.; Aschkenasi, C.J.; Lee, C.E.; Chemelli, R.M.; Saper, C.B.; Yanagisawa, M.; Elmquist, J.K. Differential expression of orexin receptors 1 and 2 in the rat brain. *J. Comp. Neurol.* **2001**, *435*, 6–25. [CrossRef] [PubMed]

13. Suzuki, H.; Takemoto, Y.; Yamamoto, T. Differential distribution of orexin-A-like and orexin receptor 1 (OX_1R)-like immunoreactivities in the Xenopus pituitary. *Tissue Cell* **2007**, *39*, 423–430. [CrossRef] [PubMed]

14. Mahler, S.V.; Smith, R.J.; Moorman, D.E.; Sartor, G.C.; Aston-Jones, G. Multiple roles for orexin/hypocretin in addiction. *Prog. Brain Res.* **2012**, *198*, 79–121. [PubMed]

15. Ida, T.; Nakahara, K.; Murakami, T.; Hanada, R.; Nakazato, M.; Murakami, N. Possible involvement of orexin in the stress reaction in rats. *Biochem. Biophys. Res. Commun.* **2000**, *270*, 318–323. [CrossRef]

16. Furlong, T.M.; Vianna, D.M.; Liu, L.; Carrive, P. Hypocretin/orexin contributes to the expression of some but not all forms of stress and arousal. *Eur. J. Neurosci.* **2009**, *30*, 1603–1614. [CrossRef]

17. Stachulski, A.V.; Baillie, T.A.; Park, B.K.; Obach, R.S.; Dalvie, D.K.; Williams, D.P.; Srivastava, A.; Regan, S.L.; Antoine, D.J.; Goldring, C.E.; et al. The generation, detection, and effects of reactive drug metabolites. *Med. Res. Rev.* **2013**, *33*, 985–1080. [CrossRef]

18. Bettica, P.; Squassante, L.; Zamuner, S.; Nucci, G.; Danker-Hopfe, H.; Ratti, E. The orexin antagonist SB-649868 promotes and maintains sleep in men with primary insomnia. *Sleep* **2012**, *35*, 1097–1104. [CrossRef]

19. Hirose, M.; Egashira, S.; Goto, Y.; Hashihayata, T.; Ohtake, N.; Iwaasa, H.; Hata, M.; Fukami, T.; Kanatani, A.; Yamada, K. N-acyl 6,7-dimethoxy-1,2,3,4-tetrahydroisoquinoline: The first orexin-2 receptor selective non-peptidic antagonist. *Bioorg. Med. Chem. Lett.* **2003**, *13*, 4497–4499. [CrossRef]

20. Whitman, D.B.; Cox, C.D.; Breslin, M.J.; Brashear, K.M.; Schreier, J.D.; Bogusky, M.J.; Bednar, R.A.; Lemaire, W.; Bruno, J.G.; Hartman, G.D.; et al. Discovery of a potent, CNS-penetrant orexin receptor antagonist based on an n,n-disubstituted-1,4-diazepane scaffold that promotes sleep in rats. *Chem. Med. Chem.* **2009**, *4*, 1069–1074. [CrossRef]

21. Malherbe, P.; Borroni, E.; Pinard, E.; Wettstein, J.G.; Knoflach, F. Biochemical and electrophysiological characterization of almorexant, a dual orexin 1 receptor (OX1)/orexin 2 receptor (OX2) antagonist: Comparison with selective OX1 and OX2 antagonists. *Mol. Pharmacol.* **2009**, *76*, 618–631. [CrossRef] [PubMed]

22. McAtee, L.C.; Sutton, S.W.; Rudolph, D.A.; Li, X.; Aluisio, L.E.; Phuong, V.K.; Dvorak, C.A.; Lovenberg, T.W.; Carruthers, N.I.; Jones, T.K. Novel substituted 4-phenyl-[1,3]dioxanes: Potent and selective orexin receptor 2 ($OX_{(2)}R$) antagonists. *Bioorg. Med. Chem. Lett.* **2004**, *14*, 4225–4229. [CrossRef] [PubMed]

23. Porter, R.A.; Chan, W.N.; Coulton, S.; Johns, A.; Hadley, M.S.; Widdowson, K.; Jerman, J.C.; Brough, S.J.; Coldwell, M.; Smart, D.; et al. 1,3-Biarylureas as selective non-peptide antagonists of the orexin-1 receptor. *Bioorg. Med. Chem. Lett.* **2001**, *11*, 1907–1910. [CrossRef]

24. Smart, D.; Sabido-David, C.; Brough, S.J.; Jewitt, F.; Johns, A.; Porter, R.A.; Jerman, J.C. SB-334867-A: The first selective orexin-1 receptor antagonist. *Br. J. Pharmacol.* **2001**, *132*, 1179–1182. [CrossRef] [PubMed]

25. Srinivasan, S.; Simms, J.A.; Nielsen, C.K.; Lieske, S.P.; Bito-Onon, J.J.; Yi, H.; Hopf, F.W.; Bonci, A.; Bartlett, S.E. The dual orexin/hypocretin receptor antagonist, almorexant, in the ventral tegmental area attenuates ethanol self-administration. *PLoS ONE* **2012**, *7*, e44726. [CrossRef]

26. Gao, M.; Wang, M.; Zheng, Q.H. Synthesis of [(^{11}C)]MK-1064 as a new PET radioligand for imaging of orexin-2 receptor. *Bioorg. Med. Chem. Lett.* **2016**, *26*, 3694–3699. [CrossRef]

27. Wang, C.; Moseley, C.K.; Carlin, S.M.; Wilson, C.M.; Neelamegam, R.; Hooker, J.M. Radiosynthesis and evaluation of [^{11}C]EMPA as a potential PET tracer for orexin 2 receptors. *Bioorg. Med. Chem. Lett.* **2013**, *23*, 3389–3392. [CrossRef]

28. Liu, F.; Majo, V.J.; Prabhakaran, J.; Castrillion, J.; Mann, J.J.; Martinez, D.; Kumar, J.S. Radiosynthesis of [11C]BBAC and [^{11}C]BBPC as potential PET tracers for orexin2 receptors. *Bioorg. Med. Chem. Lett.* **2012**, *22*, 2172–2174. [CrossRef]
29. Wang, C.; Wilson, C.M.; Moseley, C.K.; Carlin, S.M.; Hsu, S.; Arabasz, G.; Schroeder, F.A.; Sander, C.Y.; Hooker, J.M. Evaluation of potential PET imaging probes for the orexin 2 receptors. *Nucl. Med. Biol.* **2013**, *40*, 1000–1005. [CrossRef]
30. Cox, C.D.; McGaughey, G.B.; Bogusky, M.J.; Whitman, D.B.; Ball, R.G.; Winrow, C.J.; Renger, J.J.; Coleman, P.J. Conformational analysis of N,N-disubstituted-1,4-diazepane orexin receptor antagonists and implications for receptor binding. *Bioorg. Med. Chem. Lett.* **2009**, *19*, 2997–3001. [CrossRef]
31. Steiner, M.A.; Gatfield, J.; Brisbare-Roch, C.; Dietrich, H.; Treiber, A.; Jenck, F.; Boss, C. Discovery and characterization of ACT-335827, an orally available, brain penetrant orexin receptor type 1 selective antagonist. *Chem. Med. Chem.* **2013**, *8*, 898–903. [CrossRef] [PubMed]
32. Betschart, C.; Hintermann, S.; Behnke, D.; Cotesta, S.; Fendt, M.; Gee, C.E.; Jacobson, L.H.; Laue, G.; Ofner, S.; Chaudhari, V.; et al. Identification of a novel series of orexin receptor antagonists with a distinct effect on sleep architecture for the treatment of insomnia. *J. Med. Chem.* **2013**, *56*, 7590–7607. [CrossRef]
33. Callander, G.E.; Olorunda, M.; Monna, D.; Schuepbach, E.; Langenegger, D.; Betschart, C.; Hintermann, S.; Behnke, D.; Cotesta, S.; Fendt, M.; et al. Kinetic properties of "dual" orexin receptor antagonists at OX_1R and OX_2R orexin receptors. *Front. Neurosci.* **2013**, *7*, 230. [CrossRef] [PubMed]
34. Zhang, L.; Villalobos, A.; Beck, E.M.; Bocan, T.; Chappie, T.A.; Chen, L.; Grimwood, S.; Heck, S.D.; Helal, C.J.; Hou, X.; et al. Design and selection parameters to accelerate the discovery of novel central nervous system positron emission tomography (PET) ligands and their application in the development of a novel phosphodiesterase 2A PET ligand. *J. Med. Chem.* **2013**, *56*, 4568–4579. [CrossRef] [PubMed]
35. Seo, Y.J.; Kang, Y.; Muench, L.; Reid, A.; Caesar, S.; Jean, L.; Wagner, F.; Holson, E.; Haggarty, S.J.; Weiss, P.; et al. Image-guided synthesis reveals potent blood-brain barrier permeable histone deacetylase inhibitors. *ACS Chem. Neurosci.* **2014**, *5*, 588–596. [CrossRef] [PubMed]
36. Besnard, J.; Ruda, G.F.; Setola, V.; Abecassis, K.; Rodriguiz, R.M.; Huang, X.P.; Norval, S.; Sassano, M.F.; Shin, A.I.; Webster, L.A.; et al. Automated design of ligands to polypharmacological profiles. *Nature* **2012**, *492*, 215–220. [CrossRef]
37. Van de Bittner, G.C.; Ricq, E.L.; Hooker, J.M. A philosophy for CNS radiotracer design. *Acc. Chem. Res.* **2014**, *47*, 3127–3134. [CrossRef]
38. Wang, C.; Placzek, M.S.; Van de Bittner, G.C.; Schroeder, F.A.; Hooker, J.M. A Novel Radiotracer for Imaging Monoacylglycerol Lipase in the Brain Using Positron Emission Tomography. *ACS Chem. Neurosci.* **2016**, *7*, 484–489. [CrossRef]
39. Wang, C.; Schroeder, F.A.; Hooker, J.M. Development of new positron emission tomography radiotracer for BET imaging. *ACS Chem. Neurosci.* **2017**, *8*, 17–21. [CrossRef]
40. Rohlfing, T.; Kroenke, C.D.; Sullivan, E.V.; Dubach, M.F.; Bowden, D.M.; Grant, K.A.; Pfefferbaum, A. The INIA19 template and neuromaps atlas for primate brain image parcellation and spatial normalization. *Front. Neuroinform* **2012**, *6*, 27. [CrossRef]
41. Bai, P.; Wey, H.-Y.; Patnaik, D.; Lu, X.; Rokka, J.; Stephanie, F.; Haggarty, S.J.; Wang, C. Positron emission tomography probes targeting bromodomain and extra-terminal (BET) domains to enable in vivo neuroepigenetic imaging. *Chem. Commun.* **2019**, *55*, 12932–12935. [CrossRef] [PubMed]

Sample Availability: Samples of the compounds (IPSU and CW24) are available from the authors.

© 2020 by the authors. Licensee MDPI, Basel, Switzerland. This article is an open access article distributed under the terms and conditions of the Creative Commons Attribution (CC BY) license (http://creativecommons.org/licenses/by/4.0/).

Article

PET Imaging of the Adenosine A_{2A} Receptor in the Rotenone-Based Mouse Model of Parkinson's Disease with [^{18}F]FESCH Synthesized by a Simplified Two-Step One-Pot Radiolabeling Strategy

Susann Schröder [1,*], Thu Hang Lai [2], Magali Toussaint [2], Mathias Kranz [3,4], Alexandra Chovsepian [5], Qi Shang [6,7], Sladjana Dukić-Stefanović [2], Winnie Deuther-Conrad [2], Rodrigo Teodoro [2], Barbara Wenzel [2], Rareş-Petru Moldovan [2], Francisco Pan-Montojo [5,6] and Peter Brust [2]

1. ROTOP Pharmaka Ltd., Department of Research and Development, Dresden 01328, Germany
2. Helmholtz-Zentrum Dresden-Rossendorf (HZDR), Department of Neuroradiopharmaceuticals, Institute of Radiopharmaceutical Cancer Research, Research Site Leipzig, Leipzig 04318, Germany; t.lai@hzdr.de (T.H.L.); m.toussaint@hzdr.de (M.T.); s.dukic-stefanovic@hzdr.de (S.D.-S.); w.deuther-conrad@hzdr.de (W.D.-C.); r.teodoro@hzdr.de (R.T.); b.wenzel@hzdr.de (B.W.); r.moldovan@hzdr.de (R.-P.M.); p.brust@HZDR.de (P.B.)
3. PET Imaging Center, University Hospital of North Norway (UNN), Tromsø 9009, Norway; mathias.kranz@uit.no
4. Nuclear Medicine and Radiation Biology Research Group, The Arctic University of Norway, Tromsø 9009, Norway
5. University Hospital Munich, Department of Psychiatry, Ludwig-Maximilians-Universität (LMU) Munich, Munich 80336, Germany; alexandra.chovsepian@med.uni-muenchen.de (A.C.); Francisco.Pan-Montojo@med.uni-muenchen.de (F.P.-M.)
6. University Hospital Munich, Department of Neurology, Ludwig-Maximilians-Universität (LMU) Munich, Munich 81377, Germany; shangqi0911@gmail.com
7. University Hospital Carl Gustav Carus, Clinic of Neurology, Technische Universität Dresden (TUD), Dresden 01307, Germany
* Correspondence: s.schroeder@hzdr.de; Tel.: +49-341-234-179-4631

Received: 24 January 2020; Accepted: 27 March 2020; Published: 2 April 2020

Abstract: The adenosine A_{2A} receptor ($A_{2A}R$) is regarded as a particularly appropriate target for non-dopaminergic treatment of Parkinson's disease (PD). An increased $A_{2A}R$ availability has been found in the human striatum at early stages of PD and in patients with PD and dyskinesias. The aim of this small animal positron emission tomography/magnetic resonance (PET/MR) imaging study was to investigate whether rotenone-treated mice reflect the aspect of striatal $A_{2A}R$ upregulation in PD. For that purpose, we selected the known $A_{2A}R$-specific radiotracer [^{18}F]**FESCH** and developed a simplified two-step one-pot radiosynthesis. PET images showed a high uptake of [^{18}F]**FESCH** in the mouse striatum. Concomitantly, metabolism studies with [^{18}F]**FESCH** revealed the presence of a brain-penetrant radiometabolite. In rotenone-treated mice, a slightly higher striatal $A_{2A}R$ binding of [^{18}F]**FESCH** was found. Nonetheless, the correlation between the increased $A_{2A}R$ levels within the proposed PD animal model remains to be further investigated.

Keywords: adenosine A_{2A} receptor; Parkinson's disease; rotenone-based mouse model; PET imaging; [^{18}F]**FESCH**; two-step one-pot radiosynthesis

1. Introduction

Parkinson's disease (PD) is characterized by the degeneration of dopaminergic neurons in the brain, especially in the substantia nigra, resulting in a decreased dopamine level in the striatum.

Consequently, stimulation of the dopamine D_2 receptor (D_2R) is reduced and, further, the morphology of striatal neurons and synaptic connections is changing. These effects lead to motor dysfunctions like bradykinesia, tremor, rigor, and postural instability [1,2]. The standard therapy of PD is based on the pharmacological increase of dopamine levels either by administration of L-DOPA as metabolic precursor of dopamine or by dopamine agonists. Further strategies to raise or maintain the dopamine levels are based on inhibitors for dopamine reuptake and dopamine-degrading enzymes (e.g., monoamine oxidase B, catechol-O-methyl transferase) [2]. These dopamine- increasing drugs have only a positive effect on motor functions at short-term treatment but cause undesirable side effects at long-term treatment like dyskinesia, nausea, and hallucinations. Due to these limitations, the medicinal research is currently focusing on non-dopaminergic therapies of PD [2,3].

The adenosine A_{2A} receptor ($A_{2A}R$) is a G-protein coupled receptor that is highly expressed in the basal ganglia of the brain, mainly in striatopallidal GABAergic neurons. Furthermore, in this brain area the $A_{2A}R$ and the D_2R are co-localized and form heteromeric receptor complexes while showing contradictory effects on motor functions [2]. For example, an L-DOPA-induced increase of dopamine stimulates the D_2R leading to enhanced motor activity. Activation of the $A_{2A}R$ by adenosine or appropriate agonists results in the inhibition of the D_2R and, thus, in reduced dopamine binding [2,4,5]. Pharmacological inhibition of the $A_{2A}R$ amplifies the D_2R-dependent signaling cascade and improves motor symptoms [2,3,5,6]. Since the $A_{2A}R$ modulates the dopamine binding affinity of the D_2R, selective $A_{2A}R$ antagonists have great potential as appropriate non-dopaminergic PD therapeutics [2,3,6]. In clinical phase II and III trials, the $A_{2A}R$ antagonists **istradefylline, preladenant,** and **tozadenant** improved motor symptoms in PD patients [7–9]. Out of these, **istradefylline** (Nourianz®) has already been approved in Japan [10] and currently by the U.S. Food and Drug Administration (FDA) as an adjunctive treatment to L-DOPA.

Furthermore, a comparative neuropathological postmortem study on brain slices from healthy subjects and PD patients showed a 2.5-fold increased striatal $A_{2A}R$ protein level already at early PD stages pointing out that $A_{2A}R$ upregulation is an initial event in the pathogenesis of the disease [11]. Though, it remains unclear whether there is a correlation between altered $A_{2A}R$ expression and motor symptoms in PD.

Molecular imaging of the $A_{2A}R$ with positron emission tomography (PET) enables the non-invasive investigation of pathological changes of the receptor level in the human brain [12].

A clinical PET study using [^{11}C]**SCH442416** (Figure 1) revealed a 70–80% increased $A_{2A}R$ availability in the striatum of PD patients with L-DOPA-induced dyskinesia [13]. This strongly indicates a key role of $A_{2A}R$ not only in the development of PD [11] but also in the progression of the disease and in association with the dopaminergic standard therapy [2,13]. Hence, $A_{2A}R$-PET is considered as an appropriate tool for early diagnosis and staging of PD as well as evaluation of potential $A_{2A}R$ antagonists for PD treatment by means of receptor occupancy studies. The main disadvantage of ^{11}C-labeled radiotracers for clinical routine applications is the short half-life of the nuclide ($t_{1/2}$ = 20.3 min). To date, there is only one ^{18}F-labeled ($t_{1/2}$ = 109.7 min) $A_{2A}R$ radiotracer, [^{18}F]**MNI-444** [14,15] (Figure 1), that has already been evaluated in a clinical PET study. In healthy volunteers [15], a high uptake of activity in the striatum has been observed. However, no follow-up studies have been published since 2015 and further data of [^{18}F]**MNI-444** regarding specificity of $A_{2A}R$ binding in humans or selectivity over the adenosine A_1 receptor (A_1R), which is also highly expressed in the brain [16,17], are not yet available. For that reason, we selected the well-studied and highly $A_{2A}R$-selective radiotracer [^{18}F]**FESCH** [18–21] ($K_i(hA_{2A}R)$ = 12.4 nM, $K_i(hA_1R)$ ~ 10 µM, former [^{18}F]**MRS5425** [18]) for our purposes, which is the [^{18}F]fluoroethoxy analog of [^{11}C]**SCH442416** (Figure 1). [^{18}F]**FESCH** has been evaluated in healthy rats and in rats with unilateral PD symptoms by PET showing an $A_{2A}R$-specific binding in the striatum and a significantly increased $A_{2A}R$-mediated uptake in the 6-hydroxydopamine-lesioned hemisphere of 9–12% [19,21]. Therefore, [^{18}F]**FESCH** has been stated as the most suitable radiotracer for quantification of the $A_{2A}R$ in the brain with PET [21].

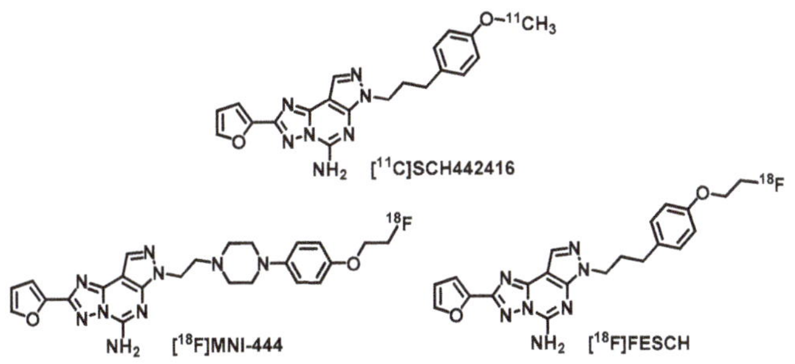

Figure 1. Molecular structures of the herein discussed radiotracers for positron emission tomography (PET) imaging of the adenosine A_{2A} receptor ($A_{2A}R$).

The rotenone-based mouse model has been well established for pre-clinical investigations of PD [22–24]. Chronic administration of the neurotoxin rotenone to mice leads to relapse of dopaminergic neurons and changes in motor functions similar to the symptoms in PD patients [24–26]. In this small animal PET/MR imaging study with [^{18}F]**FESCH**, we aimed to validate if rotenone-treated mice reflect the aspect of striatal $A_{2A}R$ upregulation in PD. In addition, we investigated [^{18}F]**FESCH** by in vitro autoradiography on mouse brain slices and in vivo metabolism as well as baseline and blocking PET studies in healthy CD-1 mice.

2. Results and Discussion

The $A_{2A}R$ antagonist **FESCH** and the required known phenol precursor desmethyl SCH442416 for ^{18}F-labeling were synthesized in three or two steps, respectively, from commercially available 2-(furan-2-yl)-7H-pyrazolo[4,3-e][1,2,4]triazolo[1,5-c]pyrimidin-5-amin and 1-(3-bromopropyl)-4-methoxybenzene according to procedures described in the literature [18,27,28]. Evaluation of **FESCH** in an established binding assay revealed high $A_{2A}R$ affinity ($K_i(hA_{2A}R) = 0.6 \pm 0.1$ nM, n = 4) and good A_1R selectivity ($K_i(hA_1R) = 203 \pm 40$ nM, n = 4).

The initially published radiosynthesis of [^{18}F]**FESCH** comprises a two-step two-pot procedure [19,20]. Briefly, the respective [^{18}F]fluoroethyl intermediate from the ^{18}F-labeling step was isolated either by semi-preparative HPLC [19] or solid-phase extraction on a cartridge [20] and, only then, reacted with the phenol precursor desmethyl SCH442416. To simplify the radiosynthesis and avoid loss of activity through purification of the [^{18}F]fluoroethyl intermediate, we developed a two-step one-pot strategy for the production of [^{18}F]**FESCH** [29] (see Scheme 1). First, ethane-1,2-diol bis(4-methylbenzenesulfonate) was ^{18}F-labeled via nucleophilic substitution of one tosylate group using anhydrous K$^+$/[^{18}F]F$^-$/K$_{222}$-carbonate complex in acetonitrile. Notably, desmethyl SCH442416 was pre-treated with aqueous tetrabutylammonium hydroxide (TBAOH) in acetonitrile to generate the activated phenolate which was directly reacted with the non-isolated 2-[^{18}F]fluoroethyl tosylate. Besides, further bases and solvents were tested for the [^{18}F]fluoroethylation of desmethyl SCH442416 resulting in decreased yields of [^{18}F]**FESCH** mainly due to partial decomposition of the phenol precursor under these conditions. Preliminary experiments using potassium carbonate or cesium carbonate (1.5–2.5 eq.) for activation of desmethyl SCH442416 (1–2 mg) and either acetonitrile or a 1:1 mixture of acetonitrile and N,N-dimethylformamide or tert-butanol as solvents revealed labeling yields of 4–14% based on TLC analysis of the reaction mixtures after 10 min at 120 °C, respectively. Replacement of the base by TBAOH (2.5 eq.) and the use of acetonitrile only resulted in a significantly increased labeling yield of 25%. Final optimization was achieved by increasing the amount of the phenol precursor to 2.5 mg leading to 46.4 ± 8.5% (n = 9) of non-isolated [^{18}F]**FESCH**.

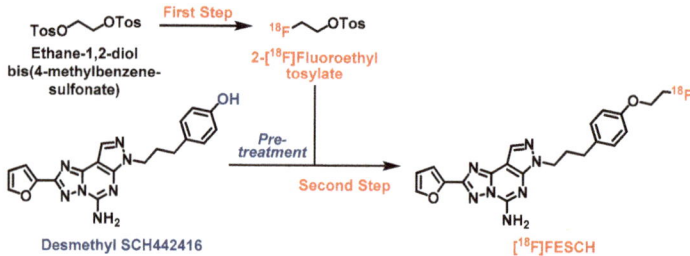

Scheme 1. Simplified two-step one-pot radiosynthesis of [^{18}F]**FESCH**. Reagents and conditions: (First step) 2 mg ethane-1,2-diol bis(4-methylbenzenesulfonate) in 100 µL MeCN, ~3 GBq azeotropically dried K$^+$/[^{18}F]F$^-$/K$_{222}$-carbonate complex in 400 µL MeCN, 90 °C, 10 min; (Second step) 2.5 mg desmethyl SCH442416 pre-treated with 10 µL TBAOH$_{aq.}$ (40%) in 490 µL MeCN at 90 °C for 10 min, directly added to 2-[^{18}F]fluoroethyl tosylate, 120 °C, 10 min.

[^{18}F]**FESCH** was purified by semi-preparative HPLC (see Figure 2) and concentrated via solid-phase extraction on a pre-conditioned reversed-phase cartridge followed by elution with absolute ethanol. After evaporation of the solvent, the radiotracer was finally formulated in sterile isotonic saline with a maximum ethanol content of 10% for better solubility. The identity of [^{18}F]**FESCH** was confirmed by analytical HPLC using an aliquot of the final product spiked with the non-radioactive reference compound (Figure 2).

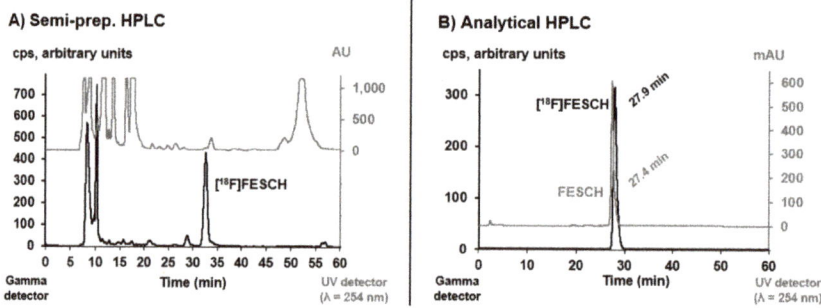

Figure 2. (**A**) Semi-preparative HPLC profile of the crude reaction mixture for isolation of [^{18}F]**FESCH** (column: Reprosil-Pur 120 C18-AQ, 250 × 10 mm, particle size: 10 µm, eluent: 50% MeCN/20 mM NH$_4$OAc$_{aq.}$, flow: 7 mL/min). (**B**) Analytical HPLC profile of the formulated radiotracer [^{18}F]**FESCH** spiked with the non-radioactive reference **FESCH** (column: Reprosil-Pur 120 C18-AQ, 250 × 4.6 mm, particle size: 5 µm; eluent: 26-90-26% MeCN/20 mM NH$_4$OAc$_{aq.}$, flow: 1 mL/min).

The herein described one-pot radiolabeling strategy provided [^{18}F]**FESCH** with a high molar activity of 116 ± 18.5 GBq/µmol (n = 7, end of synthesis) and an overall radiochemical yield of 16.1 ± 1.5% (n = 9, end of bombardment), which is a significant improvement compared to the published two-pot procedure (7 ± 2% [20]). The required total synthesis time of 114 ± 6 min and the achieved radiochemical purity of ≥ 98% were very similar for both methods.

In vitro stability of [^{18}F]**FESCH** was examined in isotonic saline and pig plasma. Samples of each medium were analyzed by radio-HPLC after 60 min incubation at 37 °C and no degradation or defluorination of the radiotracer was observed. This result is contradictory to the published stability of [^{18}F]**FESCH** especially in saline where only 85–90% of intact radiotracer have been detected by radio-TLC while multiple spots were observed [20]. Thus, Khanapur et al. [20] formulated [^{18}F]**FESCH** in phosphate-buffered saline, but Bhattacharjee et al. [19] used a saline solution of [^{18}F]**FESCH** for biological investigations. At present, we have no explanation for that and as we detected only the intact radiotracer in isotonic saline (see Figure 2), formulation of [^{18}F]**FESCH** in another medium was not required.

The distribution coefficient of [^{18}F]**FESCH** was determined by partitioning between *n*-octanol and phosphate-buffered saline (PBS, pH = 7.4) at ambient temperature using the conventional shake-flask method. The obtained logD$_{7.4}$ value of 1.97 ± 0.17 (n = 4) emphasizes the lipophilic character of the radiotracer which allows a passive diffusion through the blood–brain barrier [30–32]. However, there is a discrepancy compared to the published logD value of 3.16 ± 0.03 [20] which might be caused by different experimental setups.

In vitro autoradiography studies were accomplished by incubating sections of mouse brain with [^{18}F]**FESCH**. Non-specific binding was assessed by co-incubation with an excess of either **FESCH** or **ZM241385**, respectively. The images demonstrated an A$_{2A}$R-specific accumulation of [^{18}F]**FESCH** in the striatum (Figure 3), which is characterized by the binding parameters K_D = 4.69 ± 1.17 nM and B_{max} = 497 ± 97 fmol/mg wet weight consistent with the literature [33].

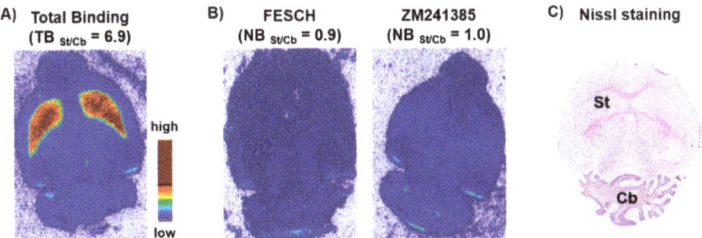

Figure 3. Representative autoradiographic images of transversal CD-1 mouse brain slices. (**A**) In vitro distribution of activity after incubation with 0.2 MBq/mL of [^{18}F]**FESCH** (TB = total binding). (**B**) Non-specific binding (NB) of [^{18}F]**FESCH** determined in the presence of 1 μM of **FESCH** or **ZM241385**, respectively, as blocking agents. (**C**) Nissl staining, St = Striatum, Cb = Cerebellum.

For in vivo imaging investigations in healthy CD-1 mice (n = 12, 10 weeks, 30–35 g), [^{18}F]**FESCH** was injected intravenously (baseline: 2.8 ± 2 MBq, vehicle: 5.8 ± 2 MBq, blocking: 6.4 ± 2.5 MBq with n = 4, respectively) and whole body scans were performed for 60 min in listmode with a Mediso nanoScan® PET/MR scanner followed by dynamic reconstruction. Time-activity curves (TACs) were generated for regions of interest such as striatum and cerebellum as reference region. The obtained PET images showed a high striatal uptake of [^{18}F]**FESCH** with a standardized uptake value ratio (SUVR) for striatum over cerebellum of > 5 (9–30 min post injection (p.i.)). In vivo selectivity for the A$_{2A}$R was highlighted by the significant reduction of the SUVR by 29–42% (9–30 min p.i. with $p < 0.05$) after pre-injection of **tozadenant** as blocking agent (Figure 4). The highest achievable dose of **tozadenant** was 2.5 mg/kg with respect to solubility and volume of injection as well as the concentrations of DMSO and Kolliphor® EL suitable for in vivo application in mice. The herein observed blocking effect is in accordance with the estimated A$_{2A}$R occupancy for **tozadenant** in rhesus monkey of about 72% at 5 mg/kg [14]. These results indicate that [^{18}F]**FESCH** is a promising radiotracer for molecular imaging of the A$_{2A}$R in the brain.

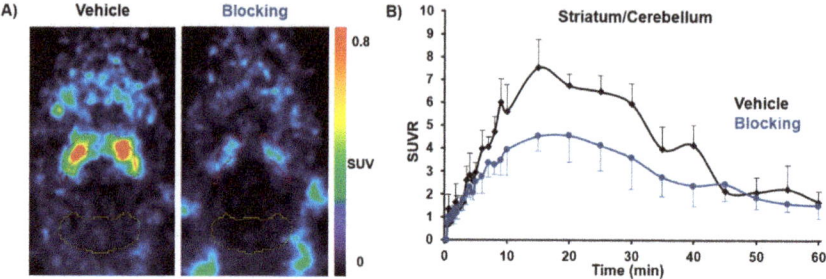

Figure 4. (**A**) Representative horizontal PET images of [^{18}F]**FESCH** uptake (average 10–30 min) in the brain of healthy CD-1 mice under vehicle (15 min pre-injection of DMSO:Kolliphor® EL:0.9% NaCl, 1:2:7, 5 µL/g) and blocking conditions (15 min pre-injection of **tozadenant** 2.5 mg/kg in DMSO:Kolliphor® EL:0.9% NaCl, 1:2:7, 5 µL/g; red = striatum; yellow = cerebellum). (**B**) Averaged TACs of [^{18}F]**FESCH** for vehicle (n = 4) and **tozadenant** pre-injected mice (n = 4) with SUVRs for striatum over cerebellum.

However, a representative metabolism study revealed only moderate in vivo stability of [^{18}F]**FESCH**. Analytical radio-HPLC (Figure 5) of the extracted mouse plasma sample showed 41% of intact radiotracer at 15 min p.i. (recovery of total activity = 84%). In the analyzed brain sample, one polar radiometabolite ([^{18}F]**M1**) was detected accounting for 29% of the total extracted activity at 15 min p.i. (recovery = 98%).

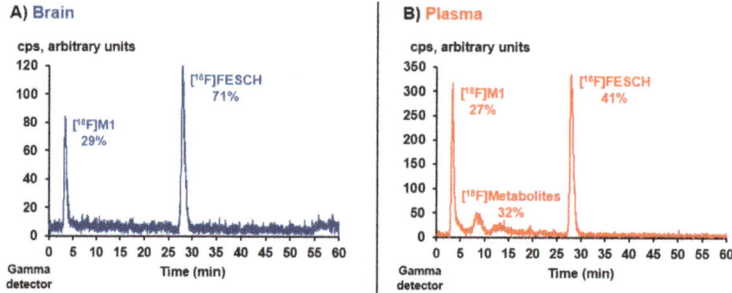

Figure 5. Representative in vivo metabolism study of CD-1 mouse plasma and brain samples at 15 min p.i. of [^{18}F]**FESCH** (~ 17 MBq): Analytical radio-HPLC profiles of extracted (**A**) brain and (**B**) plasma sample (column: Reprosil-Pur 120 C18-AQ, 250 × 4.6 mm, particle size: 5 µm; eluent: 26-90-26% MeCN/20 mM NH$_4$OAc$_{aq.}$, flow: 1 mL/min).

Compared to the published study in Wistar-Unilever rats (46% intact radiotracer in plasma at 60 min p.i. [20]) the in vivo degradation of [^{18}F]**FESCH** appears to be somewhat faster in mice. Notably and to the best of our knowledge, the formation of brain-penetrating radiometabolites of [^{18}F]**FESCH** has not been regarded before. Based on our experiences with radiotracers bearing a [^{18}F]fluoroethoxy moiety [34,35], the herein observed radiometabolite [^{18}F]**M1** is proposed to be 2-[^{18}F]fluoroethanol or the oxidized 2-[^{18}F]fluoroacetaldehyde and 2-[^{18}F]fluoroacetate, resulting from a cytochrome P450 enzyme-induced metabolic degradation [36,37], which are able to cross the blood–brain barrier [38–41].

For PET/MR studies with [^{18}F]**FESCH** in the rotenone-based mouse model of PD, the radiotracer (9.7 ± 1.3 MBq) was administrated to C57BL/6JRj mice (control: n = 5; rotenone-treated: n = 7; 16 months, 28–35 g) followed by the same imaging protocol used for the baseline and blocking studies in CD-1 mice. Although statistically not significant, the averaged TACs between 2 and 61 min p.i. revealed a slightly higher uptake of [^{18}F]**FESCH** in the striatum of rotenone-treated mice compared to controls.

An increase of the SUVR for striatum over cerebellum by 15–33% was observed, which was caused by the elevated SUV for striatum of 11–27% (21–61 min p.i., respectively; Figure 6).

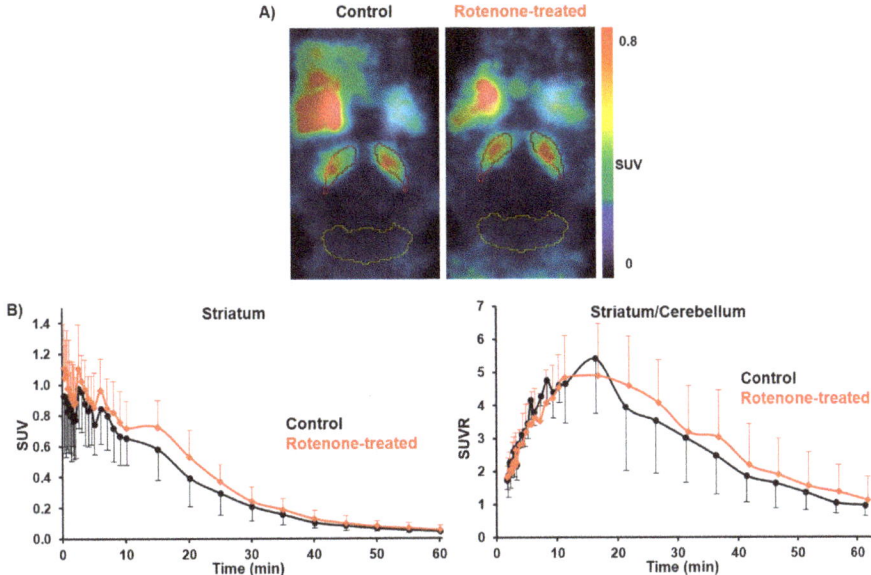

Figure 6. (**A**) Representative horizontal PET images of [^{18}F]**FESCH** uptake (average 2–61 min) in the brain of control and rotenone-treated C57BL/6JRj mice (red = striatum; yellow = cerebellum). (**B**) Averaged TACs of [^{18}F]**FESCH** for control (n = 5) and rotenone-treated mice (n = 7) with SUVs for striatum and SUVRs for striatum over cerebellum.

These results are in accordance with the determined $A_{2A}R$ levels on C57BL/6JRj mouse brain sections from a comparative in vitro immunofluorescence study. No significant difference in the fluorescence signal of the $A_{2A}R$ between control and rotenone-treated mice was observed (Figure 7).

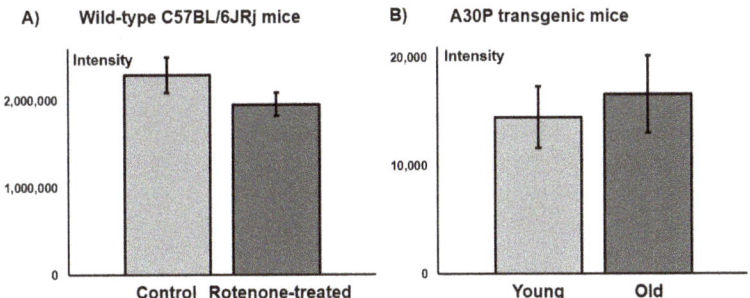

Figure 7. Mean fluorescence intensity of $A_{2A}R$ staining on the striatum of coronal C57BL/6JRj mouse brain sections: (**A**) Control vs. rotenone-treated wild-type mice (n = 3, respectively); (**B**) Young (39 weeks) vs. old (71–76 weeks) A30P transgenic mice (n = 3, respectively).

The increased $A_{2A}R$ availability in the striatum of PD patients appears to be related to L-DOPA-induced dyskinesia. In the rotenone-treated mice, we did not identify any signs of dyskinesia and, thus, it is not surprising that in this PD mouse model no significant increase in the $A_{2A}R$ density was detected either by PET imaging or by immunofluorescence staining. With regard to dyskinesia, the rotenone model is comparable to another neurotoxin-based PD mouse model using 1-methyl-4-phenyl-1,2,3,6-tetrahydropyridine (MPTP), where only weak signs of dyskinesia were observed and this exclusively under L-DOPA treatment [42]. To date, unilateral injection of 6-hydroxydopamine (6-OHDA) is the only effective neurotoxin-related approach to replicate evidently some forms of dyskinesia in rats and mice. In the 6-OHDA mouse model, dyskinesia is detected as abnormal involuntary movements (AIMs) with a more simplified range than in rats, presenting more prominent rotational locomotion with less dystonic features, i.e., axial AIMs [42]. None of the above-mentioned AIMs were detected in the rotenone-treated mice. Consequently, it remains to be examined whether treatment with L-DOPA would lead to dyskinesia in the rotenone model as shown for the MPTP model and if this would cause a quantifiable increase in striatal $A_{2A}R$ levels.

In contrast to the above-discussed neurotoxin-based PD animal models, the A30P mouse line represents a transgenic model of PD. This mouse line overexpresses the human A30P mutation in the α-synuclein gene, where the 30th amino acid residue alanine is replaced by proline, in all neurons. The A30P mutation is associated with rare familial cases of PD. Since α-synuclein is a main constituent of Lewy bodies, the A30P transgenic mice are characterized by Lewy body formation in neuronal cell bodies and neurites throughout the brain [43]. The α-synuclein pathology and PD-associated symptoms become prominent at about 69 weeks of age [43,44]. In an ongoing study, we detected slightly, but due to the rather small and highly variable data set statistically non-significant, higher striatal mean fluorescence intensities of the $A_{2A}R$ signal in older symptomatic A30P transgenic mice compared to younger ones without symptoms (see Figure 7). We will further investigate this PD mouse model by small animal PET/MR imaging studies using a novel derivative of [^{18}F]**FESCH** currently under development.

3. Summary and Conclusions

In this study, we investigated the suitability of the known radiotracer [^{18}F]**FESCH** for in vitro and in vivo imaging of the $A_{2A}R$ in the mouse brain by autoradiography and small animal PET/MR. Contrary to previous estimations, the herein observed brain-penetrating radiometabolite might limit the applicability of [^{18}F]**FESCH** for valid quantification of $A_{2A}R$ levels in the striatum with PET. One promising strategy to decelerate the degradation of ^{18}F-labeled radiotracers is the deuterium-hydrogen exchange at the metabolic labile position in the molecular structure of the compound [45]. Thus, our recent efforts are focused on a deuterated derivative of [^{18}F]**FESCH** with enhanced in vivo stability particularly at the [^{18}F]fluoroethoxy side chain.

In the investigated rotenone-based mouse model of PD, no significant difference in the striatal $A_{2A}R$ density between rotenone-treated mice and controls was detectable by PET imaging or immunofluorescence staining. These results indicate that the rotenone model does not reflect upregulation of striatal $A_{2A}R$ in PD, which appears to be related to dyskinesia. Therefore, it remains to be further examined whether treatment with L-DOPA could cause a detectable increase in $A_{2A}R$ levels in the striatum, which would induce observable dyskinesia in this PD model. In a parallel and ongoing study to characterize the A30P mouse model, initial findings reveal that this transgenic model might be more suitable with regard to displaying PD-like pathological changes in $A_{2A}R$ availability.

4. Materials and Methods

4.1. General Information

Chemicals and solvents were purchased from standard commercial sources in analytical grade and were used without further purification. Radio-TLCs were performed on pre-coated silica gel plates (Alugram® Xtra SIL G/UV$_{254}$, Polygram® SIL G/UV$_{254}$; Carl Roth, Karlsruhe, Germany). The compounds were localized at 254 nm (UV lamp). Radio-TLC was recorded using a bioimaging analyzer system (BAS-1800 II, Fuji Photo Film, Co. Ltd., Tokyo, Japan) and images were evaluated with Aida 2.31 software (raytest Isotopenmessgeräte GmbH, Straubenhardt, Germany). Column chromatography was conducted on silica gel (0.06–0.20 mm; Carl Roth, Karlsruhe, Germany). HPLC separations were performed on JASCO systems equipped with UV detectors from JASCO Deutschland GmbH (Pfungstadt, Germany) and activity detectors from raytest Isotopenmessgeräte GmbH (GABI Star, Straubenhardt, Germany).

Semi-preparative HPLC conditions were: Column: Reprosil-Pur C18-AQ, 250 × 10 mm; particle size: 10 µm; eluent: 50% MeCN/20 mM NH$_4$OAc$_{aq.}$; flow: 7 mL/min; ambient temperature; UV detection at 254 nm.

Analytical HPLC conditions were: Column: Reprosil-Pur C18-AQ, 250 × 4.6 mm; particle size: 5 µm; gradient: 0–10 min: 26% MeCN, 10–35 min: 26% → 90% MeCN, 35–45 min: 90% MeCN, 45–50 min: 90% → 26% MeCN, 50–60 min: 26% MeCN/20 mM NH$_4$OAc$_{aq.}$; isocratic: 40% MeCN/20 mM NH$_4$OAc$_{aq.}$; flow: 1 mL/min; ambient temperature; UV detection at 254 nm. Molar activity was determined on the base of a calibration curve (0.05–2.0 µg **FESCH**) carried out under isocratic HPLC conditions (36% MeCN/20 mM NH$_4$OAc$_{aq.}$) using chromatograms obtained at 264 nm as the maximum of UV absorbance.

No-carrier-added (n.c.a.) [^{18}F]fluoride ($t_{1/2}$ = 109.8 min) was produced via the [^{18}O(p,n)^{18}F] nuclear reaction by irradiation of [^{18}O]H$_2$O (Hyox 18 enriched water, Rotem Industries Ltd, Arava, Israel) on a Cyclone®18/9 (IBA RadioPharma Solutions, Louvain-la-Neuve, Belgium) with fixed energy proton beam using Nirta® [^{18}F]fluoride XL target.

4.2. Radiosynthesis

The aqueous solution of no-carrier-added [^{18}F]fluoride (~ 3 GBq) was trapped on a Chromafix® 30 PS-HCO$_3^-$ cartridge (MACHEREY-NAGEL GmbH & Co. KG, Düren, Germany). The activity was eluted with 300 µL of an aqueous K$_2$CO$_3$-solution (1.78 mg, 12.9 µmol) into a 4-mL V-vial containing Kryptofix 2.2.2 (K$_{2.2.2}$, 5.6 mg, 14.9 µmol) in 1 mL MeCN. The K$^+$/[^{18}F]F$^-$/K$_{2.2.2}$-carbonate complex was azeotropically dried under vacuum and argon flow within 7–10 min using a Discover PETwave Microwave CEM® (75 W, 50–60 °C, power cycling mode, CEM GmbH, Kamp-Lintfort, Germany). Two aliquots of MeCN (2 × 1 mL) were added during the drying procedure and the final complex was dissolved in 400 µL MeCN ready for radiolabeling.

The aliphatic radiolabeling of ethane-1,2-diol bis(4-methylbenzenesulfonate) (Sigma-Aldrich, Munich, Germany; 2 mg in 100 µL MeCN) was performed under conventional heating at 90 °C for 10 min. After pre-heating of the phenol precursor desmethyl SCH442416 (2.5 mg in 490 µL MeCN) with 10 µL TBAOH$_{aq.}$ (40%) at 90 °C for 10 min, this solution was directly added to the crude 2-[^{18}F]fluoroethyl tosylate and the reaction mixture was stirred for 10 min at 120 °C. Aliquots of the reaction mixtures from both steps were analyzed by radio-TLC (EtOAc/petroleum ether; first step = 1:1; second step = 6:1) or radio-HPLC (isocratic mode, see General Information) to determine the radiochemical yields of crude 2-[^{18}F]fluoroethyl tosylate (R_f = 0.7, 87.0 ± 2.3%; n = 9) and [^{18}F]**FESCH** (R_f = 0.5, 46.4 ± 8.5%; t_R = 29.4 min, 34.4 ± 4.1%, decay corrected for t_R = 0 min; n = 9), respectively.

After dilution with water (1:1), the crude reaction mixture was applied to an isocratic semi-preparative HPLC (see General Information) for isolation of the desired radiotracer [^{18}F]**FESCH** (t_R = 32–33 min). The collected fractions were diluted with water (total volume = 40 mL), passed through a Sep-Pak® C18 Plus cartridge (Waters, Milford, MA, USA; pre-conditioned with 5 mL of absolute EtOH and 60 mL water), and eluted with 1.5 mL of absolute EtOH. Evaporation of the solvent at 70 °C under a gentle argon stream and subsequent formulation of the radiotracer in sterile isotonic saline containing 10% EtOH afforded a [^{18}F]**FESCH** solution usable for biological investigations.

The identity of the radiotracer was proved by analytical radio-HPLC (see General Information) of samples of [^{18}F]**FESCH** spiked with the non-radioactive reference compound **FESCH** using a gradient and an isocratic mode.

4.3. Investigation of In Vitro Stability and Lipophilicity (logD$_{7.4}$)

The in vitro stability of [^{18}F]**FESCH** (4–6 MBq) was studied by incubation in isotonic saline and pig plasma (500 μL each) at 37 °C. Samples were taken at 60 min and analyzed by radio-HPLC (see General Information).

The lipophilicity of [^{18}F]**FESCH** was examined by partitioning between *n*-octanol and phosphate-buffered saline (PBS, pH = 7.4) at ambient temperature using the conventional shake-flask method. The radiotracer (65 μL, ~ 8 MBq) was added to a tube containing the *n*-octanol/PBS-mixture (6 mL, 1:1, four-fold determination). The tubes were shaken for 20 min using a mechanical shaker (HS250 basic, IKA Labortechnik GmbH & Co. KG, Staufen, Germany) followed by centrifugation (5500 rpm for 5 min) and separation of the phases. Aliquots were taken from the organic and the aqueous phase (1 mL each) and activity was measured with an automated gamma counter (1480 WIZARD, Fa. Perkin Elmer, Waltham, MA, USA). The distribution coefficient (D$_{7.4}$) was calculated as [activity (cpm/mL) in *n*-octanol]/[activity (cpm/mL) in PBS, pH = 7.4] stated as the decade logarithm (logD$_{7.4}$).

4.4. Binding Assay

Membrane homogenates prepared from stably transfected CHO-K1 cells (Chinese hamster ovary cells, clone K1) with the human A$_1$R or A$_{2A}$R, provided by Prof. Klotz, University of Würzburg, Würzburg, Germany, were used for the binding experiments. The following radioligands were employed: [^{3}H]**ZM241385** for the A$_{2A}$R and [^{3}H]**DPCPX** for the A$_1$R. For the determination of A$_{2A}$R and A$_1$R binding affinity of reference compound, frozen cell suspensions were thawed, homogenized by a 27-gauge needle, and diluted with 50 mM TRIS-HCl buffer (pH = 7.4, 100 mM NaCl, 5 mM MgCl$_2$, 1 mM EDTA) containing 1 μU/mL adenosine deaminase (ADA). Membrane suspension was incubated with 1.5 nM [^{3}H]**ZM241385** or 1 nM [^{3}H]**DPCPX** and various concentrations of the test compound. Non-specific binding was determined by co-incubation with 10 μM **ZM241385** or 1 μM **DPCPX**. The incubation was performed at room temperature for 90 min and terminated by rapid filtration using Whatman GF/B glass-fiber filters, pre-soaked in 0.3% polyethyleneimine and a 48-channel harvester (Biomedical Research and Development Laboratories, Gaithersburg, MD, USA) followed by washing four times with ice-cold TRIS-HCl buffer. Filter-bound radioactivity was quantified by liquid scintillation counting. At least two separate experiments were performed for determination of K$_i$ values. All data were analyzed with GraphPad Prism, Version 4.1 (GraphPad Inc., La Jolla, CA, USA), according to the Cheng–Prusoff equation.

4.5. Animal Experiments

All experimental work including animals was conducted in accordance with the national legislation on the use of animals for research (Tierschutzgesetz (TierSchG), Tierschutz-Versuchstierverordnung (TierSchVersV)) and was approved by the responsible animal care committees of the Free State of Bavaria and the Free State of Saxony (TVV 08/13, 24-9168.11/18/8, June 12th, 2013 and TVV 18/18, DD24.1-5131/446/19, June 20th, 2018; Landesdirektion Sachsen). Female CD-1 mice, 10–12 weeks, were obtained from the Medizinisch-Experimentelles Zentrum at University of Leipzig, Leipzig, Germany.

4.5.1. In Vitro Autoradiography

Cryosections of brains obtained from female CD-1 mice (10–12 weeks) were thawed, dried in a stream of cold air, and pre-incubated in 50 mM TRIS-HCl buffer (pH = 7.4, 100 mM NaCl, 5 mM MgCl$_2$, 1 mM EDTA) containing 1 μU/mL adenosine deaminase (ADA) for 15 min at ambient temperature. Afterwards, brain sections were incubated with 0.1–0.2 MBq/mL [^{18}F]**FESCH** in buffer for 90 min at room temperature. Non-specific binding was determined in the presence of 1 μM of **FESCH** or **ZM241385**, respectively. Subsequently, the sections were washed twice for 5 min in ice-cold TRIS-HCl buffer, and dipped for 5 s in ice-cold deionized water. The sections were rapidly dried in a stream of cold air before being exposed overnight on an imaging plate (Fujifilm Corporation, Tokyo, Japan). Developed autoradiographs were analyzed in a phosphor imager (HD-CR 35, Duerr NDT GmbH, Bietigheim Bissingen, Germany). The quantification was performed by using 2D-densitometric analysis (AIDA 2.31 software, raytest Isotopenmessgeräte GmbH, Straubenhardt, Germany). Further data analysis was performed with GraphPad Prism, Version 4.1 (GraphPad Inc., La Jolla, CA, USA).

4.5.2. Rotenone-Based and A30P Transgenic Mouse Models of Parkinson's Disease

Animal Housing

Male wild-type C57BL/6JRj mice (Janvier Labs, Le Genest-Saint-Isle, France) and male (Thy1)-h[A30P]α-syn transgenic C57BL/6JRj mice (generated according to the protocol by Kahle et al. [43], hereafter "A30P transgenic mice") were housed at room temperature under a 12:12 h dark:light cycle. Food and water was provided ad libidum.

Oral Rotenone Administration

Wild-type C57BL/6JRj mice (12 months) were divided into two groups and treated 5 days a week for 4 months. A 1.2 mm × 60 mm gavage (Unimed, Lausanne, Switzerland) was used to administer 0.01 mL/g animal weight of rotenone (Sigma-Aldrich, Munich, Germany) solution corresponding to a 5 mg/kg dose. Controls were treated only with the vehicle solution (2% carboxymethyl cellulose (Sigma-Aldrich, Munich, Germany) and 1.25% chloroform (Carl Roth, Karlsruhe, Germany)).

4.5.3. Immunofluorescence Staining and Imaging

Immunostaining

Cryosections of brains obtained from wild-type C57BL/6JRj mice (16 months, vehicle and rotenone-treated) or A30P transgenic mice (young group = 39 weeks; old group = 71–76 weeks) were immunostained using the free-floating technique and all procedures took place in a well of a 12-well plate. Briefly, brain sections were washed with PBS (pH = 7.4) for 10 min, then blocked with blocking solution (PBS, 0.02% Triton-X and 5% donkey serum) for 1 h, and incubated in the first antibody (1:100, rabbit anti-adenosine receptor A_{2A}, ab3461, Abcam, Cambridge, UK) in blocking solution at 4 °C overnight. On the next day, sections were washed in PBS (4 × 15 min) and incubated with the secondary antibody (1:500, donkey Alexa® 555 anti-rabbit, Invitrogen/Thermo Fisher Scientific Inc., Waltham, MA, USA) for three hours at room temperature. Sections were washed in PBS (4 × 15 min), transferred to the slide, and mounted with Mowiol mounting medium (50% PBS, 25% ethylenglycol and 25% glycerol).

Microscopy and Image Analysis

Images were obtained with an Apotome.2 fluorescence microscope (Carl Zeiss Microscopy GmbH, Jena, Germany). Mean fluorescence intensity (MFI) in the striatum was measured using Fiji-ImageJ, a free-software for image analysis (ImageJ v1.52p, National Institutes of Health (NIH), Bethesda, MD, USA). MFI values were normalized to background signal.

4.5.4. Small Animal PET/MR Studies

For the time of the experiments, female CD-1 mice (n = 12; age = 10 weeks; weight = 30–35 g) and male C57BL/6JRj mice (control with n = 5; rotenone-treated with n = 7; age = 16 months; weight = 28–35 g) were kept in a dedicated climatic chamber with free access to water and food under a 12:12 h dark:light cycle at a constant temperature of 24 °C. The animals received an injection of [^{18}F]**FESCH** into the tail vein (5.0 ± 2.5 MBq for CD-1 mice) or retro-orbital (9.7 ± 1.3 MBq for C57BL/6JRj mice) followed by a 60-min PET/MR scan (Mediso nanoScan®, Budapest, Hungary). Each PET image was corrected for random coincidences, dead time, scatter, and attenuation (AC), based on a whole body (WB) MR scan. The reconstruction parameters for the list mode data were the following: 3D-ordered subset expectation maximization (OSEM), 4 iterations, 6 subsets, energy window = 400–600 keV, coincidence mode = 1–5, ring difference = 81. The mice were positioned prone in a special mouse bed (heated up to 37 °C), with the head fixed to a mouth piece for the anesthetic gas supply with isoflurane in 40% air and 60% oxygen (anesthesia unit: U-410, Agnthos, Lidingö, Sweden; gas blender: MCQ, Rome, Italy). The PET data were collected by a continuous WB scan during the entire investigation. Following the 60-min PET scan, a T1-weighted WB gradient echo sequence (GRE, repetition time = 20 ms, echo time = 6.4 ms) was performed for AC and anatomical orientation. Image registration and evaluation of the region of interest (ROI) were done with PMOD (PMOD Technologies LLC, v. 3.9, Zurich, Switzerland).

The respective brain regions were identified using the mouse brain atlas template Ma–Benveniste–Mirrione–FDG. The activity data are expressed as mean standardized uptake value (SUV) of the overall ROI.

4.5.5. In Vivo Metabolism Study

The radiotracer [^{18}F]**FESCH** (~ 17 MBq in 150 μL isotonic saline) was injected into a female CD-1 mouse via the tail vein. Brain and blood samples were obtained at 15 min p.i., plasma separated by centrifugation (14,000 × g, 1 min), and brain homogenized in ~ 1 mL isotonic saline on ice (10 strokes of a polytetrafluoroethylene (Teflon®) plunger at 1000 rpm in a borosilicate glass cylinder; Potter S Homogenizer, B. Braun Melsungen AG, Melsungen, Germany).

Two consecutive extractions were performed as duplicate (plasma) or triplicate (brain) determinations. Plasma (50 μL) and brain samples (250 μL) were added to an ice-cold acetone/water mixture (4:1; plasma or brain sample/organic solvent = 1:4). The samples were vortexed for 1 min, incubated on ice for 10 min (first extraction) or 5 min (second extraction) and centrifuged at 10,000 rpm for 5 min. Supernatants were collected and the precipitates were re-dissolved in 100 μL of the ice-cold acetone/water mixture (4:1) for the second extraction. Activity of aliquots from supernatants of each extraction step and of the precipitates was quantified using an automated gamma counter (1480 WIZARD, Fa. Perkin Elmer). The supernatants from both extractions were combined, concentrated at 70 °C under argon stream, and analyzed by radio-HPLC (gradient mode, see General Information). By this protocol, [^{18}F]**FESCH** was quantitatively extracted from the biological material as proven by in vitro incubation of the radiotracer in pig plasma.

Author Contributions: S.S., T.H.L., and R.-P.M. designed and performed organic syntheses; S.S., T.H.L., R.T., and B.W. designed and performed radiosyntheses; S.S., T.H.L., A.C., S.D.-S., W.D.-C., F.P.-M., and P.B. designed and performed in vitro and in vivo studies; M.T., M.K., Q.S., W.D.-C., and P.B. designed and performed PET/MR studies; S.S., T.H.L., M.T., M.K., A.C., Q.S., S.D.-S., W.D.-C., F.P.-M., and P.B. analyzed the data. All authors read and approved the final manuscript.

Funding: This work (Project No. 100226753) was funded by the European Regional Development Fund (ERDF) and Sächsische Aufbaubank (SAB).

Acknowledgments: We thank Karsten Franke and Steffen Fischer, HZDR, for providing [^{18}F]fluoride.

Conflicts of Interest: The authors declare no conflict of interest.

References

1. Gerlach, M.; Reichmann, H.; Riederer, P. *Die Parkinson-Krankheit: Grundlagen, Klinik, Therapie*; Springer: Wien, Austria; New York, NY, USA, 2007; p. 453.
2. De Lera Ruiz, M.; Lim, Y.-H.; Zheng, J. Adenosine A$_{2A}$ receptor as a drug discovery target. *J. Med. Chem.* **2014**, *57*, 3623–3650. [CrossRef] [PubMed]
3. Jorg, M.; Scammells, P.J.; Capuano, B. The dopamine D$_2$ and adenosine A$_{2A}$ receptors: Past, present and future trends for the treatment of Parkinson's disease. *Curr. Med. Chem.* **2014**, *21*, 3188–3210. [CrossRef] [PubMed]

4. Ferre, S.; von Euler, G.; Johansson, B.; Fredholm, B.B.; Fuxe, K. Stimulation of high-affinity adenosine A$_2$ receptors decreases the affinity of dopamine D$_2$ receptors in rat striatal membranes. *Proc. Natl. Acad. Sci. USA* **1991**, *88*, 7238–7241. [CrossRef] [PubMed]
5. Ferré, S.; Bonaventura, J.; Tomasi, D.; Navarro, G.; Moreno, E.; Cortés, A.; Lluís, C.; Casadó, V.; Volkow, N.D. Allosteric mechanisms within the adenosine A$_{2A}$–dopamine D$_2$ receptor heterotetramer. *Neuropharmacology* **2016**, *104*, 154–160. [CrossRef]
6. Nazario, L.R.; da Silva, R.S.; Bonan, C.D. Targeting adenosine signaling in Parkinson's disease: From pharmacological to non-pharmacological approaches. *Front. Neurosci.* **2017**, *11*, 658. [CrossRef]
7. LeWitt, P.A.; Guttman, M.; Tetrud, J.W.; Tuite, P.J.; Mori, A.; Chaikin, P.; Sussman, N.M.; Group, U.S.S. Adenosine A$_{2A}$ receptor antagonist istradefylline (KW-6002) reduces "off" time in Parkinson's disease: A double-blind, randomized, multicenter clinical trial (6002-US-005). *Ann. Neurol.* **2008**, *63*, 295–302. [CrossRef]
8. Hauser, R.A.; Cantillon, M.; Pourcher, E.; Micheli, F.; Mok, V.; Onofrj, M.; Huyck, S.; Wolski, K. Preladenant in patients with Parkinson's disease and motor fluctuations: A phase 2, double-blind, randomised trial. *Lancet Neurol.* **2011**, *10*, 221–229. [CrossRef]
9. Hauser, R.A.; Olanow, C.W.; Kieburtz, K.D.; Pourcher, E.; Docu-Axelerad, A.; Lew, M.; Kozyolkin, O.; Neale, A.; Resburg, C.; Meya, U.; et al. Tozadenant (SYN115) in patients with Parkinson's disease who have motor fluctuations on levodopa: A phase 2b, double-blind, randomised trial. *Lancet Neurol.* **2014**, *13*, 767–776. [CrossRef]
10. Dungo, R.; Deeks, E.D. Istradefylline: First global approval. *Drugs* **2013**, *73*, 875–882. [CrossRef]
11. Villar-Menéndez, I.; Porta, S.; Buira, S.P.; Pereira-Veiga, T.; Díaz-Sánchez, S.; Albasanz, J.L.; Ferrer, I.; Martín, M.; Barrachina, M. Increased striatal adenosine A$_{2A}$ receptor levels is an early event in Parkinson's disease-related pathology and it is potentially regulated by miR-34b. *Neurobiol. Dis.* **2014**, *69*, 206–214. [CrossRef]
12. Khanapur, S.; Waarde, A.V.; Ishiwata, K.; Leenders, K.L.; Dierckx, R.A.J.O.; Elsinga, P.H. Adenosine A$_{2A}$ receptor antagonists as positron emission tomography (PET) tracers. *Curr. Med. Chem.* **2014**, *21*, 312–328. [CrossRef]
13. Ramlackhansingh, A.F.; Bose, S.K.; Ahmed, I.; Turkheimer, F.E.; Pavese, N.; Brooks, D.J. Adenosine 2A receptor availability in dyskinetic and nondyskinetic patients with Parkinson disease. *Neurology* **2011**, *76*, 1811–1816. [CrossRef]
14. Barret, O.; Hannestad, J.; Alagille, D.; Vala, C.; Tavares, A.; Papin, C.; Morley, T.; Fowles, K.; Lee, H.; Seibyl, J.; et al. Adenosine 2A receptor occupancy by tozadenant and preladenant in rhesus monkeys. *J. Nucl. Med.* **2014**, *55*, 1712–1718. [CrossRef]
15. Barret, O.; Hannestad, J.; Vala, C.; Alagille, D.; Tavares, A.; Laruelle, M.; Jennings, D.; Marek, K.; Russell, D.; Seibyl, J.; et al. Characterization in humans of ^{18}F-MNI-444, a novel PET radiotracer for brain adenosine A$_{2A}$ receptors. *J. Nucl. Med.* **2015**, *56*, 586–591. [CrossRef]
16. Chen, J.-F.; Eltzschig, H.K.; Fredholm, B.B. Adenosine receptors as drug targets—What are the challenges? *Nat. Rev. Drug Discov.* **2013**, *12*, 265–286. [CrossRef] [PubMed]
17. Svenningsson, P.; Hall, H.; Sedvall, G.; Fredholm, B.B. Distribution of adenosine receptors in the postmortem human brain: An extended autoradiographic study. *Synapse* **1997**, *27*, 322–335. [CrossRef]
18. Shinkre, B.A.; Kumar, T.S.; Gao, Z.-G.; Deflorian, F.; Jacobson, K.A.; Trenkle, W.C. Synthesis and evaluation of 1,2,4-triazolo[1,5-c]pyrimidine derivatives as A$_{2A}$ receptor-selective antagonists. *Bioorg. Med. Chem. Lett.* **2010**, *20*, 5690–5694. [CrossRef] [PubMed]
19. Bhattacharjee, A.K.; Lang, L.; Jacobson, O.; Shinkre, B.; Ma, Y.; Niu, G.; Trenkle, W.C.; Jacobson, K.A.; Chen, X.; Kiesewetter, D.O. Striatal adenosine A$_{2A}$ receptor-mediated positron emission tomographic imaging in 6-hydroxydopamine-lesioned rats using [^{18}F]-MRS5425. *Nucl. Med. Biol.* **2011**, *38*, 897–906. [CrossRef]
20. Khanapur, S.; Paul, S.; Shah, A.; Vatakuti, S.; Koole, M.J.B.; Zijlma, R.; Dierckx, R.A.J.O.; Luurtsema, G.; Garg, P.; van Waarde, A.; et al. Development of [^{18}F]-labeled pyrazolo[4,3-e]-1,2,4- triazolo[1,5-c]pyrimidine (SCH442416) analogs for the imaging of cerebral adenosine A$_{2A}$ receptors with positron emission tomography. *J. Med. Chem.* **2014**, *57*, 6765–6780. [CrossRef]
21. Khanapur, S.; van Waarde, A.; Dierckx, R.A.; Elsinga, P.H.; Koole, M.J. Preclinical evaluation and quantification of ^{18}F-fluoroethyl and ^{18}F-fluoropropyl analogs of SCH442416 as radioligands for PET imaging of the adenosine A$_{2A}$ receptor in rat brain. *J. Nucl. Med.* **2017**, *58*, 466–472. [CrossRef]

22. Pan-Montojo, F.; Anichtchik, O.; Dening, Y.; Knels, L.; Pursche, S.; Jung, R.; Jackson, S.; Gille, G.; Spillantini, M.G.; Reichmann, H.; et al. Progression of Parkinson's disease pathology is reproduced by intragastric administration of rotenone in mice. *PLoS ONE* **2010**, *5*, e8762. [CrossRef] [PubMed]
23. Arnhold, M.; Dening, Y.; Chopin, M.; Arévalo, E.; Schwarz, M.; Reichmann, H.; Gille, G.; Funk, R.H.W.; Pan-Montojo, F. Changes in the sympathetic innervation of the gut in rotenone-treated mice as possible early biomarker for Parkinson's disease. *Clin. Auton. Res.* **2016**, *26*, 211–222. [CrossRef]
24. Pan-Montojo, F.; Schwarz, M.; Winkler, C.; Arnhold, M.; O'Sullivan, G.A.; Pal, A.; Said, J.; Marsico, G.; Verbavatz, J.M.; Rodrigo-Angulo, M.; et al. Environmental toxins trigger PD-like progression via increased alpha-synuclein release from enteric neurons in mice. *Sci. Rep.* **2012**, *2*, 898. [CrossRef] [PubMed]
25. Moon, Y.; Lee, K.H.; Park, J.-H.; Geum, D.; Kim, K. Mitochondrial membrane depolarization and the selective death of dopaminergic neurons by rotenone: Protective effect of coenzyme Q10. *J. Neurochem.* **2005**, *93*, 1199–1208. [CrossRef] [PubMed]
26. Betarbet, R.; Sherer, T.B.; MacKenzie, G.; Garcia-Osuna, M.; Panov, A.V.; Greenamyre, J.T. Chronic systemic pesticide exposure reproduces features of Parkinson's disease. *Nat. Neurosci.* **2000**, *3*, 1301–1306. [CrossRef] [PubMed]
27. Baraldi, P.G.; Cacciari, B.; Romagnoli, R.; Spalluto, G.; Monopoli, A.; Ongini, E.; Varani, K.; Borea, P.A. 7-Substituted 5-amino-2-(2-furyl)pyrazolo[4,3-*e*]-1,2,4-triazolo[1,5-*c*]pyrimidines as A_{2A} adenosine receptor antagonists: A study on the importance of modifications at the side chain on the activity and solubility. *J. Med. Chem.* **2002**, *45*, 115–126. [CrossRef]
28. Damont, A.; Hinnen, F.; Kuhnast, B.; Schöllhorn-Peyronneau, M.-A.; James, M.; Luus, C.; Tavitian, B.; Kassiou, M.; Dollé, F. Radiosynthesis of [^{18}F]DPA-714, a selective radioligand for imaging the translocator protein (18 kDa) with PET. *J. Label. Compd. Rad.* **2008**, *51*, 286–292. [CrossRef]
29. Schröder, S.; Lai, T.H.; Kranz, M.; Toussaint, M.; Shang, Q.; Dukic-Stefanovic, S.; Pan-Montojo, F.; Brust, P. Investigation of [^{18}F]FESCH for PET imaging of the adenosine A_{2A} receptor in a rotenone-based mouse model of Parkinson´s disease and development of a two-step one-pot radiolabeling strategy. *J. Label. Compd. Rad.* **2019**, *62*, S183.
30. Tavares, A.A.S.; Lewsey, J.; Dewar, D.; Pimlott, S.L. Radiotracer properties determined by high performance liquid chromatography: A potential tool for brain radiotracer discovery. *Nucl. Med. Biol.* **2012**, *39*, 127–135. [CrossRef]
31. Clark, D.E. *In silico* prediction of blood–brain barrier permeation. *Drug Discov. Today* **2003**, *8*, 927–933. [CrossRef]
32. Waterhouse, R.N. Determination of lipophilicity and its use as a predictor of blood–brain barrier penetration of molecular imaging agents. *Mol. Imaging Biol.* **2003**, *5*, 376–389. [CrossRef] [PubMed]
33. Sihver, W.; Schulze, A.; Wutz, W.; Stusgen, S.; Olsson, R.A.; Bier, D.; Holschbach, M.H. Autoradiographic comparison of *in vitro* binding characteristics of various tritiated adenosine A_{2A} receptor ligands in rat, mouse and pig brain and first *ex vivo* results. *Eur. J. Pharmacol.* **2009**, *616*, 107–114. [CrossRef] [PubMed]
34. Schröder, S.; Wenzel, B.; Deuther-Conrad, W.; Teodoro, R.; Kranz, M.; Scheunemann, M.; Egerland, U.; Höfgen, N.; Briel, D.; Steinbach, J.; et al. Investigation of an ^{18}F-labelled imidazopyridotriazine for molecular imaging of cyclic nucleotide phosphodiesterase 2A. *Molecules* **2018**, *23*, 556. [CrossRef] [PubMed]
35. Liu, J.; Wenzel, B.; Dukic-Stefanovic, S.; Teodoro, R.; Ludwig, F.-A.; Deuther-Conrad, W.; Schröder, S.; Chezal, J.-M.; Moreau, E.; Brust, P.; et al. Development of a new radiofluorinated quinoline analog for PET imaging of phosphodiesterase 5 (PDE5) in brain. *Pharmaceuticals* **2016**, *9*, 22. [CrossRef] [PubMed]
36. Zoghbi, S.S.; Shetty, H.U.; Ichise, M.; Fujita, M.; Imaizumi, M.; Liow, J.-S.; Shah, J.; Musachio, J.L.; Pike, V.W.; Innis, R.B. PET imaging of the dopamine transporter with [^{18}F]FECNT: A polar radiometabolite confounds brain radioligand measurements. *J. Nucl. Med.* **2006**, *47*, 520–527.
37. Evens, N.; Vandeputte, C.; Muccioli, G.G.; Lambert, D.M.; Baekelandt, V.; Verbruggen, A.M.; Debyser, Z.; Van Laere, K.; Bormans, G.M. Synthesis, *in vitro* and *in vivo* evaluation of fluorine-18 labelled FE-GW405833 as a PET tracer for type 2 cannabinoid receptor imaging. *Bioorgan. Med. Chem.* **2011**, *19*, 4499–4505. [CrossRef]
38. Lear, J.L.; Ackermann, R.F. Evaluation of radiolabeled acetate and fluoroacetate as potential tracers of cerebral oxidative metabolism. *Metab. Brain Dis.* **1990**, *5*, 45–56. [CrossRef]
39. Mori, T.; Sun, L.-Q.; Kobayashi, M.; Kiyono, Y.; Okazawa, H.; Furukawa, T.; Kawashima, H.; Welch, M.J.; Fujibayashi, Y. Preparation and evaluation of ethyl[^{18}F]fluoroacetate as a proradiotracer of [^{18}F]fluoroacetate for the measurement of glial metabolism by PET. *Nucl. Med. Biol.* **2009**, *36*, 155–162. [CrossRef]

40. Muir, D.; Berl, S.; Clarke, D.D. Acetate and fluoroacetate as possible markers for glial metabolism *in vivo*. *Brain Res.* **1986**, *380*, 336–340. [CrossRef]
41. Ponde, D.E.; Dence, C.S.; Oyama, N.; Kim, J.; Tai, Y.-C.; Laforest, R.; Siegel, B.A.; Welch, M.J. ^{18}F-Fluoroacetate: A potential acetate analog for prostate tumor imaging—*In vivo* evaluation of ^{18}F-fluoroacetate versus ^{11}C-acetate. *J. Nucl. Med.* **2007**, *48*, 420–428.
42. Peng, Q.; Zhong, S.; Tan, Y.; Zeng, W.; Wang, J.; Cheng, C.; Yang, X.; Wu, Y.; Cao, X.; Xu, Y. The rodent models of dyskinesia and their behavioral assessment. *Front. Neurol.* **2019**, *10*, 1016. [CrossRef] [PubMed]
43. Kahle, P.J.; Neumann, M.; Ozmen, L.; Müller, V.; Jacobsen, H.; Schindzielorz, A.; Okochi, M.; Leimer, U.; van der Putten, H.; Probst, A.; et al. Subcellular localization of wild-type and Parkinson's disease-associated mutant α-synuclein in human and transgenic mouse brain. *J. Neurosci.* **2000**, *20*, 6365–6373. [CrossRef] [PubMed]
44. Wagner, J.; Ryazanov, S.; Leonov, A.; Levin, J.; Shi, S.; Schmidt, F.; Prix, C.; Pan-Montojo, F.; Bertsch, U.; Mitteregger-Kretzschmar, G.; et al. Anle138b: A novel oligomer modulator for disease-modifying therapy of neurodegenerative diseases such as prion and Parkinson's disease. *Acta Neuropathol.* **2013**, *125*, 795–813. [CrossRef] [PubMed]
45. Kuchar, M.; Mamat, C. Methods to increase the metabolic stability of ^{18}F-radiotracers. *Molecules* **2015**, *20*, 16186–16220. [CrossRef]

Sample Availability: Not available.

© 2020 by the authors. Licensee MDPI, Basel, Switzerland. This article is an open access article distributed under the terms and conditions of the Creative Commons Attribution (CC BY) license (http://creativecommons.org/licenses/by/4.0/).

Article

Synthesis and In Vitro Evaluation of 8-Pyridinyl-Substituted Benzo[e]imidazo[2,1-c][1,2,4]triazines as Phosphodiesterase 2A Inhibitors

Rien Ritawidya [1,2,*], Friedrich-Alexander Ludwig [1], Detlef Briel [3], Peter Brust [1] and Matthias Scheunemann [1,*]

1. Department of Neuroradiopharmaceuticals, Institute of Radiopharmaceuticals Cancer Research, Helmholtz-Zentrum Dresden-Rossendorf, 04318 Leipzig, Germany
2. Center for Radioisotope and Radiopharmaceutical Technology, National Nuclear and Energy Agency (BATAN), Puspiptek Area, Serpong, South Tangerang 15310, Indonesia
3. Pharmaceutical/Medicinal Chemistry, Institute of Pharmacy, Faculty of Medicine, Leipzig University, Brüderstraße 34, 04103 Leipzig, Germany
* Correspondence: r.ritawidya@hzdr.de or rienrita@batan.go.id (R.R.); m.scheunemann@hzdr.de (M.S.); Tel.: +49-341-234-179-4611 (R.R.); +49-341-234-179-4618 (M.S.)

Academic Editor: Diego Muñoz-Torrero
Received: 24 May 2019; Accepted: 26 July 2019; Published: 31 July 2019

Abstract: Phosphodiesterase 2A (PDE2A) is highly expressed in distinct areas of the brain, which are known to be related to neuropsychiatric diseases. The development of suitable PDE2A tracers for Positron Emission Tomography (PET) would permit the in vivo imaging of the PDE2A and evaluation of disease-mediated alterations of its expression. A series of novel fluorinated PDE2A inhibitors on the basis of a Benzoimidazotriazine (BIT) scaffold was prepared leading to a prospective inhibitor for further development of a PDE2A PET imaging agent. BIT derivatives (**BIT1–9**) were obtained by a seven-step synthesis route, and their inhibitory potency towards PDE2A and selectivity over other PDEs were evaluated. **BIT1** demonstrated much higher inhibition than other BIT derivatives (82.9% inhibition of PDE2A at 10 nM). **BIT1** displayed an IC_{50} for PDE2A of 3.33 nM with 16-fold selectivity over PDE10A. This finding revealed that a derivative bearing both a 2-fluoro-pyridin-4-yl and 2-chloro-5-methoxy-phenyl unit at the 8- and 1-position, respectively, appeared to be the most potent inhibitor. In vitro studies of **BIT1** using mouse liver microsomes (MLM) disclosed **BIT1** as a suitable ligand for ^{18}F-labeling. Nevertheless, future in vivo metabolism studies are required.

Keywords: Phosphodiesterase 2A (PDE2A); Positron Emission Tomography (PET); Benzoimidazotriazine (BIT); fluorinated; Mouse Liver Microsomes (MLM)

---

1. Introduction

Cyclic nucleotide Phosphodiesterases (PDEs) represent a class of enzymes catalyzing the hydrolysis of the intracellular second messengers, cyclic Adenosine Monophosphate (cAMP) and cyclic Guanosine Monophosphate (cGMP) [1]. cAMP and cGMP are involved in a great variety of cellular functions related to physiological and pathophysiological processes in brain and periphery [2–6].

The 11 family members of PDEs are encoded by 21 genes and classified by their substrate specificity [7–10]. PDEs 4, 7, and 8 are cAMP-specific and PDEs 5, 6, and 9 cGMP-specific, and others (PDEs 1, 2, 3, 10, and 11) hydrolyze both substrates [7,9,11].

The dual substrate enzyme PDE2A is highly expressed in some brain areas, such as nucleus accumbens, cortex, hippocampus [7], striatum, amygdala [12], substantia nigra, and olfactory neurons [13–15], thus being involved in complex neuronal processes like learning, concentration, memory, emotion, and related diseases [8,9]. The inhibition of PDE2A will increase the intracellular levels of cGMP and cAMP in PDE2A-abundant tissues and may result in improvement of neuroplasticity and memory function [8].

The activity of PDE2A inhibitors related to cognitive improvement has been evaluated in animal models [1,7,16,17]. These results suggest the use of PDE2A inhibitors for treatment of neuropsychiatric diseases such as Alzheimer's disease and schizophrenia [18].

Up to now, there have been several PDE2A inhibitors reported. **BAY 60-7550** (Figure 1), the first highly-selective PDE2A inhibitor, has been widely used to evaluate PDE2A activity. However, this compound shows a poor pharmacokinetic profile, as well as poor ability to cross the blood-brain barrier [11,16]. The finding of **BAY 60-7550** triggered many pharmaceutical companies to discover potent PDE2A inhibitors for potential use in treating a variety of brain diseases [11].

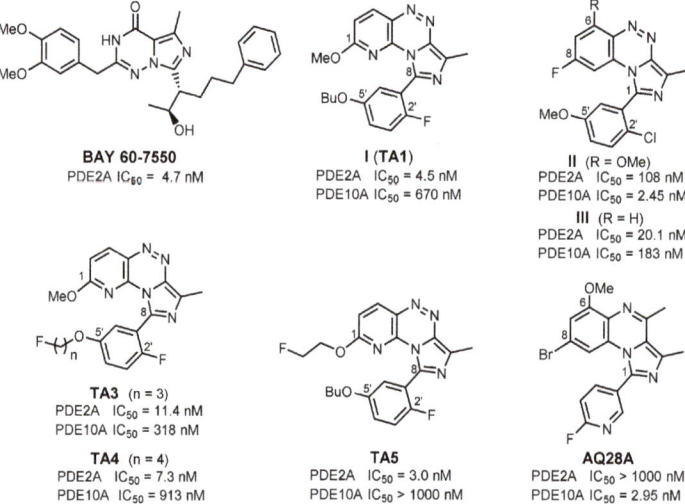

Figure 1. Different PDE inhibitors [15,19–27].

In 2010, Biotie Therapies and Wyeth claimed a series of 1,2,4-triazine- and pyrazine-containing tricyclic compounds exhibiting dual inhibition against both PDE2A and PDE10A as therapeutic targets [25–28]. The triazine series comprises benzo- and pyridine-fused imidazo[5,1-c][1,2,4]triazine derivatives [27], some of which are depicted in Figure 1 (Compounds **I (TA1)**, **II**, and **III**), along with their binding affinities. In 2015, our group reported on the first PDE2A PET tracers on the basis of a Pyridoimidazotriazine (PIT) scaffold starting from **I (TA1**, Figure 1), as the lead compound. Two fluoroalkyl derivatives, **TA3** and **TA4** (Figure 1), demonstrated high affinity towards PDE2A with 28-fold and 125-fold selectivity over PDE10A, respectively [15,21]. More recently, we gained a further improvement in terms of in vitro PDE2A binding via replacement of the 1-methoxy in **TA1** by 1-(2-fluoroethoxy), resulting in the compound **TA5**, (Figure 1) [15,21,22]. However, after successful ^{18}F-labeling, the obtained tracers, [^{18}F]**TA3** and [^{18}F]**TA4**, did not entirely succeed and were proven to be suitable only for in vitro autoradiographic PDE2A imaging. Beyond that, [^{18}F]**TA5** failed to demonstrate specific binding in vitro [22]. The formation of brain-penetrating radiometabolites due to the O-defluoroalkylation limited their application for in vivo PDE2A imaging. Therefore, further structural modifications of tricyclic lead are inevitably required to allow appropriate in vivo PDE2A imaging [21,22]. A further tracer, [^{18}F]**AQ28A**, obtained as a developmental compound from our PDE10 program (**AQ28A**, Figure 1), was proven to be sufficiently metabolically stable and to enable in vivo imaging of PDE10A in rodents by PET [24]. As one feature, the fluorinated aromatic ring of this compound was ascribed to its performance as a promising tracer.

As part of our ongoing commitment in ^{18}F-PET tracer development devoted to PDE imaging (2A, 5A, and 10A) [21,22,24,29], we focused on Benzoimidazotriazine (BIT) as the scaffold, which differs in a benzo ring formally replacing the pyrido ring of our hitherto investigated **TA** compounds. Because of our findings from **AQ28A**, we felt encouraged to combine a 2-fluorpyridine moiety as a fluorine-bearing group with structural features required for PDE2A binding (Figure 2). Herein, we report the synthesis and in vitro evaluation of fluorinated derivatives containing a BIT scaffold and the selection of a promising ligand for the development of a ^{18}F-labeled ligand for PDE2A imaging with PET. The use of a benzo fused imidazo[2,1-c][1,2,4]triazine allows us to readily perform structural modification in the six-membered benzo ring (8-position) while retaining the structural modifiability of the imidazole portion (1-position) of the tricycle (Figure 2). An example of this BIT series, compound **III** (Figures 1 and 2), bearing no substituent at the 6-position, was more potent to PDE2A in comparison to **II** and served as the lead in our studies [26].

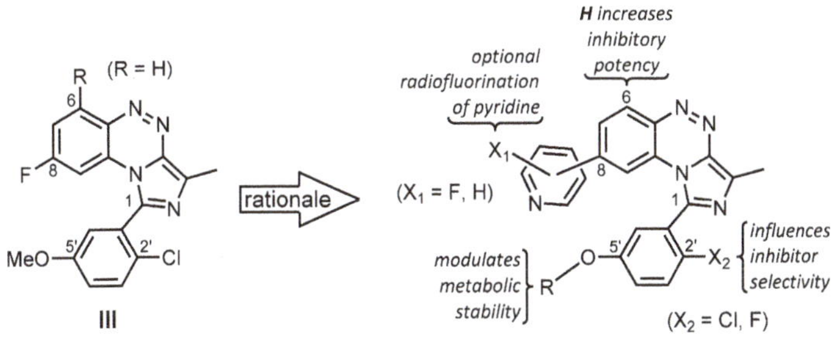

Figure 2. Approach for structural modification of the lead structure **III**.

Therefore, our strategy was to modify this structure by changing the substituents at 1-(phenyl) and the 8-position (pyridinyl instead of F) in order to obtain potent and selective PDE2A inhibitors (Figure 2). These can be established via a standard palladium (Pd)-catalyzed Suzuki coupling reaction [30]. Our main interest was to obtain fluorinated compounds for the development of PDE2A PET tracers. Therefore, the introduction of fluorine-containing groups particularly at the 8-position (pyridinyl) is required. According to our previous work, several fluorinated pyridinyl substituents may enhance the potency and selectivity to PDE2A, and the incorporation of the pyridine moiety is tolerable. Besides, pyridine, as an electron-deficient aromatic ring, offers a convenient way to perform radiofluorination via nucleophilic aromatic substitution [31]. Furthermore, fluorinated pyridinyl substituents are relatively more stable towards metabolic degradation compared with fluoroalkyl groups, which were found to be more prone to defluoroalkylation [15,21,22,24].

2. Results and Discussion

2.1. Chemistry

The first of our synthetic approaches focused on bromo-3-methylbenzo[e]imidazo[5,1-c][1,2,4]triazine (Scheme 1, Compound **5**) as the key intermediate for further conversions into the desired final products (BIT derivatives). The 8-bromo-substituted tricycle **5** was synthesized in four steps starting from 4-bromo-2-fluoro-aniline (**1**) as depicted in Scheme 1.

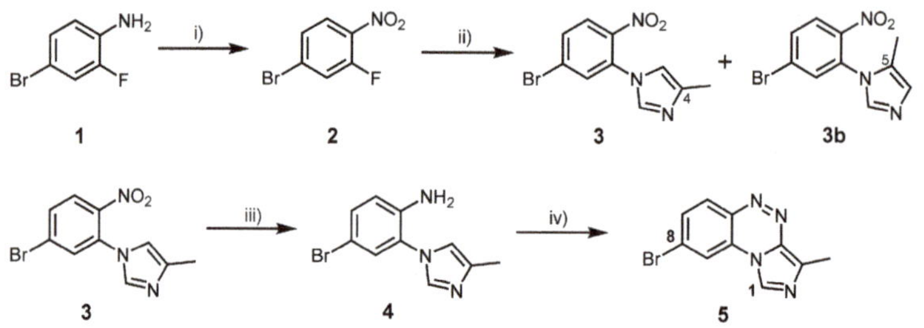

Scheme 1. Synthesis of Compound **5**. (i) Oxidation, NaBO$_3$·4H$_2$O, acetic acid, 65 °C, 2–3 d; (ii) 4-methyl-1H-imidazole, K$_2$CO$_3$, DMF, room temperature (rt), 1–2 d; (iii) Fe (5 eq), HOAc/EtOH, reflux, 2.5 h; (iv) NaNO$_2$, acetic acid, H$_2$O, rt, 1 h.

The oxidation of the aromatic amino group with sodium perborate tetrahydrate (NaBO$_3$·4H$_2$O) in acetic acid afforded nitro compound **2** in a 51% yield. To avoid significant by-product formation, the reaction temperature was kept below 65 °C, and the aniline **1** was slowly added to the oxidizing agent [23]. The obtained 4-bromo-2-fluoro-nitrobenzene (**2**) was reacted with 4-methyl-1H-imidazole to provide the corresponding N-aryl-4- and N-aryl-5-methylimidazoles **3** and **3b**. This nucleophilic aromatic substitution (S$_N$Ar) employed 1.2 eq of 4-Me-imidazole and 2.0 eq of K$_2$CO$_3$ in DMF and formed a mixture of **3** and **3b** in a ratio of ~4:1 [32,33]. However, the main regioisomer **3** could be purified via repeated recrystallization from methanol (50% yield). The reduction of **3** using iron afforded aniline **4** in high yield (92%). The iron powder represents a preferred reducing agent for nitro compounds bearing sensitive functional groups such as halides and/or other reducible groups [34]. A subsequent one-pot diazotation-intramolecular azo coupling of compound **4** by use of aqueous sodium nitrite in acetic acid gave the intermediate **5** in a satisfactory yield of 93%. The overall yield was 22%.

The synthesis of diaryl-substituted compounds (BIT derivatives) starting from **5** is depicted in Scheme 2. Positions 1 and 8 were substituted with different aryl moieties by two independent Suzuki couplings, possible through an intermediate bromination on position 1. The first Suzuki coupling was used to introduce a substituted pyridine ring to the 8-position. Six different Boronic Acids (BAs), including fluoropyridinyl boronic acids (**8a–8d**) and pyridinyl boronic acids (**8e** and **8f**) as well, were coupled in the presence of K$_2$CO$_3$ as a base and Pd(Ph$_3$)$_4$ as a catalyst under standard conditions (Scheme 1, Table 1).

Cross-coupling products **6a–6f** were obtained after flash column chromatography purification in yields from 28%–98%, as depicted in Table 1. Afterwards, the bromination of the imidazo fused ring in the 1-position was carried out using N-bromosuccinimide (NBS) in acetonitrile at room temperature and provided Compounds **7a–7f** in moderate to high yields (Table 1).

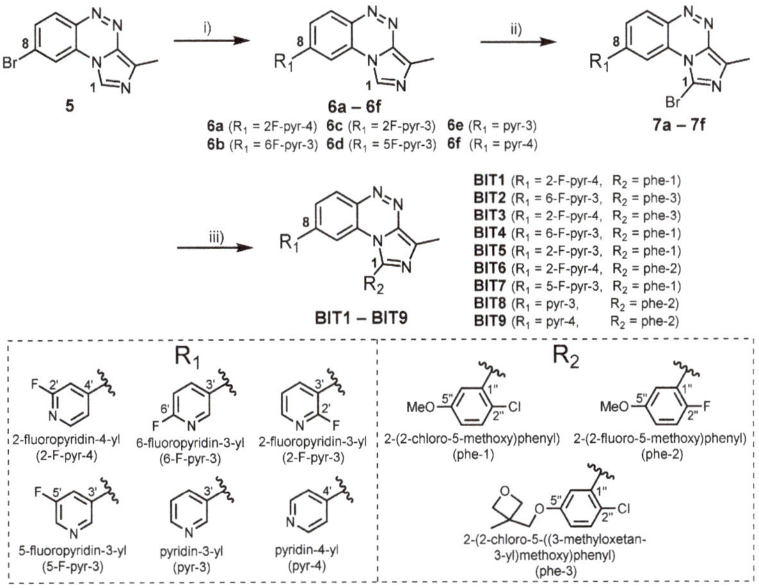

Scheme 2. Approach to achieve a series of nine BIT derivatives. (i) R₁B(OH)₂ (**8a–8f**), K₂CO₃, Pd(PPh₃)₄, 1,4-dioxane:H₂O (4:1), reflux, 1–2 d; (ii) NBS, MeCN, rt 4–6 h; (iii) R₂B(OH)₂ (**9a** [phe-1-B(OH)₂], **9b** [phe-2-B(OH)₂]), or R₂B(pin) (**9c**, [phe-3-B(pin)]), K₂CO₃, Pd(PPh₃)₄, dioxane:H₂O (4:1), reflux, 1–2 d.

Table 1. Yields of Suzuki coupling with compound **5** (products: **6a–6f**) and of bromination (products: **7a–7f**).

Entry	Starting Boronic Acid (BA)	Product of First Suzuki Coupling	Yield	Product of Bromination	Yield
1	2-fluoro-pyridine-4-yl-BA (**8a**)	6a	76%	7a	79%
2	6-fluoro-pyridine-3-yl-BA (**8b**)	6b	28%	7b	63%
3	2-fluoro-pyridine-3-yl-BA (**8c**)	6c	98%	7c	88%
4	3-fluoro-pyridine-5-yl-BA (**8d**)	6d	56%	7d	98%
5	pyridine-3-yl-BA (**8e**)	6e	77%	7e	87%
6	pyridine-4-yl-BA (**8f**)	6f	87%	7f	92%

By a second Suzuki coupling, the five-membered imidazole portion was functionalized (Scheme 2). Initially, two *ortho*-chlorophenyl boronic acid (BA) derivatives **9a** and **9c** were used for coupling with **7a** and **7b**, as well. In addition, to include the 5″-methoxy substituted BA **9a**, we investigated a derivative with a modified propoxy substitution by using **9c** with a (3-methyloxetan-3-yl)methoxy moiety at the C-5″-position (Scheme 2, **9c** [phe-3-B(pin)]). It has been reported that an oxetane residue may modulate and positively influence physicochemical, as well as biological properties of a drug candidate [35,36]. Besides, an oxetane moiety is regarded to induce stability towards metabolic attack similar to a corresponding geminal dimethyl unit (gem-Me₂) [37], but in contrast will not increase the lipophilicity of the compound [35,36]. To make this oxetane derivative available, we prepared boronic ester **9c** in three steps, starting from 3-bromo-phenol (**10**) (Scheme 3).

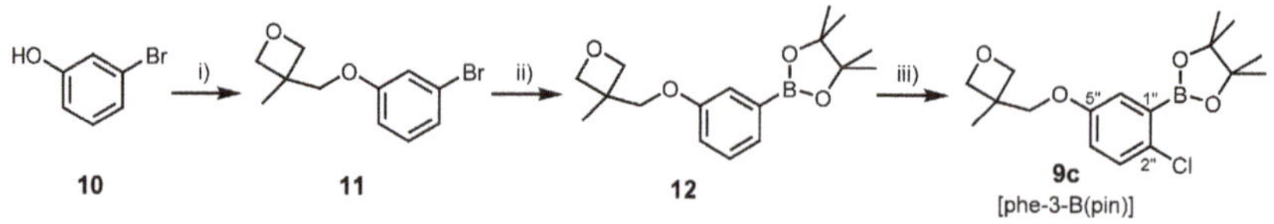

Scheme 3. Preparation of **9c**. (i) 3-methyloxetan-3-yl-methyl-mesylate, Cs_2CO_3, K_2CO_3, MeCN, 75–79 °C, 18 h; (ii) B_2pin_2, $PdCl_2(dppf)$, KOAc, 2-Me-THF, Ar; (iii) NCS, DMF, rt.

First, phenol **10** was reacted with 3-methyloxetan-3-yl-methyl-mesylate to provide aryl ether **11** (85% yield). Afterwards, Miyaura-borylation was performed to react **11** with bis(pinacolato)diboron (B_2pin_2) in the presence of $PdCl_2(dppf)$ and KOAc as a catalyst and base, respectively, providing boronate ester **12** in a good yield of 81%. Finally, the chlorination of **12** ortho to the boronate ester by means of N-chloro-succinimide (NCS) in DMF at room temperature afforded **9c** in a yield of 56% [38].

Boronic acid derivatives **9a** and **9c** were reacted with **7a** and **7b** via cross-coupling under conditions similar to those described for the reactions of BAs **8a–8f** with compound **5**. Reaction products **BIT1**, **BIT2**, and **BIT3** were obtained in satisfactory yields, as depicted in Table 2 (Entries 1–3).

Table 2. Series of BIT derivatives according to Scheme 2.

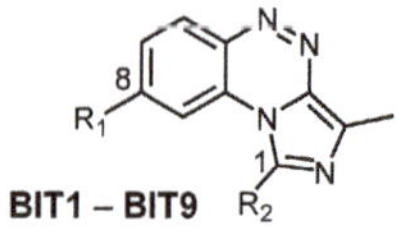

Entry	Bromo Compound	Boronic Acid Derivative	R_1/R_2	Product	Yield
1	7a	9a	2-F-pyr-4/phe-1	BIT1	80%
2	7b	9c	6-F-pyr-3/phe-3	BIT2	79%
3	7a	9c	2-F-pyr-4/phe-3	BIT3	68%
4	7b	9a	6-F-pyr-3/phe-1	BIT4	64%
5	7c	9a	2-F-pyr-3/phe-1	BIT5	69%
6	7a	9b	2-F-pyr-4/phe-2	BIT6	72%
7	7d	9a	5-F-pyr-3/phe-1	BIT7	56%
8	7e	9b	pyr-3/phe-2	BIT8	82%
9	7f	9b	pyr-4/phe-2	BIT9	39%

In vitro evaluation data of the first series of final compounds revealed **BIT1** to have a higher inhibitory potency towards PDE2A in contrast to **BIT2** and **BIT3** (see the In Vitro Evaluation section). Therefore, on the basis of this first finding, we prepared further BIT derivatives and investigated the effect of an alteration of the pyridinyl substitution in 8-position on the PDE2A inhibition. For that purpose, we synthesized **BIT4–BIT9** with the methoxy group in the 5″-position, since this substitution appeared to be the most promising for the inhibition of PDE2A. Of the newly-synthesized compounds, **BIT6**, **BIT8**, and **BIT9** possessed the fluoro substitution in position 2″ of the phenyl residue. **BIT4–BIT9** were obtained in moderate to good yields as shown in Table 2 (Entries 4–9). After characterization by one- and two-dimensional NMR spectroscopy and HR-MS (see Supplementary Materials), the new derivatives were evaluated in vitro (see the In Vitro Evaluation Section).

2.2. In Vitro Evaluation: Structure-Activity Relationships of BIT Derivatives

Final products (BIT derivatives) were evaluated in vitro towards PDEs by means of radioligand binding assays. Inhibition of PDE2A was measured at an inhibitor concentration of **10 nM** and those of other PDEs at a concentration of **1 µM** for each BIT derivative. The inhibitory potencies of all BIT compounds are summarized in Table 3.

Table 3. Percentage inhibition of PDE sub-types from synthesized compounds; for PDE2A3, the compounds were measured at a concentration of 10 nM and for PDE4A1, PDE5A1, PDE9A1, and PDE10A1 at 1 µM.

	Inhibition of PDE Sub-Types (%)				
	2A3	4A1	5A1	9A1	10A1
BIT1	82.9	58.6	29.3	NI	32.4
BIT2	8.52	64.1	NI	NI	2.46
BIT3	13.2	72.7	NI	NI	28.2
BIT4	2.63	NI	NI	NI	10.8
BIT5	NI	NI	NI	NI	22.1
BIT6	56.6	54.2	13.6	28.4	74.0
BIT7	50.6	65.7	25.1	11.5	87.0
BIT8	NI	13.9	36.2	NI	55.5
BIT9	80.5	86.8	57.7	17.5	96.6
TA1	93.5	24.9	3.08	NI	41

NI: No Inhibition

We first focused our attention on modifications at the 1-position while keeping the 2-fluoropyridin-4-yl (2-F-pyr-4) fixed at the 8-position as for compounds **BIT1** and **BIT3**. According to our previous work [21], a longer alkyl side chain length (3-fluoropropoxy to 4-fluorobutoxy, as **TA3–TA4**) contributes to improved potency, as well as selectivity towards PDE2A, as was also shown for **TA1** (Figure 1) [15,21]. However, the substitution of the 5″-methoxy group (**BIT1**) by a (3-methyloxetan-3-yl)methoxy moiety (**BIT3**) in the 1-phenyl moiety resulted in a strong reduction of inhibitory potency towards PDE2A3 from 82.9% to 13.2% (at 10 nM). One possible explanation for that result might be the increased polarity of the chosen oxetanyl-methoxy residue compared with alkoxy groups, which possibly is unfavorable for key interactions towards PDE2A. It was reported that the phenylpropyl portion in the imidazole ring of **BAY 60-7550** interacts with Leu770, one of the amino acids that is known to be responsible for hydrophobic pocket (H-pocket) formation in the PDE2A active site [39]. Moreover, **BAY 60-7550** binding to the PDE2A active site suggested that the selectivity for PDE2A was generated from the ability of the ligand to induce a conformational change of Leu770 [1,39]. Therefore, it is hypothesized that **BIT3** may not induce the conformational change of the Leu770 amino acid that generates the selectivity pocket formation, thus leading to potency losses towards PDE2A [1,39].

In the case of a pair of 6-F-pyr-3-substituted products, **BIT2** and **BIT4**, a weak inverse effect of PDE2A binding was observed [40], depending on the nature of the 5″ substituent and in comparison to **BIT3**/**BIT1**. Strikingly, the much lower inhibition displayed by **BIT4** (2.63%) in contrast to **BIT1** (82.9%), bearing the same phe-1 substituent, points to the impact of the pyridine substitution pattern in exerting inhibition on PDE2A.

In this connection, another factor that contributes to the selectivity of **BAY 60-7550** is a glutamine-switch mechanism [39], which is generally proposed for dual substrate-specific PDEs, such as PDE2A [1,39]. We assume that triazine N-5 in our scaffold might accept a hydrogen bond from Gln859 in a similar mode as described for the pyrazine portion in bioisosteric triazoloquinoxaline-based PDE2A inhibitors [16,20]. We further investigated the influence of different pyridinyl substituents at the 8-position on PDE2A inhibition while keeping the phe-1 fixed at the 1-position. In addition to **BIT4**, containing a 6-F-pyr-3, the PDE2A inhibition also dropped significantly with **BIT5**, containing a 2-F-pyr-3 at the 8-position. Interestingly, a potency of at least 50% remained with the introduction of the 5-F-pyr-3 residue (**BIT7**) at the 8-position. It is not unambiguously clear whether this was caused only by the position of the pyridine nitrogen, which may affect rotational freedom of the ring *via* active site H-bonding to the pyridine N [16], or in the case of **BIT1** and **BIT7**, additionally by X-H⋯F-C interactions with fluorine [41,42]. Considering this, we found an influence of substituent variation at the 8-position while keeping phe-1 at the 1-position in decreasing order of activity towards PDE2A as 2-F-pyr-4 > 5-F-pyr-3 > 6-F-pyr-3 > 2-F-pyr-3. It was supposed that 2-F-pyr-4 and 5-F-pyr-3 likely maintained the conformational locking by the H-bond, resulting in higher PDE2A potency of **BIT1** and **BIT7**. Moreover, compared to **BIT7**, both **BIT4** and **BIT5** demonstrated eight- and four-fold selectivity over PDE10A, respectively.

We then directed our attention to investigate the influence of ortho-fluorophenyl (phe-2) at the 1-position. In the case of **BIT6**, having a 2-F-pyr-4 at the 8-position, a weakly lower potency towards PDE2A of 56% was observed, when compared to **BIT1**, which is in accordance with the known positive PDE2A inhibitory effect of 2″-Cl in comparison to 2″-F [20,28]. A non-fluorinated pyridine (pyr-4 in **BIT9**) also maintained the inhibitory activity towards PDE2A of 80.5%, close to **BIT1** (82.9%), however at the expense of increasing inhibition towards PDE10A and PDE4A (96.6% and 86.8% at a 1 µM inhibitor concentration) in contrast to **BIT1** (32.4% and 58.6% at 1 µM). Incorporation of pyr-3 resulted in a significant loss of PDE2A inhibition, which is consistent with the result of **BIT2**, **BIT4**, and **BIT5**, having also a substituted pyridin-3-yl at the 8-position, but not with **BIT7**, which may be due to reasons discussed above. Again, the strong effect of the N-atom position in the pyridine ring on PDE2A inhibition may be explained by differing conformational preferences between pyr-4 and pyr-3 [11]. It was assumed that pyr-3 allows more rotational freedom, resulting in energy loss of binding [16]. The selectivity towards certain PDEs was decreased when exchanging phe-1 with phe-2. Thus, in addition to lower potency towards PDE2A, **BIT6** displayed higher inhibition of PDE10A when compared to **BIT1** (74% vs. 32%, at 1 µM). Therefore, these results suggest that the ortho-chlorophenyl (phe-1) at 1-position was useful for PDE2A potency and selectivity.

Three compounds, **BIT1**, **BIT6**, and **BIT9**, were selected for estimation of IC_{50} values of PDE2A and PDE10A inhibition, to determine the potency in more detail and also the PDE10A/PDE2A selectivity ratio. The related IC_{50} values are shown in Table 4.

Table 4. Affinity and selectivity of three new fluorinated compounds towards PDE2A and PDE10A.

Compounds	IC_{50} PDE2A (nM)	IC_{50} PDE10A (nM)	Selectivity Ratio
BIT1	3.33	53.23	16
BIT6	65.06	168.4	2.5
BIT9	17.7	20.24	1.1

Finally, the potencies of **BIT1** towards a range of other PDEs were evaluated as depicted in Table 5. **BIT1** showed a good selectivity over other PDEs (12.6%–23.3% for other PDEs, measured at 1 µM). Summarizing, **BIT1** had a good selectivity over PDE10A (16-fold), while **BIT6** and **BIT9** had 2.5-fold and 1.1-fold selectivity over PDE10A, respectively. Applying similar assay conditions, **TA1** and **BAY 60-7550** displayed an IC_{50} value towards PDE2A of 10.8 nM and 0.66 nM, respectively. Thus, **BIT1**, possessing phe-1 and 2-F-pyr-4 at the 1- and 8-positions, proved to be a prospective ligand for labeling with ^{18}F for PDE2A PET imaging.

Table 5. Percentage inhibition of PDE sub-types by **BIT1** (measured at 1 µM).

PDE Subtypes	%Inhibition
PDE1A3	12.6
PDE3A	20
PDE6AB	16.8
PDE7A	16.4
PDE8A1	23.3
PDE11A	14.5

2.3. Incubations with Mouse Liver Microsomes

To gain first information about whether **BIT1** is prone to metabolic degradation, the compound was investigated in an in vitro test system. Briefly, **BIT1** was incubated with Mouse Liver Microsomes (MLM), in the presence of NADPH, at 37 °C for 90 min [43]. After protein precipitation by addition of acetonitrile and subsequent centrifugation, the supernatant was examined by HPLC-UV-MS.

After incubation with MLM for 90 min, unchanged **BIT1** could still be detected in high amounts (Figure 3). The main fractions of in vitro metabolites were products of a mono-oxygenation of **BIT1**, namely Metabolites **M3**, **M5**, and **M6** (see Table 6).

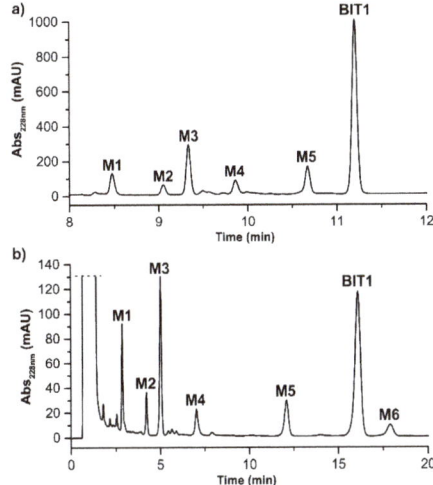

Figure 3. HPLC-UV chromatogram after incubation of **BIT1** with MLM in the presence of NADPH (for conditions, see the experimental part): (**a**) UV chromatogram (partial) recorded during gradient elution (Method A); (**b**) UV chromatogram recorded during isocratic elution (Method B).

Table 6. Overview of HPLC and MS data of in vitro metabolites of **BIT1** formed by MLM in the presence of NADPH.

Metabolite	t_R (min) Method A Gradient	t_R (min) Method B Isocratic	m/z Found	m/z Theoret.	Identification
BIT1	11.28	16.10	420.7	420.1	parent $(M + H)^+$
M1	8.47	2.85	422.7	422.1	reduction $(M + 2H + H)^+$
M2	9.05	4.24	452.8	452.1	di-oxygenation $(M + 2O + H)^+$
M3	9.33	5.02	436.7	436.1	mono-oxygenation $(M + O + H)^+$
M4	9.86	7.03	406.6	406.1	demethylation $(M-CH_3 + H)^+$
M5	10.67	12.04	436.7	436.1	mono-oxygenation $(M + O + H)^+$
M6	not detected	17.89	436.7	436.1	mono-oxygenation $(M + O + H)^+$

The metabolite **M6** was detectable only by an HPLC method with isocratic elution as shown in Figure 3b. Furthermore, one metabolite resulting from a two-fold oxygenation (**M2**) was found. Due to the limitations of a single quadrupole MS, it was not possible to obtain information regarding the sites of functionalization in the molecule, neither to distinguish products from N-oxidation or C-hydroxylation [44]. In contrast, for the formation of **M1** and **M4** by reduction and demethylation, respectively, the chloro(methoxy)-phenyl moiety is expected to undergo these reactions under the conditions applied.

The considerable metabolic stability of **BIT1** in vitro supports the potential suitability of the corresponding ^{18}F-labelled compounds as a radioligand for PET. However, due to the retention properties of the metabolite **M6** in HPLC this metabolite may have a similar or higher lipophilicity in comparison to **BIT1**. Therefore, of the metabolites detected, **M6** most likely bears the risk of passing the blood-brain barrier and influencing brain PET results. Hence, future in vivo metabolism studies should pay attention to the occurrence of this metabolite, as well as to distinguish it chromatographically from the parent compound.

3. Materials and Methods

3.1. General Information

Chemicals were purchased from following suppliers: Manchester Organics, abcr, VWR, Fluka, Acros, Roth, ChemPure, Merck, Sigma Aldrich, Apollo scientific, and Fluorochem. Solvents were dried before use, if required. Air- and moisture-sensitive reactions were carried out under argon atmosphere. Room temperature (rt) refers to 20–25 °C. The progress of a reaction was monitored by thin layer chromatography using pre-coated TLC sheets POLYGRAM® SIL G/UV254 purchased from Macherey-Nagel. Detected spots were observed under UV light at λ 254 nm and 365 nm. Flash chromatography was performed with silica gel 40–63 μm from VWR Chemicals.

NMR spectra (^{1}H, ^{13}C, ^{19}F) were recorded on Mercury 300/Mercury 400 (Varian, Palo Alto, CA, USA) or Fourier 300/Avance DRX 400 Bruker (Billerica, MA, USA) instruments. Signals of solvents were used as internal standards for ^{1}H-NMR (CDCl$_3$, δ_H = 7.26; DMSO-d_6, δ_H = 2.50) and ^{13}C-NMR (CDCl$_3$, δ_C = 77.16; DMSO-d_6, δ_C = 39.52). The chemical shifts (δ) are reported in ppm as follows: s, singlet; d, doublet; t, triplet; m, multiplet; and the coupling constants (J) are reported in Hz. Mass spectra were recorded on an ESQUIRE 3000 Plus (ESI, low resolution) and a 7 T APEX II (ESI, high resolution) from Bruker Daltonics.

3.2. Syntheses

4-Bromo-2-fluoro-nitrobenzene (**2**): A suspension of sodium perborate tetrahydrate (20.24 g, 0.13 mol) in glacial acetic acid (80 mL) was stirred at 65 °C. 4-Bromo-2-fluoro-aniline (5 g, 26.31 mmol, 1 eq) in 35 mL acetic acid was dropwise added over 5 h. The reaction was heated overnight, and subsequently, another portion of NaBO$_3$·4H$_2$O (12.2 g, 78.9 mmol) was added. After full consumption of the starting material, the reaction mixture was cooled to room temperature, the solid filtered off, and the filtrate quenched with ice-cold water (600 mL). Then, the precipitate was filtrated and purified with column chromatography (hexane/chloroform, 2:1) to give the product as yellow solid 2 (2.96 g, 51%). TLC (hexane/CHCl$_3$ (2:1)): R$_f$ = 0.32. ^{1}H-NMR (400 MHz, CDCl$_3$) δ_H = 8.01–7.93 (m, 1H), 7.50 (dd, J = 10.1, 1.9 Hz, 1H), 7.48–7.44 (m, 1H). ^{13}C-NMR (101 MHz, CDCl$_3$) δ_C = 155.48 (d, J = 269.9 Hz), 136.60 (d, J = 7.2 Hz), 129.56 (d, J = 8.9 Hz), 128.24 (d, J = 4.3 Hz), 127.28 (d, J = 2.5 Hz), 122.21 (d, J = 23.6 Hz). LR-MS (EI): m/z = 219 (calcd. 219 for C$_6$H$_3$^{79}BrFNO$_2$$^+$ [M]$^+$)

4-Bromo-2-(4-methyl-1H-imidazol-1-yl) nitrobenzene (**3**): A mixture of compound **2** (9.46 g, 43 mmol) and K$_2$CO$_3$ (11.9 g, 86 mmol) in DMF (40 mL) was stirred at 4 °C, while a solution of 4-methyl imidazole (3.78 g, 46 mmol) in DMF (10 mL) was slowly added in the course of 2 h. Afterwards, stirring was continued for 10 h at room temperature. The reaction mixture was poured into water (250 mL), and the formed precipitate was filtered off, washed, and dried to give a brown yellow solid (10.47 g), which was found to be an impure mixture of regioisomeric Products **3** and **3b** in a ratio of ~4:1, according to ^{1}H-NMR. The solid was dissolved in CHCl$_3$ (80 mL) and filtered through a plug of silica gel (2 g). The solvent was evaporated, and the remaining solid was recrystallized twice from aqueous EtOH

(80%) to give pure 3, as bright yellow crystals (7.09 g, 58%). Mp. 151–152.5 °C; TLC (CHCl$_3$/MeCN (10:1)]: R$_f$ = 0.33. ^{1}H-NMR (400 MHz, CDCl$_3$) δ$_H$ = 7.85 (d, J = 8.7 Hz, 1H), 7.70 (dd, J = 8.7, 2.0 Hz, 1H), 7.61 (d, J = 2.0 Hz, 1H), 7.52 (d, J = 0.9 Hz, 1H), 6.75 (s, 1H), 2.27 (d, J = 0.9 Hz, 3H). ^{13}C-NMR (101 MHz, CDCl$_3$) δ$_C$ = 143.91, 140.24, 136.31, 132.37, 131.95, 131.47, 127.95, 126.68, 116.35, 13.71. LR-MS (ESI+): m/z = 282, (calcd. 282 for C$_{10}$H$_9$^{79}BrN$_3$O$_2$$^+$ [M + H]$^+$).

The minor regioisomer was isolated from a sample (0.36 g) of the filtrate obtained from the first recrystallization. It was purified via flash purification applying a gradient from CHCl$_3$ (100%) to CHCl$_3$/MeCN (30:1) to give pure 3b, as a yellow solid (0.11 g). Mp. 107.5–109 °C; TLC [CHCl$_3$/MeCN (10:1)]: R$_f$ = 0.26. ^{1}H-NMR (400 MHz, CDCl$_3$) δ$_H$ = 7.97 (d, J = 8.7 Hz, 1H), 7.81 (dd, J = 8.7, 2.1 Hz, 1H), 7.59 (d, J = 2.1 Hz, 1H), 7.46 (s, 1H), 6.91 (s, 1H), 2.04 (s, 3H). ^{13}C-NMR (101 MHz, CDCl$_3$) δ$_C$ = 145.38, 136.78, 133.60, 133.54, 131.16, 128.99, 128.22, 127.82, 126.78, 9.24.

4-Bromo-2-(4-methyl-1 H-imidazol-1-yl) aniline (**4**): Nitro compound 3 (4.42 g, 15.7 mmol, 1 eq) was dissolved under argon in a 1:1 mixture of ethanol (50 mL) and acetic acid (50 mL). Iron powder (5.25 g, 94 mmol) was added, and the reaction mixture was stirred under reflux for 2.5 h (bath temperature 115–120 °C). The mixture was filtered through celite, and the filtrate was adjusted to pH 8 with Na$_2$CO$_3$ until a greenish solid formed. The mixture was extracted with Ethyl Acetate (EE, 3 × 100 mL), and the organic phase was washed with water (2 × 100 mL) and saturated NaCl solution (100 mL), and dried (Na$_2$SO$_4$). The solvent was evaporated to give a beige solid (3.64 g, 92%). TLC (CHCl$_3$/MeOH/30% NH$_3$ (10:1:0.1)): R$_f$ = 0.41. ^{1}H-NMR (400 MHz, CDCl$_3$) δ$_H$ = 7.52 (d, J = 1.0 Hz, 1H), 7.30 (dd, J = 8.6, 2.3 Hz, 1H), 7.23 (d, J = 2.2 Hz, 1H), 6.80 (s, 1H), 6.71 (t, J = 8.0 Hz, 1H), 3.68 (d, J = 48.0 Hz, 2H), 2.30 (d, J = 0.6 Hz, 3H).^{13}C-NMR (101 MHz, CDCl$_3$) δ$_C$ = 141.22, 139.48, 136.72, 132.47, 129.50, 124.52, 117.94, 116.32, 109.38, 13.77. LR-MS (ESI+): m/z = 274, (calcd. 274 for C$_{10}$H$_{10}$^{79}BrN$_3$Na$^+$ [M + Na]$^+$).

8-Bromo-3-methylbenzo[e]imidazo[5,1-c][1,2,4]triazine (**5**): A solution of NaNO$_2$ (1.28 g, 18.6 mmol) in H$_2$O (10 mL) was added to a stirred solution of 4 (3.13 g, 12.4 mmol) in acetic acid (60 mL) at room temperature. A yellow precipitate formed immediately. After 5 h of stirring, the reaction mixture was evaporated to half of its original volume and diluted with H$_2$O (50 mL). The pH was carefully adjusted to pH 8 by the addition of solid NaHCO$_3$. The yellow solid was filtered off and dried to give 5 (3.24 g, 98% sufficiently pure for the next step). The solid was dissolved in CHCl$_3$ (150 mL) and filtered through a plug of silica gel (2 g). The filtrate was concentrated and triturated with CHCl$_3$/heptane mixtures to afford a yellow powder (2.9 g, 89%). TLC (hexane/ethyl acetate (1:3)): R$_f$ = 0.45. ^{1}H-NMR (400 MHz, DMSO-d$_6$) δ$_H$ = 9.16 (s, 1H), 8.70 (d, J = 2.0 Hz, 1H), 8.29 (d, J = 8.6 Hz, 1H), 7.88 (dd, J = 8.6, 2.0 Hz, 1H), 2.74 (s, 3H). ^{13}C-NMR (101 MHz, DMSO-d$_6$) δ$_C$ = 137.71, 135.87, 134.37, 131.40, 130.55, 127.23, 126.01, 123.04, 118.35, 12.39. LR-MS (ESI+): m/z = 263 (calcd. 263 for C$_{10}$H$_8$^{79}BrN$_4$$^+$ [M + H]$^+$).

3.2.1. General Procedure A for the Suzuki Coupling of Bromo Derivatives **5, 7a, 7b, 7c, 7d, 7e**, and **7f**

Brominated compound (1 eq), boronic acid derivative (BA, **8a–8f, 9a–9c**, 1.1–1.3 eq), and K$_2$CO$_3$ (2–3 eq) were combined in a mixture of 1,4-dioxane and water (4:1). The suspension was degassed with argon for 5–10 min, and Pd(PPh$_3$)$_4$ (5–10 mol%) was added. The mixture was refluxed (bath temperature 102–108 °C) for 1–2 days until completion of the reaction (as indicated by TLC). The solvent was removed, and the residue was partitioned between CHCl$_3$ and water. The aqueous layer was extracted twice with CHCl$_3$, and the combined organic layers were dried (Na$_2$SO$_4$) and the solvent evaporated. Unless stated otherwise, the obtained residue was purified by flash chromatography on silica gel using CHCl$_3$/MeOH (10:1) as the eluent. The following products were isolated:

8-(2-Fluoropyridin-4-yl)-3-methylbenzo[e]imidazo[5,1-c][1,2,4]triazine (**6a**): Based on General Procedure A of Suzuki coupling, a mixture of compound 5 (101 mg, 0.38 mmol,1 eq), BA 8a (65 mg,0.46 mmol, 1.2 eq), and K$_2$CO$_3$ (131 mg, 0.95 mmol, 2.5 eq), in dioxane/water (5 mL) were reacted in the presence of Pd(PPh$_3$)$_4$ (44 mg, 10 mol%) to give after purification compound 6a (81 mg, 76%) as a yellow solid. TLC (hexane/ethyl acetate (1:1)): R$_f$ = 0.25. ^{1}H-NMR (400 MHz, DMSO-d$_6$) δ$_H$ = 9.28 (s, 1H), 8.92

(d, J = 1.9 Hz, 1H), 8.52 (d, J = 8.5 Hz, 1H), 8.46 (d, J = 5.3 Hz, 1H), 8.25 (dd, J = 8.5, 1.9 Hz, 1H), 7.97 (dt, J = 5.3, 1.8 Hz, 1H), 7.82 (s, 1H), 2.79 (s, 3H). LR-MS (ESI+): m/z = 280, (calcd. 280 for $C_{15}H_{11}FN_5^+$ [M + H]$^+$).

8-(6-Fluoropyridin-3-yl)-3-methylbenzo[e]imidazo[5,1-c][1,2,4]triazine (**6b**): According to General Procedure A, compound **5** (500 mg, 1.9 mmol, 1 eq), BA **8b** (321 mg, 2.28 mmol, 1.2 eq), K_2CO_3 (788 mg, 5.7 mmol, 3 eq), and Pd(PPh$_3$)$_4$ (220 mg, 10 mol%) were reacted in dioxane/water (24 mL) to yield after purification compound **6b** (150 mg, 28%), as a yellow powder. TLC (CH$_3$Cl$_2$/MeOH (10:1)): R$_f$ = 0.38. ^{1}H-NMR (400 MHz, CDCl$_3$) δ_H = 8.62–8.52 (m, 3H), 8.12 (ddd, J = 8.4, 7.4, 2.7 Hz, 1H), 7.98 (d, J = 1.8 Hz, 1H), 7.82 (dd, J = 8.4, 1.8 Hz, 1H), 7.13 (dd, J = 8.5, 3.1 Hz, 1H), 2.91 (s, 3H). LR-MS (ESI+): m/z = 280, (calcd. 280 for $C_{15}H_{11}FN_5^+$ [M + H]$^+$).

8-(2-Fluoropyridin-3-yl)-3-methylbenzo[e]imidazo[5,1-c][1,2,4]triazine (**6c**): According to General Procedure A, compound **5** (501 mg, 1.9 mmol, 1 eq), BA **8c** (325 mg, 2.28 mmol, 1.2 eq), K_2CO_3 (661 mg, 4.75 mmol, 2.5 eq), and Pd(PPh$_3$)$_4$ (219 mg, 10 mol%) were reacted in dioxane/water (15 mL) to afford after purification compound **6c** (0.52 g, 98%), as a yellow solid. TLC (CHCl$_3$/MeOH (10:1)): R$_f$ = 0.38. ^{1}H-NMR (400 MHz, CDCl$_3$) δ_H = 8.58–8.52 (m, 1H), 8.34 (dt, J = 5.0, 1.5 Hz, 1H), 8.10–7.99 (m, 1H), 7.85 (dd, J = 8.4, 1.6 Hz, 1H), 7.71–7.62 (m, 1H), 7.58–7.37 (m, 2H), 2.92 (s, 3H). LR-MS (ESI+): m/z = 280 (calcd. 280 for $C_{15}H_{11}FN_5^+$ [M + H]$^+$).

8-(5-Fuoropyridin-3-yl)-3-methylbenzo[e]imidazo[5,1-c][1,2,4]triazine (**6d**): According to General Procedure A, compound **5** (200 mg, 0.76 mmol, 1 eq), BA **8d** (118 mg, 0.84 mmol, 1.1 eq), K_2CO_3 (315 mg, 2.28 mmol, 3.0 eq), and Pd(PPh$_3$)$_4$ (44 mg, 5 mol%) were reacted in dioxane/water (10 mL) to give after purification Compound **6d** (0.12 g, 56%), as a yellow powder. TLC (CHCl$_3$/MeOH (10:1)): R$_f$ = 0.29. ^{1}H-NMR (400 MHz, DMSO-d_6) δ_H = 9.26 (s, 1H), 9.09 (d, J = 1.9 Hz, 1H), 8.88 (d, J = 1.8 Hz, 1H), 8.71 (d, J = 2.7 Hz, 1H), 8.49 (d, J = 8.5 Hz, 1H), 8.36 (dt, J = 10.3, 2.3 Hz, 1H), 8.19 (dd, J = 8.5, 1.9 Hz, 1H), 2.77 (s, 3H).

3-Methyl-8-(pyridin-3-yl)benzo[e]imidazo[5,1-c][1,2,4]triazine (**6e**): According to General Procedure A, compound **5** (1.0 g, 3.8 mmol, 1 eq), BA **8e** (0.61 mg, 4.94 mmol, 1.3 eq), K_2CO_3 (1.58 g, 11.4 mmol, 3 eq), and Pd(PPh$_3$)$_4$ (220 mg, 5 mol%) were reacted in dioxane/water (50 mL) to give after purification by flash chromatography compound **6e** (0.76 g, 77%), as a yellow solid. TLC (CHCl$_3$/MeOH (10:1)): R$_f$ = 0.20. ^{1}H-NMR (400 MHz, DMSO-d_6) δ_H = 9.23 (s, 1H), 9.16 (d, J = 2.1 Hz, 1H), 8.77 (d, J = 1.8 Hz, 1H), 8.69 (dd, J = 4.8, 1.6 Hz, 1H), 8.45 (d, J = 8.4 Hz, 1H), 8.35–8.30 (m, 1H), 8.11 (dd, J = 8.5, 1.8 Hz, 1H), 7.60 (ddd, J = 8.0, 4.8, 0.9 Hz, 1H), 2.75 (s, 3H). LR-MS (ESI+): m/z = 262, (calcd. 262 for $C_{15}H_{12}N_5^+$ [M+H]$^+$).

3-Methyl-8-(pyridin-4yl)benzo[e]imidazo[5,1-c][1,2,4]triazine (**6f**): According to General Procedure A, compound **5** (200 mg, 0.76 mmol, 1 eq), K_2CO_3 (265 mg, 1.92 mmol, 2.52 eq), BA **8f** (119 mg, 0.968 mmol, 1.27 eq), and Pd(PPh$_3$)$_4$ (44 mg, 0.038 mmol, 5 mol%) were reacted in dioxane/water (5 mL) to afford after purification compound **6f** (173 mg, 87%), as a yellow solid. TLC (CHCl$_3$/MeOH (10:1)): R$_f$ = 0.23. ^{1}H-NMR (300 MHz, DMSO-d_6) δ_H = 9.26 (s, 1H), 8.82 (d, J = 1.8 Hz, 1H), 8.78–8.75 (m, 2H), 8.47 (d, J = 8.5 Hz, 1H), 8.15 (dd, J = 8.5, 1.9 Hz, 1H), 7.96–7.92 (m, 2H), 2.75 (s, 3H).

3.2.2. General Procedure B for Bromination of Compounds **6a, 6b, 6c, 6d, 6e**, and **6f**

N-Bromosuccinimide (NBS, ~1.5 eq) was added to a suspension of compounds **6a–6f** in acetonitrile. The reaction mixture was protected from light and stirred at room temperature for 1–2 d. After the full conversion of the starting material, the mixture was partitioned between CHCl$_3$ and H$_2$O. The organic layer was washed with aqueous NaHCO$_3$ (5%), water, and saturated NaCl solution. After drying (Na$_2$SO$_4$) and evaporation, the residue was purified by flash chromatography, eluting with CHCl$_3$/MeOH (10:1).

1-Bromo-8-(2-fluoropyridin-4-yl)methylbenzo[e]imidazo[5,1-c][1,2,4]triazine (**7a**): According to General Procedure B, for bromination, compound **6a** (300 mg, 1.07 mmol) and NBS (287 mg, 1.61 mmol) were reacted in acetonitrile (20 mL) to give after purification compound **7a**, as a yellow solid (306 mg, 79%). TLC (CHCl$_3$/MeOH (10:1)): R$_f$ = 0.74. ^{1}H-NMR (400 MHz, CDCl$_3$) δ_H = 9.25 (t, J = 10.2 Hz, 1H), 8.59 (d, J = 8.4 Hz, 1H), 8.42 (t, J = 8.7 Hz, 1H), 7.94 (dd, J = 8.4, 1.8 Hz, 1H), 7.55–7.48 (m, 1H), 7.25 (s, 1H), 2.89 (d, J = 3.6 Hz, 3H). LR-MS (ESI+): m/z = 358 (calcd. 358 for C$_{15}$H$_{10}$^{79}BrFN$_5$$^+$ [M + H]$^+$)

1-Bromo-8-(6-fluoropyridin-3-yl)-3-methylbenzo[e]imidazo[5,1-c][1,2,4]triazine (**7b**): According to General Procedure B, compound **6b** (354 mg, 1.27 mmol) and NBS (338 mg, 1.9 mmol) were reacted in acetonitrile (16 mL) to give after purification Compound 7b, as a yellow solid (290 mg, 63%). TLC (CHCl$_3$/MeOH (10:1)): R$_f$ = 0.71. ^{1}H-NMR (400 MHz, CDCl$_3$) δ_H = 9.15 (d, J = 1.8 Hz, 1H), 8.59 (d, J = 2.6 Hz, 1H), 8.56 (d, J = 8.4 Hz, 1H), 8.12 (ddd, J = 8.5, 7.4, 2.7 Hz, 1H), 7.87 (dd, J = 8.4, 1.8 Hz, 1H), 7.13 (dd, J = 8.5, 3.0 Hz, 1H), 2.88 (s, 3H). LR-MS (ESI+): m/z = 358 (calcd. 358 for C$_{15}$H$_{10}$^{79}BrFN$_5$$^+$, [M + H]$^+$).

1-Bromo-8-(2-fluoropyridin-3-yl)-3-methylbenzo[e]imidazo[5,1-c][1,2,4]triazine (**7c**): According To General Procedure B, compound **6c** (169 mg, 0.61 mmol) and NBS (162 mg, 0.91 mmol) were reacted in acetonitrile (6 mL) to give after purification compound **7c**, as a yellow solid (192 mg, 88%). TLC (CHCl$_3$/MeOH (10:1)), R$_f$ = 0.66. ^{1}H-NMR (400 MHz, CDCl$_3$) δ_H = 9.23 (t, J = 1.5 Hz, 1H), 8.54 (d, J = 8.4 Hz, 1H), 8.34 (dt, J = 4.9, 1.5 Hz, 1H), 8.08–8.00 (m, 1H), 7.89 (dt, J = 8.4, 1.6 Hz, 1H), 7.46–7.37 (m, 1H), 2.88 (s, 3H). LR-MS (ESI+): m/z = 358, (calcd. 358 for C$_{15}$H$_{10}$^{79}BrFN$_5$$^+$ [M + H]$^+$).

1-Bromo-8-(5-fluoropyridin-3-yl)-3-methylbenzo[e]imidazo[5,1-c][1,2,4]triazine (**7d**): According to General Procedure B, compound **6d** (120 mg, 0.43 mmol) and NBS (115 mg, 0.65 mmol) were reacted in acetonitrile (4 mL) to afford after purification Compound **7d**, (151 mg, 98%), as a yellow solid. TLC (CHCl$_3$/MeOH (10:1)), R$_f$ = 0.71. ^{1}H-NMR (300 MHz, CDCl$_3$) δ_H = 9.20–9.19 (m, 1H), 8.83–8.81 (m, 1H), 8.61–8.59 (m, 1H), 8.56 (d, J = 0.4 Hz, 1H), 7.90 (dd, J = 8.4, 1.8 Hz, 1H), 7.73 (ddd, J = 9.1, 2.7, 1.9 Hz, 1H), 2.88 (s, 3H). LR-MS (ESI+): m/z = 358, (calcd. 358 for C$_{15}$H$_{10}$^{79}BrFN$_5$$^+$ [M + H]$^+$).

1-Bromo-3-methyl-8-(pyridin-3-yl)benzo[e]imidazo[5,1-c][1,2,4]triazine (**7e**): According to General Procedure B, compound **6e** (300 mg, 1.15 mmol, 1 eq) and NBS (306 mg, 1.72 mmol) were reacted in acetonitrile (8 mL) to afford after purification compound **7e** (343 mg, 87%), as a yellow solid. TLC (CHCl$_3$/MeOH (10:1)): R$_f$ = 0.47. ^{1}H-NMR (400 MHz, CDCl$_3$) δ_H = 9.19 (d, J = 1.7 Hz, 1H), 9.00 (dd, J = 2.5, 0.9 Hz, 1H), 8.73 (dd, J = 4.8, 1.6 Hz, 1H), 8.55 (d, J = 8.4 Hz, 1H), 8.03 (ddd, J = 8.0, 2.5, 1.6 Hz, 1H), 7.91 (dd, J = 8.4, 1.8 Hz, 1H), 7.50 (ddd, J = 8.0, 4.8, 0.9 Hz, 1H), 2.87 (s, 3H). LR-MS (ESI+): m/z = 340, (calcd. 340 for C$_{15}$H$_{11}$^{79}BrN$_5$$^+$ [M + H]$^+$).

1-Bromo-3-methyl-8-(pyridin-4-yl)benzo[e]imidazo[5,1-c][1,2,4]triazine (**7f**): According to General Procedure B, compound **6f** (100 mg, 0.383 mmol, 1 eq) and NBS (102 mg, 0.573 mmol) were reacted in MeCN (4 mL) to give after purification compound **7f** (120 mg, 92%), as a yellow powder. TLC (CHCl$_3$/MeOH (10:1)): R$_f$ = 0.54. ^{1}H-NMR (400 MHz, DMSO-d_6) δ_H = 9.27 (d, J = 1.8 Hz, 1H), 8.85–8.78 (m, 2H), 8.58 (d, J = 8.4 Hz, 1H), 8.24 (dd, J = 8.4, 1.8 Hz, 1H), 7.97–7.91 (m, 1H), 2.76 (s, 3H). LR-MS (ESI+): m/z = 340, (calcd. 340 for C$_{15}$H$_{11}$^{79}BrN$_5$$^+$ [M + H]$^+$).

3.2.3. Synthesis of Oxetanyl Building Block **9c**

3-((3-Bromophenoxy)methyl)-3-methyloxetane (**11**): A mixture of (3-methyloxetan-3-yl) methanol (2.09 g, 20.4 mmol) and triethylamine (TEA, 3.2 mL, 23 mmol) in MeCN (8 mL) was stirred and cooled at 15–25 °C, while a solution of methanesulfonyl chloride (1.55 mL, 20 mmol) in MeCN (3 mL) was dropwise added in the course of 20 min. The mixture was stirred at 22 °C for 4 h and at 0–2 °C for 30 min. The precipitated TEA hydrochloride was filtered off and washed with MeCN (2 × 2.5 mL). To the filtrate was successively added 3-bromophenol (**10**, 3.46 g, 20 mmol) and Cs$_2$CO$_3$ (3.26 g, 10 mmol) along with K$_2$CO$_3$ (2.07 g, 15 mmol). The resulting suspension was stirred and heated at 75–79 °C for 18 h (progress monitored by TLC). Upon completion, the suspension was stirred with methyl *tert*-butyl ether (MTBE, 50 mL) for 2 h. The solid was filtered off and the filtrate washed with sodium hydroxide solution (0.8 M, 1 × 20, 2 × 10 mL) and dried (Na_2CO_3). The solvent was evaporated and the remaining oil bulb-to-bulb distilled (4 mbar, air bath 150–190 °C) to afford compound **11**, as a colorless oil (4.35 g, 85%). TLC (heptane/MTBE (3:2)): R_f = 0.4. ^{1}H-NMR (400 MHz, $CDCl_3$) δ_H = 7.18–7.12 (m, 1H), 7.12-7.08 (m, 2H), 6.87 (ddd, *J* = 8.1, 2.5, 1.3 Hz, 1H), 4.61 (d, *J* = 6.0 Hz, 2H), 4.46 (d, *J* = 6.0 Hz, 2H), 4.00 (s, 2H), 1.43 (s, 3H). ^{13}C-NMR (101 MHz, $CDCl_3$) δ_C = 159.86, 130.70, 124.24, 122.94, 117.97, 113.62, 79.81 (2C), 73.06, 39.74, 21.33.

4,4,5,5-Tetramethyl-2-(3-((3-methyloxetan-3-yl) methoxy) phenyl-1,3,2-dioxaborolane (**12**): A mixture of 3-bromophenylether **11** (1.75 g, 6.8 mmol), KOAc (1.5 g, 15.3 mmol), and bis(pinacolato)diboron (1.77 g, 15 mmol, in 2-methyltetrahydrofuran (24 mL)) was degassed with argon for 10 min. After addition of Pd(dppf)Cl_2 (0.07 g, 0.096 mmol), the mixture was refluxed for 7 h. Upon completion (monitored by TLC), CH_2Cl_2 (25 mL) was added, the resulting suspension stirred for 1 h, and the solid filtered off. The filtrate was evaporated, and the viscous residue (3.2 g) was dissolved in MTBE (70 mL) and subsequently filtered through a short silica gel column (H 2 cm × D 2 cm). The filtrate was extracted with aqueous NaOH (0.75 M, 4 × 12 mL). The alkaline extract was neutralized by slow addition of aqueous HCl (4 M) at 0–2 °C (→ pH 6–7). The separated oil was extracted with MTBE (3 × 15 mL), and after drying ($MgSO_4$), the solvent was evaporated to leave a viscous residue (2.15 g). Crystallization was achieved by trituration with MTBE/heptane to yield **12**, as a colorless solid (1.67 g, 81%). TLC (heptane/MTBE (2:1)): R_f = 0.38. ^{1}H-NMR (300 MHz, $CDCl_3$) δ_H = 7.42 (dt, *J* = 7.3, 1.1 Hz, 1H), 7.38–7.35 (m, 1H), 7.31 (t, *J* = 7.7 Hz, 1H), 7.04 (ddd, *J* = 8.2, 2.8, 1.2, 30 Hz, 1H), 4.63 (d, *J* = 5.9 Hz, 2H), 4.45 (d, *J* = 5.9 Hz, 2H), 4.06 (s, 2H), 1.44 (s, 3H), 1.35 (s, 12H). ^{13}C-NMR (75 MHz, $CDCl_3$) δ_C = 158.65, 129.15, 127.59, 119.86, 118.27, 84.01, 80.01, 72.87, 39.86, 25.01, 21.47 (the boron-substituted carbon atom was not detectable).

2-(2-Chloro-5-((3-methyloxetan-5 3-yl) methoxy) phenyl)-4,4,5,5-tetramethyl-1,3,2-dioxaborolane (**9c**): N-Chlorosuccinimide (NCS, 0.4 g, 3.0 mmol) was added to a solution of compound **12** (0.73 g, 2.4 mmol) in DMF (7 mL). The mixture was stirred at room temperature for 20 h (monitored by TLC). The solvent was removed in vacuo, and the viscous residue was dissolved in MTBE (20 mL). The solution was successively washed with aqueous $Na_2S_2O_3$ (10%, 9 mL) and water (6 mL) and extracted with aqueous NaOH (0.75 M, 3 × 6 mL). The alkaline extract was neutralized (→ pH 7–8) by slow addition of aqueous HCl (4 M) at 0–2 °C. The separated oil was extracted with MTBE (3 × 10 mL), and after drying ($MgSO_4$), the solvent was evaporated to leave a viscous residue. Crystallization was achieved by trituration with MTBE/heptane to yield **9c**, as a colorless solid (0.45 g, 56%). TLC (heptane/MTBE (2:1)): R_f = 0.30. ^{1}H-NMR (300 MHz, $CDCl_3$) δ_H = 7.25 (d, *J* = 8.8 Hz, 1H), 7.23 (d, *J* = 3.2 Hz, 1H), 6.90 (dd, *J* = 8.8, 3.2 20 Hz, 1H), 4.61 (d, *J* = 5.9 Hz, 2H), 4.44 (d, *J* = 5.9 Hz, 2H), 4.01 (s, 2H), 1.42 (s, 3H), 1.37 (s, 12H). ^{13}C-NMR (75 MHz, $CDCl_3$) δ_C = 157.11, 131.36, 130.49, 121.90, 121.88, 118.33, 84.39, 79.90, 79.88, 73.15, 39.82, 24.95, 21.40.

3.2.4. Synthesis of Compounds **BIT1**–**BIT9** via Suzuki Coupling According to General Procedure A

1-(2-Chloro-5-methoxyphenyl)-8-(2-fluoropyridin-4-yl)-3methylbenzo[e]imidazo[5,1-c][1,2,4]triazine (**BIT1**): Based on General Procedure A for Suzuki coupling, a mixture of compound **7a** (100 mg, 0.28 mmol, 1 eq), BA **9a** (63.4 mg,0.34 mmol, 1.2 eq), and K_2CO_3 (97 mg, 0.7 mmol, 2.5 eq) in dioxane/water (5 mL) was reacted in the presence of Pd(PPh$_3$)$_4$ (33 mg, 10 mol%) to give after purification by flash chromatography (gradient: hexane/ethyl acetate, 9:1 → ethyl acetate, 100%) compound **BIT1** (92 mg, 80%) as a yellow powder. TLC (hexane/ethyl acetate (1:1)): R_f = 0.33. ^{1}H-NMR (400 MHz, $CDCl_3$) δ_H = 8.55 (d, *J* = 8.4 Hz, 1H), 8.27 (d, *J* = 5.2 Hz, 1H), 7.85 (dd, *J* = 8.4, 1.7 Hz, 1H), 7.56 (d, *J* = 8.8 Hz, 1H), 7.48 (d, *J* = 1.6 Hz, 1H), 7.21 (d, *J* = 3.0 Hz, 1H), 7.19 (d, *J* = 3.1 Hz, 1H), 7.16 (d, *J* = 3.0 Hz, 1H), 6.88 (s, 1H), 3.87 (s, 3H), 2.99 (s, 3H). ^{13}C-NMR (75 MHz, $CDCl_3$) δ_C = 164.59 (d, *J* = 239.3 Hz), 159.09, 151.71 (d, *J* = 8.1 Hz), 148.68 (d, *J* = 15.4 Hz), 140.20 (d, *J* = 3.3 Hz), 139.84, 137.41, 136.98, 136.18, 131.51, 131.44, 130.81, 126.00, 124.05, 119.38 (d, *J* = 4.1 Hz), 118.56, 117.68, 113.90, 107.44 (d, *J* = 38.7 Hz), 56.01,

12.89. ^{19}F-NMR (282 MHz, CDCl$_3$) δ$_F$ = −66.98 (s, br). HR-MS (ESI+): m/z = 420.1025 and 442.0848 (calcd. 420.1022 for C$_{22}$H$_{16}$ClFN$_5$O$^+$ [M+H]$^+$ and 442.0841 for C$_{22}$H$_{15}$ClFN$_5$NaO$^+$ [M + Na]$^+$).

1-(2-Chloro-5-((3-methyloxetan-3-yl)methoxy)phenyl)-8-(6-fluoropyridin-3-yl)-3-methylbenzo[e]imidazo[5,1-c] [1,2,4]triazine (**BIT2**): According to General Procedure A, compound **7b** (144 mg, 0.40 mmol,1 eq), BA **9c** (163 mg, 0.48 mmol, 1.2 eq), K$_2$CO$_3$ (157 mg, 1.14 mmol, 2.84 eq), and Pd(PPh$_3$)$_4$ (23 mg, 5 mol%) were reacted in dioxane/water (5 mL) to give a raw product, which was purified by two-times column chromatography (CHCl$_3$/MeOH (10:1) to yield compound **BIT2** (155 mg, 79%), as a yellow solid. TLC (hexane/ethyl acetate (1:3)): R$_f$ = 0.29. ^{1}H-NMR (400 MHz, CDCl$_3$) δ$_H$ = 8.52 (d, J = 8.3 Hz, 1H), 8.21 (d, J = 2.5 Hz, 1H), 7.90–7.81 (m, 1H), 7.77 (dd, J = 8.3, 1.8 Hz, 1H), 7.55 (d, J = 8.8 Hz, 1H), 7.45 (d, J = 1.8 Hz, 1H), 7.27 (s, 1H), 7.17 (dd, J = 8.9, 2.9 Hz, 1H), 7.02 (dd, J = 8.5, 3.0 Hz, 1H), 4.57 (dd, J = 6.0, 2.0 Hz, 2H), 4.44 (d, J = 5.9 Hz, 2H), 4.17–3.99 (m, 2H), 2.96 (s, 3H), 1.41 (s, 3H). ^{13}C-NMR (101 MHz, CDCl$_3$) δ$_C$ = 163.85 (d, J = 242.3 Hz), 158.37, 146.27 (d, J = 15.2 Hz), 140.19, 139.79 (d, J = 8.1 Hz), 139.45, 137.37, 136.39, 135.92, 133.01 (d, J = 4.7 Hz), 131.74, 131.46, 130.85, 126.38, 125.94, 124.10, 118.73 (d, J = 41.2 Hz), 113.32, 110.38, 110.00, 79.66 (d, J = 3.4 Hz), 73.58, 39.77, 21.28, 12.87. ^{19}F-NMR (282 MHz, CDCl$_3$) δ$_F$ = −67.82 (dd, J = 7.4, 3.2 Hz). HR-MS (ESI+): m/z = 490.1442 and 512.1260 (calcd. 490.1441 for C$_{26}$H$_{22}$ClFN$_5$O$_2^+$ [M+H]$^+$ and 512.1260 for C$_{26}$H$_{21}$ClFN$_5$NaO$_2^+$ [M + Na]$^+$).

1-(2-Chloro-5-((3-methyloxetan-3-yl)methoxy)phenyl)-8-(2-fluoropyridin-4-yl)-3-methylbenzo[e]imidazo[5,1-c] [1,2,4]triazine (**BIT3**): According to General Procedure A, compound **7a** (100 mg, 0.279 mmol, 1 eq), BA **9c** (104 mg, 0.307 mmol, 1.1 eq), K$_2$CO$_3$ (93 mg, 0.642 mmol, 2.3 eq), and Pd(PPh$_3$)$_4$ (16 mg, 5 mol%) were reacted in dioxane/water (5 mL) to give after purification by flash chromatography (gradient: hexane/ethyl acetate, 1:2 → ethyl acetate, 100%) compound **BIT3** (93 mg, 68%), as a yellow solid. TLC (hexane/ethyl acetate (1:3)): R$_f$ = 0.29. ^{1}H-NMR (400 MHz, CDCl$_3$) δ$_H$ = 8.55 (d, J = 8.4 Hz, 1H), 8.27 (d, J = 5.2 Hz, 1H), 7.86 (dd, J = 8.4, 1.8 Hz, 1H), 7.58 (d, J = 8.9 Hz, 1H), 7.52 (d, J = 1.8 Hz, 1H), 7.28 (d, J = 3.0 Hz, 1H), 7.23–7.18 (m, 2H), 6.89 (d, J = 1.6 Hz, 1H), 4.60–4.55 (m, 2H), 4.44 (d, J = 6.0 Hz, 2H), 4.15–4.03 (m, 2H), 2.97 (s, 3H), 1.42 (s, 3H). ^{13}C-NMR (101 MHz, CDCl$_3$) δ$_C$ = 164.60 (d, J = 239.3 Hz), 158.47, 151.79 (d, J = 8.2 Hz), 148.72 (d, J = 15.5 Hz), 140.25 (d, J = 3.4 Hz), 139.92, 137.45, 137.02, 136.10, 131.68, 131.49, 130.91, 126.37, 125.90, 124.06, 119.44 (d, J = 4.1 Hz), 118.91, 118.56, 114.00, 107.49 (d, J = 38.7 Hz), 79.63 (d, J = 4.1 Hz), 73.61, 39.77, 21.26, 12.90. ^{19}F-NMR (282 MHz, CDCl$_3$) δ$_F$ = −66.93 (s, br). HR-MS (ESI+) m/z = 490.1443 and 512.1261 (calcd. 490.1441 for C$_{26}$H$_{22}$ClFN$_5$O$_2^+$ [M + H]$^+$ and 512.1260 for C$_{26}$H$_{21}$ClFN$_5$NaO$_2^+$ [M + Na]$^+$).

1-(2-Chloro-5-methoxyphenyl)-8-(6-fluoropyridin-3-yl)-3 methylbenzo[e]imidazo[5,1-c][1,2,4]triazine (**BIT4**): According to General Procedure A, compound **7b** (111 mg, 0.308 mmol, 1 eq), BA **9a** (69 mg, 0.370 mmol, 1.2 eq), K$_2$CO$_3$ (107 mg, 0.77 mmol, 2.5 eq), and Pd(PPh$_3$)$_4$ (36 mg, 10 mol%) were reacted in dioxane/water (5 mL) to give after purification by flash chromatography (gradient: CHCl$_3$/MeOH, 50:1 → CHCl$_3$/MeOH, 30:1) compound **BIT4** (83 mg, 64%), as a yellow solid. TLC (hexane/ethyl acetate (1:1)): R$_f$ = 0.37. ^{1}H-NMR (400 MHz, CDCl$_3$) δ$_H$ = 8.51 (d, J = 8.4 Hz, 1H), 8.20 (d, J = 2.3 Hz, 1H), 7.82 (ddd, J = 8.6, 7.3, 2.7 Hz, 1H), 7.76 (dd, J = 8.4, 1.8 Hz, 1H), 7.52 (d, J = 8.9 Hz, 1H), 7.39 (d, J = 1.8 Hz, 1H), 7.19 (d, J = 3.1 Hz, 1H), 7.13 (dd, J = 8.9, 3.0 Hz, 1H), 7.01 (dd, J = 8.5, 3.1 Hz, 1H), 3.84 (s, 3H), 2.96 (s, 3H).^{13}C-NMR (75 MHz, CDCl$_3$) δ$_C$ = 163.82 (d, J = 242.0 Hz), 159.00, 146.24 (d, J = 15.0 Hz), 140.19, 139.76 (d, J = 8.1 Hz), 139.38, 137.33, 136.36, 136.00, 132.98 (d, J = 4.7 Hz), 131.58, 131.43, 130.77, 126.04, 125.87, 124.10, 118.64, 117.62, 113.26, 110.16 (d, J = 37.5 Hz), 55.99, 12.86. ^{19}F-NMR (282 MHz, CDCl$_3$) δ$_F$ = −67.90 (dd, J = 7.4, 3.1 Hz). HR-MS (ESI+): m/z = 420.1010 (calcd. 420.1022 for C$_{22}$H$_{16}$ClFN$_5$O$^+$ [M + H]$^+$).

1-(2-Chloro-5-methoxyphenyl)-8-(2-fluoropyridin-3-yl)-3 methylbenzo[e]imidazo[5,1-c][1,2,4]triazine (**BIT5**): According to General Procedure A, compound **7c** (115 mg, 0.322 mmol, 1 eq), BA **9a** (72 mg, 0.386 mmol, 1.2 eq), K$_2$CO$_3$ (115 mg, 0.832 mmol, 2.58 eq), and Pd(PPh$_3$)$_4$ (37 mg, 10 mol%) were reacted in dioxane/water (5 mL) to give after purification by flash chromatography (gradient: light petroleum ether/ethyl acetate, 1:1 → light petroleum ether/ethyl acetate, 1:2) compound **BIT5** (93 mg, 69%), as a yellow solid. TLC (light petroleum ether/ethyl acetate (1:2)): R_f = 0.43. ^{1}H-NMR (300 MHz, CDCl$_3$) δ_H = 8.51 (d, J = 8.4 Hz, 1H), 8.22 (dd, J = 4.8, 0.7 Hz, 1H), 7.83–7.73 (m, 2H), 7.52–7.45 (m, 2H), 7.32–7.27 (m, 1H), 7.17 (d, J = 3.0 Hz, 1H), 7.10 (dd, J = 8.9, 3.0 Hz, 1H), 3.83 (s, 3H), 2.97 (s, 3H). ^{13}C-NMR (75 MHz, CDCl$_3$) δ_C = 160.19 (d, J = 242.0 Hz), 158.95, 147.84 (d, J = 15.0 Hz), 140.67 (d, J = 3.5 Hz), 139.44, 137.51, 137.42 (d, J = 2.1 Hz), 136.37, 136.21, 131.33, 131.01, 130.93, 127.58 (d, J = 2.8 Hz), 126.05, 123.76, 122.28 (d, J = 4.6 Hz), 121.95 (d, J = 27.2 Hz), 118.72, 117.03, 115.51 (d, J = 5.2 Hz), 55.95, 12.88. ^{19}F-NMR (282 MHz, CDCl$_3$) δ_F = −70.11 (d, J = 9.8 Hz). HR-MS (ESI+): m/z = 420.1014 (calcd. 420.1022 for $C_{22}H_{16}ClFN_5O^+$ [M + H]$^+$).

1-(2-Fluoro-5-methoxyphenyl)-8-(2-fluoropyridin-4-yl)-3-methylbenzo[e]imidazo[5,1-c][1,2,4]triazine (**BIT6**): According to General Procedure A, compound **7a** (100 mg, 0.279 mmol, 1 eq), BA **9b** (62 mg, 0.365 mmol, 1.3 eq), K$_2$CO$_3$ (97 mg, 0.698 mmol, 2.58 eq), and Pd(PPh$_3$)$_4$ (32 mg, 10 mol%) were reacted in dioxane/water (5 mL) to give after purification by flash chromatography (gradient: light petroleum ether/ethyl acetate, 1:1 → light petroleum ether/ethyl acetate, 1:2) compound **BIT6** (81 mg, 72%), as an orange solid. TLC (petroleum ether/ethyl acetate (1:2)): R_f = 0.54. ^{1}H-NMR (300 MHz, CDCl$_3$) δ_H = 8.55 (d, J = 8.4 Hz, 1H), 8.27 (dd, J = 5.3, 0.7 Hz, 1H), 7.86 (dd, J = 8.4, 1.9 Hz, 1H), 7.73 (dd, J = 3.5, 1.8 Hz, 1H), 7.30–7.27 (m, 1H), 7.25–7.14 (m, 3H), 6.94–6.91 (m, 1H), 3.88 (s, 3H), 2.97 (s, 3H). ^{13}C-NMR (75 MHz, CDCl$_3$) δ_C = 164.61 (d, J = 239.5 Hz), 156.62 (d, J = 7.1 Hz), 153.86 (d, J = 246.3 Hz), 153.58, 151.80, 148.73 (d, J = 15.4 Hz), 140.32, 140.24 (d, J = 2.9 Hz), 137.50 (d, J = 62.8 Hz), 133.85, 131.56, 125.94, 124.28, 120.00 (d, J = 16.3 Hz), 119.46 (d, J = 4.1 Hz), 118.76 (d, J = 7.9 Hz), 117.03 (d, J = 22.7 Hz), 116.44, 113.90 (d, J = 2.2 Hz), 107.56 (d, J = 38.8 Hz), 56.21, 12.88. ^{19}F-NMR (282 MHz, CDCl$_3$) δ_F = −66.90—66.97 (m), −121.68–121.78 (m). HR-MS (ESI+): m/z = 404.1325 (calcd. 404.1317 for $C_{22}H_{16}F_2N_5O^+$ [M + H]$^+$).

1-(2-Chloro-5-methoxyphenyl)-8-(3-fluoropyridin-5-yl)-3 methylbenzo[e]imidazo[5,1-c][1,2,4]triazine (**BIT7**): According to General Procedure A, compound **7d** (100 mg, 0.279 mmol, 1 eq), BA **9a** (62 mg, 0.335 mmol, 1.2 eq), K$_2$CO$_3$ (116 mg, 0.839 mmol, 3.0 eq), and Pd(PPh$_3$)$_4$ (17 mg, 5 mol%) were reacted in dioxane/water (5 mL) to give after purification by flash chromatography (gradient: hexane/ethyl acetate, 1:2 → hexane/ethyl acetate, 1:3) compound **BIT7** (79 mg, 67%), as a yellow solid. TLC (hexane/ethyl acetate (1:2)): R_f = 0.38. ^{1}H-NMR (400 MHz, CDCl$_3$) δ_H = 8.53 (d, J = 8.4 Hz, 1H), 8.48 (d, J = 2.7 Hz, 1H), 8.46–8.45 (m, 1H), 7.81 (dd, J = 8.4, 1.9 Hz, 1H), 7.55 (d, J = 8.9 Hz, 1H), 7.44–7.38 (m, 2H), 7.20 (d, J = 3.0 Hz, 1H), 7.14 (dd, J = 8.9, 3.0 Hz, 1H), 3.85 (s, 3H), 2.97 (s, 3H). ^{13}C-NMR (101 MHz, CDCl$_3$) δ_C = 159.71 (d, J = 258.4 Hz), 159.05, 143.99 (d, J = 4.0 Hz), 139.77, 139.56, 138.21 (d, J = 23.1 Hz), 137.37, 136.58, 136.23 (d, J = 3.8 Hz), 136.09, 131.52, 131.48, 130.84, 126.02, 125.99, 124.10, 121.21 (d, J = 19.0 Hz), 118.59, 117.70, 113.68, 55.99, 12.87. ^{19}F-NMR (377 MHz, CDCl$_3$) δ_F = −125.95 (d, J = 9.2 Hz). HR-MS (ESI+): m/z = 420.1024 (calcd. 420.1022 for $C_{22}H_{16}ClFN_5O^+$ [M + H]$^+$).

1-(2-Fluoro-5-methoxyphenyl)-3-methyl-8-(pyridin-3-yl)benzo[e]imidazo[5,1-c][1,2,4]triazine (**BIT8**): According to General Procedure A, compound **7e** (100 mg, 0.294 mmol, 1 eq), BA **9b** (60 mg, 0.353 mmol, 1.2 eq), K$_2$CO$_3$ (110 mg, 0.796 mmol, 2.7 eq), and Pd(PPh$_3$)$_4$ (17 mg, 5 mol%) were reacted in dioxane/water (5 mL) to give after purification by flash chromatography (gradient: hexane/ethyl acetate, 1:3 → ethyl acetate, 100%) compound **BIT8** (93 mg, 82%), as an orange solid. TLC (hexane/ethyl acetate (1:3)): R_f = 0.35. ^{1}H-NMR (300 MHz, CDCl$_3$) δ_H = 8.69–8.62 (m, 2H), 8.53 (d, J = 8.4 Hz, 1H), 7.83 (dd, J = 8.4, 1.8 Hz, 1H), 7.77 (ddd, J = 8.0, 2.4, 1.6 Hz, 1H), 7.70 (dd, J = 3.5, 1.8 Hz, 1H), 7.39 (ddd, J = 8.0, 4.8, 0.8 Hz, 1H), 7.30–7.23 (m, 2H), 7.18–7.11 (m, 1H), 3.86 (s, 3H), 2.97 (s, 3H). ^{13}C-NMR (75 MHz, CDCl$_3$) δ_C = 156.52 (d, J = 1.8 Hz), 154.85 (d, J = 243.0 Hz), 149.93, 148.39, 140.58 (d, J = 125.3 Hz), 137.76, 136.60, 134.74, 134.53, 133.64, 131.43, 128.62 (d, J = 12.4 Hz), 126.12, 124.28, 123.94, 120.12 (d, J = 16.6 Hz), 118.83 (d, J = 7.9 Hz), 117.01 (d, J = 22.9 Hz), 116.29 (d, J = 2.0 Hz), 113.35 (d, J = 1.9 Hz), 56.18, 12.86. ^{19}F-NMR (282 MHz, CDCl$_3$) δ_F = −121.71 (td, J = 9.0, 4.0 Hz). HR-MS (ESI+): m/z = 386.1400 (calcd. 386.1412 for $C_{22}H_{17}FN_5O^+$ [M + H]$^+$).

1-(2-Fluoro-5-methoxyphenyl)-3-methyl-8-(pyridin-4-yl)benzo[e]imidazo[5,1-c][1,2,4]triazine (**BIT9**): According to General Procedure A, compound **7f** (100 mg, 0.294 mmol, 1 eq), BA **9b** (61 mg, 0.366 mmol, 1.2 eq), K_2CO_3 (103 mg, 0.745 mmol, 2.53 eq), and $Pd(PPh_3)_4$ (19 mg, 5 mol%) were reacted in dioxane/water (5 mL) to give a raw product, which was purified by two-times flash chromatography: $CHCl_3$/MeOH (10:1) was used first, followed by hexane/ethyl acetate (gradient 1:3 → ethyl acetate, 100%) to obtain compound **BIT9** (44 mg, 39%), as a yellow solid. TLC (hexane/ethyl acetate (1:1)): R_f = 0.29. ^{1}H-NMR (300 MHz, $CDCl_3$) δ_H = 8.68–8.65 (m, 2H), 8.53 (d, J = 8.5 Hz, 1H), 7.87 (dd, J = 8.4, 1.8 Hz, 1H), 7.74 (ddd, J = 3.5, 1.9, 0.5 Hz, 1H), 7.33–7.30 (m, 2H), 7.29–7.22 (m, 2H), 7.19–7.13 (m, 1H), 3.87 (s, 3H), 2.96 (s, 3H). ^{13}C-NMR (101 MHz, $CDCl_3$) δ_C = 156.57 (d, J = 2.0 Hz), 154.85 (d, J = 243.4 Hz), 150.60, 146.38, 141.51, 140.07, 137.81, 137.04, 133.73, 131.46, 125.98, 124.28, 121.65, 120.11 (d, J = 16.6 Hz), 118.86 (d, J = 7.9 Hz), 117.01 (d, J = 22.6 Hz), 116.24 (d, J = 2.0 Hz), 113.67 (d, J = 2.3 Hz), 56.21, 12.87. ^{19}F-NMR (282 MHz, $CDCl_3$) δ_F = −121.63—121.71 (m). HR-MS (ESI+): m/z = 386.1404 (calcd. 386.1412 for $C_{22}H_{17}FN_5O^+$ $[M + H]^+$).

3.3. Biology

3.3.1. In Vitro Evaluation of BIT Derivatives Towards PDEs

The inhibitory potency of BIT derivatives against the human recombinant PDE subtype was determined in enzyme assays, conducted by SB Drug Discovery (Scotland, U.K.). The phosphodiesterase assays were performed using recombinant human PDE enzymes expressed in a baculoviral system. This had similarity to PDE enzymes taken from human tissue using known inhibitor standards where available. The radiometric assay was a modification of the two-step method of Thompson and Appleman [45], which was adapted for 96-well plate format.

Firstly, tritium-labeled cyclic AMP or cyclic GMP was hydrolyzed to 5'-AMP or 5'-GMP by phosphodiesterase. The 5'-AMP or 5'-GMP was then further hydrolyzed to adenosine or guanosine by nucleotidase in snake venom. An anion-exchange resin bound all charged nucleotides and left [^{3}H]adenosine or [^{3}H]guanosine as the only labeled compound to be counted by liquid scintillation. Briefly, 50 µL of diluted human PDE enzyme were incubated with 50 µL of [^{3}H]-cAMP or [^{3}H]-cGMP and 11 µL of 50% DMSO (or compound dilution) for 20 min (at 30 °C). Reactions were carried out in a Greiner 96-deep well 1-mL master-block. The enzyme was diluted in 20 mM Tris HCl pH 7.4 and [^{3}H]-cAMP or [^{3}H]-cGMP are diluted in 10 mM $MgCl_2$, 40 mM Tris HCl pH 7.4. The reaction is terminated by denaturing the PDE enzyme (at 70 °C for 2 min), after which 25 µL of snake venom nucleotidase were added and incubated for 10 min (at 30 °C). Plates were centrifuged for 9 seconds before incubation. After incubation, 200 µL of Dowex resin were added, and the plate was shaken for 20 min then centrifuged for 3 min at 2500 r.p.m. Fifty microliters of supernatant were removed and added to 200 µL of MicroScint-20 in white plates (Greiner 96-well Optiplate) and shaken for 30 min before reading on a Perkin Elmer TopCount Scintillation Counter.

Compounds were tested at a concentration of 10 nM against human PDE2A3 and at 1 µM for PDE4A1, PDE5A, PDE6AB, PDE9A1, and PDE10A1. The percentage of inhibition of the compounds and standard inhibitors was determined and compared to historical assay data to ensure that they fell within acceptable ranges. For IC_{50} values' measurements, **BIT1** was tested at concentrations of 0.1, 0.5, 1.0, 5.0, 10.0, and 100 nM, and **BIT6** and **BIT9** were tested at concentrations of 0.5, 1.0, 5.0, 10.0, 100, and 1000 nM against PDE2A3, while the concentration against PDE10A1 at 0.25, 5.0, 10, 100, 250, and 1000 nM (n = 2).

Data generated were analyzed using Prism software (GraphPad Inc.).

3.3.2. Incubations with Mouse Liver Microsomes

For microsome experiments, the following instruments were used: BioShake iQ (QUANTIFOIL Instruments, Jena, Germany), Centrifuge 5424 (Eppendorf, Hamburg, Germany), DB-3D TECHNE Sample Concentrator (Biostep, Jahnsdorf, Germany), and UltiMate 3000 UHPLC System (Thermo Scientific,

Germering, Germany) including a DAD detector (DAD-3000RS) coupled to an MSQ Plus Single Quadrupole Mass Spectrometer (Thermo Scientific, Austin, Texas, USA).

NADPH (Nicotinamide Adenine Dinucleotide Phosphate) and testosterone were purchased from Sigma-Aldrich (Steinheim, Germany). GIBCO Mouse Liver Microsomes (MLM, 20 mg/mL) were purchased from Life Technologies (Darmstadt, Germany). Dulbecco's Phosphate-Buffered Saline (PBS) (without Ca^{2+}, Mg^{2+}) was purchased from Biochrom (Berlin, Germany).

Incubations had a final volume of 250 µL and were performed in PBS (pH 7.4). **BIT1** was freshly dissolved in DMSO to provide a stock solution of 2 mM that was used for the experiments and resulted in an amount of DMSO of 1% in each incubation mixture. In the following, final concentrations are provided in brackets [43]. PBS, MLM (1 mg/mL), and **BIT1** (20 µM) were mixed and preincubated at 37 °C for 5 min. Analogously, preincubated NADPH (2 mM) was added, and the mixtures were gently shaken at 37 °C. After 90 min, portions of 1.0 mL of cold acetonitrile (−20 °C) were added, respectively, followed by vigorous shaking (30 s), cooling on ice (4 min), and centrifugation at 14,000 rpm (10 min). Supernatants were concentrated at 50 °C under a flow of nitrogen to provide residual volumes of 40–60 µL, which were reconditioned by adding acetonitrile/water 1:1 (*v/v*) to obtain samples of 100 µL, which were stored at 4 °C until analyzed by HPLC-UV-MS. As a positive control, testosterone was used as the substrate and incubated at an appropriate concentration, similar to the protocol described above, to give complete conversion confirmed by HPLC. Furthermore, incubations without NADPH, microsomal protein, and **BIT1**, respectively, were performed as negative controls.

HPLC-UV-MS analyses were performed on a ReproSil-Pur 120 C18-AQ-column, 125 mm × 3 mm, 3 µm (Dr. Maisch GmbH, Ammerbuch, Germany) equipped with an appropriate precolumn at 25 °C and a flow rate of 0.7 mL/min. The solvent system consisted of Eluent A: water, containing 2 mM ammonium acetate, and Eluent B: water/acetonitrile 20/80 (v/v), containing 2 mM ammonium acetate. Two methods were applied. Method A (gradient elution, % acetonitrile): 0–1.5 min, 10%; 1.5–10 min, 10–80%; 10-13 min, 80%; 13–16 min 10%. Method B (isocratic elution, % acetonitrile): 0–25 min, 42%. For both methods, UV detection was performed at a wavelength of 228 nm (maximum UV absorbance of **BIT1**) and a bandwidth of 20 nm. MSQ Plus single quadrupole mass spectrometer was operated in positive electrospray ionization mode: probe temperature 500 °C, needle voltage 5 V, cone voltage 75 V.

4. Conclusions

A series of novel fluorinated PDE2A inhibitors based on a **BIT** (benzoimidazotriazine) scaffold was successfully prepared and evaluated in vitro. The binding results revealed that a modification with a 2,2-oxetanyl propoxy portion on the phenyl (phe-3) at the 1-position led to a potency loss towards PDE2A. However, a small methoxy group at the same position (phe-1) led to a significant increase of potency towards PDE2A. It appeared that a higher inhibition towards PDE2A was obtained by pyridine-4-yl residues in this position, either fluorinated or non-fluorinated. Furthermore, the introduction of ortho-fluorophenyl at the 1-position, instead of o-chlorophenyl, showed a lack of selectivity towards PDE2A. The introduction of the 2-fluoropyridin-4-yl residue at the 8-position provided us a promising candidate for future labeling with fluorine-18.

In vitro metabolism studies of **BIT1** with MLM showed that **BIT1** was sufficiently stable and suitable for the development of an ^{18}F-labeled radioligand. This should further be proven by in vivo investigations in the future. Taken together, **BIT1** might be a prospective ligand to be developed as a radioligand for PDE2A-PET imaging.

Supplementary Materials: Supplementary Materials including NMR spectra of final compounds **BIT1–BIT9**, and dose response curves (**BIT1**, **BIT6**, and **BIT9**) can be accessed at.

Author Contributions: R.R., D.B., and M.S. designed and performed the organic syntheses; R.R., P.B., and M.S. designed the in vitro evaluation; R.R. and F.-A.L. designed and performed the microsomal incubations; R.R., F.-A.L, D.B., P.B., and M.S. wrote the paper. All authors read and approved the final manuscript.

Funding: This research was funded by the Deutsche Forschungsgemeinschaft (DFG), Project Number: SCHE 1825/3-1.

Acknowledgments: The authors would like to thank the Indonesia Ministry of Research and Technology-Program for Research and Innovation in Science and Technology Project (RISET-Pro), World Bank Loan No. 8245-ID, for supporting the PhD thesis of Rien Ritawidya. We also thank the staff of the Institute of Analytical Chemistry, Department of Chemistry and Mineralogy of University of Leipzig, for recording and processing the NMR and LR/HR-MS spectra and Tina Spalholz (HZDR) for technical assistance.

Conflicts of Interest: The authors declare no conflict of interest.

References

1. Helal, C.J.; Arnold, E.P.; Boyden, T.L.; Chang, C.; Chappie, T.A.; Fennell, K.F.; Forman, M.D.; Hajos, M.; Harms, J.F.; Hoffman, W.E.; et al. Application of Structure-Based Design and Parallel Chemistry to Identify a Potent, Selective, and Brain Penetrant Phosphodiesterase 2A Inhibitor. *J. Med. Chem.* **2017**, *60*, 5673–5698. [CrossRef] [PubMed]
2. Fajardo, A.M.; Piazza, G.A.; Tinsley, H.N. The role of cyclic nucleotide signaling pathways in cancer: Targets for prevention and treatment. *Cancers* **2014**, *6*, 436–458. [CrossRef] [PubMed]
3. Maurice, D.H.; Ke, H.; Ahmad, F.; Wang, Y.; Chung, J.; Manganiello, V.C. Advances in targeting cyclic nucleotide phosphodiesterases. *Nat. Rev. Drug Discov.* **2014**, *13*, 290–314. [CrossRef] [PubMed]
4. Conti, M.; Beavo, J. Biochemistry and physiology of cyclic nucleotide phosphodiesterases: Essential components in cyclic nucleotide signaling. *Annu. Rev. Biochem.* **2007**, *76*, 481–511. [CrossRef] [PubMed]
5. Francis, S.H.; Blount, M.A.; Corbin, J.D. Mammalian cyclic nucleotide phosphodiesterases: Molecular mechanisms and physiological functions. *Physiol. Rev.* **2011**, *91*, 651–690. [CrossRef] [PubMed]
6. Keravis, T.; Lugnier, C. Cyclic nucleotide phosphodiesterase (PDE) isozymes as targets of the intracellular signalling network: Benefits of PDE inhibitors in various diseases and perspectives for future therapeutic developments. *Br. J. Pharmacol.* **2012**, *165*, 1288–1305. [CrossRef]
7. Lakics, V.; Karran, E.H.; Boess, F.G. Quantitative comparison of phosphodiesterase mRNA distribution in human brain and peripheral tissues. *Neuropharmacology* **2010**, *59*, 367–374. [CrossRef] [PubMed]
8. Gomez, L.; Massari, M.E.; Vickers, T.; Freestone, G.; Vernier, W.; Ly, K.; Xu, R.; McCarrick, M.; Marrone, T.; Metz, M.; et al. Design and Synthesis of Novel and Selective Phosphodiesterase 2 (PDE2a) Inhibitors for the Treatment of Memory Disorders. *J. Med. Chem* **2017**, *60*, 2037–2051. [CrossRef] [PubMed]
9. Gomez, L.; Breitenbucher, J.G. PDE2 inhibition: Potential for the treatment of cognitive disorders. *Bioorg. Med. Chem. Lett* **2013**, *23*, 6522–6527. [CrossRef]
10. Redrobe, J.P.; Rasmussen, L.K.; Christoffersen, C.T.; Bundgaard, C.; Jørgensen, M. Characterisation of Lu AF33241: A novel, brain-penetrant, dual inhibitor of phosphodiesterase (PDE) 2A and PDE10A. *Eur. J. Pharmacol.* **2015**, *761*, 79–85. [CrossRef]
11. Mikami, S.; Sasaki, S.; Asano, Y.; Ujikawa, O.; Fukumoto, S.; Nakashima, K.; Oki, H.; Kamiguchi, N.; Imada, H.; Iwashita, H.; et al. Discovery of an Orally Bioavailable, Brain-Penetrating, in Vivo Active Phosphodiesterase 2A Inhibitor Lead Series for the Treatment of Cognitive Disorders. *J. Med. Chem.* **2017**, *60*, 7658–7676. [CrossRef] [PubMed]
12. Kehler, J. Targeting Phosphodiesterases in the CNS. In *Comprehensive Medicinal Chemistry*, 3rd ed; Chackalamannil, S., Rotela, D., Ward, S.E., Eds.; Elsevier: Oxford, UK, 2017; Volume 7, pp. 1–24.
13. Stephenson, D.T.; Coskran, T.M.; Kelly, M.P.; Kleiman, R.J.; Morton, D.; O'Neill, S.M.; Schmidt, C.J.; Weinberg, R.J.; Menniti, F.S. The distribution of phosphodiesterase 2A in the rat brain. *Neuroscience* **2012**, *226*, 145–155. [CrossRef] [PubMed]
14. Stephenson, D.T.; Coskran, T.M.; Wilhelms, M.B.; Adamowicz, W.O.; O'Donnell, M.M.; Muravnick, K.B.; Menniti, F.S.; Kleiman, R.J.; Morton, D. Immunohistochemical localization of phosphodiesterase 2A in multiple mammalian species. *J. Histochem. Cytochem* **2009**, *57*, 933–949. [CrossRef] [PubMed]
15. Schroder, S.; Wenzel, B.; Deuther-Conrad, W.; Scheunemann, M.; Brust, P. Novel Radioligands for Cyclic Nucleotide Phosphodiesterase Imaging with Positron Emission Tomography: An Update on Developments Since 2012. *Molecules* **2016**, *21*, 650. [CrossRef] [PubMed]
16. Buijnsters, P.; de Angelis, M.; Langlois, X.; Rombouts, F.J.R.; Sanderson, W.; Tresadern, G.; Ritchie, A.; Trabanco, A.A.; VanHoof, G.; van Roosbroeck, Y.; et al. Structure-Based Design of a Potent, Selective, and Brain Penetrating PDE2 Inhibitor with Demonstrated Target Engagement. *ACS Med. Chem. Lett.* **2014**, *5*, 1049–1053. [CrossRef]

17. Blokland, A.; Menniti, F.S.; Prickaerts, J. PDE inhibition and cognition enhancement. *Expert. Opin. Ther. Pat.* **2012**, *22*, 349–354. [CrossRef] [PubMed]
18. Helal, C.J.; Arnold, E.; Boyden, T.; Chang, C.; Chappie, T.A.; Fisher, E.; Hajos, M.; Harms, J.F.; Hoffman, W.E.; Humphrey, J.M.; et al. Identification of a Potent, Highly Selective, and Brain Penetrant Phosphodiesterase 2A Inhibitor Clinical Candidate. *J. Med. Chem.* **2018**, *61*, 1001–1018. [CrossRef]
19. Malamas, M.S.; Ni, Y.; Erdei, J.J.; Egerland, U.; Langen, B. The Substituted Imidazo [1,5-a]Quinoxalines as Inhibitors of Phosphodiesterase 10. WO 2010/138833 A1, 2 December 2010.
20. Am Ende, C.W.; Kormos, B.L.; Humphrey, J.M. The State of the Art in Selective PDE2A Inhibitor Design. In *Phosphodiesterases and their inhibitors*; Liras, S., Bell, A.S., Eds.; Wiley-VCH: Weinheim an der Bergstrasse, Germany, 2014; Volume 61, pp. 83–104.
21. Schroder, S.; Wenzel, B.; Deuther-Conrad, W.; Teodoro, R.; Egerland, U.; Kranz, M.; Scheunemann, M.; Hofgen, N.; Steinbach, J.; Brust, P. Synthesis, ^{18}F-Radiolabelling and Biological Characterization of Novel Fluoroalkylated Triazine Derivatives for in Vivo Imaging of Phosphodiesterase 2A in Brain via Positron Emission Tomography. *Molecules* **2015**, *20*, 9591–9615. [CrossRef]
22. Schröder, S.; Wenzel, B.; Deuther-Conrad, W.; Teodoro, R.; Kranz, M.; Scheunemann, M.; Egerland, U.; Höfgen, N.; Briel, D.; Steinbach, J.; et al. Investigation of an ^{18}F-labelled Imidazopyridotriazine for Molecular Imaging of Cyclic Nucleotide Phosphodiesterase 2A. *Molecules* **2018**, *23*, 556. [CrossRef]
23. Wagner, S.; Scheunemann, M.; Dipper, K.; Egerland, U.; Hoefgen, N.; Steinbach, J.; Brust, P. Development of highly potent phosphodiesterase 10A (PDE10A) inhibitors: Synthesis and in vitro evaluation of 1,8-dipyridinyl- and 1-pyridinyl-substituted imidazo[1,5-a]quinoxalines. *Eur. J. Med. Chem.* **2016**, *107*, 97–108. [CrossRef]
24. Wagner, S.; Teodoro, R.; Deuther-Conrad, W.; Kranz, M.; Scheunemann, M.; Fischer, S.; Wenzel, B.; Egerland, U.; Hoefgen, N.; Steinbach, J.; et al. Radiosynthesis and biological evaluation of the new PDE10A radioligand [^{18}F]AQ28A. *J. Label. Compd. Radiopharm.* **2017**, *60*, 36–48. [CrossRef] [PubMed]
25. Stange, H.; Langen, B.; Egerland, U.; Hoefgen, N.; Priebs, M.; Malamas, M.S.; Erdei, J.; Ni, Y. Triazine Derivatives as Inhibitors of Phosphodiesterases. Patent US 2010/0120762 A1, 13 May 2010.
26. Stange, H.; Langen, B.; Egerland, U.; Hoefgen, N.; Priebs, M.; Malamas, M.S.; Erdei, J.J.; Ni, Y. Imidazo[5,1-C][1,2,4]Benzotriazine Derivatives As Inhibitors of. Patent US 2010/0120763 A1, 13 May 2010.
27. Malamas, M.S.; Stange, H.; Schindler, R.; Lankau, H.-J.; Grunwald, C.; Langen, B.; Egerland, U.; Hage, T.; Ni, Y.; Erdei, J.; et al. Novel triazines as potent and selective phosphodiesterase 10A inhibitors. *Bioorg. Med. Chem. Lett.* **2012**, *22*, 5876–5884. [CrossRef] [PubMed]
28. Trabanco, A.A.; Buijnsters, P.; Rombouts, F.J.R. Towards selective phosphodiesterase 2A (PDE2A) inhibitors: A patent review (2010 - present). *Expert. Opin. Ther. Pat.* **2016**, *26*, 933–946. [CrossRef] [PubMed]
29. Liu, J.; Wenzel, B.; Dukic-Stefanovic, S.; Teodoro, R.; Ludwig, F.-A.; Deuther-Conrad, W.; Schröder, S.; Chezal, J.-M.; Moreau, E.; Brust, P.; et al. Development of a new radiofluorinated quinoline analog for PET imaging of phosphodiesterase 5 (PDE5) in brain. *Pharmaceuticals* **2016**, *9*, 22. [CrossRef] [PubMed]
30. Gujral, S.S.; Khatri, S.; Riyal, P. Suzuki Cross Coupling Reaction- A Review. *Indo Glob. J. Pharm. Sci.* **2012**, *2*, 351–367.
31. Dollé, F. [^{18}F]Fluoropyridines: From Conventional Radiotracers to the Labeling of Macromolecules Such as Proteins and Oligonucleotides. In *PET Chemistry The Driving Force in Molecular Imaging*; Schubinger, P.A., Lehmann, L., Friebe, M., Eds.; Springer: Berlin, Germany, 2006; Volume 62, pp. 113–157.
32. Goldstein, S.W.; Bill, A.; Dhuguru, J.; Ghoneim, O. Nucleophilic Aromatic Substitution—Addition and Identification of an Amine. *J. Chem. Educ.* **2017**, *94*, 1388–1390. [CrossRef]
33. Artamkina, G.A.; Egorov, M.P.; Beletskaya, I.P. Some aspects of anionic sigma complexes. *Chem. Rev.* **1982**, *82*, 427–459. [CrossRef]
34. Gamble, A.B.; Garner, J.; Gordon, C.P.; O'Conner, S.M.J.; Keller, P.A. Aryl Nitro Reduction with Iron Powder or Stannous Chloride under Ultrasonic Irradiation. *Synth. Commun.* **2007**, *37*, 2777–2786. [CrossRef]
35. Wuitschik, G.; Rogers-Evans, M.; Muller, K.; Fischer, H.; Wagner, B.; Schuler, F.; Polonchuk, L.; Carreira, E.M. Oxetanes as promising modules in drug discovery. *Angew. Chem. Int. Ed.* **2006**, *45*, 7736–7739. [CrossRef]
36. Bull, J.A.; Croft, R.A.; Davis, O.A.; Doran, R.; Morgan, K.F. Oxetanes: Recent Advances in Synthesis, Reactivity, and Medicinal Chemistry. *Chem. Rev.* **2016**, *116*, 12150–12233. [CrossRef]
37. Talele, T.T. Natural-Products-Inspired Use of the gem-Dimethyl Group in Medicinal Chemistry. *J. Med. Chem.* **2018**, *61*, 2166–2210. [CrossRef] [PubMed]

38. Kamei, T. Metal-free halogenation of arylboronate with N-halosuccinimide. *Tetrahedron Lett.* **2014**, *55*, 4245–4247. [CrossRef]
39. Zhu, J.; Yang, Q.; Dai, D.; Huang, Q. X-ray crystal structure of phosphodiesterase 2 in complex with a highly selective, nanomolar inhibitor reveals a binding-induced pocket important for selectivity. *J. Am. Chem. Soc.* **2013**, *135*, 11708–11711. [CrossRef] [PubMed]
40. Gomez, L.; Xu, R.; Sinko, W.; Selfridge, B.; Vernier, W.; Ly, K.; Truong, R.; Metz, M.; Marrone, T.; Sebring, K.; et al. Mathematical and Structural Characterization of Strong Nonadditive Structure-Activity Relationship Caused by Protein Conformational Changes. *J. Med. Chem.* **2018**, *61*, 7754–7766. [CrossRef] [PubMed]
41. Schneider, H.-J. Hydrogen bonds with fluorine. Studies in solution, in gas phase and by computations, conflicting conclusions from crystallographic analyses. *Chem. Sci.* **2012**, *3*, 1381–1394. [CrossRef]
42. Taylor, R. The hydrogen bond between N-H or O-H and organic fluorine: favourable yes, competitive no. *Acta. Crys.* **2017**, *B73*, 474–488. [CrossRef] [PubMed]
43. Ludwig, F.-A.; Smits, R.; Fischer, S.; Donat, C.K.; Hoepping, A.; Brust, P.; Steinbach, J. LC-MS Supported Studies on the in Vitro Metabolism of both Enantiomers of Flubatine and the in Vivo Metabolism of (+)-[^{18}F]Flubatine-A Positron Emission Tomography Radioligand for Imaging α4β2 Nicotinic Acetylcholine Receptors. *Molecules* **2016**, *21*, 1200. [CrossRef] [PubMed]
44. Testa, B.; Kraemer, S.D. The Biochemistry of Drug Metabolism—An The Biochemistry of Drug Metabolism-Introduction Part 2. Redox Reaction and Their Enzymes. *Chem. Biodivers.* **2007**, *4*, 257–405. [CrossRef]
45. Russell, T.R.; Thompson, W.J.; Schneider, F.; Schneider, F.W.; Appleman, M.M. 3':5'-Cyclic Adenosine Monophosphate Phosphodiesterase: Negative Cooperativity. *Proc. Natl. Acad. Sci. USA* **1972**, *69*, 1791–1795. [CrossRef]

Sample Availability: **BIT2**, and **BIT6** are avaliable from the authors.

© 2019 by the authors. Licensee MDPI, Basel, Switzerland. This article is an open access article distributed under the terms and conditions of the Creative Commons Attribution (CC BY) license (http://creativecommons.org/licenses/by/4.0/).

Article

Radiosynthesis and Biological Investigation of a Novel Fluorine-18 Labeled Benzoimidazotriazine-Based Radioligand for the Imaging of Phosphodiesterase 2A with Positron Emission Tomography

Rien Ritawidya [1,2,*], Barbara Wenzel [1], Rodrigo Teodoro [1], Magali Toussaint [1], Mathias Kranz [3,4], Winnie Deuther-Conrad [1], Sladjana Dukic-Stefanovic [1], Friedrich-Alexander Ludwig [1], Matthias Scheunemann [1] and Peter Brust [1]

- [1] Helmholtz-Zentrum Dresden-Rossendorf, Institute of Radiopharmaceutical Cancer Research, Research Site Leipzig, Department of Neuroradiopharmaceuticals, 04318 Leipzig, Germany; b.wenzel@hzdr.de (B.W.); r.teodoro@hzdr.de (R.T.); m.toussaint@hzdr.de (M.T.); w.deuther-conrad@hzdr.de (W.D.-C.); s.dukic-stefanovic@hzdr.de (S.D.-S.); f.ludwig@hzdr.de (F.-A.L.); m.scheunemann@hzdr.de (M.S.); p.brust@hzdr.de (P.B.)
- [2] Center for Radioisotope and Radiopharmaceutical Technology, National Nuclear Energy Agency (BATAN), Puspiptek Area, Serpong, South Tangerang 15314, Indonesia
- [3] Tromsø PET Center, University Hospital of North Norway, 9009 Tromsø, Norway; Mathias.Kranz@unn.no
- [4] Nuclear Medicine and Radiation Biology Research Group, The Arctic University of Norway, 9009 Tromsø, Norway
- * Correspondence: r.ritawidya@hzdr.de or rienrita@batan.go.id; Tel.: +49-341-234-179-4611 or +62-21-756-3141

Received: 24 October 2019; Accepted: 12 November 2019; Published: 15 November 2019

Abstract: A specific radioligand for the imaging of cyclic nucleotide phosphodiesterase 2A (PDE2A) via positron emission tomography (PET) would be helpful for research on the physiology and disease-related changes in the expression of this enzyme in the brain. In this report, the radiosynthesis of a novel PDE2A radioligand and the subsequent biological evaluation were described. Our prospective compound 1-(2-chloro-5-methoxy phenyl)-8-(2-fluoropyridin-4-yl)-3-methylbenzo[e]imidazo[5,1-c][1,2,4]triazine, benzoimidazotriazine (**BIT1**) (IC_{50} PDE2A = 3.33 nM; 16-fold selectivity over PDE10A) was fluorine-18 labeled via aromatic nucleophilic substitution of the corresponding nitro precursor using the $K[^{18}F]F-K_{2.2.2}$-carbonate complex system. The new radioligand $[^{18}F]$**BIT1** was obtained with a high radiochemical yield (54 ± 2%, n = 3), a high radiochemical purity (≥99%), and high molar activities (155–175 GBq/µmol, n = 3). In vitro autoradiography on pig brain cryosections exhibited a heterogeneous spatial distribution of $[^{18}F]$**BIT1** corresponding to the known pattern of expression of PDE2A. The investigation of in vivo metabolism of $[^{18}F]$**BIT1** in a mouse revealed sufficient metabolic stability. PET studies in mouse exhibited a moderate brain uptake of $[^{18}F]$**BIT1** with a maximum standardized uptake value of ~0.7 at 5 min p.i. However, in vivo blocking studies revealed a non-target specific binding of $[^{18}F]$**BIT1**. Therefore, further structural modifications are needed to improve target selectivity.

Keywords: cyclic nucleotide phosphodiesterase; PDE2A radioligand; nitro-precursor; fluorine-18; in vitro autoradiography; PET imaging

1. Introduction

The cyclic nucleotide phosphodiesterases (PDEs) are a superfamily of enzymes catalyzing the hydrolysis of the intracellular secondary messengers, cyclic adenosine monophosphate (cAMP) and cyclic guanosine monophosphate (cGMP) [1–3]. These secondary messengers are involved in a great variety of cellular functions associated with normal and pathophysiological processes in the brain and periphery [3–6].

The 11 family members of PDEs are encoded by 21 genes and classified by their substrate specificity [3,7]. PDE 4, 7, and 8 are only cAMP-specific, PDE 5, 6, and 9 are cGMP-specific, and the remaining PDE 1, 2, 3, 10, and 11 hydrolyze both substrates [3,6,7].

The dual-substrate enzyme PDE2A is abundantly expressed in brain, in particular, in caudate, nucleus accumbens, cortex, hippocampus [3,8,9], amygdala [10,11], substantia nigra, as well as olfactory tubercle [11,12], while the expression in the midbrain, hindbrain, and cerebellum is comparatively low [8,11,12]. This specific distribution indicates a role of PDE2A in the modulation of complex neuronal processes, such as learning, concentration, memory, emotion, depression, anxiety, and CNS related disorder [10,13,14]. The pharmacological inhibition of PDE2A has been evaluated in preclinical studies, thus suggesting PDE2A inhibitors as a potential treatment for neurodegenerative diseases, such as Alzheimer's disease, schizophrenia, and dementia [15,16]. PDE2A inhibition has the potential to prolong the duration of cAMP- and cGMP-dependent signaling pathways, eventually improving neural plasticity and memory function [3,13,17].

Positron emission tomography (PET) is a molecular imaging modality that enables the visualization, characterization, and measurement of molecular targets and biochemical processes in living systems [18]. Accordingly, a PDE2A specific radiotracer would allow quantification of PDE2A expression, as well as disease-related changes thereof.

The so far most developed PDE2A radioligands are shown in Figure 1. The first highly potent PDE2A compound [^{18}F]**B-23** (IC$_{50}$ hPDE2A = 1 nM; IC$_{50}$ rPDE10A = 11 nM) was developed by Janssen Pharmaceutica NV (Beerse, Belgium) [10,19]. Biodistribution and microPET imaging studies in rats demonstrated high uptake of activity in the striatum; however, brain-penetrating radio-metabolites limited further evaluation. Pfizer Inc. (New York, NY, USA) also reported on a highly affine PDE2A radioligand, [^{18}F]**PF-05270430** (IC$_{50}$ hPDE2A = 0.5 nM; IC$_{50}$ hPDE10A >3000 nM) [10,20], which has been evaluated in monkeys [20] and translated to clinical trials [21,22]. [^{18}F]**PF-05270430** showed high target-specific accumulation in the putamen, caudate, and nucleus-accumbens, as well as good metabolic stability and a favorable kinetic profile, pointing out [^{18}F]**PF-05270430** as a promising PDE2A PET ligand [21]. However, using the cerebellum as a reference region, the estimated binding potential of [^{18}F]**PF-05270430** was low, and the authors concluded that further studies are required to validate the suitability of the cerebellum as a reference region [21].

In parallel to Pfizer, our group developed further PDE2A radioligands on the basis of pyridoimidazotriazine. The recently published compounds [^{18}F]**TA3** (IC$_{50}$ PDE2A = 11.4 nM; IC$_{50}$ PDE10A = 318 nM) and [^{18}F]**TA4** (IC$_{50}$ PDE2A = 7.3 nM; IC$_{50}$ PDE10A = 913 nM) (Figure 1) were characterized by high potency and selectivity towards PDE2A [24]. Notably, both radiotracers were found to be suitable radioligands for in vitro imaging of PDE2A; however, in vivo metabolism studies revealed a high fraction of polar radio-metabolites in the brain limiting their use for in vivo PDE2A imaging. The latest radioligand out of this series, [^{18}F]**TA5** ((IC$_{50}$ PDE2A = 3.0 nM; IC$_{50}$ PDE10A >1000 nM, Figure 1), exhibited the highest potency towards PDE2A and selectivity over PDE10A. However, autoradiographic studies of [^{18}F]**TA5** showed a homogenous and non-displaceable binding in rat and pig brain cryosections, indicating an insufficient specificity of this radioligand [25]. In addition, [^{18}F]**TA5** is degraded extremely fast in the mouse. Therefore, we decided to perform further structural modifications to develop fluorine-18 labeled PET tracers with improved metabolic stability for the molecular imaging of PDE2A.

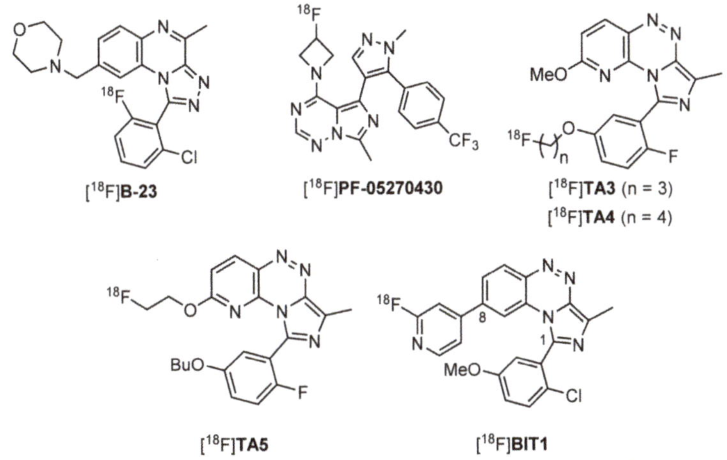

Figure 1. Selected phosphodiesterase 2A (PDE2A) PET radioligands [10,19,20,23–25].

In our continuous effort to develop fluorine-18 labeled PET tracers dedicated to molecular imaging of PDE2A, we selected the benzoimidazotriazine (BIT) scaffold (Figure 1) as lead structure by replacing the pyrido ring of the **TA** compounds with a benzo ring [26,27]. As a first result, we very recently reported on the synthesis and inhibitory potency of nine new fluorinated derivatives based on this BIT scaffold [28]. In order to increase the metabolic stability of the corresponding PDE2A radioligands, the fluorine was introduced by substituting a fluoropyridine ring at the benzene moiety. Out of this series, derivative **BIT1** (Figure 1) was selected as the most suitable candidate for ^{18}F-labeling and biological investigation (IC$_{50}$ PDE2A = 3.33 nM; 16-fold selectivity over PDE10A) [28].

Herein, we reported on the development and the evaluation of [^{18}F]**BIT1**, including the synthesis of the corresponding nitro precursor, the manual radiosynthesis, and the transfer to an automated synthesis module, and the subsequent in vitro and in vivo investigations.

2. Results and Discussion

2.1. Precursor Synthesis and Radiochemistry

2.1.1. Synthesis of the Labeling Precursor 5

The synthesis of the reference compound **BIT1** has been reported previously [28]. For an efficient radiosynthesis of [^{18}F]**BIT1**, we selected the nitro precursor **5** because of its higher reactivity to nucleophilic aromatic substitutions in comparison to a bromo-substituted precursor [29,30]. As depicted in Scheme 1, **5** was prepared in four steps starting from the BIT key intermediate **1** [28]. Firstly, Miyaura borylation of **1** with bis(pinacolato)diboron, using potassium acetate and [Pd(dppf)Cl$_2$] as base and catalyst, afforded the pinacol boronic ester **2** in a satisfactory yield of 85% [31]. By palladium-catalyzed Suzuki coupling with 4-bromo-2-nitropyridine, compound **3** was obtained in 63% yield. This 4-bromo-2-nitropyridine was synthesized according to a slightly modified procedure, already described in the literature [32]. Thereafter, the bromination at 1-position of the imidazole ring using N-bromo-succinimide (NBS) afforded compound **4** in an excellent yield of 97%. Finally, the subsequent Suzuki coupling with 2-chloro-5-methoxy-phenyl boronic acid gave the nitro precursor **5** in 68% yield (NMR spectrums of precursor **5** see in the Supplementary Materials).

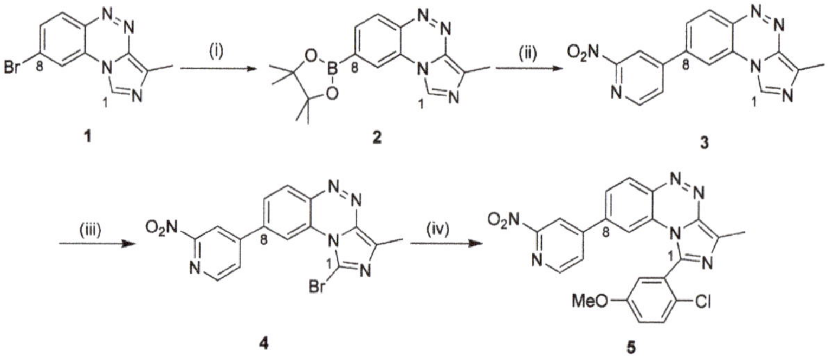

Scheme 1. Synthesis of precursor **5**. Reagents and conditions: (i) Bis(pinacolato)diboron [Pd(dppf)Cl$_2$], KOAc, 2-methyltetrahydrofuran (2-MeTHF), 100 °C (6 h), 85 °C (12 h), 85%; (ii) 4-bromo-2-nitropyridine, K$_2$CO$_3$, Pd(PPh$_3$)$_4$, 1,2-dimethoxyethane, 63%; (iii) N-bromo-succinimide (NBS), acetonitrile (MeCN), room temperature (RT), 4–6 h, 97%; (iv) 2-chloro-5-methoxy-phenyl-1-yl-boronic acid, K$_2$CO$_3$, [Pd(PPh$_3$)$_4$], 1,4-dioxane/H$_2$O (4:1), reflux, 1–2 d, 68%.

2.1.2. Radiosynthesis and Characterization of [^{18}F]**BIT1**

Manual Radiosynthesis

The new PDE2A radioligand [^{18}F]**BIT1** was prepared by heteroaromatic nucleophilic substitution of **5** using a classical anhydrous K[^{18}F]F-K$_{2.2.2}$ carbonate complex. The radiolabeling process of [^{18}F]**BIT1** was carried out under thermal heating applying a constant amount of 1.0 mg of precursor **5** by using different (i) polar aprotic solvents dimethylsulfoxide (DMSO), N,N-dimethylformamide (DMF), and acetonitrile (MeCN), (ii) temperatures, and (iii) reaction times.

As shown in Figure 2, after 5 min (min) reaction time, a radiochemical yield (RCY) of ~57% was obtained when DMF was used at 150 °C. According to radio-TLC, besides [^{18}F]fluoride, three radioactive by-products were observed, which accounted for 10% of total radioactivity. When the temperature was reduced to 120 °C, an increase of the RCY up to 93% was observed at this early time point. However, the RCY decreased with increasing the reaction time, indicating a decomposition of [^{18}F]**BIT1** under these conditions. The highest RCY in DMF (~94%) was achieved when the reaction temperature was further reduced to 100 °C. At this temperature, the product remained stable over 20 min reaction time. By contrast, almost no radiofluorination could be observed when MeCN was used (RCY ≤1%). With DMSO as a solvent, high RCYs (>85%) were obtained at 150 °C and 120 °C, which remained constant up to 20 min reaction time in contrast to the findings with DMF. A further decrease to 100 °C resulted in an increase of the RCY (>95%) after 10 min of reaction time. The precursor **5** was stable throughout the time of analysis, as proven by HPLC (data not shown). Based on these results, DMSO was selected for the production of [^{18}F]**BIT1**.

Due to the similarity of the chromatographic behavior of the nitro precursor and the corresponding ^{18}F-radiotracer, the use of a low amount of precursor is beneficial for subsequent isolation of the radiotracer via semi-preparative HPLC [33,34]. Accordingly, the further reduction of the amount of the nitro precursor **5** up to 0.5 mg was investigated, achieving an excellent RCY of ≥95% (DMSO, 100 °C, 5 min reaction time). The selection of a suitable column for semi-preparative HPLC was also somewhat crucial in order to have a separation of the radiotracer and its nitro precursor in a reasonable time [33]. According to our previous experiences [33], a slightly polar C18 phase turned out to be most appropriate for the isolation of [^{18}F]**BIT1** using a mixture of MeCN/water and ammonium acetate as a buffer.

[^{18}F]**BIT1** was successfully isolated under the aforementioned conditions and further purified by solid-phase extraction (SPE). After formulation in 0.9% saline containing 10% ethanol, [^{18}F]**BIT1** was obtained with an RCY of 38% (decay corrected to the end of the bombardment, EOB), and molar activities in the range of 38 GBq/µmol (at the end of synthesis, EOS).

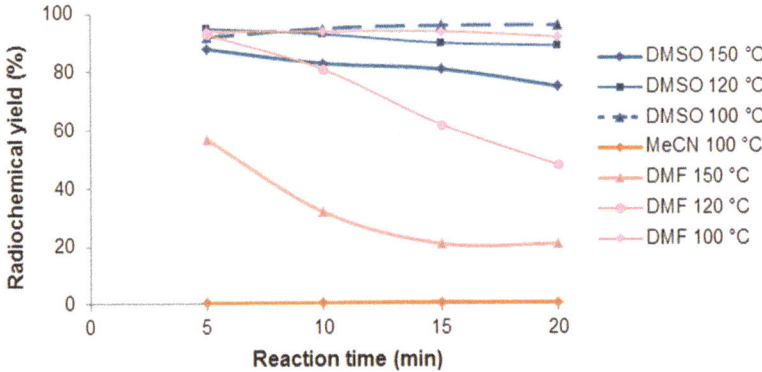

Figure 2. The radiochemical yield of [^{18}F]**BIT1** depending on reaction time, temperature, and solvent. (Conditions: K[^{18}F]F-K$_{2.2.2}$, thermal heating, 1.0 mg of **5**). BIT, benzoimidazotriazine.

Automated Radiosynthesis of [^{18}F]**BIT1**

By using the most appropriate conditions of the manual procedure, the automated radiosynthesis of [^{18}F]**BIT1** (Scheme 2) was established using a TRACERlab FX2 N synthesis module (GE Healthcare, Waukesha, WI, USA). The detailed configuration is shown in the experimental part. Briefly, the [^{18}F]F$^-$ was firstly trapped on an anion exchange cartridge and then eluted into the reactor using an aqueous potassium carbonate solution. The reaction took place with the [^{18}F]F$^-$/K$_{2.2.2}$/K$_2$CO$_3$ system and the nitro precursor **5** (0.5 mg) in DMSO at 100 °C for 5 min. After the isolation of [^{18}F]**BIT1** via semi-preparative RP-HPLC (Figure 3A), the product was purified via SPE on an RP cartridge and formulated in sterile isotonic saline containing 10% of EtOH. The total synthesis time of [^{18}F]**BIT1** was approximately 75 min. Analytical radio- and UV-HPLC of the final product, spiked with the non-labeled reference compound **BIT1**, confirmed the identity of [^{18}F]**BIT1** (Figure 3B). Finally, the radiotracer was obtained with a radiochemical purity of ≥99%, an RCY of 54 ± 2% (EOB, n = 3), and molar activities in the range of 155–175 GBq/µmol (EOS, n = 3).

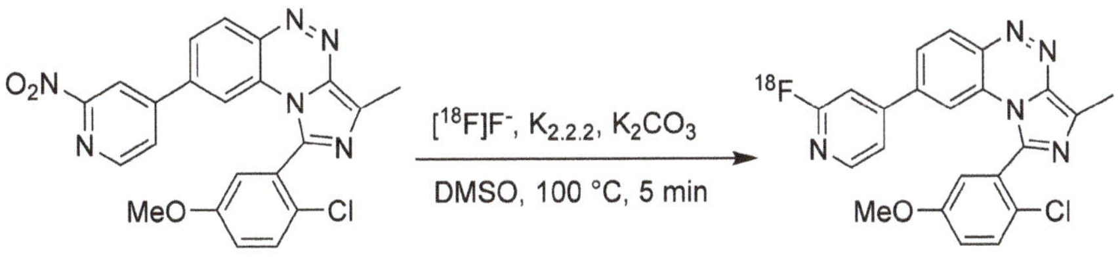

Scheme 2. ^{18}F-labeling of [^{18}F]**BIT1**.

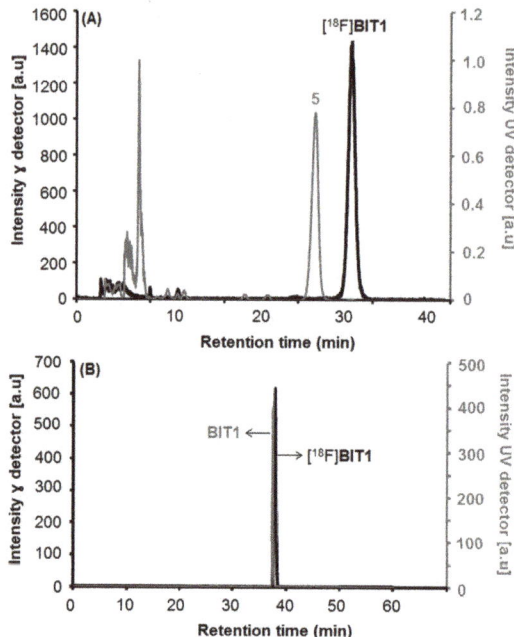

Figure 3. (**A**) Semi-preparative radio- and UV-HPLC chromatograms of [^{18}F]**BIT1** (conditions: Reprosil-Pur C18-AQ, 250 × 10 mm, 46% MeCN/20 mM NH$_4$OAc$_{aq.}$, 5.5 mL/min); (**B**) Analytical radio- and UV-HPLC chromatograms of the final product of [^{18}F]**BIT1** spiked with the non-radioactive reference **BIT1** (conditions: Reprosil-Pur C18-AQ, 250 × 4.6 mm, gradient system with eluent mixture of MeCN/20 mM NH$_4$OAc$_{aq.}$, 1.0 mL/min).

In vitro Stability and Lipophilicity of [^{18}F]**BIT1**

The radioligand [^{18}F]**BIT1** was stable (radiochemical purity ≥99%) in phosphate-buffered saline (PBS), pig plasma, *n*-octanol, as well as in the saline formulation containing 10% ethanol at 40 °C for up to 1 h. The lipophilicity was determined by the shake-flask method in the *n*-octanol/PBS system. With a logD$_{7.4}$ value of 1.81 ± 0.05, [^{18}F]**BIT1** falls within the range of radiotracers with optimal brain passive diffusion [35–37]. However, the calculated coefficient distribution value (ACD/Labs, Version 12.0, Advanced Chemistry Development, Inc.) displayed value of 4.03. The significant deviations between the calculation and experimental methods were observed in particular when the pattern of connectivity and non-bonded intramolecular interactions are not included in the applied database [38–41]. Moreover, the big discrepancy between the experimental and the calculated logD$_{7.4}$ value was already observed for our PDE2 tracer [^{18}F]**TA5** and has been previously discussed [25]. It is assumed that the experimentally determined higher hydrophilicity of [^{18}F]**TA5** might be due to the solvation effect related to hydrogen bonding and ionization of the radioligand in the aqueous buffered system [25,35]. Whereas with the software-based determination, this effect may be underestimated [25]. Therefore, the calculated logD value is often higher than the experimentally determined value [25,35]. Since **TA5** and **BIT1** are rather structurally similar, we assume the apparent strong discrepancy of logD for [^{18}F]**BIT1** might also be caused by the same reasons as those for [^{18}F]**TA5**.

2.2. In Vitro and In Vivo Characterization of [^{18}F]BIT1

2.2.1. In Vitro Autoradiography of [^{18}F]BIT1

To investigate the distribution of binding sites of [^{18}F]**BIT1** in the brain, in vitro autoradiographic studies using cryosections of pig brain were performed. As depicted in Figure 4B, the distribution pattern of [^{18}F]**BIT1** corresponds to the known spatial distribution of PDE2A with a high density of binding sites in the caudate nucleus (Cd), nucleus accumbens (Acb), cortex (Cx), and hippocampus (Hip) (Nissl staining of the corresponding slice is shown in Figure 4A). However, [^{18}F]**BIT1** also binds to the non-PDE2A specific regions cerebellum (Cb) and thalamus (Th), indicating binding of [^{18}F]**BIT1** to other targets.

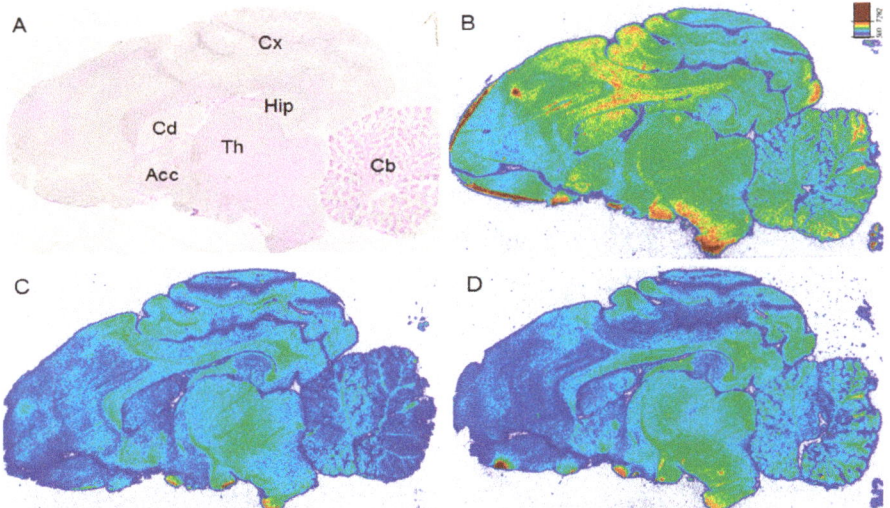

Figure 4. In vitro autoradiography of [^{18}F]**BIT1** in pig brain slices, (**A**) Nissl staining, (**B**) Total binding of 1.72 nM of [^{18}F]**BIT1**, (**C**) Blocking studies (10 μM of **TA1**), (**D**) Self-blocking (10 μM of **BIT1**). Abbreviations: (Acb: nucleus accumbens; Cb: cerebellum; Cd: nucleus caudatus; Cx: cortex; Hip: hippocampus; Th: thalamus).

To verify these findings, blocking studies with 10 μM of **TA1** (a potent PDE2 ligand) [42] (Figure 4C) and **BIT1** (Figure 4D) were performed. The decrease of [^{18}F]**BIT1** binding of ~50% and ~30% in PDE2A specific regions Cd and Acb, respectively, observed for both **TA1** and **BIT1** indicated in vitro specificity of the radiotracer. However, the simultaneous decrease of [^{18}F]**BIT1** binding in the range of 20–30% in Cb by **TA1**, as well as **BIT1**, suggested further non-specific binding of the radiotracer. We hypothesized the high non-specific binding of [^{18}F]**BIT1** could be related to the moderate selectivity of **BIT1** over PDE10A. Accordingly, these results limited the suitability of [^{18}F]**BIT1** for in vitro molecular imaging of PDE2A.

2.2.2. In Vivo Metabolism of [^{18}F]BIT1

The in vivo metabolism of [^{18}F]**BIT1** was investigated in plasma and brain homogenate obtained from mice at 30 min p.i.. Prior to the analysis with RP-HPLC, the samples were treated with a mixture of MeCN/H$_2$O (4:1, v/v) to precipitate the proteins. The recovery of radioactivity in plasma and brain samples was 96% and 99%, respectively. Intact tracer accounted for 43% and 78% of total activity in plasma and brain, respectively, indicating higher metabolic stability of [^{18}F]**BIT1** in comparison to our previous PDE2A radioligands [^{18}F]**TA3**, [^{18}F]**TA4**, and [^{18}F]**TA5** [23]. As shown in the chromatograms

(Figure 5), the two radio-metabolites ([^{18}F]**M1** and [^{18}F]**M2**) found in the brain were also observed in the corresponding plasma sample, indicating their ability to penetrate the blood-brain barrier. For further clarification, samples obtained as previously described in our in vitro metabolism study with **BIT1** using mouse liver microsomes [28] were investigated similarly by HPLC, but with UV detection (Figure 5C). On the basis of the in vitro metabolites **M1** and **M2**, both elucidated by LC-MS in the mentioned study, we can conclude about the brain penetrating radio-metabolites [^{18}F]**M1** and [^{18}F]**M2** as products of N-oxidation or C-hydroxylation. Besides, some of the more polar radio-metabolites detected in plasma could only tentatively be assigned as formed by mono-oxygenation, di-oxygenation, reduction, and demethylation, but were not investigated further.

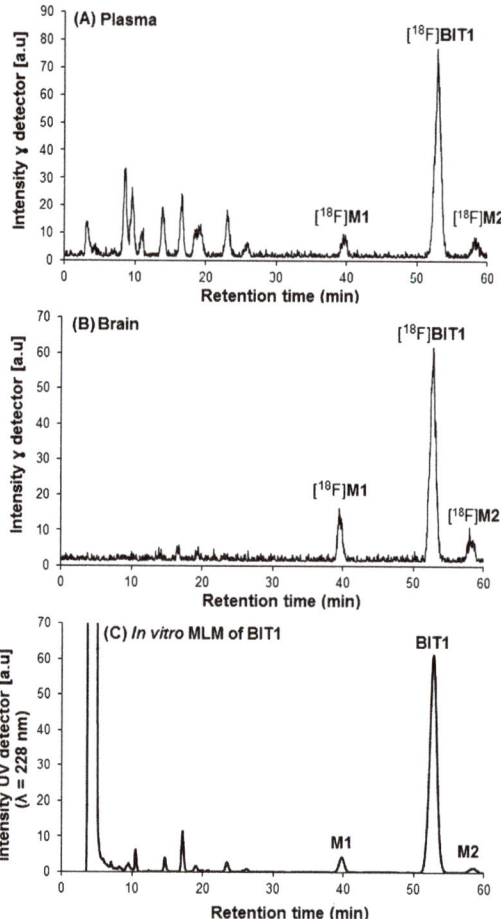

Figure 5. Representative analytical HPLC chromatogram: (**A**) plasma and (**B**) brain samples at 30 min p.i of [^{18}F]**BIT1** in a CD-1 mouse; (**C**) sample from incubation of **BIT1** with mouse liver microsomes (MLM) in the presence of NADPH for 90 min at 37 °C (UV detection: 228 nm; axis of retention time was adjusted). Conditions: isocratic system, column: Reprosil-Pur C18-AQ, 250 × 4.6 mm; eluent: 42% MeCN in 20 mM NH$_4$OAc$_{aq.}$, flow: 1.0 mL/min.

2.3. In Vivo PET-MRI Studies of [^{18}F]BIT1

Dynamic PET-MRI studies were performed in female CD-1 mice after intravenous administration of [^{18}F]**BIT1**. As reflected by the time-activity curves (TACs) shown in Figure 6, the radioactivity uptake in the whole brain, striatum, and cerebellum reached standardized uptake values (SUVs) of about 0.7 at 5 min p.i., indicating blood-brain barrier penetration. Since there is no difference in the TACs between the whole brain, striatum, and cerebellum, [^{18}F]**BIT1** is assumed to possess low specific binding also in vivo.

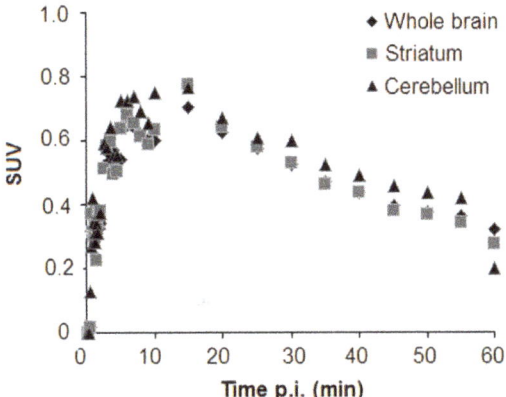

Figure 6. Averaged time-activity curves of [^{18}F]**BIT1** in CD-1 mice (n = 2) with standard uptake values (SUV) in the whole brain, striatum, and cerebellum.

To further investigate this hypothesis, blocking studies were performed by concomitant injection of **TA1** and [^{18}F]**BIT1**. However, since no significant reduction of radioactivity uptake in the PDE2A-specific region striatum was detectable (data not shown), the high non-specific binding of [^{18}F]**BIT1** already observed in vitro was confirmed.

Overall, the herein reported potential PDE2A radioligand, selected from a new class of 8-pyridinyl-BIT compounds, demonstrated sufficient blood-brain barrier (BBB) permeability. However, [^{18}F]**BIT1** was found to be insufficient for in vivo imaging of PDE2A with PET. Further structural modifications are needed to obtain PDE2A selective radioligands for in vitro and in vivo research. Extensive structure-activity relationship (SAR) studies could lead to the improvement of the selectivity and specificity of compounds. In particular, modifications of the set substituents at 1- and 8-positions of the BIT scaffold might result in additive as well as nonadditive effects in compound potency [43].

3. Materials and Methods

3.1. General Methods

All chemicals were purchased from commercial sources and used without further purification. Solvents were dried before used if required. Air and moisture-sensitive reactions were carried out under argon atmosphere. Room temperature (RT) refers to 20–25 °C. Reaction mixtures were monitored by thin-layer chromatography (TLC) using pre-coated TLC-plates POLYGRAM® SIL G/UV254 (Macherey-Nagel GmbH & Co. KG, Düren, Germany). The spots were detected under UV light at λ 254 nm and 365 nm. For the purification of final products, flash chromatography was performed with silica gel 40–63 μm (Merck KGaA, Darmstadt, Germany). Radio-TLC was performed on Polygram SIL G/UV$_{254}$ plates (Macherey-Nagel GmbH & Co. KG, Düren, Germany) pre-coated plates with a mixture of chloroform/methanol (10:1, v/v) as eluent. The radio-TLC plates were exposed to storage phosphor screens (BAS-IP MS 2025, FUJIFILM Co., Tokyo, Japan) and recorded using the AmershamTyphoon RGB Biomolecular Imager (GE Healthcare Life Sciences, Freiburg, Germany), and the images were quantified with ImageQuant TL8.1 software (GE Healthcare Life Sciences, Freiburg, Germany).

NMR spectra (^{1}H and ^{13}C) were recorded on Mercury 300/Mercury 400 (Varian, Palo Alto, CA, USA) or Fourier 300/Avance DRX 400 Bruker (Billerica, MA, USA) instruments. Residual solvent proton (^{1}H) and carbon (^{13}C) resonances were used as internal standards for ^{1}H-NMR (CDCl$_3$, δ_H = 7.26; DMSO-d_6, δ_H = 2.50) and ^{13}C-NMR (CDCl$_3$, δ_C = 77.16; DMSO-d_6, δ_C = 39.52). The chemical shifts (δ) were reported in ppm as follows: s, singlet; d, doublet; t, triplet; m, multiplet, and the coupling constants (J) are reported in Hz. Mass spectra were recorded on an ESQUIRE 3000 Plus (ESI, low resolution) and a 7 T APEX II (ESI, high resolution) (Bruker Daltonics, Bruker Corporation, Billerica, MA, USA).

For semi-preparative HPLC, the following conditions were used. Column: Reprosil-Pur C18-AQ, 250 × 10 mm, particle size: 10 µm; eluent: 46% MeCN/20 mM NH$_4$OAc$_{aq.}$; flow: 5.5 mL/min; ambient temperature; UV detection at 254 nm. The semi-preparative HPLC separations in the manual radiosynthesis were performed on a JASCO LC-2000 system, using a PU-2080-20 pump, an UV/VIS-2075 detector coupled with a radioactivity HPLC flow monitor (Gabi Star, raytest Isotopenmessgeräte GmbH, Straubenhardt, Germany) and a fraction collector (Advantec CHF-122SC, Dublin, CA, USA). Data recording was performed with the Galaxie chromatography software (Agilent Technologies, Santa Clara, USA).

Analytical chromatographic separations were performed on a JASCO LC-2000 system, incorporating a PU-2080*Plus* pump, AS-2055*Plus* auto-injector (100 µL sample loop), and a UV-2070 *Plus* detector coupled with a gamma radioactivity HPLC detector (Gabi Star, raytest Isotopenmessgeräte GmbH, Straubenhardt, Germany). Data analysis was performed with the Galaxie chromatography software (Agilent Technologies, Santa Clara, USA) using the chromatogram obtained at 254 nm. A Reprosil-Pur C18-AQ column (250 × 4.6 mm; 5 µm; Dr. Maisch HPLC GmbH, Ammerbuch-Entringen, Germany) with MeCN/20 mM NH$_4$OAc$_{aq.}$ (pH 6.8) as eluent mixture and a flow of 1.0 mL/min was used (gradient: eluent A 10% MeCN/20 mM NH$_4$OAc$_{aq.}$; eluent B 90% MeCN/20 mM NH$_4$OAc$_{aq.}$; 0–10 min 100% A, 10–40 min up to 100% B, 40–45 min 100% B, 45–50 min up to 100% A, 50–60 min 100% A; isocratic system 42% MeCN/20 mM NH$_4$OAc$_{aq.}$; flow: 1.0 mL/min; ambient temperature).

The molar activities were determined on the basis of a calibration curve (0.1–6 µg of **BIT1**) performed under isocratic HPLC conditions (46% MeCN/20 mM NH$_4$OAc$_{aq.}$) using chromatograms obtained at 228 nm as the maximum of UV absorbance of compound **BIT1**.

No-carrier-added (n.c.a.) [^{18}F]fluoride ($t_{1/2}$ = 109.8 min) was produced by irradiation of [^{18}O]H$_2$O (Hyox 18 enriched water, Rotem Industries Ltd., Arava, Israel) via [^{18}O(p,n)^{18}F] nuclear reaction by irradiation of on a Cyclone®18/9 (iba RadioPharma Solutions, Louvain-la-Neuve, Belgium).

Remote-controlled automated syntheses were performed using a TRACERlab FX2 N synthesizer (GE Healthcare, USA) equipped with a N810.3FT.18 pump (KNF Neuburger GmbH, Freiburg, Germany), a BlueShadow UV detector 10D (KNAUER GmbH, Berlin, Germany), NaI(Tl)- counter, and the TRACERlab FX Software.

3.2. Precursor Synthesis and Radiochemistry

The final compounds described in this manuscript meet the purity requirement (>95%) determined by UV-HPLC.

3.2.1. Synthesis of Precursor

3-Methyl-8-(4,4,5,5-tetramethyl-1,3,2-dioxaborolan-2-yl)benzo[e]imidazo[5,1-c][1,2,4]triazine (**2**)

A mixture of bromo derivative **1** [28] (1.32 g, 5 mmol), potassium acetate (1.1 g, 11.2 mmol), and bis(pinacolato)diboron (1.3 g, 5.11 mmol) in 2-MeTHF (20 mL) was degassed with argon for 10 min. After the addition of Pd(dppf)Cl$_2$ (0.055 g, 0.075 mmol), the mixture was refluxed at 100 °C for 6 h and at 85 °C for 12 h. Upon completion of the reaction (monitored by TLC), CH$_2$Cl$_2$ (25 mL) was added, and the reaction mixture was stirred for 1 h. The solid was filtered off, and the filtrate was evaporated. The dark brown semi-solid residue (2.5 g) was dissolved in MTBE (60 mL) and filtered through a short silica gel column (H: 3 cm × D: 2 cm). Heptane (30 mL) was added to the eluate, followed by concentration to a volume of approximately 20 mL. The precipitated solid was filtered and dissolved in MTBE (120 mL), and the solution obtained was filtered again through a short silica gel column (H: 2 cm × D: 2 cm). The yellow eluate was concentrated (→ ~15 mL) and treated with *n*-heptane (30 mL). The solid obtained was filtered and dried under reduced pressure to afford a yellow solid **2** (1.32 g, 85%). TLC [CHCl$_3$/MeCN (10:1)]: R$_f$ = 0.26 (strong tailing). ^{1}H NMR (400 MHz, CDCl$_3$) δ$_H$ = 8.57 (s, 1H), 8.41 (d, J = 8.0 Hz, 1H), 8.25 (s, 1H), 8.05 (dd, J = 8.0, 1.1 Hz, 1H), 2.89 (s, 3H), 1.40 (s, 13H).

3-Methyl-8-(2-nitropyridin-4-5 yl)benzo[e]imidazo[5,1-c][1,2,4]triazine (**3**)

A mixture of boronic ester **2** (470 mg, 1.5 mmol, 1 eq) and 4-bromo-2-nitropyridine (243 mg, 1.2 mmol) in 1,2-dimethoxyethane (6 mL) and an aqueous solution of K$_2$CO$_3$ (2 M, 1.75 mL, 3.5 mmol) was degassed with argon for 20 min. Pd(PPh$_3$)$_4$ (50 mg, 0.043 mmol) was added, and the reaction mixture was heated at 95 °C for 1 h and 88 °C for 14 h. Upon full conversion of starting material (monitored by TLC), the solvent was evaporated, and the residue was partitioned between water (30 mL) and CHCl$_3$ (50 mL). After separation of the organic layer, the aqueous layer was extracted with CHCl$_3$/MeCN (8:1, 6 × 30) and CHCl$_3$ (5 × 30 mL). The combined organic layers were dried over Na$_2$CO$_3$, and the solvent was evaporated to leave a solid residue (370 mg). The residue was refluxed in MeCN (50 mL) for 5 min, and after cooling to 0–2 °C, the product was filtered and dried to give **3** as yellow solid (233 mg, 63%). TLC [CHCl$_3$/MeOH/30% NH$_3$ (10:1:0.1)]: R$_f$ = 0.35. ^{1}H NMR (300 MHz, DMSO-d6) δ$_H$ = 9.26 (s, 1H), 8.96 (dd, J = 2.0, 0.5 Hz, 1H), 8.85 (dt, J = 5.1, 0.8 Hz, 1H), 8.82–8.80 (m, 1H), 8.52 (dd, J = 8.7, 1.5 Hz, 1H), 8.42 (dd, J = 5.1, 1.7 Hz, 1H), 8.28 (dd, J = 8.5, 2.0 Hz, 1H), 2.78 (d, J = 1.7 Hz, 3H). ^{13}C NMR (75 MHz, DMSO-d6) δ$_C$ = 157.67, 149.59, 149.28, 138.82, 138.14, 136.18, 135.67, 130.31, 127.22, 127.17, 126.02, 122.45, 115.55, 114.35, 12.18.

1-Brom-3-methyl-8-(2-nitropyridin-4-yl)benzo[e]imidazo[5,1-c][1,2,4]triazine (**4**)

N-Bromosuccinimide (NBS) (48 mg, 1.58 eq) was added to a suspension of compound **3** (52 mg, 1 eq) in 2.5 mL MeCN. The reaction mixture was protected from light and stirred at RT for 1–2 d. After the complete conversion of starting material (monitored by TLC), the mixture was partitioned between CHCl$_3$ (15 mL) and water (15 mL). The organic layer was washed with aqueous NaHCO$_3$ (5%, 15 mL) and Na$_2$SO$_3$ (5%, 15 mL). The aqueous layer was extracted again with CHCl$_3$ (4x5 mL). The combined organic layers were dried (Na$_2$SO$_4$) and evaporated. The crude product was purified by column chromatography (CHCl$_3$/MeOH, 20:1) to afford compound **4** as a yellow solid (64 mg, 97%). TLC [CHCl$_3$/MeOH (20:1)]: R$_f$ = 0.59. ^{1}H NMR (400 MHz, CDCl$_3$) δ$_H$ = 9.31 (d, J = 1.8 Hz, 1H), 8.84 (d, J = 5.0 Hz, 1H), 8.65 (d, J = 8.4 Hz, 1H), 8.59 (d, J = 1.6 Hz, 1H), 8.04–7.98 (m, 2H), 2.90 (s, 3H).

1-(2-Chloro-5-methoxyphenyl)-3-methyl-8-(2-nitropyridin-4-yl)benzo[e]imidazo[5,1-c][1,2,4]triazine (**5**)

Brominated compound **4** (43 mg, 0.112 mmol, 1 eq), 2-chloro-5-methoxyphenyl boronic acid (27 mg, 0.145 mmol, 1.29 eq), and K$_2$CO$_3$ (44 mg, 0.318 mmol, 2.8 eq) were combined in a mixture of 1,4-dioxane and H$_2$O (4:1). Under argon atmosphere, [Pd(PPh$_3$)$_4$] (13 mg, 10 mol%) was added. After refluxing at 101–105 °C for 3 h, the reaction mixture was stirred at RT for 16 h. Afterward, the residue was partitioned between CHCl$_3$ (20 mL) and water (20 mL). The organic layer was washed with water (20 mL) and brine (20 mL). The aqueous layer was extracted with CHCl$_3$ (3 × 10 mL), and the combined organic layers were dried over Na$_2$SO$_4$, and the solvent was evaporated under reduced pressure. The crude product was purified by column chromatography using a gradient system (*n*-hexane/ethyl acetate from 1:1 to 1:3) to get the corresponding precursor **5** as a yellow solid (34 mg, 68%). TLC [*n*-hexane/ethyl acetate (1:2)]: R$_f$ = 0.38. ^{1}H NMR (400 MHz, CDCl$_3$) δ$_H$ = 8.71 (dd, J = 5.0, 0.6 Hz, 1H), 8.60 (d, J = 8.4 Hz, 1H), 8.15 (dd, J = 1.6, 0.6 Hz, 1H), 7.95 (dd, J = 8.4, 1.9 Hz, 1H), 7.77 (dd, J = 5.0, 1.6 Hz, 1H), 7.67–7.64 (m, 1H), 7.58 (d, J = 1.9 Hz, 1H), 7.24–7.20 (m, 2H), 3.87 (s, 3H), 3.00 (s, 3H). ^{13}C NMR (101 MHz, CDCl$_3$) δ$_C$ = 159.24, 157.85, 150.52, 149.96, 140.32, 138.66, 137.53, 137.22, 136.47, 131.76, 131.27, 131.11, 126.50, 125.79, 125.61, 124.31, 118.59, 117.99, 115.77, 114.21, 56.06, 12.94. ESI (+): m/z = 447.10 (calcd. 447.10 for C$_{22}$H$_{16}$ClN$_6$O$_3^+$ [M + H]$^+$).

3.2.2. In Vitro Metabolism of **BIT1**

BIT1 was incubated with mouse liver microsomes in the presence of NADPH, including positive and negative controls, for 90 min at 37 °C, as previously described [28]. Obtained samples were investigated by analytical HPLC with an isocratic system (42% MeCN/20 mM NH_4OAc_{aq}; flow rate of 1.0 mL/min) and UV detection (228 nm).

3.2.3. Radiochemistry

Manual Radiosynthesis

No carrier added [^{18}F]fluoride in 1.5 mL water and was trapped on a Chromafix® 30 PS-HCO_3^- cartridge (Macherey-Nagel GmbH & Co. KG, Düren, Germany). The activity was eluted with 300 µL of an aqueous solution of potassium carbonate solution (K_2CO_3, 1.8 mg, 13 µmol) into a 4 mL V-vial, and Kryptofix$_{2.2.2}$. ($K_{2.2.2}$) (11 mg, 29 µmol) in 1.0 mL MeCN was added. The aqueous [^{18}F]fluoride was azeotropically dried under vacuum and nitrogen flow within 7–10 min using a single-mode microwave (75 W, at 50–60 °C, power cycling mode). Two aliquots of MeCN (2 × 1.0 mL) were added during the drying procedure, and the final complex was dissolved in 500 µL of solvent (DMSO, DMF, and MeCN) ready for labeling. Thereafter, 0.5–1.0 mg of precursor in 500 µL labeling solvent was added, and the ^{18}F-labeling was performed at different temperatures (100 °C, 120 °C, and 150 °C). To analyze the reaction mixture and to determine radiochemical yields, samples were taken for radio-HPLC and radio-TLC at different time points (5, 10, 15, and 20 min).

After cooling to <30 °C, the reaction mixture was diluted with 1.0 mL water and 2.0 mL of MeCN/H_2O (1:1) and directly applied to an isocratic semi-preparative RP-HPLC for isolation of [^{18}F]**BIT1** (46% MeCN/20 mM $NH_4OAc_{aq.}$, 5.5 mL/min, Reprosil-Pur C18-AQ column, 250 × 10 mm; 10 µm; Dr. Maisch HPLC GmbH, Ammerbuch-Entringen, Germany). The collected radiotracer fraction was diluted with 35 mL water to perform purification by sorption on a Sep-Pak® C18 light cartridge (Waters, Milford, MA, USA) and successive elution with 0.75 mL of ethanol.

Automated Radiosynthesis of [^{18}F]**BIT1**

The automated radiosynthesis was performed in a TRACERlab FX2 N synthesis module (GE Healthcare, Waukesha, WI, USA). About 2–4.5 GBq of aqueous n.c.a [^{18}F]fluoride was trapped on a Chromafix® 30 PS-HCO_3^- cartridge (Figure 7, Macherey-Nagel GmbH & Co. KG, Düren, Germany, entry 1) and eluted with K_2CO_3 (1.8 mg, 13 µmol, entry 2) in 400 µL water. Kryptofix $K_{2.2.2}$ (11 mg, 29 µmol, in 1.0 mL MeCN, entry 3) was directly added. Afterward, azeotropic drying was performed at 60 °C and 85 °C under a vacuum. Thereafter, 0.5 mg of **5** in 1.0 mL DMSO (entry 4) was transferred to the reactor, and subsequently, the reaction was performed at 100 °C for 5 min. After cooling the reactor to 30 °C, 1.0 mL of H_2O and 2.0 mL of MeCN/H_2O (1:1) (entry 5) were added and transferred into the injection vial (entry 6). The semi-preparative HPLC was performed using a Reprosil-Pur C-18 AQ column (250 × 10 mm; 10 µM) using 46% MeCN in aqueous 20 mM ammonium acetate at a flow rate of 5.5 mL/min (entry 7). The [^{18}F]**BIT1** fraction was pooled in 35 mL H_2O (entry 8) and then loaded on a preconditioned Sep-Pak® C18 light cartridge (Waters, Milford, MA, USA, entry 9), and washed with 2.0 mL of H_2O (entry 10). Afterward, the trapped [^{18}F]**BIT1** was eluted with 1.2 mL EtOH (entry 11) into the product vial (entry 12). The product was transferred out of the hot cell, and the solvent was reduced under a gentle argon stream at 70 °C to a final volume of 10–50 µL. Afterward, the radiotracer was diluted in isotonic saline to obtain a final product containing 10% of EtOH (v/v). The identity of the final product was confirmed using radioanalytical HPLC.

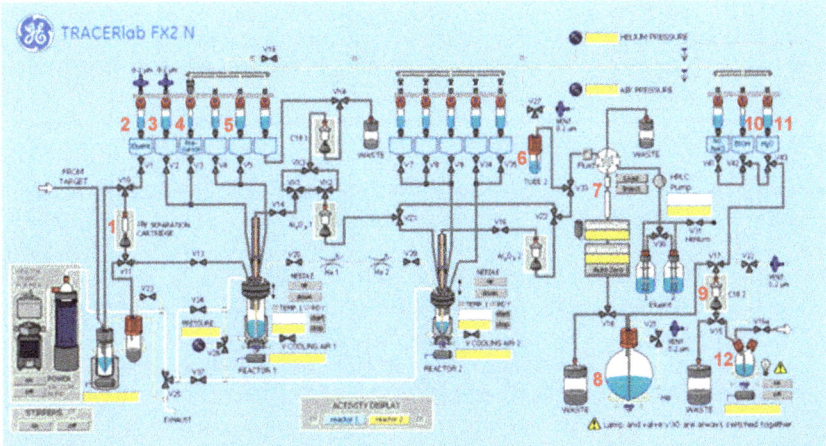

Figure 7. Scheme of the synthesis module TRACERlab FX2 N for the radiosynthesis of [^{18}F]**BIT1**. (1) Chromafix® 30 PS-HCO$_3$−, (2) K$_2$CO$_3$ (1.8 mg in 400 μL water), (3) K$_{2.2.2}$ (11 mg, 29 μmol in 1.0 mL MeCN), (4) precursor (0.5 mg of **5** in 1.0 mL DMSO), (5) 1.0 mL H$_2$O and 2.0 mL MeCN/H$_2$O (1:1), (6) injection vial, (7) Reprosil-Pur C18-AQ column (46% MeCN/20mM NH$_4$OAc$_{aq.}$, flow 5.5 mL/min), (8) 35 mL water, (9) Sep-Pak® C18 light, (10) 2.0 mL water, (11) 1.2 mL EtOH, (12) product vial.

The molar activity of [^{18}F]**BIT1** was determined by analytical HPLC (isocratic mode: 46% MeCN/20 5mM NH$_4$OAc$_{aq.}$, Reprosil-Pur C18-AQ column 250 × 4.6 mm, flow rate: 1.0 mL/min). The UV/Vis detection was carried out at 228 nm as the absorption maximum of **BIT1**.

Determination of Stability

In vitro stability of [^{18}F]**BIT1** was investigated by incubation of the radioligand in phosphate-buffered saline (PBS, pH 7.4), *n*-octanol, and pig plasma at 40 °C for 60 min (~5 MBq of the radioligand was added to 500 μL of each medium). Samples were taken at 30 and 60 min and analyzed by radio-TLC and radio-HPLC.

Determination of log D

The lipophilicity of [^{18}F]**BIT1** was determined by partitioning between *n*-octanol and phosphate-buffered saline (PBS, pH 7.4) at ambient temperature using the conventional shake-flask method. An aliquot of 10 μL of the formulated solution containing ~500 kBq of [^{18}F]**BIT1** was added to a tube containing 6 mL of the *n*-octanol/PBS-mixture (1:1, v/v, four-fold determination). The tubes were shaken for 20 min using a mechanical shaker (HS250 basic, IKA Labortechnik GmbH & Co. KG, Staufen, Germany) followed by centrifugation (5000 rpm for 5 min) and separation of the phases. Aliquots of 1 mL of the organic and the aqueous phase were taken, and the activity was measured using an automated gamma counter (1480 WIZARD, Fa. Perkin Elmer, Waltham, MA, USA). The distribution coefficient (D) was calculated as [activity (cpm/mL) in *n*-octanol]/[(activity (cpm/mL) in PBS], specified as the decadic logarithm (logD).

3.3. Animal Studies

All experimental work, including animals, was conducted in accordance with the national legislation on the use of animals for research (Tierschutzgesetz (TierSchG), Tierschutz-Versuchstierverordnung (TierSchVersV)) and was approved by the responsible research ethics committee (TVV 18/18, DD24.1-5131/446/19, Landesdirektion Sachsen, 20th June 2018). Female CD-1 mice, 10–12 weeks, were obtained from the Medizinisch-Experimentelles Zentrum at Universität Leipzig.

For the time of the experiments, the animals were kept in a dedicated climatic chamber with free access to water and food under a 12:12 h dark:light cycle at a constant temperature of 24 °C. Piglet brains were obtained from anesthetized and euthanized juvenile female German landrace pigs (Lehr- und Versuchsgut Oberholz, Universität Leipzig).

3.3.1. In Vitro Autoradiography of [^{18}F]**BIT1**

Pig brain cryosections (16 μm; Microm HM560 Cryostat, FischerScientific GmbH, Schwerte, Germany) were thawed, dried in a stream of cold air, and preincubated for 10 min with buffer (50 mM TRIS-HCl, pH 7.4, 120 mM NaCl, 5 mM KCl, 2 mM $MgCl_2$) at room temperature. Afterward, brain sections were incubated with 1.72 nM of [^{18}F]**BIT1** in the buffer for 60 min at room temperature. The sections were then washed twice with ice-cold 50 mM TRIS-HCl (pH 7.4), dipped in ice-cold deionized water, dried in a stream of cold air, and exposed for 60 min to an image plate. The analysis was performed using an image plate scanner (HD-CR 35; Duerr NDT GmbH, Bietigheim Bissingen, Germany). Non-specific binding was determined using 10 μM **TA1** and **BIT1** as blocking compounds.

3.3.2. In Vivo Metabolism of [^{18}F]**BIT1**

[^{18}F]**BIT1** (~19 MBq) was injected in female CD-1 mice (10–12 weeks old) via the tail vein. Brain and blood samples were obtained at 30 min p.i., plasma separated by centrifugation (14,000 rpm, 1 min), and brain homogenized in ~1 mL isotonic saline on ice (10 strokes of a PTFE plunge at 1000 rpm) in borosilicate glass.

Protein precipitation of plasma and brain samples was performed in duplicate with ice-cold $MeCN/H_2O$ (8:2, v/v), which was added to the samples (sample/solvent, 1:4, v/v). The samples were vortexed for 1 min, followed by resting on ice for 10 min. Afterward, the samples were centrifuged for 5 min at 10,000 g. Supernatants were collected, and 100 μL ice-cold $MeCN/H_2O$ (8:2, v/v) was added to the pellets for the second extraction, applying the same treatment as before. The combined supernatants were concentrated at 70 °C under nitrogen stream until a remaining volume of 100 μL and subsequently quantified by analytical radio-HPLC with an isocratic system (42% MeCN/20 mM $NH_4OAc_{aq.}$; flow rate: 1.0 mL/min). The activity recovery was determined by measuring the radioactivity of aliquots taken from supernatants and the pellets using a gamma counter.

3.4. PET-MRI Studies of [^{18}F]BIT1

PET/MRI scans were performed using a preclinical PET/MRI system (nanoScan, Mediso Medical Imaging Systems, Budapest, Hungary). Animals were anesthetized with a mixture of air/isoflurane (2.0%, 200 mL/min) (Anesthesia Unit U-410, agntho's, Lidingö, Sweden), and their body temperature maintained at 37 °C with a thermal bed system. [^{18}F]**BIT1** (2 MBq, A_m: 155 GBq/μmol, EOS) was injected via the tail vein, either without (baseline study, n = 2) or with **TA1** (2 mg/kg, DMSO/NaCl (1:25); blocking study, n = 2). A dynamic 60-min PET scan was started 20 sec before radioligand injection. The list-mode data were reconstructed into 33 frames (12 × 10 sec, 6 × 30 sec, 5 × 1 min, 10 × 5 min) with 3D-ordered subset expectation maximization, 4 iterations, and 6 subsets, using an energy window of 400 to 600 KeV and coincidence mode of 1–5. After the PET scan, an MRI scan was performed for anatomical orientation and attenuation correction on a 1T magnet using a T1-weighted gradient-echo sequence (TR = 15 ms, TE = 2.6 ms). The data analysis was done with PMOD (PMOD Technologies Ltd., Zurich, Switzerland, v. 3.6), and an atlas-based method was used to obtain SUV time-activity curves for the striatum, cerebellum, and the whole brain.

4. Conclusions

In our efforts to develop a specific radioligand for PET imaging of PDE2A in the brain on the basis of a benzoimidazotriazine scaffold, we successfully prepared the novel PDE2A radioligand [^{18}F]**BIT1** with a high radiochemical yield, radiochemical purity, as well as molar activity. Our findings showed that this class of compound demonstrated good brain penetration and sufficient in vivo metabolic stability. However, [^{18}F]**BIT1** was not suitable for the molecular imaging of PDE2A in the brain because of a high non-specific binding. Further structural modifications are required to obtain more satisfactory PDE2A specific radioligands.

Supplementary Materials: The following are available online, Figures S1–S2: NMR spectra of precursor **5** (^{1}H and ^{13}C), Figure S3. The calibration curve of molar activity of reference **BIT1**.

Author Contributions: B.W., M.S., and P.B. designed the study; M.S. and R.R. designed and performed organic syntheses; B.W., R.R., and R.T. performed radiosynthesis; W.D.-C., B.W., R.T., F.-A.L., and R.R. performed in vivo metabolism studies; S.D.-S., W.D.-C., and P.B. performed in vitro autoradiographic studies; M.T., M.K., W.D.-C., and P.B. designed and performed PET-MR studies; R.R., B.W., R.T., M.T., M.K., W.D.-C., S.D.-S, F.-A.L., M.S., and P.B. wrote the paper. All authors read and approved the final manuscript.

Acknowledgments: This work was financially supported by the Deutsche Forschungsgemeinschaft (DFG), Project Number: SCHE 1825/3-1. We further thank the Research and Innovation in Science and Technology Project (RISET-Pro), the Ministry of Research, Technology, and Higher Education, Republic of Indonesia, World Bank Loan No. 8245-ID, for supporting the Ph.D. thesis of Rien Ritawidya, the staff of the Institute of Analytical Chemistry (University of Leipzig) for measuring NMR and LR/HR-MS spectra, Karsten Franke and Steffen Fischer for the production of [^{18}F]fluoride, and Tina Spalholz for technical assistance.

Conflicts of Interest: The authors declare no conflict of interest.

References

1. Helal, C.J.; Arnold, E.P.; Boyden, T.L.; Chang, C.; Chappie, T.A.; Fennell, K.F.; Forman, M.D.; Hajos, M.; Harms, J.F.; Hoffman, W.E.; et al. Application of structure-based design and parallel chemistry to identify a potent, selective, and brain penetrant phosphodiesterase 2A inhibitor. *J. Med. Chem.* **2017**, *60*, 5673–5698. [CrossRef] [PubMed]
2. Zhu, J.; Yang, Q.; Dai, D.; Huang, Q. X-ray crystal structure of phosphodiesterase 2 in complex with a highly selective, nanomolar inhibitor reveals a binding-induced pocket important for selectivity. *J. Am. Chem. Soc.* **2013**, *135*, 11708–11711. [CrossRef] [PubMed]
3. Wu, Y.; Li, Z.; Huang, Y.-Y.; Wu, D.; Luo, H.-B. Novel phosphodiesterase inhibitors for cognitive improvement in Alzheimer's disease. *J. Med. Chem.* **2018**, *61*, 5467–5483. [CrossRef] [PubMed]
4. Svensson, F.; Bender, A.; Bailey, D. Fragment-based drug discovery of phosphodiesterase inhibitors. *J. Med. Chem.* **2018**, *61*, 1415–1424. [CrossRef] [PubMed]
5. Jäger, R.; Schwede, F.; Genieser, H.-G.; Koesling, D.; Russwurm, M. Activation of PDE2 and PDE5 by specific GAF ligands: Delayed activation of PDE5. *Br. J. Pharmacol.* **2010**, *161*, 1645–1660. [CrossRef] [PubMed]
6. Maurice, D.H.; Ke, H.; Ahmad, F.; Wang, Y.; Chung, J.; Manganiello, V.C. Advances in targeting cyclic nucleotide phosphodiesterases. *Nat. Rev. Drug Discov.* **2014**, *13*, 290–314. [CrossRef]
7. Conti, M.; Beavo, J. Biochemistry and physiology of cyclic nucleotide phosphodiesterases: Essential components in cyclic nucleotide signaling. *Annu. Rev. Biochem.* **2007**, *76*, 481–511. [CrossRef]
8. Lakics, V.; Karran, E.H.; Boess, F.G. Quantitative comparison of phosphodiesterase mRNA distribution in human brain and peripheral tissues. *Neuropharmacology* **2010**, *59*, 367–374. [CrossRef]
9. Menniti, F.S.; Faraci, W.S.; Schmidt, C.J. Phosphodiesterases in the CNS: Targets for drug development. *Nat. Rev. Drug Discov.* **2006**, *5*, 660–670. [CrossRef]
10. Gomez, L.; Breitenbucher, J.G. PDE2 inhibition: Potential for the treatment of cognitive disorders. *Bioorganic Med. Chem. Lett.* **2013**, *23*, 6522–6527. [CrossRef]
11. Stephenson, D.T.; Coskran, T.M.; Wilhelms, M.B.; Adamowicz, W.O.; O'Donnell, M.M.; Muravnick, K.B.; Menniti, F.S.; Kleiman, R.J.; Morton, D. Immunohistochemical localization of phosphodiesterase 2A in multiple mammalian species. *J. Histochem. Cytochem.* **2009**, *57*, 933–949. [CrossRef] [PubMed]
12. Stephenson, D.T.; Coskran, T.M.; Kelly, M.P.; Kleiman, R.J.; Morton, D.; O'Neill, S.M.; Schmidt, C.J.; Weinberg, R.J.; Menniti, F.S. The distribution of phosphodiesterase 2A in the rat brain. *Neuroscience* **2012**, *226*, 145–155. [CrossRef] [PubMed]
13. Gomez, L.; Massari, M.E.; Vickers, T.; Freestone, G.; Vernier, W.; Ly, K.; Xu, R.; McCarrick, M.; Marrone, T.; Metz, M.; et al. Design and synthesis of novel and selective phosphodiesterase 2 (PDE2a) inhibitors for the treatment of memory disorders. *J. Med. Chem.* **2017**, *60*, 2037–2051. [CrossRef] [PubMed]

14. Sierksma, A.S.R.; Rutten, K.; Sydlik, S.; Rostamian, S.; Steinbusch, H.W.M.; van den Hove, D.L.A.; Prickaerts, J. Chronic phosphodiesterase type 2 inhibition improves memory in the APPswe/PS1dE9 mouse model of Alzheimer's disease. *Neuropharmacology* **2013**, *64*, 124–136. [CrossRef] [PubMed]
15. Umar, T.; Hoda, N. Selective inhibitors of phosphodiesterases: Therapeutic promise for neurodegenerative disorders. *Med. Chem. Commun.* **2015**, *6*, 2063–2080. [CrossRef]
16. Mikami, S.; Sasaki, S.; Asano, Y.; Ujikawa, O.; Fukumoto, S.; Nakashima, K.; Oki, H.; Kamiguchi, N.; Imada, H.; Iwashita, H.; et al. Discovery of an orally bioavailable, brain-penetrating, in vivo active phosphodiesterase 2A inhibitor lead series for the treatment of cognitive disorders. *J. Med. Chem.* **2017**, *60*, 7658–7676. [CrossRef]
17. Reneerkens, O.A.H.; Rutten, K.; Steinbusch, H.W.M.; Blokland, A.; Prickaerts, J. Selective phosphodiesterase inhibitors: A promising target for cognition enhancement. *Psychopharmacology* **2009**, *202*, 419–443. [CrossRef]
18. Brust, P.; van den Hoff, J.; Steinbach, J. Development of ^{18}F-labeled radiotracers for neuroreceptor imaging with positron emission tomography. *Neurosci. Bull.* **2014**, *30*, 777–811. [CrossRef]
19. Andrés-Gil, J.I.; Rombouts, F.J.R.; Trabanco, A.A.; Vanhoof, G.C.P.; De Angelis, M.; Buijnsters, P.J.J.A.; Guillemont, J.E.G.; Bormans, G.M.R.; Celen, S.J.L.; Vliegen, M. 1-aryl-4-methyl-[1,2,4]triazolo[4,3-a]quinoxaline Derivatives. Patent No. WO 2013/000924 A1, 3 January 2013.
20. Zhang, L.; Villalobos, A.; Beck, E.M.; Bocan, T.; Chappie, T.A.; Chen, L.; Grimwood, S.; Heck, S.D.; Helal, C.J.; Hou, X.; et al. Design and selection parameters to accelerate the discovery of novel central nervous system positron emission tomography (PET) ligands and their application in the development of a novel phosphodiesterase 2A PET ligand. *J. Med. Chem.* **2013**, *56*, 4568–4579. [CrossRef]
21. Naganawa, M.; Waterhouse, R.N.; Nabulsi, N.; Lin, S.-F.; Labaree, D.; Ropchan, J.; Tarabar, S.; DeMartinis, N.; Ogden, A.; Banerjee, A.; et al. First-in-human assessment of the novel PDE2A PET radiotracer ^{18}F-PF-05270430. *J. Nucl. Med.* **2016**, *57*, 1388–1395. [CrossRef]
22. Naganawa, M.; Nabulsi, N.; Waterhouse, R.; Lin, S.-F.; Zhang, L.; Cass, T.; Ropchan, J.; McCarthy, T.; Huang, Y.; Carson, R. Human PET studies [^{18}F]PF-05270430, a PET radiotracer for imaging phosphodiesterase-2A. *J. Nucl. Med.* **2013**, *54*, 201.
23. Schröder, S.; Wenzel, B.; Deuther-Conrad, W.; Scheunemann, M.; Brust, P. Novel radioligands for cyclic nucleotide phosphodiesterase imaging with positron emission tomography: An update on developments since 2012. *Molecules* **2016**, *21*, 650. [CrossRef] [PubMed]
24. Schröder, S.; Wenzel, B.; Deuther-Conrad, W.; Teodoro, R.; Egerland, U.; Kranz, M.; Scheunemann, M.; Höfgen, N.; Steinbach, J.; Brust, P. Synthesis, ^{18}F-radiolabelling and biological characterization of novel fluoroalkylated triazine derivatives for in vivo imaging of phosphodiesterase 2A in brain via positron emission tomography. *Molecules* **2015**, *20*, 9591–9615. [CrossRef] [PubMed]
25. Schröder, S.; Wenzel, B.; Deuther-Conrad, W.; Teodoro, R.; Kranz, M.; Scheunemann, M.; Egerland, U.; Höfgen, N.; Briel, D.; Steinbach, J.; et al. Investigation of an ^{18}F-labelled imidazopyridotriazine for molecular imaging of cyclic nucleotide phosphodiesterase 2A. *Molecules* **2018**, *23*, 556. [CrossRef]
26. Stange, H.; Langen, B.; Egerland, U.; Hoefgen, N.; Priebs, M.; Malamas, M.S.; Erdei, J.J.; Ni, Y. Imidazo[5,1-c][1,2,4]benzotriazine Derivatives as Inhibitors of Phosphodiesterases. Patent US 2010/0120763 A1, 13 May 2010.
27. Malamas, M.S.; Stange, H.; Schindler, R.; Lankau, H.-J.; Grunwald, C.; Langen, B.; Egerland, U.; Hage, T.; Ni, Y.; Erdei, J.; et al. Novel triazines as potent and selective phosphodiesterase 10A inhibitors. *Bioorganic Med. Chem. Lett.* **2012**, *22*, 5876–5884. [CrossRef]
28. Ritawidya, R.; Ludwig, F.-A.; Briel, D.; Brust, P.; Scheunemann, M. Synthesis and in vitro evaluation of 8-pyridinyl-substituted benzo[e]imidazo[2,1-c][1,2,4]triazines as phosphodiesterase 2A inhibitors. *Molecules* **2019**, *24*, 2791. [CrossRef]
29. Dolci, L.; Dolle, F.; Valette, H.; Vaufrey, F.; Fuseau, C.; Bottlaender, M.; Crouzel, C. Synthesis of a fluorine-18 labeled derivative of epibatidine for in vivo nicotinic acetylcholine receptor PET imaging. *Bioorganic Med. Chem.* **1999**, *7*, 467–479. [CrossRef]
30. Wagner, S.; Teodoro, R.; Deuther-Conrad, W.; Kranz, M.; Scheunemann, M.; Fischer, S.; Wenzel, B.; Egerland, U.; Hoefgen, N.; Steinbach, J.; et al. Radiosynthesis and biological evaluation of the new PDE10A radioligand [^{18}F]AQ28A. *J. Label. Compd. Radiopharm.* **2017**, *60*, 36–48. [CrossRef]
31. Ishiyama, T.; Murata, M.; Miyaura, N. Palladium(0)-catalyzed cross-coupling reaction of alkoxydiboron with haloarenes: A direct procedure for arylboronic esters. *J. Org. Chem.* **1995**, *60*, 7508–7510. [CrossRef]
32. Wiley, R.H.; Hartman, J.L. Oxidation of aminopyridines to nitropyridines. *J. Am. Chem. Soc.* **1951**, *73*, 494.

33. Wenzel, B.; Günther, R.; Brust, P.; Steinbach, J. A fluoro versus a nitro derivative-a high-performance liquid chromatography study of two basic analytes with different reversed phases and silica phases as basis for the separation of a positron emission tomography radiotracer. *J. Chromatogr. A* **2013**, *1311*, 98–105. [CrossRef] [PubMed]
34. Lindemann, M.; Hinz, S.; Deuther-Conrad, W.; Namasivayam, V.; Dukic-Stefanovic, S.; Teodoro, R.; Toussaint, M.; Kranz, M.; Juhl, C.; Steinbach, J.; et al. Radiosynthesis and in vivo evaluation of a fluorine-18 labeled pyrazine based radioligand for PET imaging of the adenosine A_{2B} receptor. *Bioorganic Med. Chem.* **2018**, *26*, 4650–4663. [CrossRef] [PubMed]
35. Waterhouse, R.N. Determination of lipophilicity and its use as a predictor of blood–brain barrier penetration of molecular imaging agents. *Mol. Imaging Biol.* **2003**, *5*, 376–389. [CrossRef] [PubMed]
36. Pike, V.W. PET radiotracers: Crossing the blood-brain barrier and surviving metabolism. *Trends Pharmacol. Sci.* **2009**, *30*, 431–440. [CrossRef] [PubMed]
37. van de Waterbeemd, H.; Camenisch, G.; Folkers, G.; Chretien, J.R.; Raevsky, O.A. Estimation of blood-brain barrier crossing of drugs using molecular size and shape, and H-bonding descriptors. *J. Drug Target.* **1998**, *6*, 151–165. [CrossRef] [PubMed]
38. Testa, B.; Crivori, P.; Reist, M.; Carrupt, P.-A. The influence of lipophilicity on the pharmacokinetic behavior of drugs: Concepts and examples. *Perspect. Drug Discov. Des.* **2000**, *19*, 179–211. [CrossRef]
39. Bodor, N.; Gabanyi, Z.; Wong, C.-K. A new method for the estimation of partition coefficient. *J. Am. Chem. Soc.* **1989**, *111*, 3783–3786. [CrossRef]
40. Donovan, S.F.; Pescatore, M.C. Method for measuring the logarithm of the octanol–water partition coefficient by using short octadecyl–poly(vinyl alcohol) high-performance liquid chromatography columns. *J. Chromatogr. A* **2002**, *952*, 47–61. [CrossRef]
41. Vraka, C.; Nics, L.; Wagner, K.-H.; Hacker, M.; Wadsak, W.; Mitterhauser, M. LogP, a yesterday's value? *Nucl. Med. Biol.* **2017**, *50*, 1–10. [CrossRef]
42. Stange, H.; Langen, B.; Egerland, U.; Hoefgen, N.; Priebs, M.; Malamas, M.S.; Erdei, J.; Ni, Y. Triazine Derivatives as Inhibitors of Phosphodiesterases. Patent No. WO 2010/054253 A1, 14 May 2010.
43. Gomez, L.; Xu, R.; Sinko, W.; Selfridge, B.; Vernier, W.; Ly, K.; Truong, R.; Metz, M.; Marrone, T.; Sebring, K.; et al. Mathematical and Structural characterization of strong nonadditive structure-activity relationship caused by protein conformational changes. *J. Med. Chem.* **2018**, *61*, 7754–7766. [CrossRef]

Sample Availability: Samples of the compounds are not available from the authors..

© 2019 by the authors. Licensee MDPI, Basel, Switzerland. This article is an open access article distributed under the terms and conditions of the Creative Commons Attribution (CC BY) license (http://creativecommons.org/licenses/by/4.0/).

Review

A Survey of Molecular Imaging of Opioid Receptors

Paul Cumming [1,2,*], János Marton [3], Tuomas O. Lilius [4], Dag Erlend Olberg [5] and Axel Rominger [1,*]

1. Department of Nuclear Medicine, University of Bern, Inselspital, Freiburgstraße 18, 3010 Bern, Switzerland
2. School of Psychology and Counselling and IHBI, Queensland University of Technology, QLD 4059, Brisbane, Australia
3. ABX Advanced Biochemical Compounds, Biomedizinische Forschungsreagenzien GmbH, Heinrich-Glaeser-Strasse 10-14, D-1454 Radeberg, Germany; marton@abx.de
4. Center for Translational Neuromedicine, Faculty of Health and Medical Sciences, University of Copenhagen, 2200 Copenhagen N, Denmark; tuomas.lilius@sund.ku.dk
5. School of Pharmacy, University of Oslo, Norwegian Medical Cyclotron Centre, N-0372 Oslo, Norway and Norwegian Medical Cyclotron Centre Ltd., Sognsvannsveien 20, N-0372 Oslo, Norway; Dag.Erlend.Olberg@syklotronsenteret.no
* Correspondence: paul.cumming@insel.ch (P.C.); axel.rominger@insel.ch (A.R.); Tel.: +41-31-664-04-98 (P.C.); +41-31-632-26-10 (A.R.)

Academic Editor: Peter Brust
Received: 21 October 2019; Accepted: 13 November 2019; Published: 19 November 2019

Abstract: The discovery of endogenous peptide ligands for morphine binding sites occurred in parallel with the identification of three subclasses of opioid receptor (OR), traditionally designated as μ, δ, and κ, along with the more recently defined opioid-receptor-like (ORL1) receptor. Early efforts in opioid receptor radiochemistry focused on the structure of the prototype agonist ligand, morphine, although N-[methyl-^{11}C]morphine, -codeine and -heroin did not show significant binding *in vivo*. [^{11}C]Diprenorphine ([^{11}C]DPN), an orvinol type, non-selective OR antagonist ligand, was among the first successful PET tracers for molecular brain imaging, but has been largely supplanted in research studies by the μ-preferring agonist [^{11}C]carfentanil ([^{11}C]Caf). These two tracers have the property of being displaceable by endogenous opioid peptides in living brain, thus potentially serving in a competition-binding model. Indeed, many clinical PET studies with [^{11}C]DPN or [^{11}C]Caf affirm the release of endogenous opioids in response to painful stimuli. Numerous other PET studies implicate μ-OR signaling in aspects of human personality and vulnerability to drug dependence, but there have been very few clinical PET studies of μORs in neurological disorders. Tracers based on naltrindole, a non-peptide antagonist of the δ-preferring endogenous opioid enkephalin, have been used in PET studies of δORs, and [^{11}C]GR103545 is validated for studies of κORs. Structures such as [^{11}C]NOP-1A show selective binding at ORL-1 receptors in living brain. However, there is scant documentation of δ-, κ-, or ORL1 receptors in healthy human brain or in neurological and psychiatric disorders; here, clinical PET research must catch up with recent progress in radiopharmaceutical chemistry.

Keywords: opioid receptors; positron emission tomography; radiotracers; μOR-, δOR-, κOR- and ORL1-ligands; epilepsy; movement disorders; pain; drug dependence

---

1. Introduction

The analgesic and soporific properties of opium have been known since antiquity, perhaps first attested in the detached reveries of Homer's Lotophagi. The sinister side of opium dreams is depicted in Tennyson's version of that story, and more distinctly in the memoires of Thomas de Quincy, who may have had the distinction of establishing a genre of literature, the addiction diary. A key active constituent of the sap of *Papaver somniferum* was first isolated in 1804 by the apothecary Friedrich

Wilhelm Sertürner, who named it morphium, later morphine (**1**). Chemists identified its elemental composition in the 19th century, and efforts to determine its structure were rewarded in 1925, when Gulland and Robinson [1] recommended a structure consistent with the characteristics of morphine and codeine and their degradation products. Subsequent investigations confirmed the correctness of the analytically deduced structure of morphine, culminating in its total synthesis, achieved in the 1950s by Gates and Tschudi [2,3]. The absolute stereochemistry of morphine's five chiral carbons (5, 6, 9, 13 and 14) was reported by Bentley and Cardwell [4] in 1955, and the first practically realizable morphine total synthesis with reasonable yields was reported by Rice in 1980 [5]. To this day, it is more economical to allow the poppy plant to do the main work of morphine (**1**) synthesis, although chemists have since produced so many structural variants that one might consider opioid pharmacology to be a discipline in its own right. There have been several reviews of opioid receptor imaging in the past decade [6–8], but we now present a comprehensive update on the the main classes of opioid receptor (OR) ligands used for positron emission tomography (PET), and review clinical findings with this technology. Relevant chemical structures of endogenous opioid peptides and representative small molecule opioid receptor ligands are depicted in Figure 1.

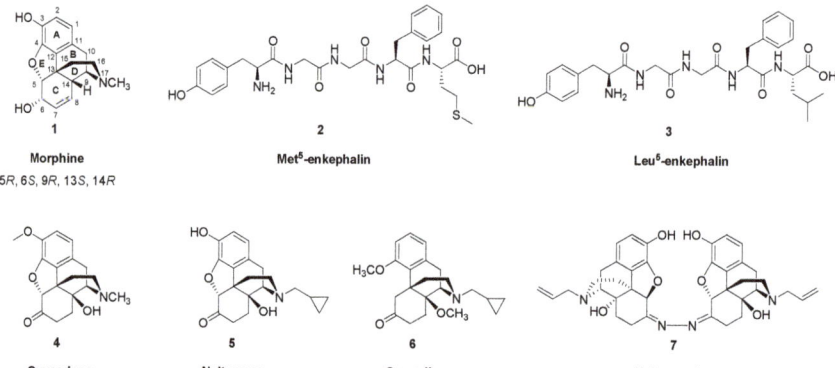

Figure 1. Chemical structures of endogenous opioid peptides and selected opioid receptor ligands.

The modern era of opioid pharmacology began with the identification of an opioid binding site in brain tissue in studies with tritiated naloxone [9]. Soon thereafter, opioid peptides were isolated from pig brain [10], which famously involved whisky as an emolument for the slaughterhouse workers. The pentapeptides Met^5-enkephalin (**2**) and Leu^5-enkephalin (**3**) both had morphine-like effects in inhibiting the electrically stimulated contraction of the *vas deferens*, with the latter compound being somewhat less potent. The enkephalins were most abundant in striatum and hypothalamus of rat, guinea pig and calf, and Met^5-enkephalin (**2**) was generally 3–6 times more abundant than Leu^5-enkephalin (**3**) [11,12]. An additional higher molecular weight opioid (β-endorphin) isolated and sequenced from camel pituitary extracts proved to be a 31 amino acid polypeptide possessing homology with Met^5-enkephalin (**2**) [13]. A trypsin-sensitive opioid activity (dynorphin-A) isolated from pituitary is a 17 amino acid polypeptide possessing N-terminal homology with Leu^5-enkephalin (**3**), with the shorter peptide dynorphin-B [1-13] having almost 1000-fold higher potency than Leu^5-enkephalin (**3**) in the guinea pig ileum muscle preparation [14]. Soon after these discoveries, bovine DNA sequences were cloned for the β-endorphin precursor corticotropin-β-lipotropin [15], the Met^5/Leu^5-enkephalin (**2**, **3**) precursor preproenkephalin-A [16] (which proved to contain four copies of Met^5-enkephalin (**2**) and one copy of Leu^5-enkephalin (**3**), consistent with the ratio of their tissue concentrations), and the closely related preproenkephalin-B [17]. Other researchers cloned porcine preprodynorphin, the precursor for dynorphin A [1-17], dynorphin-A [1-8], dynorphin B [1-13], and other opioid peptides [18].

It was soon apparent that the endogenous opioid peptides bound to at least three distinct sites in the brain and peripheral tissues, known as μ-, δ-, and κORs. For a time, the orphaned σ receptors were thought to comprise another type of OR [20], due to the analgesic (and hallucinogenic) action of pentazocine at κORs in brain. However, binding of σOR ligands is not displacable by the opioid antagonist naloxone, nor do σORs bind opioid peptides with high affinity, such that the σ- receptor is now recognized as a pharmacological class in its own right. The μOR gene has at least 14 exons that can give rise to diverse splice variants, and at least three pharmacologically distinct subtypes are recognized: $μ_1$, $μ_2$ and $μ_3$. Displacement studies *in vitro* and *in vivo* with the μ-selective competitive antagonist cyprodime (**6**) and the $μ_1$-specific competitor naloxonazine (**7**) showed that [^{11}C]carfentanil ([^{11}C]Caf, **8**) binds predominantly to the $μ_1$ subtype [21]. The $μ_3$ subtype is alkaloid sensitive and opioid peptide insensitive; it couples to nitric oxide generation, and, extraordinarily, its endogenous agonist in amygdala seems to be morphine (**1**) [22]. There is also an opioid receptor-like receptor 1 (ORL1), which is activated by the 17 amino acid polypeptide known as nociceptin/ orphanin FQ (N/OFQ) [23].

Opioid signaling has an important function in the modulation of pain processing at the spinal level. ORs synthetized in the dorsal root ganglion are transported to peripheral nerve endings and to the superficial layers of the spinal cord dorsal horn. In the dorsal horn, μORs are the most densely expressed subtype, followed by δ-, and κORs. Over 70% of the ORs there are located on the central terminals of small-diameter (mostly C and A-delta fibres) primary afferent neurons. A main mechanism of opioid analgesia lies in the activation of presynaptic μORs in the spinal cord, leading to decreased release of excitatory transmitters and nociceptive transmission. Unforunately, PET methods do not suffice for detecting ORs in the human spinal cord.

Figure 2 shows PET images of the distributions in human telencephalon of binding sites for the four main classes of ORs, μ, δ, κ, and ORL1. The pattern of μORs in Figure 2A encompasses the telencephalic pain pathway of limbic brain regions. Supraspinal μORs in the nucleus accumbens and amygdala have a role in the analgesic and reinforcing properties of opioids. The thalamus, especially the medial structures, relay nociceptive spinothalamic input from the spinal cord to higher structures. μORs also have a prominent distribution in the brainstem, with high density in several structures associated with analgesia, such as the periaqueductal gray, rostroventral medulla, the reticular formation, and locus coeruleus [24–26]. From these structures, efferent outflow descends to the spinal cord where it acts to inhibit nociceptive transmission in afferent fibres. μORs are also abundant in the hypothalamus, where they might affect hormonal regulation. Receptors in the medullary vagal complex, area postrema, and nucleus tractus solitarius, can mediate endocrine actions and nausea.

As seen in Figure 2B, the δOR has high expression in the cerebral cortex, nucleus accumbens, and the caudate putamen. This receptor is involved in analgesic activity at both spinal and supraspinal sites. Similarly as μORs, agonists of central δORs contribute to respiratory depression, whereas receptors in the gut mediate constipation, an important side effect of morphine. The δORs receptors localize presynaptically where they inhibit the release of excitatory neurotransmitters [27]. Despite these properties, δ-selective drugs have not yet found clinical application.

The κORs have wide expression in rat brain, with highest levels in the ventral tegmental area, substantia nigra, nucleus accumbens, caudate putamen, claustrum, endoperiform nucleus, various hypothalamic nuclei, and the amygdala [28]. A similar expression profile occurs in the human brain [29,30], as seen in Figure 2C. Activation of κORs does not produce respiratory depression, but typical adverse effects include sedation and dysphoria, limiting the clinical use of κOR targeting drugs [27]. Despite generally dysphoric effects in humans, the κOR agonist Salvinorin-A, which is obtained from the leaves of *Salvia divinorum*, finds a niche market in the drug subculture for those seeking to briefly experience a dissociative state.

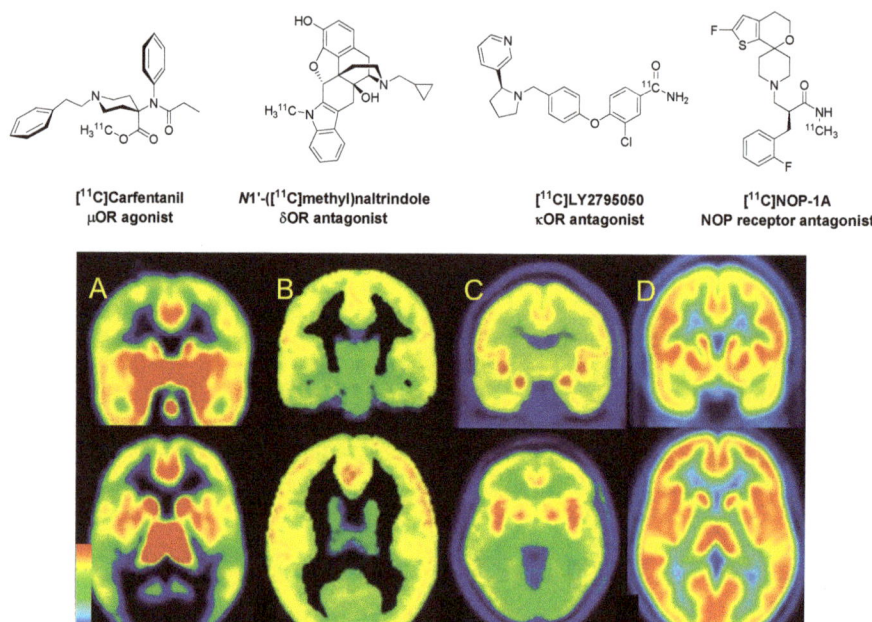

Figure 2. Human brain opioid receptor PET images in coronal (upper row) and axial (lower row) planes. Modified and reproduced with permission from Peciña et al. [19]. From left to right, we see (**A**) the μOR agonist [^{11}C]carfentanil, binding most abundantly in the caudate nucleus, anterior cingulate cortex, thalamus, and pituitary gland; (**B**) the δOR antagonist N1'-([^{11}C]methyl)naltrindole, which has diffuse binding throughout neocortex; (**C**) the κOR antagonist [^{11}C]LY2795050, which has high binding in the insular cortex, lateral frontal cortex and amygdala; (**D**) the NOP antagonist [^{11}C]NOP-1A, which binds abundantly throughout the brain. Binding sites of μ-, κ- and NOP-OR ligands are expressed as binding potential relative to the cerebellum (BP_{ND}), whereas binding of the δ-ligand (which has no non-binding reference region) is expressed as net influx (K_i,) in units of perfusion (mL cm^{-3} min^{-1}). The color scale in the lower right indicates (for **A**, **C**, and **D**) BP_{ND} ranging from 0 to 2, or (**B**) K_i ranging from 0–0.1 mL cm^{-3} min^{-1}

Figure 2D shows a widespread and abundant expression of NOP binding sites in human brain. Agonists of the NOP receptor, unlike μOR agonists, are devoid of reinforcing or motivational properties, but are implicated in homeostatic functions such as feeding and body weight, as well as anxiety, stress, and alcohol dependence [31].

2. Radiotracers for the PET Imaging of ORs

2.1. μOR Ligands and Non-Selective Ligands

The era of OR PET imaging was preceded by a phase of studies *ex vivo* with tritiated ligands such as the antagonist [^{3}H]diprenorphine ([^{3}H]DPN), which accumulated in striatum, *locus coeruleus*, *substantia nigra pars compacta*, and *substantia gelatinosa* of the living rat [32]. A similar pattern was revealed with the antagoninst [^{3}H]naloxone, which showed sodium-dependent saturable binding *ex vivo*, with a B_{max} close to that seen *in vitro* [33]. In contrast, the agonist [^{3}H]Foxy bound with low nM affinity at μORs *in vitro*, but failed to accumulate in brain of living rats, a property which was attributed to the presence of high sodium in the living organism. The presence of sodium in the biding medium enhanced antagonist binding *in vitro* but decreases agonist binding [34]. However, addition of

Na^+ to the incubation medium had little effect on the affinity of the morphiceptin analog μOR agonist Tyr-Pro-(*N*Me)Phe-D-Pro-NH_2 *in vitro* ([^{3}H]PL017) [35]. Unpredictable sensitivity of binding to the presence of sodium in the medium seems mainly to be a property of agonist ligands.

OR PET imaging began with the introduction of 3-*O*-acetyl-[^{18}F]cyclofoxy (3-*O*-Ac-[^{18}F]FcyF), (**10**). 3-*O*-Ac-[^{18}F]FcyF (**10**, Figure 3) is an opioid antagonist radiotracer, which was prepared from 3-*O*-acetyl-6α-naltrexol triflate *via* direct nucleophilic substitution with tetraethylammonium [^{18}F]fluoride in anhydrous acetonitrile at 80 °C for 15 min [36,37]. Based on displacement studies with CyF, binding of 3-*O*-Ac-[^{18}F]FcyF (**10**) is likely to reveal the composite of μ- and κOR binding [38], despite the qualitatively μOR-like binding pattern reported in living baboon brain examined with 3-*O*-Ac-[^{18}F]FcyF (**10**) [36], and the pattern of [^{3}H]cyclofoxy retention in rat brain analysed *ex vivo* [39]. However, in a rat study, the increased [^{3}H]cyclofoxy binding provoked by chronic treatment with morphine (**1**) could be attributed to upregulation of μOR sites [40]. Saturation binding PET studies with [^{18}F]FcyF (**11**) in awake rat indicated a single binding site with apparent affinity of 2 nM and B_{max} ranging from 15 pmol/g in white matter to 74 pmol/g in striatum; these results matched closely the corresponding results obtained *in vitro* [41].

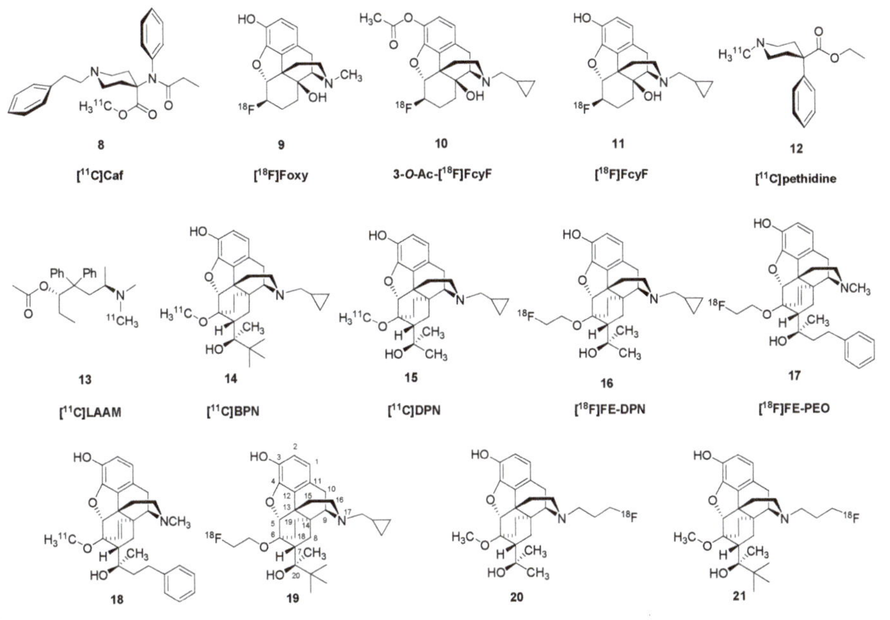

8 [^{11}C]Caf; 9 [^{18}F]Foxy; 10 3-*O*-Ac-[^{18}F]FcyF; 11 [^{18}F]FcyF; 12 [^{11}C]pethidine; 13 [^{11}C]LAAM; 14 [^{11}C]BPN; 15 [^{11}C]DPN; 16 [^{18}F]FE-DPN; 17 [^{18}F]FE-PEO; 18 [^{11}C]PEO; 19 [^{18}F]FE-BPN; 20 [^{18}F]FP-*nor*DPN; 21 [^{18}F]FP-*nor*BPN

Figure 3. Structures of μ-selective and non-selective opioid receptor radioligands.

PET studies with *N*-[methyl-^{11}C]-labelled morphine, codeine, heroin and pethidine indicated distinct differences in uptake and kinetics in rhesus brain [42], all seemingly in relation to lipophilicity of the various drugs. A more detailed kinetics analysis of *N*-[methyl-^{11}C]pethidine (**12**) in brain of rhesus monkey indicated a very low binding potential [43]. The methadone analogue [^{11}C]*L*-α-acetylmethadol ([^{11}C]LAAM, **13**) had moderate uptake in brain of mice, but its specific binding was not reported [44]. These observations are a case in point supporting the generalization that effective pharmaceuticals do not necessary make good ligands for molecular imaging of their targets.

The displaceability of OR ligands by competitors *in vivo* is a complex matter, and one increasingly relevant given the current opioid abuse crisis in some countries. Whereas antagonists such as naloxone are effective in rescuing addicts from death by overdose, naloxone also finds experimental use in molecular imaging studies to confirm binding of PET tracers to ORs. Thus, the BP$_{ND}$ of [^{11}C]Caf (**8**) to μORs in human brain, which ranged from 1.0 in cerebellum to 2.7 in caudate nucleus, was nearly completely displaced throughout in brain by a 50-mg dose of naltrexone (NTX, **5**) [45], closely matching the dose used for resuce from opioid agonist overdose. In humans, intranasal naloxone administration caused a rapid displacement of [^{11}C]Caf (**8**), in accordance with the rapid response seen in treatment for overdose [46].

Buprenorphine (BPN) [47–49] is a narcotic analgesic used since the 1970s in the low dose management of post-operative pain. Since 2002, BPN has approval in the United States at higher dose or in combination with naloxone (Suboxone®) for substitution therapy in the management of opiate addiction. BPN contains the same 6,14-ethenomorphinan skeleton as diprenorphine (DPN), and both compounds have an N^{17}-cyclopropylmethyl substituent, although BPN contains a *tert*-butyl group in position-20 instead of methyl. Interestingly, BPN has a completely different pharmacological profile than DPN. Whereas DPN is a mixed antagonist, BPN is a partial μOR agonist and κOR antagonist, and displays some affinity for the NOP receptor [50] (Table 1).

Table 1. Binding profile of selected ligands at the human opioid receptors [51].

Ligand	Ki [nM]				Action	Compound Class	Ref.
	μ-OR	δ-OR	κ-OR	NOP			
Met^5-enkephalin	261	9.9	-	-	agonist δOR	EOP	Stefanucci [52]
Leu^5-enkephalin	513	10.7	-	-	agonist δOR	EOP	Stefanucci [52]
β-Endorphin	2.1	2.4	96	-	agonist μOR, κOR	EOP	Corbett [53]
Dinorphin A	1.6	1.25	0.05	386	agonist κOR	EOP	Zhang [54]
Nociceptin	437	2846	147	0.08	agonist	EOP	Cami-Cobeci [50]
Morphine	2.06	>10,000	134	>10,000	agonist μOR	EM	Valenzano [55]
Oxycodone	16	7680	43,000	-	agonist μOR	EM	Miyazaki [56]
Naltrexone	0.62	12.3	1.88	-	antagonist	EM	Zheng [57]
Carfentanil^a	0.07	-	-	-	agonist μOR	4-AP	Henriksen [58]
Carfentanil^a	0.051	4.7	13	-	agonist μOR	4-AP	Frost [59]
Carfentanil^b	0.024	3.28	43.1	-	agonist μOR	4-AP	Cometta [60]
Cyclofoxy^a	2.62	89	9.3	-	antagonist μOR, κOR	EM	Rothman [38]
DPN	0.07	0.23	0.02	-	antagonist	orvinol	Raynor [61]
BPN	1.5	6.1	2.5	77.4	partial μOR agonist, κOR antagonist	orvinol	Cami-Cobeci [50]
PEO	0.18	5.1	0.12	-	full agonist	orvinol	Marton [62]
FE-DPN	0.24	8.00	0.20	-	antagonist	orvinol	Schoultz [63]
FE-BPN	0.24	2.10	0.12	-	mixed agonist/antagonist	orvinol	Schoultz [63]
FE-PEO	0.10	0.49	0.08	-	full agonist	orvinol	Schoultz [63]
NTI^b	3.8	0.03	332	-	antagonist δOR	EM	Portoghese [64]
MeNTI^b	14	0.02	65	-	antagonist δOR	EM	Portoghese [64]
GR103545	16.2	536	0.02	-	agonist κOR	ArAP	Schoultz [65]
LY2459989^a	7.68	91.3	0.18	-	antagonist κOR	APPB	Zheng [66]
LY2795050	25.8	153	0.72	-	antagonist κOR	APPB	Zheng [57]
FEKAP	7.4	139	0.43	-	agonist κOR	ArAP	Li [67]
EKAP	8.6	386	0.28	-	agonist	ArAP	Li [68]
MeJDTic	8.88	118	1.01	-	antagonist κOR	JDTic	Poisnel [69]
Salvinorin A	>1000	5790	1.9	-	agonist κOR	NND	Harding [70]
NOP-1A	-	-	-	0.15	antagonist NOP	FDPTP	Pike [71]
MK-0911	94	-	-	0.6	antagonist NOP	SPB	Hostetler [72]

EOP: Endogenous opioid peptide, EM: 4,5-Eopxy-morphinan, 4-AP: 4-Anilidopiperidine, orvinol: 6,14-ethenomorphinan, Bentley-compound, ArAP: Arylacetamidopiperazine, APPB: Aryl-phenylpyrrolidinylmethyl-phenoxy-benzamide, JDTic: *trans*-3,4-dimethyl-4-(3-hydroxyphenyl)-piperidine, **NND**: "non nitrogenous" diterpene, FDPTP: 2′-fluoro-4′,5′-dihydrospiro[piperidine- 4,7′-thieno [2,3-c]pyran]- derivative, SPB: [[spiro[2.5]octan-8-yl]-methyl]piperidin-4-yl] benzimidazol-2-one, ^{a}: in the rat brain ^{b}: in guinea pig brain membranes.

Luthra et al. [73] synthesized N^{17}-cyclopropyl[^{11}C]methyl-buprenorphine starting from N^{17}-*nor*-buprenorphine. Analogously to the N^{17}-cyclopropyl[^{11}C]methyl-diprenorphine synthesis [74] the corresponding precursor, *nor*-BPN, was reacted with cyclopropyl[^{11}C]carbonyl chloride and the carbonyl functional group of the resulting intermediate was reduced with LiAlH$_4$. Lever et al. [75] developed a metabolically stable radiotracer, 6-O-(methyl-^{11}C)-BPN, at Johns Hopkins University in 1990 in a two-step synthesis from 3-O-TBDMS-6-O-desmethyl-BPN. The precursor was selectively alkylated in position-6 with [^{11}C]iodomethane/NaH in DMF at 80 °C for two min. Following desilylation, [^{11}C]BPN (**14**) was produced in 10% radiochemical yield with molar activity of 41 GBq/μmol. Subsequently, Luthra et al. [76], aiming to avoid the formation of 3-O-alkylated by-products, introduced the base-stable, acid labile trityl protecting group to protect the phenolic hydroxyl in position-3. Applying 3-O-trityl-6-O-desmethyl-BPN in a two-step, fully-automated radiosynthesis (^{11}C-methylation/deprotection), yielded [^{11}C]BPN (**14**) in 15% radiochemical yield and with a molar activity of 13–22 GBq/µmol. In 2014, Schoultz et al. [63] reported a procedure for the radiosynthesis of 6-O-(2-[^{18}F]fluoroethyl)-6-O-desmethyl-BPN ([^{18}F]FE-BPN (**19**)) via ^{18}F-fluoro-alkylation of 3-O-trityl-6-O-desmethyl-BPN (TDBPN) precursor with [^{18}F]fluoroethyl tosylate and subsequent trityl deprotection. The decay corrected formulated product yield was 26 % and the molar activity 50–300 GBq/µmol.

DPN, a semisynthetic thebaine/oripavine derivative with a methyl group amenable for labelling in position-20, belongs structurally to the ring-C bridged morphinans (6,14-ethenomorphinans, orvinols, Bentley-compounds) [48]. DPN is a nonselective OR antagonist with affinity in the nanomolar range (Table 1), 100 times more potent than nalorphine. Indeed, DPN is used in the veterinary medicine as an antidote/reversing agent/antagonist for remobilizing large African animals (rhinos/elephants, Revivon®), which had been immobilized with the astonishingly potent agonists etorphine or carfentanil.

The first attempt at labelling of DPN with carbon-11 in postition-20 was reported by Burns et al. [77], who used N-cyclopropylmethyl-dihydronororvinone as precursor. The reaction of the precursor bearing an acetyl group in position-7-alpha with [^{11}C]methyllithium yielded 20-[^{11}C]methyl-DPN ([^{11}C]DPN (**15**)). Luthra et al. [74] developed cyclopropyl[^{11}C]methyl-DPN by alkylating N^{17}-nor-DPN with cyclopropane[^{11}C]carbonyl chloride and then reducing the N-cyclopropyl[^{11}C]carbonyl intermediate with LiAlH$_4$ in THF. The radiochemical yield of the corresponding radioligands were low in both cases [74,77].

In 1987, Lever et al. [78] developed a [^{11}C]DPN (**15**) synthesis by alkylating the precursor 3-O-TBDMS-6-O-desmethyl-DPN in position-6 with [^{11}C]iodomethane in DMF containing sodium hydride at 80 °C for two min. After cleavage of the TBDMS protecting group, [^{11}C]DPN (**15**) was obtained with 10 % radiochemical yield and 64 GBq/µmol molar activity. In 1994, Luthra and her associates at the Hammersmith Hospital developed a new precursor for the radiosynthesis of [^{11}C]DPN (**15**) [76]. Selective alkylation of 3-O-trityl-6-O-desmethyl-DPN (TDDPN) with [^{11}C]iodomethane in the presence of NaH/DMF (95 °C, five min). Upon deprotection with 2 M hydrochloric acid (95 °C, two min) the radiotracer **15** was obtained with a radiochemical yield of 13–19% and a molar activity of 16–24 GBq/µmol. Recently, Fairclough et al. [79] at the University of Manchester reported a modified synthetic method also starting from TDDPN, yielding [^{11}]DPN (**15**) with ten times higher molar activity (240 GBq/µmol) [76] and a radiochemical yield of 32%. The non-selective OR partial mixed agonist/antagonist 6-O-(methyl-^{11}C)-BPN ([^{11}C]BPN, **14**) accumulated in striatum, thalamus and cingulate cortex in living baboon brain. Analysis of the dynamic PET data with a model assuming irreversible trapping gave a net blood-brain clearance (K_i) of about 0.064 mL cm^{-3} min^{-1}, which was halved by administration of naloxone, indicating substantial displaceability [80]. In a study in heroin addicts, the BPN occupancy at [^{11}C]Caf (**8**) binding sites was estimated relative to the drug-free baseline. An oral dose of 2 mg BPN had an occupancy of about 50% throughout brain, whereas 16 mg had 85% global occupancy [81]. On the other hand, therapeutic methadone (18–90 mg/day) did not provoke any discernible occupancy at 6-O-(methyl-^{11}C)-diprenorphine ([^{11}C]DPN, **15**) binding sites, neither in human opioid addicts, nor in mice, a phenomenon attributed to high agonist potency of methadone, such that withdrawl effects are averted with a rather low occupancy [82]. In further preclinical studies from the same research group, binding of [^{11}C]DPN (**15**) in mouse brain was unaltered by treatment with oxycodone (**4**) or morphine (**1**) (full agonists at µORs), but was reduced by approximately 90% by BPN (partial agonist at µORs and antagonist at the δ- and κORs).

A comparative OR PET study in humans compared the distributions of the µOR-selective agonist [^{11}C]Caf (**8**) and the mixed antagonist [^{11}C]DPN (**15**) [83]. Qualitatively, [^{11}C]DPN (**15**) binding in the striatum, cingulate and frontal cortex exceeded that of [^{11}C]Caf (**8**) (which had highest binding in the µOR-rich thalamus), consistent with labeling of additional non-µOR sites by [^{11}C]DPN (**15**). An investigation of ORs in human cerebellum showed abundant binding of a µOR-specific ligand in the molecular layer, moderate binding of a κOR-selective ligand, but a near absence of δOR binding sites, which was consistent with the observations of [^{11}C]DPN (**15**) binding *in vivo* [84]. The presence of binding sites in cerebellum can be an obstacle to the valid use of reference tissue methods of PET quantitation.

As a longer-lived alternative to [^{11}C]DPN, (**15**), Wester et al. developed 6-O-(2-[^{18}F]fluoroethyl)-6-O-desmethyl-DPN ([^{18}F]FE-DPN (**16**)) [85], which contains a 2-fluoroethoxy group in position-6 instead of an OCH$_3$. [^{18}F]FE-DPN (**16**) was synthesized from TDDPN, the so called "*Luthra-precursor*" [76], the same precursor as for [^{11}C]DPN (**15**). For the synthesis of **16**, TDDPN was reacted with [^{18}F]fluoroethyl tosylate ([^{18}F]FE-Tos) in DMF in the presence of sodium hydride for five min at 100 °C. The trityl protecting group was removed with 2 N hydrochloric acid, yielding [^{18}F]FE-DPN (**16**) with a radiochemical yield of 22 ± 7% and the molar activity was 37 GBq/µmol [85]. In 2013, Schoultz et al. [86] reported an automated radiosynthesis of **16** from TDDPN with a decay-corrected radiochemical yield of 25 ± 7%. [^{18}F]FE-DPN (**16**) has similar uptake as [^{11}C]DPN (**15**) in mouse brain, and obtained a BP$_{ND}$ in human thalamus of about 2 relative to occipital cortex, versus only 0.3 in somatosensory cortex [87]. Women showed faster plasma metabolism [^{18}F]FE-DPN (**16**) than men, which might contribute to apparent gender differences in binding [88].

The first instance of full compartmental analysis of an opioid PET ligand in living brain was for the case of [^{11}C]Caf, as described below. This fentanyl analogue belongs to the 4-anilidopiperidine (4AP) class of OR ligands, which are potent µOR-selective agonists. Since 1960, numerous 4AP-type OR ligands were synthesized and their structure-activity-relationship at ORs were recently summarized [89,90]. Caf is structurally different from fentanyl in that it contains an additional carboxymethyl group in position-4 of the piperidine ring. Caf is a µOR-selective full agonist of extreme potency, being almost 10,000 times more potent than morphine (**1**) [58,91]. In 1985, [^{11}C]Caf (**8**) radiotracer was applied in the first human PET study [59,92]. For the radiosynthesis of [^{11}C]Caf, desmethyl-Caf sodium carboxylate was alkylated with [^{11}C]iodomethane in DMF at 35 °C for five min [92]. This procedure gave molar activity at the end of synthesis of 122 GBq/µmol, which would correspond to mass dose of about 500 pg in a human PET study, which is too low to have any effect on particpants. According to a novel version of the radiosynthesis, desmethyl-Caf free acid serves as precursor in a reaction performed in dimethylsulfoxide with [^{11}C]methyl triflate in the presence of tetrabutylammonium hydroxide [93]. This procedure gave a molar activity of 5 GBq/µmol, which would correspond to a mass dose of 100 µg, certainly intruding into the range causing some pharmacological effects. Risk of toxicity is a serious matter in PET imaging with potent agonists, and for society in general, given the weaponization [94] (figurative and literal) that is possible with Caf.

For the compartmental analysis of [^{11}C]Caf (**8**), two models were fitted to dynamic time activity curves (TACs) measured by PET in human brain relative to a metabolite-corrected arterial input function [95]; this approach is the gold standard for PET quantitation. The input function obtained by HPLC analysis of plasma extracts showed that untransformed parent fractions declined to 50% at 25 min post injection. The authors estimated microparameters for the reversible transfer of the tracer across the blood-brain barrier (K_1/k_2), the reversible binding to a receptor compartment (k_3/k_4), and the reversible association to a non-specific compartment in brain (k_5/k_6). The mean binding potential (BP; k_3/k_4) was 1.8 in frontal cortex and 3.4 in thalamus at baseline, versus only 0.16 and 0.26 after treatment with naloxone (0.1 mg/kg). This study set a very high standard for quantitative PET analysis, although the molar activity of the tracer may not have been completely adequate. The amount of mass injected corresponded to about 5 µg Caf per scan, which could cause some analgesia in humans, although being less than 10% of the dose causing loss of consciousness. Nonetheless, this again raises the issue of safety in PET studies with the using of high-potency full agonist ligands, as noted above. Unless the the highest possible molar activity is obtained, pharmacologically significant occupancy can occur, bringing a risk of toxicity. Test-retest studies with [^{11}C]Caf (**8**) showed admirable low variability (< 6%) and high intraclass correlation coefficients (ICC > 0.90) of the total distribution volume (V_T) relative to the metabolite-corrected arterial input, and likewise BP$_{ND}$ relative to a reference tissue [96].

Phenethyl-orvinol (PEO) [97] shares the same in ring-C bridged morphinan scaffold as DPN and BPN. PEO contains a 6,14-*etheno*-bridge, an N^{17}-methyl substituent and a 2-phenethyl group in position-20. It is a full agonist with higher affinity at µOR (0.18 nM) and κORs (0.12 nM) than to δORs (5.1 nM). The radiosynthesis of 6-O-(methyl-^{11}C)-phenethyl-orvinol ([^{11}C]PEO, **18**) was reported by Marton et al. [62] in 2009. The *Luthra-type* trityl-protected precursor (3-O-trityl-6-O-desmethyl-phenethyl-orvinol, TDPEO) was alkylated in position-6 with [^{11}C]iodomethane in the presence of 8-10 equiv. sodium hydride. The protecting group was removed with 1 M hydrochloric acid in ethanol, yielding [^{11}C]PEO (**18**) with a radiochemical yield of 57 ± 16% and a molar activity of 60 GBq/µmol.

In 2012, Marton and Henriksen [98] reported the preliminary results of the synthesis of 6-O-(2-[^{18}F]fluoroethyl)-6-O-desmethyl-phenethyl-orvinol ([^{18}F]FE-PEO, **17**) starting from 6-O-(2-tosyloxyethyl)-6-O-desmethyl-phenylethyl-orvinol (TE-TDPEO) *via* direct nucleophilic fluorination and subsequent deprotection. This procedure gave [^{18}F]FE-PEO (**17**) in an isolated preparative yield of 35 ± 8% with a molar activity of 55–130 GBq/µmol. In 2013, a research group of the University of Cambridge [99] investigated [^{18}F]FE-PEO (**17**) as a candidate OR PET-ligand, obtained by an automated cGMP-compliant method the [^{18}F]FE-PEO at 28 ± 15% yield and 52–224 GBq/µmol molar activity. In 2014, Schoultz et al. [63] reported the synthesis and biological evaluation of three structurally-related 6-O-(2-[^{18}F]fluoroethyl)-6-O-desmethy-orvinols, i.e. [^{18}F]FE-DPN (**16**), [^{18}F]FE-BPN (**19**), and [^{18}F]FE-PEO (**17**). The production of these ^{18}F-fluoroethyl-orvinol radiotracers (**16,17,19**) was accomplished from 3-O-trityl-6-O-desmethyl-orvinol precursors (TDDPN, TDBPN, TDPEO) in a two-pot, three-step synthesis. [^{18}F]FE-PEO (**17**) had a molar activity at end of synthesis of 50–250 GBq/µmol [99], corresponding to an injected mass <1 µg in human PET studies. The total distribution volume (V_T) in rat brain ranged from 1 mL cm^{-3} in cerebellum to 8 mL cm^{-3} in thalamus; displacement studies *in vitro* with the µOR-selective agonist DAMGO indicated high specificity in certain brain regions. [^{18}F]FE-DPN (**16**) had a molar activity of 37 GBq/µmol [85].

2.2. Delta Ligands

N1'-Methylnaltrindole (MeNTI, Figure 4) is a highly selective δOR antagonist (Table 1). MeNTI was prepared from naltrexone (**5**) in a Fischer-indol synthesis with N-methyl-N-phenylhydrazine [64]. The radiosynthesis of [^{11}C]MeNTI was reported by Lever et al. in 1995 [100]. In the first step, 3-O-benzyl-naltrindole was reacted with [^{11}C]iodomethane in DMF in the presence of either sodium hydride or tetrabutylammonium hydroxide at 80 °C for two min. The next step was hydrogenolysis of the formed 3-O-benzyl-N1'-(methyl-^{11}C)-naltrindole under heterogenous catalytic conditions (H$_2$, 10% Pd/C, DMF/ethanol, 80 °C, four min), or alternatively catalytic transfer hydrogenation (HCOONH$_4$, 10% Pd/C, MeOH). This gave [^{11}C]MeNTI (**22**) with 6% radiochemical yield and a molar activity of 76 GBq/µmol.

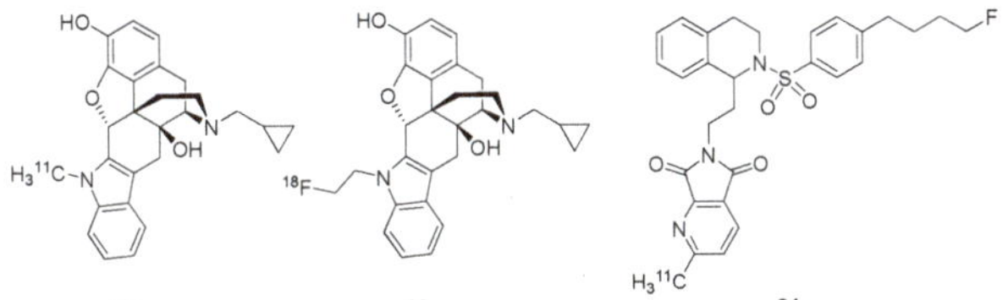

Figure 4. Labeled δ-opioid receptor ligands.

Human PET studies with [^{11}C]MeNTI (**22**) confirmed earlier demonstrations in living mice of δOR-selectivity *in vivo* [101]. The binding ratio relative to cerebellum ranged from 1.1 in hippocampus to 1.7 in striatum and insular cortex, regional values correlated rather precisely with known density of δORs *in vitro*, and showed 50% displacement after administration of 50 mg NTX (**5**). The tracer showed pseudo-irreversible binding characteristics over the course of 90 min, with net blood-brain clearance (K_i) ranging from 0.04 in cerebellum to 0.11 mL cm^{-3} min^{-1} in putamen [102]. The K_i for [^{11}C]MeNTI (**22**) in human brain declined by only about 20% after treatment with naloxone at a dose completely displacing μOR sites [45]. The authors of that study suggested that incomplete and variable δOR blockade might contribute to the success of NTX (**5**) as a treatment for alcoholism.

N1'-(2-[^{18}F]fluoroethyl)naltrindole (**23**, [^{18}F]FE-NTI, BU97001) was developed by Matthews et al. in 1999 [103]. The precursor, N1'-[2-(tosyloxyethyl)]-3-O-benzyl-naltrindole, was prepared in four consecutive transformations from naltrexone (**5**). In the first step, naltrexone (**5**) was reacted in a Fischer-indol synthesis with 2-(N1-phenylhydrazino)acetic acid ethyl ester. The resulting indolomorphinanyl-acetic ester was reacted with benzyl bromide to yield the 3-O-benzyl protected NTI derivative, which was reduced with LiAlH$_4$ in THF-toluene to afford the corresponding indolomorphinanyl ethanol intermediate. This compound was reacted with tosyl chloride to provide the appropriate precursor with a tosyloxy leaving group. For the radiosynthesis of [^{18}F]FE-NTI (**23**), the precursor was reacted in a direct nucleophilic reaction with potassium [^{18}F]fluoride/K$_2$CO$_3$/Kryptofix[2.2.2] in DMF to yield N1'-(2-[^{18}F]fluoroethyl)-3-O-benzyl-naltrindole. Final debenzylation by hydrogenolysis under heterogenous catalytical conditions (H$_2$, Pd/C, N,N-dimethyl formamide) gave [^{18}F]FE-NTI (**23**) with a radiochemical yield of 10% and molar activity of 31 GBq/μmol. [^{18}F]FE-NTI (**23**) was an antagonist in mouse *vas deferens* with high selectivity over μ- and κOR sites, and its tritiated version bound to rat whole brain as a single site with K$_D$ of 0.42 nM and B$_{max}$ of 3 pmol g^{-1} [104].

In 2007, Bourdier et al., [105] reported the radiosynthesis of a 2 [^{11}C]methyl pyrrolo[3,4-b]pyridine-5,7-dione derivative (N-substituted-[^{11}C]quinolinimide) (**24**). The radiotracer containing a [^{11}C]methyl-group on the pyridine ring was synthesized from a tributylstannyl precursor, with introduction of the [^{11}C]methyl group by the Stille reaction using [^{11}C]iodomethane in the presence of *tris*(dibenzylideneacetone)dipalladium, tri-*o*-tolylphosphine, K$_2$CO$_3$, and CuCl in DMF, heated at 90 °C for five min. The labelled compound (**24**) was synthesized with a radiochemical yield of 60 ± 10% and a molar activity of 30–56 GBq/μmol. The unlabelled version of the N-substituted quinolinimide had higher δOR-selectivity than MeNTI, but its ^{11}C-derivative (**24**) failed to label ORs in mouse brain, due either to excessively rapid metabolism [105], or its only moderate affinity.

2.3. Kappa Ligands

GR89696 ((±)-methyl 4[(3,4-dichlorophenyl)acetyl]-3-(1-pyrrolidinylmethyl)- 1-piperazine-carboxylate [106–108] (Glaxo Group Research Ltd.) is a κOR-selective agonist with an arylacetamidopiperazine/diacylpiperazine structural core. GR103545, the biologically active (R)-(−)-enantiomer of GR89696, displays subnanomolar affinity and 1000-fold selectivity for human κOR (K_i = 0.02 nM), [65]. Ravert et al. [109] reported synthesis of both enantiomers of [^{11}C]GR89696 (**26**, Figure 5) from the corresponding chiral normethylcarbamoyl precursor [108,110]. The radiosynthesis was accomplished by acylation of the secondary amine with [^{11}C]methyl chloroformate in dichloromethane in the presence of trimethylamine, giving product with molar radioactivity of 69 GBq/μmol. Mouse brain distribution of the synthesized enantiomers, ((R)-(−)-[^{11}C]GR103545 (**25**) and the (S)-(+)-enantiomer [^{11}C]SGR) was determined *in vivo*, which showed the (S)-enantiomer to be inactive. The low radiochemical yield of the radiosynthesis (2–14%) [110,111] motivated the development of elaborate new radiochemical methods. In 2008, Schoultz et al. [112] developed a simple [^{11}C]methyl triflate mediated methylation of carbamino adducts. Normethylcarbamoyl-GR103545 was converted to [^{11}C]GR103545 (**25**) with [^{11}C]CH$_3$OTf under mild conditions in 64–91% radiochemical yield. Wilson et al. [113] developed a method for preparing [^{11}C-carbonyl]-methylcarbamates directly from primary or secondary amines, applying either DBU or BEMP and cyclotron-produced [^{11}C]CO$_2$. [^{11}C-carbonyl]-GR103545 (**25**) was synthesized with high radiochemical purity (>98%) and molar activity of 108–162 GBq/μmol. In 2011, Nabulsi et al. [114] reported an automated two-step, one-pot procedure for the synthesis of [^{11}C]GR103545 (**25**) from normethylcarbamoyl-GR103545 *via* transcarboxylation using the zwitterionic carbamic complex DBU-CO$_2$ and [^{11}C]CH$_3$OTf.

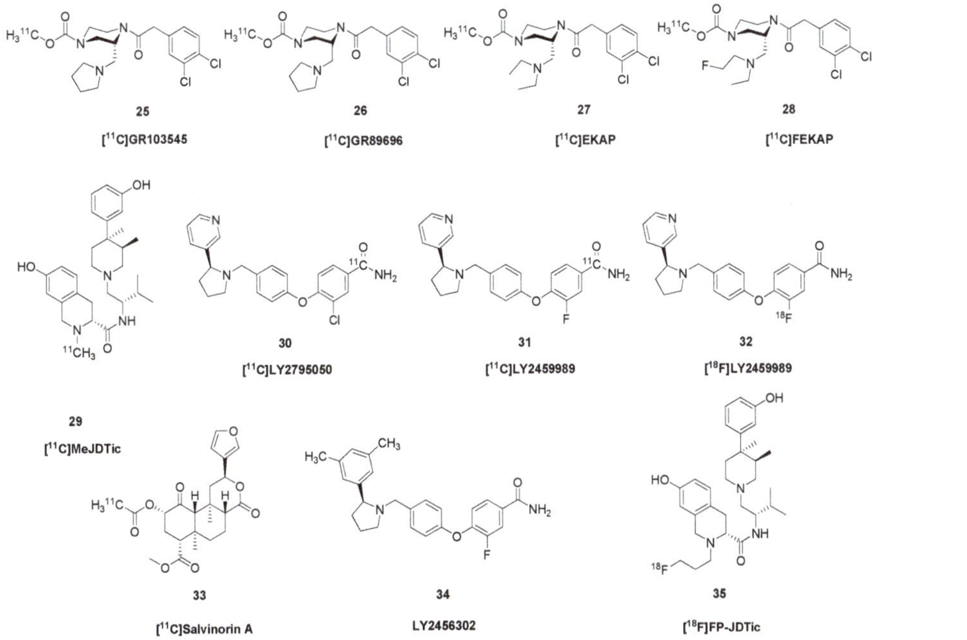

Figure 5. Selected κOR ligands.

In PET studies, the κOR-agonist ligand [^{11}C]GR103545 (**25**) had a V_T in baboon brain ranging from 3 mL cm^{-3} in cerebellum to 10 mL cm^{-3} in striatum and cingulate cortex [111]. Naloxone homogeneously displaced tracer throughout brain, but had no effect on V_T in cerebellum, which would support use of that brain region as a reference tissue. The tracer had >100-fold selectivity for κ- over δ- and μORs *in vitro* [65]. Much as in baboons, PET studies in rhesus monkey showed V_T ranging from 8 mL cm^{-3} in cerebellum to 21 mL cm^{-3} in striatum. Other monkey studies showed BP_{ND} ranging from 0.3 in amygdala to 2.2 in putamen [115]. This bolus plus infusion study with increasing mass dose in monkeys indicated an *in vivo* K_D of 2 nM and B_{max} of 1–6 pmol g^{-1}. In 2014, Naganawa et al. [116] reported the first-in-human PET study with [^{11}C]GR103545 (**25**). Test-retest variability of the quantitative endpoint V_T was about 15%, and binding ranged from 8 mL cm^{-3} in cerebellum to 41 mL cm^{-3} in amygdala; Lassen plots with naltrexone blocking indicated a non-specific uptake (V_{ND}) of only 3.4 mL cm^{-3}, thus emphasizing the absence of any non-binding reference region.

Based on the substituted-diacylpiperazine scaffold of GR103545, researchers at Yale University developed the new κOR agonist radiotracers [^{11}C]EKAP (**27**) [68] and [^{11}C]FEKAP (**28**) [67] with improved pharmacological and PET-imaging profile compared with the native compound. In the open-ring analogs of GR103545, the pyrrolidinyl-methyl group of the original molecule in position-3 was replaced by a diethylamino-methyl in EKAP and a ((ethyl)2-fluoroethyl)amino)methyl group in FEKAP. Imaging studies [^{11}C]EKAP (**27**) in rhesus monkey showed rapid metabolism *in vivo* and fast, reversible binding kinetics in brain that was blockable with specific competitors. The BP_{ND} ranged from 0.8 in frontal cortex to 1.8 in globus pallidus.

Researchers at Eli Lilly, in cooperation with the Yale University, developed κOR antagonist radiotracers with the 3-pyridinyl-1-pyrrolidinylmethyl structural scaffold. Along these lines, in 2013, Zheng et al. [57] synthesized the selective κOR antagonist radiotracer [^{11}C]LY2795050 (**30**) from an iodophenyl precursor in a two-step procedure. The precursor was converted by transition metal-mediated cyanation using H^{11}CN and Pd$_2$(dba)$_3$/dppf to a [^{11}C]nitrile intermediate. This latter was partially hydrolysed with NaOH/H$_2$O$_2$ in DMF at 80 °C for five min, giving a 12% radiochemical yield with molar activity of 23.6 GBq/µmol. [^{11}C]LY2459989 (**31**) was prepared in a two-step one-pot radiosynthesis. In the first step, an aryl-iodide-type precursor was transformed in a palladium catalyzed reaction (Pd$_2$dba$_3$/dppf) with H^{11}CN to the corresponding [^{11}C]nitrile, which was reacted with H$_2$O$_2$ under basic condition to afford **31** with 7.4% radiochemical yield and 23 GBq/µmol molar activity.

LY2459989 is the fluorine-containing analogue of LY2795050. Li et al. [117] synthesised the ^{18}F-fluorine-labelled version of LY2459989 using two different methods. Using the nitro precursor, the radiochemical yield was too low, but applying the iodonium ylide precursor, [^{18}F]LY2459989 (**30**) was prepared with 36% radiochemical yield and 1,175 GBq/µmol molar activity. While admirably high, this molar activity falls far short of the theoretical maximum for ^{18}F-, which is 63,000 GBq/µmol. Where does all that non-radioactive fluoride come from?

As noted above, salvinorin A [118–120] is a naturally occurring non-alkaloid neo-clerodane diterpenoid, isolated from *Salvia divinorum*. Also as noted above, smokings Salvinorin-A can provoke a dissociative hallucinogenic experience distinct from that of the classical hallucinogens. It has a unique structure with seven chiral carbons and is potent and highly selective κOR agonist; salvinorin A does not display any significant activity at other OR subtypes. In 2008, Hooker et al. [121] synthesized the carbon-11 labelled version of salvinorin A (**33**), using salvinorin B as precursor for the radiosynthesis. The 2-alpha-hydroxyl group of the precursor was acylated with [^{11}C]acetyl chloride in DMF in the presence of DMAP at 0 °C for 7–10 min, giving a radiochemical yield of 3.5–10% with molar activity 7.4–28 GBq/µmol. PET studies in baboon brain showed rapid uptake and washout of [^{11}C]salvinorin A (**33**), matching the brief duration of the hallucinatory/dissociative experience reported by humans users. Rat studies showed that acute doses of salvinorin A caused dose-dependent occupancy at brain κORs labelled *in vivo* with [^{11}C]GR103545 (**25**). Pretreatment with a high dose in the hours before PET examination caused persistent reductions in receptor availability, despite the brief plasma half-life of the drug, and despite the rather brief duration of the hallucinogenic experience. This suggests that κOR activation by agonists such as salvinorin A provokes a delayed and persistent receptor internalization [122].

In 2001, Thomas et al. [123] identified the first κOR-selective antagonist ligand, *trans*-3,4-dimethyl-4-(3-hydroxyphenyl)piperidine (JDTic, > 200-fold selective, Table 1), with non-opiate structure. MeJDTic is a derivative of JDTic that is ring methylated on the nitrogen of the tetrahydroisoquinoline. Poisnel et al. [69] prepared [^{11}C]MeJDTic (**29**) from JDTic by methylation with [^{11}C]methyl triflate in acetonitrile at room temperature for three min. The radiochemical yield was 78–98% and the molar activity 1.5–4.4 GBq/µmol. Recently, Schmitt et al. [124] synthesized N-[^{18}F]fluoropropyl-JDTic ([^{18}F]FP-JDTic) (**35**)) from JDTic with [^{18}F]fluoropropyl-tosylate in DMSO in the presence of DIPEA and LiI at 150 °C for 30 min. *In vivo* studies in mouse showed accumulation of [^{18}F]FP-JDTic (**35**) in peripheral organs rich in κORs. [^{11}C]MeJDTic (**29**) entered mouse brain *in vivo*, albeit attaining a concentration of only 0.2–0.3% ID/g. Its binding was substantially reduced by treatment with the κOR agonist U50,488, but was unaffected by morphine (**1**) or naltrindole (NTI), thus attesting to its high selectivity for κOR sites [69]. The κOR-agonist [^{11}C]LY2795050 (**30**) had an *in vitro* binding affinity of 1 nM, and 25 fold selectivity over µORs (Table 1). It readily entered monkey brain, and was substantially displacement by naloxone [57]. Displacement studies showed scant specific binding in monkey cerebellum, which supports its use as a reference region for quantitation. The tracer had a BP$_{ND}$ as high as 1.0 in parts of the basal ganglia [125]. Dual tracer studies in monkey showed the LY2795050 displaced [^{11}C]Caf (**8**) from µOR sites with an ED$_{50}$ of 119 µg/kg, whereas the ED$_{50}$ at κOR sites was 16 µg/kg, indicating 8-fold selectivity *in vivo* [125,126]. Corresponding human studies with kinetic modelling showed V$_T$ ranging from 2 mL cm^{-3} in cerebellum to 4 mL cm^{-3} in amygdala, and Lasson plots with partial NTX (**5**) blocking indicated a V$_{ND}$ (non-specific binding) close to 1.6 mL cm^{-3}, thus giving a BP$_{ND}$ of 1.5 in amygdala versus only 0.2 in cerebellum [126]. The test-retest reliability in human brain was about 10% [127]. PET with [^{11}C]LY2795050 (**30**) has revealed the dose-occupancy relationship in human brain for the experimental κOR antagonist LY2456302 (**34**), which is under development as a treatment of alcoholism [128].

[^{11}C]LY2459989 (**31**) had sub-nM affinity at κOR sites *in vitro*, with 30-fold selectivity over μOR and 400-fold selectivity over δOR sites [66]. Preliminary PET studies in monkey showed rapid kinetics and substantial displaceability *in vivo*, with BP$_{ND}$ ranging from 0.5 in thalamus to 2.2 in globus pallidus. A comparison of κOR ligands in rat showed that the agonist [^{11}C]GR103545 (**25**) and the antagonist [^{11}C]LY2459989 (**31**) had similar displacement by various κOR antagonists. However, and of great significance, the κOR agonists salvinorin A and U-50488, while displacing [^{11}C]GR103545 (**25**) binding *in vivo*, did not alter [^{11}C]LY2459989 (**31**) binding [129], which may indicate an allosteric binding mechanism. The novel κOR agonist tracer [^{11}C]-EKAP (**27**) showed fast uptake kinetics and high specific binding in monkey brain, with V$_T$ ranging from 12 mL cm^{-3} in cerebellum to mL cm^{-3} in globus pallidus, corresponding to a BP$_{ND}$ of 1.8, its binding was 95% displaced by pre-blocking with the antagonists naloxone or LY2795050 [68].

The highly selective and potent κOR-ligand U-50488 served as a scaffold for developing fluoro-alkylated PET ligands, but proved inappropriate due to 100-fold loss of affinity relative to the starting compound [130]. The novel fluorinated κ-ligand [^{18}F]LY2459989 (**32**) had similar kinetic properties in monkey PET studies to those of [^{11}C]LY2459989 (**31**) [117].

2.4. Nociceptin and Opioid-like 1 Receptors (ORL1)

Emerging evidence supports the use of agonists for the nociceptin/orphanin FQ peptide receptor (NOP) in the clinical management of pain and for substance abuse [131], thus presenting an attractive target for molecular imgaing A series of NOP ligands based on a 2′-fluoro-4′,5′-dihydrospiro[piperidine-4,7′-thieno[2,3-c]pyran]-scaffold were screened in rats [71]. Uptake in monkey brain in a baseline condition contrasted with a blocking condition indicated specific binding of several of the [^{11}C]-labelled compounds, of which [^{11}C]NOP-1A ((2S)-2-[(2-fluorophenyl)methyl]- 3-(2-fluorospiro[4,5-dihydrothieno[2,3-c]pyran- 7,4′-piperidine-1′-yl)-N-methyl-propanamide (**36**, Figure 6) was selected for further investigations. In the synthesis developed by Pike et al. [71], [^{11}C]NOP-1A (**36**) was prepared from a primary-amide type precursor by methylation with [^{11}C]iodomethane in DMSO basified with potassium hydroxide at 80 °C for 5 min. PET imaging experiments with **36** showed a V$_T$ in monkey brain ranging from 13 mL cm^{-3} in cerebellum to 21 mL cm^{-3} in amygdala. This fell globally to 8 mL cm^{-3} after blocking with the antagonist SB-612111, indicating a BP$_{ND}$ of 1–2 [132]. A somewhat lower V$_T$ range was detected in human brain [133], where the test-retest reliability was about 12% [134].

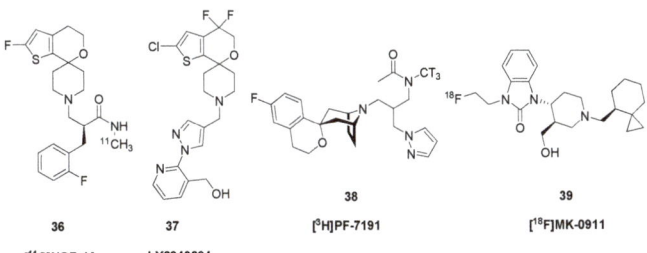

Figure 6. Chemical structures of selected ORL1 receptor ligands.

The NOP/ORL1 antagonist LY2940094 (**37**) exerted a dose-dependent reduction in immobility in the forced swim test, matching that provoked by imipramine, consistent with a potential antidepressant action [23]. Changes in [^{11}C]NOP-1A (**36**) binding in brain of living rats revealed the ORL1 occupancy of orally administered LY2940094 (**37**) [135]. MK-0911 (1-(2-fluoroethyl)-3-[(3R,4R)-3-(hydroxymethyl)-1-[[(8S)-spiro[2.5]octan-8-yl]methyl]piperidin-4-yl]benzimidazol-2- one) is a high affinity, selective NOP receptor antagonist developed by Merck Pharmaceuticals. The fluorine-18 labelled version [^{18}F]MK-0911 (**39**) had a V$_T$ in human brain ranging from 5 mL cm^{-3}

in cerebellum to 7 mL cm^{-3} in temporal cortex, with excellent test-retest stability [72]. Displacement studies with antagonists revealed the presence of a small specific binding component in cerebellum, again raising a red flag about reference tissue quantitation. Studies with the nociceptine ligand [^{3}H]PF-7191 (**38**) showed sub-nM binding displacement *in vitro* (K_i = 0.1 nM) and high selectivity over other OR types, as well as promising displaceable binding in rat brain measured *ex vivo* [136].

3. Clinical Studies

3.1. Age and Gender

The BP$_{ND}$ of [^{11}C]Caf (**8**) relative to occipital cortex was 20% lower in thalamus and amygdala of healthy, aged women compared with young women, but tended to increase with age in frontal cortex, whereas increases were more consistently seen in aged men [137]. This finding seems relevant to the age-dependent changes in sensitivity to μOR agonists, compounded by possible gender differences in hepatic tracer metabolism, noted above. Preliminary results with [^{11}C]LY2795050 (**30**) PET did not indicate any change in κOR availability with age in humans [138]. Another κ-OR PET study with that ligand showed slightly higher (5–10%) V_T in widespread brain regions of male subjects than that seen in age-matched women [139].

3.2. Epilepsy

A dual tracer PET study of patients with temporal lobe epilepsy showed increased binding of [^{11}C]Caf (**8**) to μORs in the temporal neocortex and decreased binding in the amygdala ipsilateral to the epileptic focus [140]. However, binding of [^{11}C]DPN (**15**) to μ- and other OR subtypes did not differ between affected and unaffected cerebral hemispheres, emphasizing the importance of subtype selectivity in PET studies. Another multi-tracer PET study in temporal lobe epilepsy showed increased μOR binding ([^{11}C]Caf (**8**)) in medial inferior temporal cortex and a more widespread increase in δOR binding ([^{11}C]MeNTI (**22**)) in the affected temporal lobe [141]. Increased [^{11}C]DPN (**15**) binding in temporal pole and fusiform gyrus of epilepsy patients declined with time since last seizure, indicating a transient response of the opioid system [142]. Applying a partial volume correction revealed small post-ictal increases in [^{11}C]DPN (**15**) V_T in the (sclerotic) hippocampus relative to the interictal state [143], possibly indicating reduced competition from endogenous opioids. Thus, there may be reduced opioid transmission in the post-ictal period.

A study of five patients with reading epilepsy (i.e. seizures provoked by reading text) revealed very circumscribed 10% reductions in [^{11}C]DPN binding the temporal parietal cortex in the activation condition compared to non-reading baseline, whereas no such changes was seen in control subjects [144]. In the context of a competition model, the authors interpreted this to indicate task-dependent release of opioid peptides in the patients, but it is difficult to determine the causal relationship between this release and the seizures.

3.3. Movement Disorders

In an MPTP model of acquired parkinsonism, a substantial striatal dopamine depletion to FDOPA–PET was associated with a 20–30% reduction in the V_T of [^{18}F]FcyF, (**11**) in opioid-receptor rich areas, i.e., caudate, anterior putamen, thalamus, and amygdala relative to intact animals [145]. These animals had recovery of their motor function, suggesting that the μOR changes were an adaptive response to dopamine depletion. The same group reported that this effect was (paradoxically) bilateral in animals with unilateral dopamine lesions [146]. In patients with Parkinson's disease, the [^{11}C]DPN (**15**) binding relative to occipital cortex was 20–30% reduced in striatum and thalamus only in those patients with iatrogenic DOPA-dyskinesia, but was unaffected in nondyskinetic Parkinson's disease patients [147]. This observation seems to merit further investigation, given the disabling effect of dyskinesias often encountered in the treatment of Parkinson's disease. On the other hand, there was no difference in [¹¹C]DPN (**15**) binding in symptomatic DYT1 primary torsion dystonia patients relative to controls [148].

Regional [¹¹C]DPN (**15**) binding was unaffected in patients with restless legs syndrome (which, like, Parkinson's disease, is responsive to levodopa treatment). However, there was a negative correlation between V_T and motor symptom severity, and a negative correlation between severity of pain and ligand binding in the medial pain system (medial thalamus, amygdala, caudate nucleus, anterior cingulate gyrus, insular and orbitofrontal cortex). The authors interpreted this result in relation to pain-induced release of endogenous peptides, rather than a primary aspect of restless legs syndrome [149].

Binding of the non-selective OR ligand [¹¹C]DPN PET was reduced by 31% in the caudate nucleus and 26% in putamen of a small group of symptomatic Huntington's disease patients compared to age-matched healthy controls [150]. This effect was less pronounced than was the loss of dopamine transporters seen in the same patients, suggesting that the reduction in ORs may partially accommodate the nigrostriatal degeneration. Despite this finding, there has been no indication for opioid medications in the symptomatic treatment of HD.

3.4. Pain

A qualitative study of post pontine infarct central pain showed a reduction in [¹¹C]DPN (**15**) uptake in the lateral cerebral cortex on the side contralateral to the main symptoms [151]. A more detailed study indicated [¹¹C]DPN (**15**) binding reductions in contralateral thalamus, parietal, secondary somatosensory, insular and lateral prefrontal cortices; these reductions were similar irrespective of the site of the lesion causing the central pain syndrome [152]. Indeed, this network of brain regions is recognized as comprising a central pain pathway. Another [¹¹C]DPN (**15**) study of central neuropathic pain showed 15–30% lower OR-binding within the medial pain system (cingulate, insula and thalamus), as well as the inferior parietal cortex of the lateral system (Brodman area 40). Patients with peripheral neuropathic pain had bilateral and symmetrical decreases in [¹¹C]DPN (**15**) binding in contrast to the hemispheric changes seen in central pain patients [153]. As always, these binding reductions are ambiguous, perhaps due to reduced receptor expression, increased occupancy, or internalization.

Poor sleep quality in relation to topical application of 10% capsaicin cream (which directly activates cutaneous pain receptors) was associated with higher baseline [¹¹C]Caf (**8**) BP_{ND} in the frontal lobe [154]. Thus, we suppose that baseline cortical binding may reflect a tradeoff between pain sensitivity and some cognitive or resilience function subserved by μ-ORs. Capsaicin-induced pain provoked a decrease in [¹¹C]Caf (**8**) binding in the contralateral thalamus by as much as 50%, increasing as the subjective severity of the pain [155]. Heat pain reduced the [¹⁸F]FE-DPN (**16**) binding in limbic and paralimbic brain areas including the rostral ACC and insula [156]. Application of sustained painful stimulus of the jaw muscle with saline injection provoked bilateral reductions in [¹¹C]Caf (**8**) binding in the ipsilateral amygdala (5%) and contralateral ventrolateral thalamus (7%) [157]. The same painful stimulus that provoked 5–10% decreases in [¹¹C]Caf (**8**) binding in healthy young men tended to increase binding in women; this gender difference was most pronounced in the ventral striatum ipsilateral to the pain [158]. A small group of patients with trigeminal neuralgia had reduced [¹¹C]Caf (**8**) binding in left nucleus accumbens, a brain region earlier implicated in pain modulation and response to reward or aversive stimuli [159]. It would be interesting to test if this phenomenon correlated with individual differences in affective state or trait neuroticism.

In another [¹¹C]Caf (**8**) study, the reduced binding provoked soon after administration of a sustained pain of moderate intensity had normalized in the hours after cessation of the stimulus [160]. In general, painful stimuli do not desensitize with time, so the relationship between temporal dynamics of opioid signaling and pain perception must be complex. Indeed, pain researchers and clinicians alike are familiar with the phenomenon of allodynia, which is a decrease in pain threshold, or the conversion of previously non-painful stimuli to pain. In a sciatic nerve stimulation model, pain conditioning some hours after stimulation was associated with increased C-fibre response and reduced C-fibre threshold, as well as supraspinal changes marked by increased binding of [^{11}C]PEO (**18**) in ipsilateral and bilateral structures of the rat brain [161]. Thus, allodynia may indicate inactivation of pain-mediated opioid release in brain, resulting in greater OR availability. On the other hand, an [^{18}F]FE-DPN (**16**) PET study in athletes contrasting receptor availability at rest with the condition immediately after running a half marathon showed reduced binding in various paralimbic and prefontal cortical structures, to an extent correlation with post-running euphoria ("runner's high") [162]. Similarly, a [^{11}C]Caf (**8**) study in recreationally active men showed a relationship between post-exercising euphoria with decreased μOR binding in widespread cortical areas after high intensity exercise, although effects were less clear after moderate intensity exercise [163].

Prolonged electrical stimulation of the motor cortex for relief of neuropathic pain caused reductions in [^{11}C]DPN (**15**) binding in part of the cingulate cortex, prefrontal cortex, the periaqueductal gray prefrontal cortex, and cerebellum [164]. Some of these changes correlated with the extent of pain relief. In a case study, a single session of motor cortex stimulation improved the cold pain threshold, while decreasing [^{11}C]Caf (**8**) binding in pain-network brain regions [165]. Other studies showed that low cerebral binding of [^{11}C]DPN (**15**) predicted for poor response to motor cortex stimulation for the treatment of neuropathic pain [166]. Stimulation of the central grey for treatment of phantom limb pain provoked a focal decrease in midbrain [^{11}C]DPN (**15**) binding, indicating endogenous opioid release [167].

Visceral pain applied by gastric inflammation was without effect on cerebral [^{11}C]Caf (**8**) binding in healthy volunteers [168], this standing in contrast to findings with somatic pain as described above. On the other hand, vestibular stimulation provoked decreased [^{18}F]FE-DPN (**16**) binding in parts of the right dominant cortical vestibular network [169].

A [^{11}C]Caf (**8**) study in which painful heat was applied after administration of supposedly analgesic cream indicated a placebo-mediated reduction in receptor availability [170]. The same group later showed that placebo transcranial direct current stimulation (tDCS) reduced [^{11}C]Caf (**8**) BP$_{ND}$ in the periaqueductal gray matter (PAG), precuneus, and thalamus, indicating endogenous opioid release [171]. This placebo effect apparently increased upon administration of verum tDCS. In another study, acupuncture administered according to an authentic analgesic procedure had only slight effects on the binding of [^{11}C]DPN (**15**) in human brain [172]. However, a study with [^{11}C]Caf (**8**) showed acupuncture therapy to provoke short-term and persistent 10–30% increases in μOR binding in pain-related brain regions; importantly the verum acupuncture condition was contrasted with a sham acupuncture condition in that study [173]. Thus, while acupuncture analgesia may be "in one's head", there seems to be a real component mediated by increased opioid transmission. Transcutaneous electrical acupoint stimulation (TEAS) is an analogue of the electrical acupuncture technique. Administration of TEAS at 2 Hz to anesthetized monkey provoked reductions in [^{11}C]Caf (**8**) binding in striatum, amygdala and ACC, whereas 100 Hz stimulation had no effect relative to baseline PET recordings [174].

Reminiscent of the findings in the study of pain-induced sleep disturbance noted above, a cross sectional study of sensory processing in healthy volunteers showed lower baseline binding of [^{18}F]FE-DPN (**16**) in regions such as insular cortex and orbitofrontal cortex of those with greater sensitivity to cold pain. In addition, there were negative correlations between regional binding and sensory thresholds for non-painful stimuli [175]. Similarly, the individual striatal binding of [^{11}C]Caf (**8**) BP$_{ND}$ predicted cold pressor pain threshold, but not cold pressor pain tolerance or tactile sensitivity [176]. A longitudinal [^{18}F]FE-DPN (**16**) PET study in neuropathic pain model rats showed lower μ+κ OR availability in the insula, caudate putamen, and motor cortex at three months after the injury [177]. These reductions occurred in association with anhedonia (disinterest in sucrose solution). Overall, these studies suggest that individual differences in OR signaling may mediate vulnerability to environmental stressors, a topic to be elaborated in Section 3.5 below.

Binding of [^{11}C]DPN (**15**) was reduced in pineal gland (but not in the brain *per se*) of patients who had been experiencing cluster headache attacks [178], said to be one of the most painful experiences.

The authors suggested that inputs from trigeminal nerve to the pineal gland might mediate this change. A group of seven spontaneous migrainers showed ictal reductions in [^{11}C]Caf (**8**) binding in the medial prefrontal cortex, which correlated with the baseline or intra-ictal binding [179]. The [^{11}C]DPN (**15**) BP$_{ND}$ in brain did not differ between migrainers and healthy controls, nor was there any effect of placebo treatment in either group [180].

3.5. Personality, Drug Dependence, and Psychiatric Disorders

Scores in the personality trait of harm avoidance in a group of 23 healthy males correlated positively with binding of [^{18}F]FE-DPN (**16**) in the bilateral ventral striatum, suggesting a link with predisposition to substance abuse [181]. It might follow that drug abuse is a kind of self-medication for those with pronounced harm avoidance trait. A comparison of [^{11}C]Caf (**8**) uptake in healthy individuals showed that high scores in the harm avoidance trait were associated with high μOR availability in frontal and insular cortex [182], again linking the hard avoidance trait with lower tonic opioid transmission. Score in a scale of behavioral activation, which conceptually guides approach behavior, and notably in a scale designated "fun-seeking", correlated positively with [^{11}C]Caf (**8**) in widespread brain regions [183]. A [^{11}C]Caf (**8**) study in 49 healthy volunteers showed an *inverse* relationship between μOR availability in various brain regions and individual scores in the avoidance dimension of interpersonal attachment [184]. Considering the harm avoidance findings, baseline μOR availability may mediate a trade-off between harm avoidance and avoidant behavior in interpersonal relationships, in a psychological analogue of pain or cold sensitivity.

In a large group of healthy women, [^{11}C]Caf (**8**) binding had a negative correlation with BOLD signal responses in amygdala, hippocampus, thalamus, and hypothalamus to viewing emotionally arousing scenes [185]. Non-sexual, albeit pleasurable social touch from a partner provoked widespread increases in [^{11}C]Caf (**8**) binding, suggesting reduced opioid signaling [186], whereas social laughter provoked by viewing comedic film clips decreased [^{11}C]Caf (**8**) binding in thalamus, caudate nucleus, and anterior insula. Furthermore, baseline μOR availability in some regions was associated with the rate of social laughter [187]. These results are difficult to reconcile, since pleasurable social experiences can seemingly have opposite effects on μOR availability. Contrasting the [^{11}C]Caf (**8**) binding in euthymic and unhappy states (provoked by autobiographical reflection) in young women showed higher μOR availability in the rostral anterior cingulate, ventral pallidum, amygdala, and inferior temporal cortex in the unhappy state [188]. This kind of sad reflection provoked greater increases in [^{11}C]Caf (**8**) binding in widespread brain regions of women with major depression [189], suggesting an exaggerated opioid response in relation to mood disorder, as distinct from ordinary sadness. A pilot PET study with the κOR-ligand [^{11}C]GR103545 (**25**) did not reveal any binding differences between healthy control and patients suffering from major depression [190]. However, a [^{11}C]EKAP (**27**) κOR study in healthy volunteers showed an inverse correlation between social status and [^{11}C]salvinorin A (**33**) binding in widespread brain areas, with a special association occurring in brain regions mediating reward or aversion [191]. Given the association between social stress and depression, one might have expected covariance κORs in the two studies.

A recent [^{11}C]Caf (**8**) PET study of 19 schizophrenia patients and 20 controls showed a 10% lower (Cohen's d = 0.7) μOR -availability in striatum of the patient group. While such a decrease can hardly be pathognomic of disease, the authors also reported considerably higher inter-regional covariance of the [^{11}C]Caf (**8**) binding in the patients, which might indicate an aberent spatial pattern of opioid signalling in schizophrenia [192]. There have been no OR PET studies in bipolar disorder.

A [^{11}C]Caf (**8**) study showed that circulating levels of the anti-nociceptive cytokine IL-1ra (which correlated with neuroticism scores) predicted for the pain response to a standard stimulus (saline infusion to the masseter muscle), and likewise the reduction in μOR availability in the basal ganglia during the painful stimulus [193]. In a group of female patients suffering from fibromyalgia, [^{11}C]Caf (**8**) binding correlated with pain-evoked BOLD signal changes in several brain regions, including dorsolateral prefrontal cortex and nucleus accumbens [194]. Overall, these studies suggest some linking between opioid transmission, mood, and inflammatory markers, which returns to the the notioin that OR signaling may mediate personality traits and vulnerability to stresses of various sorts.

[^{11}C]Caf (**8**) PET showed persistently increased µOR binding in frontal and cingulate cortex of acutely detoxified cocaine addicts, which correlated with the extent of craving [195]. Elevated [^{11}C]Caf (**8**) binding in frontal and temporal cortical regions was a significant predictor of time to relapse to cocaine use among abstinent addicts [196]. Binge cocaine users showed a significant association between [^{11}C]GR103545 (**25**) binding to κORs with the amount of drug consumed. Furthermore, a three-day cocaine binge reduced binding by about 15% [197]. The cerebral binding (V_T) of the ORL1 ligand [^{11}C]NOP-1A (**36**) was globally elevated 10% in detoxified cocaine users [198].

One [^{11}C]Caf (**8**) PET study showed persistently increased µOR binding in striatum of detoxified alcoholics, which furthermore correlated with the extent of craving [199]. Abstinent alcoholics showed significantly higher [^{11}C]Caf (**8**) binding compared to controls, but a blunting of the response to amphetamine (which indirectly displaces µOR binding), resembling that seen by the same research group in compulsive gamblers [200]. However, others saw only a small increase in [^{11}C]DPN (**15**) V_T in brain of acutely withdrawn alcoholics, although there was a correlation between individual PET results and craving scores at the time of scanning [201]; the combined (µ+κ) PET signal in that study makes difficult a comparison with [^{11}C]Caf (**8**) studies. *Post mortem* autoradiographic examination of [^{3}H]DAMGO binding in brain of a large group of alcoholics showed substantial reductions in µOR binding sites, whereas low [^{11}C]Caf (**8**) BP_{ND} in ventral striatum of acutely detoxified patients predicted high risk of relapse and poor response to naloxone in interaction with the µOR rs1799971 allele [202]. The disagreement between µOR findings *in vivo* and *post mortem* could indicate low tonic occupancy in alcohol dependent patients, since competition effects would disappear in autoradiographic studies.

In a [^{11}C]MeNTI (**22**) PET study, there was globally 10–20% higher δOR binding in brain of a large group of alcohol-dependent subjects; which reached significance upon correcting for age, gender, and smoking status; there was an inverse relationship between binding in some regions and intensity of craving [203]. The V_T of the κOR-selective ligand [^{11}C]LY2795050 (**30**) was significantly lower in amygdala and pallidum of alcohol-dependent subjects [138]. This stands in contrast to the usual finding of increased µOR binding and the single report of elevated δOR binding.

The naloxone challenge paradigm has a long history in investigations of the regulation of the neuroendocrine axis, but it has been uncertain if naloxone-induced increases in ACTH and cortisol secretion bear any relation to central OR pathways. In a [^{11}C]Caf (**8**) PET study of healthy volunteers there were negative relationships between cortisol (but not ACTH) response to naloxone and ligand BP_{ND} in ventral striatum, putamen and caudate [204]. The inverse relationship between naloxone-induced cortisol secretion and [^{11}C]Caf (**8**) BP_{ND} in various brain regions of healthy volunteers was absent in alcohol dependent subjects [205]. This suggests that central ORs exert a top-down regulation of the neuroendocrine axis, which might contribute to individual differences in stress response, and that the normal regulation of this axis is disrupted in alcohol dependence.

There was only a slight difference in [^{11}C]Caf (**8**) binding between non-smoking carriers of the µOR rs1799971 allelic variants, but this allelic difference was greater among smokers. Furthermore, the contrast in PET results between active and denicotinized cigarette conditions revealed a positive relationship between reward and altered µOR availability in the smokers [206]. An apparent re-analysis of the same data showed widespread reductions in [^{11}C]Caf (**8**) binding after smoking a nicotine-containing cigarette; this effect was moderated by the rs1799971 polymorphism, where carriers of the A allele showed greater response to active cigarette smoking, and higher baseline µOR binding [207]. The authors conceded that non-nicotinergic factors, i.e. conditioning, could be contributing to aspects of smoking related opioid transmission [208].

Smoking subjects with higher dependence, craving, and cigarette consumption rates showed lower baseline [^{11}C]Caf (**8**) BP_{ND} in limbic brain regions. There was blunting of this association during NTX (**5**) treatment [209], but there was very low residual specific binding in the NTX condition, which must have compromised the sensitivity of the method. Another [^{11}C]Caf (**8**) PET study showed no difference in BP$_{ND}$ between placebo and active nicotine cigarette conditions, and no difference between smokers and nonsmokers. However, there was a negative correlation in the smokers between BP$_{ND}$ in bilateral superior temporal cortex with scores in an index of nicotine dependence [210].

As noted above, challenge with amphetamine can indirectly provoke increased opioid peptide release. However, in a placebo-controlled, double-blinded and randomized [^{11}C]Caf (8) PET study, amphetamine challenge (0.3 mg/kg) did not alter μOR availability in healthy male volunteers [211]. This stands in contrast to another study, wherein a high dose of amphetamine (0.5 mg/kg) provoked reductions in [^{11}C]Caf (8) binding in widespread brain regions, i.e. frontal cortex, putamen, caudate, thalamus, anterior cingulate, and insula, whereas a sub-pharmacological dose was without such an effect [212,213]. Preclinical studies point to the importance of receptor internalization on the vulnerability of OR-receptor binding to challenge with amphetamine [214]. Notwithstanding this caveat, amphetamine induced reductions in [^{11}C]Caf (8) binding were blunted in compulsive gamblers compared to that in a healthy control group controlled for smoking and drinking [215]. There was a general correlation between dopamine synthesis capacity to FDOPA PET and [^{11}C]Caf (8) binding in putamen of healthy controls, and likewise in pathological gamblers, consistent with a tight relationship between dopamine and opioid systems in relation to compulsive behaviors [216].

Women with bulimia nervosa had reduced [^{11}C]Caf (8) binding in the left insula, to an extent correlating with their duration of fasting [217]. Obese patients (BMI 40) had globally 20% lower [^{11}C]Caf (8) BP$_{ND}$ compared to lean volunteers; contrary to some reports, the same obese subjects had normal dopamine D$_2$ receptor levels in striatum [218]. There were similar reductions in [^{11}C]Caf (8) binding in morbidly obese subjects and patients with binge eating disorder [219]. Weight loss after bariatric surgery for the treatment of obesity resulted in a global 25% increase of μOR binding [220]. A dual tracer study with [^{11}C]Caf (8) and the dopamine receptor ligand [^{11}C]raclopride showed a high correlation in the striatum of lean subjects, whereas this correlation was considerably weaker in the ventral (limbic) striatum of the morbidly obese, suggesting an uncoupling of opioid/dopamine interactions in that condition [221]. This finding might predict analogous results in gambling and substance abuse disorders, which likewise may involve dysregulation opioid/dopamine interactions.

Feeding, regardless of the hedonic experience (palatable versus unpalatable meal), provoked widespread decreases in [^{11}C]Caf (8) binding in non-obese healthy subjects, suggesting that OR transmission mediates some aspect of the rewarding properties of food [222]. Also in non-obese subjects, [^{11}C]Caf (8) BP$_{ND}$ in amygdala correlated inversely with BMI in the range 20–27 [223]. In that study, BP$_{ND}$ in other brain regions correlated with the BOLD signal response in orbitofrontal cortex upon viewing palatable food. In a group of lean subjects, the [^{11}C]Caf (8) BP$_{ND}$ at baseline in widespread brain regions correlated with BOLD responses to viewing palatable food [224], suggesting that low basal occupancy increases the response to cues. Interestingly, exercise increased or decreased thalamic μOR binding in these subjects; the direction of this change correlated with the individual BOLD signal in the contrast between viewing palatable and non-palatable food. This draws attention to individual differences in effects of exercise on the hedonic response to food, which may have some bearing on the relationship between exercise and weight loss, with the caveat that only intense exercise may significantly engage opioid transmission, as claimed above. Other studies show widespread reductions in μOR availability in frontolimbic regions after high intensity aerobic exercise, in correlation with negative affect. In contrast, mean binding was unaltered after moderate-intensity exercise, although there was some positive association with euphoria [225]; too much of a good thing spoils runner's high, it seems.

4. Conclusions and Outlook

The past decades have seen extraordinary progress in the development of ligands for PET studies of ORs. Early radiopharmaceutical research focused on studies with the antagonist [^{11}C]DPN (15) and the μOR-selective agonist [^{11}C]Caf (8), and the great preponderance of human PET studies have employed these and closely related tracers. While studies with non-selective tracers reveal the composite of OR binding sites, specific tracers may be more indicative of physiological changes in disease states. Attaining high molar activity is of great importance in PET studies with [^{11}C]Caf (**8**) and other highly potent agonist ligands; fortunately, most tracers described in this review have molar activities of at least 50 GBq/μmol, corresponding to an injected mass of about 1 μg of the drug. This is hardly a relevant dose in the case of antagonist ligands, and would give a comfortable 100-fold margin of safety with the potent μOR agonist [^{11}C]Caf (**8**).

The μOR ligands have the useful property of binding in competition with endogenous opioid peptides, such that changes in the uptake in PET studies can reveal altered endogenous opioid release under various physiological conditions. This model has been particularly useful in studies of pain pathways, which largely involve μORs in telencephalon, and in some pharmacological or behavioral activation studies. However, the simple competition model may be inadequate to account for all observations. Thus, one of the [^{11}C]Caf (**8**) studies noted above reports widespread reductions in μOR availability after smoking [207], despite the 30–3000-fold lower affinity of endogenous opioid peptides at μOR. This would seem to imply an implausibly enormous increase in peptide release to effect such changes by competition alone.

Pain studies have so far dominated the field of clinical PET research with OR-ligands, with relatively few reports on other models or conditions, as summarized in Table 2. For example, there in only one PET study of opioid receptors in schizophrenia, and only scant documentation in depression, or for that matter, in a range of common neurological disorders. In several human diseases noted above, the OR binding may be only 10% higher or lower than in the control group; while these small differences can have a large effect size, it is perhaps difficult to argue that such small differences can be causative of complex disorders or symptoms.

Table 2. A summary of the key results with opioid PET in clinical research.

Condition	Ligand	Main Finding	Ref.
Healthy aging	[^{11}C]Caf (**8**) (μOR)	20% decrease in frontal cortex (females)	[137]
Epilepsy	[^{11}C]Caf (**8**) (μOR)	Increased in ipsilateral temporal lobe, decreased in amygdala	[140]
Epilepsy	[^{11}C]DPN (**15**) (mixed ligand)	No change	[140]
Parkinson's disease	[^{11}C]DPN (**15**) (mixed ligand)	20–30% decrease in striatum and thalamus only in those patients with iatrogenic DOPA-dyskinesia	[147]
Huntington's disease	[^{11}C]DPN (**15**) (mixed ligand)	30% reduced in caudate/putamen	[150]
Pontine infarct central pain	[^{11}C]DPN (**15**) (mixed ligand)	Reduced throughout pain network	[151]
Capsaicin-induced acute pain	[^{11}C]Caf (**8**) (μOR)	Up to 50% decrease contralateral thalamus, in proportion to subjective severity	[155]
Sustained painful stimulus of the jaw muscle with saline injection	[^{11}C]Caf (**8**) (μOR)	Blateral decrease in binding in the ipsilateral amygdala (5%) and contralateral ventro-lateral thalamus (7%)	[157]
painful heat	[^{11}C]Caf (**8**) (μOR)	Placebo effect on binding changes	[170]
Acupuncture therapy with sham acupuncture control	[^{11}C]Caf (**8**) (μOR)	Persistent 10–30% increases in μ-OR binding in pain-related brain regions	[173]
Harm avoidance trait in healthy males	[^{18}F]FE-DPN (**16**) (mixed ligand)	Trait correlated positively with binding in vental striatum, suggesting link with substance abuse	[181]
Correlation with BOLD signal responses A large group of healthy women, binding to viewing emotionally arousing scenes	[^{11}C]Caf (**8**) (μOR)	Negative correlation in amygdala, hippocampus, thalamus, and hypothalamus	[185]
Major depressive disorder	[^{11}C]GR103545 (**25**) (κOR)	No difference from controls	[190]
Detoxified cocaine addicts	[^{11}C]Caf (**8**) (μOR)	Increased in frontal and cingulate cortex, which correlated with the extent of craving	[195]
Detoxified alcohol-dependent subjects	[^{11}C]MeNTI (**22**) (δOR)	Globally 10–20% increased binding inverse relationship in some regions with intensity of craving	[203]
Obesity (BMI > 40)	[^{11}C]Caf (**8**) (μOR)	Globally 20% lower compared to lean volunteers	[218]
Feeding, regardless of the hedonic experience in non-obese subjects	[^{11}C]Caf (**8**) (μOR)	Widespread decreases in binding	[222]

Addiction research using PET studies of ORs are so far mostly confined to alcohol, cocaine, and nicotine abuse and (strangely, perhaps), opioid addiction has hardly been a research theme, other than in a few occupancy studies. Since antagonists are relatively safe at doses provoking high occupancy (viz 50 mg naloxone for opioid overdose), we suppose that the B_{max} of ORs might be determinable in relation to opioid addiction and withdrawal by conducting serial PET studies over a range of molar activity, even in the presence of significant agonist occupancy. Indeed, chronic morphine was reported 45 years ago to increase the abundance of [^{3}H]naloxone binding sites in rat brain [226], but no such studies are reported in human opioid users, despite the catastrophe of the current opioid addiction epidemic. This kind of information might help to understand better the correlates of addiction and relapse. In addition, genetic studies of dopaminergic and opioid systems in relation to addiction [227], in conjunction with molecular imaging studies, could help to establish better the risk factors for opioid addiction. Endomorphins and other novel opioid petpides may present new avenues for obtaining opioid analgesia [228], while moderating the risk of "iatrogenic opioid addiction". The development of PET tracers with good binding properties in vivo and high selectivity for ORs other than the µ-type has accelerated in the past decade. However, there remain relatively few clinical molecular imaging studies of these important targets. Thus, developments in radioligand chemistry have for the presence to read for the present presence outpaced clinical PET imaging, a state of affairs that could enable and motivate a broad range of studies focusing on non-µORs over the coming decades. Just for example, κORs have an established role in the reinstatement of stress induced drug use in experimental animals, i.e. nicotine use [229], and very recent results indicate a relationship between κORs and stress-induced binge cocaine use [197].

Funding: This research received no external funding.

Conflicts of Interest: The authors declare no conflict of interest.

Abbreviations

Compound	Name, Synonyms
3-O-Ac-[^{18}F]FcyF (10)	3-O-[^{18}F]acetylcyclofoxy, 3-O-Acetyl-6-deoxy-6-beta-[^{18}F]fluoronaltrexone
β-Endorphin	Tyr-Gly-Gly-Phe-Met-Thr-Ser-Glu-Lys-Ser-Gln-Thr-Pro-Leu-Val-Thr-Leu-Phe-Lys-Asn-Ala-Ile-Ile-Lys-Asn-Ala-Tyr-Lys-Lys-Gly-Glu CAS RN: [61214-51-5]
BEMP	2-tert-butylimino-2-diethylamino-1,3-dimethyl-perhydro-1,3,2-diazaphosphorine CAS RN: [98015-45-3]
[^{11}C]BPN (14)	[^{11}C]buprenorphine, [6-O-methyl-^{11}C]buprenorphine
Caf	carfentanil, 1-(2-phenylethyl)-4-[(1-oxopropyl)phenylamino]-4-piperidine carboxylic acid methyl ester, R-31,833, 4-carboxymethyl-fentanyl, CAS RN: [59708-52-09]
[^{11}C]Caf (8)	[^{11}C]carfentanil
[^{3}H]DAMGO	[3,5-^{3}H]Tyr-D-Ala-Gly-(NMe)Phe-Gly-ol
DBU	1,8-diazabicyclo[5.4.0]undec-7-ene, CAS RN: [6674-22-2]
DIPEA	ethyldiisopropylamine, N,N-diisopropylamine, Hünig's base
DMF	N,N-dimethylformamide
[^{11}C]DPN (15)	[^{11}C]diprenorphine, [6-O-methyl-^{11}C]diprenorphine
dppf	1,1'-ferrocenediyl-bis(diphenylphosphine), CAS RN: [121450-46-8]
Dynorphin A [1-17]	Tyr-Gly-Gly-Phe-Leu-Arg-Arg-Ile-Arg-Pro-Lys-Leu-Lys-T rp-Asp-Asn-Gln CAS RN: [80448-90-4]
Dynorphin A [1-8]	Tyr-Gly-Gly-Phe-Leu-Arg-Arg-Ile, CAS RN: [75790-53-3]
Dynorpin B [1-13]	Tyr-Gly-Gly-Phe-Leu-Arg-Arg-Gln-Phe-Lys-Val-Val-Thr, CAS RN: [83335-41-5]
[^{11}C]EKAP (27)	(R)-(methyl-^{11}C) 4-(2-(3,4-dichlorophenyl)acetyl)-3-((diethylamino)methyl) piperazine-1-carboxylate
cyF	cyclofoxy, 6-deoxy-6β-[^{18}F]fluoro-naltrexone, N-cyclopropylmethyl-6-deoxy-6β-fluoro-noroxymorphone, 17-cyclopropylmethyl-4,5α-epoxy-6β-fluoro-morphinan-3,14-diol CAS RN: [103223-57-0]
[^{18}F]FcyF (11)	[^{18}F]cyclofoxy, 6-Deoxy-6-β-[^{18}F]fluoro-naltrexone, N-cyclopropylmethyl-6-deoxy-6β-[^{18}F]fluoro-noroxymorphone, CAS RN: [103223-58-1]
FDOPA	6-fluoro-L-dopa, 2-Fluoro-5-hydroxy-L-tyrosine, CAS RN: [144334-59-8]
[^{18}F]FE-DPN (16)	6-O-(2-[^{18}F]fluoroethyl)-6-O-desmethyl-diprenorphine
[^{11}C]FEKAP (28)	(R)-(methyl-^{11}C)4-(2-(3,4-dichlorophenyl)acetyl)-3-((ethyl(2-fluoroethyl)amino) methyl)piperazine-1-carboxylate
[^{18}F]FE-BPN (19)	6-O-(2-[^{18}F]fluoroethyl)-6-O-desmethyl-buprenorphine
[^{18}F]FP-norBPN (21)	N^{17}-(3-[^{18}F]fluoropropyl)-nor-buprenorphine
[^{18}F]FP-norDPN (20)	N^{17}-(3-[^{18}F]fluoropropyl)-nor-diprenorphine

[¹⁸F]FE-NTI (23)	[¹⁸F]fluoroethyl-naltrindole, N1'-(2-[¹⁸F]Fluoroethyl)-naltrindole, [¹⁸F]BU97001
[¹⁸F]FE-PEO (17)	6-O-(2-[¹⁸F]fluoroethyl)-6-O-desmethyl-phenethyl-orvinol
[¹⁸F]FE-Tos	[¹⁸F]fluoroethyl tosylate
Foxy	6-deoxy-6β-fluoro-oxymorphone, «fluorooxymorphone», 4,5α-Epoxy-6β-fluoro-17-methyl-morphinan-3,14-diol, CAS RN: [92593-44-7]
[³H]Foxy	[1,2-³H]-4,5α-epoxy-6β-fluoro-17-methyl-morphinan-3,14-diol, CAS RN: [96917-45-2]
[¹¹C]GR103545 (25)	[(3,4-dichlorophenyl)acetyl]-(3R)-(1-pyrrolidinylmethyl)-1-piperazine carboxylic acid methyl-¹¹C ester
[¹¹C]GR89696 (26)	4-[(3,4-dichlorophenyl)acetyl-3-(R,S)-(1-pyrrolidinylmethyl)-1-piperazine carboxylic acid methyl-¹¹C ester
HD	Huntington's disease
[¹¹C]LAAM (13)	N-(methyl-¹¹C)-L-α-acetoxymethadol, N-(methyl-¹¹C)-levo-alpha-acetyl methadol, N-(methyl-¹¹C)-levomethadyl acetate
Leu⁵-enkephalin	H-Tyr-Gly-Gly-Phe-Leu-OH, CAS RN: [58569-55-4]
LY2459989	3-fluoro-4-[4-[[(2S)-2-(3-pyridyl)pyrrolidin-1-yl]methyl]phenoxy] benzamide
LY2795050	3-chloro-4-[4-[[(2S)-2-(3-pyridyl)pyrrolidin-1-yl]methyl]phenoxy] benzamide, CAS RN: [1346133-08-1]
LY2456302 (34)	(S)-4-(4-((2-(3,5-dimethylphenyl)pyrrolidin-1-yl)methyl)phenoxy)- 3-fluorobenzamide, CERC-501, CAS RN: [1174130-61-0]
LY2940094	([2-[4-[(2-chloro-4,4-difluoro-spiro[5H-thieno[2,3-c]pyran-7,4'-piperidine]-1'-yl]methyl]-3-methyl-pyrazol-1-yl]-3-pyridyl]methanol), BTRX-246040, CAS RN: [1307245-86-0]
Met⁵-enkephalin	H-Tyr-Gly-Gly-Phe-Met-OH
[¹¹C]MeJDTic (29)	(3S)-7-hydroxy-N-((1S)-1-[[3R,4R)-4-(3-hydroxyphenyl)-3,4-dimethyl-1-piperidinyl]methyl]-2-methylpropyl)-2-[¹¹C]methyl-1,2,3,4-tetrahydro-3-isoquinolinecarboxamide
[¹¹C]MeNTI (22)	[¹¹C]methyl-naltrindole, N1'-[¹¹C]methyl-naltrindole
[¹⁸F]MK-0911 (39)	1-(2-[¹⁸F]fluoroethyl)-3-[(3R,4R)-3-(hydroxymethyl)-1-[[(8S)-spiro[2.5]octan-8-yl]methyl]piperidin-4-yl]benzimidazol-2-one
MPTP	1-methyl-4-phenyl-1,2,3,6-tetrahydropyridine, CAS RN: [28289-54-5]
Oxycodone	14-hydroxy-dihydrocodeinone, CAS RN: [76-42-6]
Naloxone	N¹⁷-allyl-14-hydroxy-dihydromorphinone, N¹⁷-Allyl-noroxymorphone, CAS RN: [465-65-6]
Naltrexone	N¹⁷-cyclopropylmethyl-14-hydroxy-dihydromorphinone, N¹⁷-cyclopropylmethyl-noroxymorphone, NTX CAS RN: [16590-41-3]
Nociceptine	Phe-Gly-Gly-Phe-Thr-Gly-Ala-Arg-Lys-Ser-Ala-Arg-Lys-Leu-Ala-Asn-Gln
N-substituted-[¹¹C]quinolinimide (24)	6-(2-{2-[4-(4-fluorobutyl)-benzenesulfonyl]-1,2,3,4-tetrahydro-isoquinolin-1-yl}-ethyl)-2-[¹¹C]methyl-pyrrolo[3,4-b]pyridine-5,7-dione
NTI	naltrexone-indole, naltrindole, CAS RN: [111555-53-4]
[¹¹C]NOP-1A (36)	(2S)-2-[(2-fluorophenyl)methyl]-3-(2-fluorospiro[4,5-dihydro-thieno[2,3-c]pyran-7,4'-piperidine]-1'-yl)-N-[¹¹C]methyl-propanamide
OR	opioid receptor
[¹¹C]PEO (18)	[¹¹C]phenethyl-orvinol, [6-O-methyl-¹¹C]phenethyl-orvinol
PET	positron emission tomography
Pd₂dba₃	Pd₂(dba)₃, tris(dibenzylideneacetone)dipalladium, CAS RN: [51364-51-3]
[³H]PL017	[3,5-³H]Tyr-Pro-(NMe)Phe-D-Pro-NH₂, 3-N-Me-Phe-[³H]morphiceptin, [³H] [MePhe³,D-Pro⁴]morphiceptin
Salvinorin A	methyl (2S,4aR,6aR,7R,9S,10aS,10bR)-9-acetyloxy-2-(furan-3-yl)-6a,10b-dimethyl-4,10-dioxo-2,4a,5,6,7,8,9,10a-octahydro-1H-benzo[f]isochromene-7-carboxylate, CAS RN: [83729-01-5]
TBDMS	tert-butyldimethylsilyl protecting group
TDBPN	3-O-trityl-6-O-desmethyl-buprenorphine, CAS RN: [157891-93-5]
TDDPN	3-O-trityl-6-O-desmethyl-diprenorphine, CAS RN: [157891-92-4], "Luthra-precursor", TDDPN was the first product of the company ABX advanced biochemical compounds Biomedizinische Forschungsreagenzien GmbH, Radeberg in 1997
TDPEO	3-O-trityl-6-O-desmethyl-phenethyl-orvinol, CAS RN: [1187551-69-4]
U-50488	3,4-dichloro-N-methyl-N-[(1R,2R)-2-(1-pyrrolidinyl)cyclohexyl]- benzene acetamide, NIH10533, CAS RN: [67198-13-4]

References

1. Gulland, J.M.; Robinson, R. Constitution of codeine and thebaine. *Mem. Proc. Manchester Lit. Phil. Soc.* **1925**, *69-86*, 79–86.
2. Gates, M.; Tschudi, G. The synthesis of morphine. *J. Am. Chem. Soc.* **1952**, *74*, 1109–1110.
3. Gates, M.; Tschudi, G. The synthesis of morphine. *J. Am. Chem. Soc.* **1956**, *78*, 1380–1393. [CrossRef]

4. Bentley, K.W.; Cardwell, H.M.E. The Morphine-Thebaine group of alkaloids. Part V. The absolute stereochemistry of the morphine, benzylisoquinoline, aporphine, and tetrahydroberberine alkaloids. *J. Chem. Soc.* **1955**, 3252–3260. [CrossRef]
5. Rice, K.C. Synthetic opium alkaloids and derivatives. A short total synthesis of (±)-dihydrothebainone, (±)-dihydrocodeinone, and (±)-nordihydrocodeinone as an approach to a practical synthesis of morphine, codeine, and congeners. *J. Org. Chem.* **1980**, *45*, 3135–3137. [CrossRef]
6. Lever, J.R. PET and SPECT imaging of the opioid system: Receptors, radioligands and avenues for drug discovery and development. *Curr. Pharm. Des.* **2007**, *13*, 33–49. [CrossRef] [PubMed]
7. Henriksen, G.; Willoch, F. Imaging of opioid receptors in the central nervous system. *Brain* **2008**, *131*, 1171–1196.
8. Dannals, R.F. Positron emission tomography radioligands for the opioid system. *J. Label. Compd. Radiopharm.* **2013**, *56*, 187–195. [CrossRef]
9. Pert, C.B.; Snyder, S.H. Properties of opiate-receptor binding in rat brain. *Proc. Natl. Acad. Sci. USA* **1973**, *70*, 2243–2247. [CrossRef]
10. Hughes, J.; Smith, T.W.; Kosterlitz, H.W.; Fothergill, L.A.; Morgan, B.A.; Morris, H.R. Identification of two related pentapeptides from the brain with potent opiate agonist activity. *Nature* **1975**, *258*, 577–579. [CrossRef]
11. Di Giulio, A.M.; Majane, E.M.; Yang, H.Y. On the distribution of [met^5]- and [leu^5]-enkephalins in the brain of the rat, guinea-pig and calf. *Br. J. Pharmacol.* **1979**, *66*, 297–301. [CrossRef] [PubMed]
12. Hughes, J.; Kosterlitz, H.W.; Smith, T.W. The distribution of methionine-enkephalin and leucine-enkephalin in the brain and peripheral tissues. *Br. J. Pharmacol.* **1977**, *61*, 639–647. [CrossRef] [PubMed]
13. Li, C.H.; Chung, D. Isolation and Structure of an Untriakontapeptide with Opiate Activity from Camel Pituitary Glands. *Proc. Natl. Acad. Sci. USA* **1976**, *73*, 1145–1148. [CrossRef] [PubMed]
14. Goldstein, A.; Tachibana, S.; Lowney, L.I.; Hunkapiller, M.; Hood, L. Dynorphin-(1-13), an extraordinarily potent opioid peptide. *Proc. Natl. Acad. Sci. USA* **1979**, *76*, 6666–6670. [CrossRef] [PubMed]
15. Nakanishi, S.; Inoue, A.; Kita, T.; Numa, S.; Chang, A.C.; Cohen, S.N.; Nunberg, J.; Schimke, T.R. Construction of bacterial plasmids that contain the nucleotide sequence for bovine corticotropin-beta-lipotropin precursor. *Proc. Natl. Acad. Sci. USA* **1978**, *75*, 6021–6025. [CrossRef]
16. Noda, M.; Furutani, Y.; Takahashi, H.; Toyosato, M.; Hirose, T.; Inayama, S.; Nakanishi, S.; Numa, S. Cloning and sequence analysis of cDNA for bovine adrenal preproenkephalin. *Nature* **1982**, *295*, 202–206. [CrossRef]
17. Horikawa, S.; Takai, T.; Toyosato, M.; Takahashi, H.; Noda, M.; Kakidani, H.; Kubo, T.; Hirose, T.; Inayama, S.; Hayashida, H.; et al. Isolation and structural organization of the human preproenkephalin B gene. *Nature* **1983**, *306*, 611–614. [CrossRef]
18. Kakidani, H.; Furutani, Y.; Takahashi, H.; Noda, M.; Morimoto, Y.; Hirose, T.; Asai, M.; Inayama, S.; Nakanishi, S.; Numa, S. Cloning and sequence analysis of cDNA for porcine β-neo-endorphin/dynorphin precursor. *Nature* **1982**, *298*, 245–249. [CrossRef]
19. Peciña, M.; Karp, J.F.; Mathew, S.; Todtenkopf, M.S.; Ehrich, E.W.; Zubieta, J.-K. Endogenous opioid system dysregulation in depression: Implications for new therapeutic approaches. *Mol. Psychiatry* **2019**, *24*, 576–587. [CrossRef]
20. Su, T.P. Evidence for sigma opioid receptor: Binding of [^{3}H]SKF-10047 to etorphine-inaccessible sites in guinea-pig brain. *J. Pharmacol. Exp. Ther.* **1982**, *223*, 284–290.
21. Eriksson, O.; Antoni, G. [^{11}C]Carfentanil binds preferentially to mu-opioid receptor subtype 1 compared to subtype 2. *Mol. Imaging* **2015**, *14*, 476–483. [CrossRef] [PubMed]
22. Zhu, W.; Ma, Y.; Bell, A.; Esch, T.; Guarna, M.; Bilfinger, T.V.; Bianchi, E.; Stefano, G.B. Presence of morphine in rat amygdala: Evidence for the mu$_3$ opiate receptor subtype via nitric oxide release in limbic structures. *Med. Sci. Monit.* **2004**, *10*, 433–439.
23. Witkin, J.M.; Rorick-Kehn, L.M.; Benvenga, M.J.; Adams, B.L.; Gleason, S.D.; Knitowski, K.M.; Li, X.; Chaney, S.; Falcone, J.F.; Smith, J.W.; et al. Preclinical findings predicting efficacy and side-effect profile of LY2940094, an antagonist of nociceptin receptors. *Pharma Res. Per.* **2016**, *4*, e00275. [CrossRef] [PubMed]
24. Janson, W.; Stein, C. Peripheral opioid analgesia. *Curr. Pharm. Biotechnol.* **2003**, *4*, 270–274. [CrossRef] [PubMed]
25. Mansour, A.; Fox, C.A.; Akil, H.; Watson, S.J. Opioid-receptor mRNA expression in the rat CNS: Anatomical and functional implications. *Trends Neurosci.* **1995**, *18*, 22–29. [CrossRef]

26. Atweh, S.F.; Kuhar, M.J. Distribution and physiological significance of opioid receptors in the brain. *Br. Med. Bull.* **1983**, *39*, 47–52. [CrossRef]
27. Benyhe, S.; Zádor, F.; Ötvös, F. Biochemistry of opioid (morphine) receptors: Binding, structure and molecular modelling. *Acta Biol. Szeged* **2015**, *59* (Suppl. 1), 17–37.
28. Meng, F.; Xie, G.X.; Thompson, R.C.; Mansour, A.; Goldstein, A.; Watson, S.J.; Akil, H. Cloning and pharmacological characterization of a rat kappa opioid receptor. *Proc. Natl. Acad. Sci. USA* **1993**, *90*, 9954–9958. [CrossRef]
29. Simonin, F.; Gavériaux-Ruff, C.; Befort, K.; Matthes, H.; Lannes, B.; Micheletti, G.; Mattéi, M.G.; Charron, G.; Bloch, B.; Kieffer, B. kappa-Opioid receptor in humans: cDNA and genomic cloning, chromosomal assignment, functional expression, pharmacology, and expression pattern in the central nervous system. *Proc. Natl. Acad. Sci. USA* **1995**, *92*, 7006–7010. [CrossRef]
30. Zhu, J.; Chen, C.; Xue, J.-C.; Kunapuli, S.; DeRiel, J.K.; Liu-Chen, L.-Y. Cloning of a human kappa opioid receptor from the brain. *Life Sci.* **1995**, *56*, 201–207. [CrossRef]
31. Witkin, J.M.; Statnick, M.A.; Rorick-Kehn, L.M.; Pintar, J.E.; Ansonoff, M.; Chen, Y.; Tucker, R.C.; Ciccocioppo, R. The biology of Nociceptin/Orphanin FQ (N/OFQ) related to obesity, stress, anxiety, mood, and drug dependence. *Pharmacol. Ther.* **2014**, *141*, 283–299. [CrossRef] [PubMed]
32. Pert, C.B.; Kuhar, M.J.; Snyder, S.H. Autoradiograhic localization of the opiate receptor in rat brain. *Life Sci.* **1975**, *16*, 1849–1853. [CrossRef]
33. Pert, C.B.; Snyder, S.H. Identification of opiate receptor binding in intact animals. *Life Sci.* **1975**, *16*, 1623–1634. [CrossRef]
34. Hu, X.; Wang, Y.; Hunkele, A.; Provasi, D.; Pasternak, G.W.; Filizola, M. Kinetic and thermodynamic insights into sodium ion translocation through the µ-opioid receptor from molecular dynamics and machine learning analysis. *PLoS Comput. Biol.* **2019**, *15*, e1006689. [CrossRef] [PubMed]
35. Blanchard, S.G.; Lee, P.H.; Pugh, W.W.; Hong, J.S.; Chang, K.J. Characterization of the binding of a morphine (mu) receptor-specific ligand: Tyr-Pro-NMePhe-D-Pro-NH$_2$, [^{3}H]-PL17. *Mol. Pharmacol.* **1987**, *31*, 326–333.
36. Pert, C.B.; Danks, J.A.; Channing, M.A.; Eckelman, W.C.; Larson, S.M.; Bennett, J.M.; Burke, T.R.J.; Rice, K.C. 3-[^{18}F]Acetylcyclofoxy: A useful probe for the visualization of opiate receptors in living animals. *FEBS Lett.* **1984**, *177*, 281–286. [CrossRef]
37. Larson, S.M.; Di Chiro, G. Comparative anatomo-functional imaging of two neuroreceptors and glucose metabolism: A PET study performed in the living baboon. *J. Comput. Assist. Tomogr.* **1985**, *9*, 676–681. [CrossRef]
38. Rothman, R.B.; Bykov, V.; Reid, A.; De Costa, B.R.; Newman, A.H.; Jacobson, A.E.; Rice, K.C. A brief study of the selectivity of norbinaltorphimine, (−)-cyclofoxy, and (+)-cyclofoxy among opioid receptor subtypes in vitro. *Neuropeptides* **1988**, *12*, 181–187. [CrossRef]
39. Ostrowski, N.L.; Burke, T.R.J.; Rice, K.C.; Pert, A.; Pert, C.B. The pattern of [^{3}H]cyclofoxy retention in rat brain after in vivo injection corresponds to the in vitro opiate receptor distribution. *Brain Res.* **1987**, *402*, 275–286. [CrossRef]
40. Rothman, R.; McLean, S.; Bykov, V.; Lessor, R.A.; Jacobson, A.E.; Rice, K.C.; Holaday, J.W. Chronic morphine upregulates a mu-opiate binding site labeled by [^{3}H]cycloFOXY: A novel opiate antagonist suitable for positron emission tomography. *Eur. J. Pharmacol.* **1987**, *142*, 73–81. [CrossRef]
41. Kawai, R.; Carson, R.E.; Dunn, B.; Newman, A.H.; Rice, K.C.; Blasberg, R.G. Regional brain measurement of Bmax and KD with the opiate antagonist cyclofoxy: Equilibrium studies in the conscious rat. *J. Cereb. Blood Flow. Metab.* **1991**, *11*, 529–544. [CrossRef] [PubMed]
42. Hartvig, P.; Bergström, K.; Lindberg, B.; Lundberg, P.O.; Lundqvist, H.; Långström, B.; Svärd, H.; Rane, A. Kinetics of ^{11}C-labeled opiates in the brain of rhesus monkeys. *J. Pharmacol. Exp. Ther.* **1984**, *230*, 250–255. [PubMed]
43. Hartvig, P.; Eckernäs, S.A.; Lindberg, B.S.; Lundqvist, H.; Antoni, G.; Rimland, A.; Långström, B. Regional distribution of the opioid receptor agonist N-(methyl-^{11}C)pethidine in the brain of the rhesus monkey studied with positron emission tomography. *Pharmacol. Toxicol.* **1990**, *66*, 37–40. [CrossRef] [PubMed]
44. Sai, K.K.; Fan, J.; Tu, Z.; Zerkel, P.; Mach, R.H.; Kharasch, E.D. Automated radiochemical synthesis and biodistribution of [^{11}C]l-α-acetylmethadol ([^{11}C]LAAM). *Appl. Radiat. Isot.* **2014**, *91*, 135–140. [CrossRef]

45. Weerts, E.M.; Kim, Y.K.; Wand, G.S.; Dannals, R.F.; Lee, J.S.; Frost, J.J.; McCaul, M.E. Differences in delta- and mu-opioid receptor blockade measured by positron emission tomography in naltrexone-treated recently abstinent alcohol-dependent subjects. *Neuropsychopharmacology* **2008**, *33*, 653–665. [CrossRef]
46. Johansson, J.; Hirvonen, J.; Lovró, Z.; Ekblad, L.; Kaasinen, V.; Rajasilta, O.; Helin, S.; Tuisku, J.; Sirén, S.; Pennanen, M.; et al. Intranasal naloxone rapidly occupies brain mu-opioid receptors in human subjects. *Neuropsychopharmacology* **2019**, *44*, 1667–1673. [CrossRef]
47. Lewis, J.W. Buprenorphine. *Drug Alcohol Depen.* **1985**, *14*, 363–372. [CrossRef]
48. Lewis, J.W.; Husbands, S.M. The orvinols and related opioids—High affinity ligands with diverse efficacy profiles. *Curr. Pharm. Des.* **2004**, *10*, 717–732. [CrossRef]
49. Husbands, S.M. Buprenorphine and related orvinols. In *Research and Development of Opioid-Related Ligands*; ACS Symposium Series; Ko, M.-C., Husbands, S.M., Eds.; American Chemical Society: Washington, DC, USA, 2013; Volume 1131, pp. 127–144.
50. Cami-Kobeci, G.; Polgar, W.E.; Khroyan, T.V.; Toll, L.; Husbands, S.M. Structural determinants of opioid and NOP receptor activity in aerivatives of buprenorphine. *J. Med. Chem.* **2011**, *54*, 6531–6537. [CrossRef]
51. Borsodi, A.; Bruchas, M.; Caló, G.; Chavkin, C.; Christie, M.J.; Civelli, O.; Connor, M.; Cox, B.M.; Devi, L.A.; Evans, C.; et al. Opioid receptors (version 2019.4) in the IUPHAR/BPS Guide to Pharmacology Database. In *IUPHAR/BPS Guide to Pharmacology CITE, 2019(4)*; 2019. [CrossRef]
52. Stefanucci, A.; Lei, W.; Pieretti, S.; Novellino, E.; Dimmito, M.P.; Marzoli, F.; Streicher, J.M.; Mollica, A. On resin click-chemistry-mediated synthesis of novel enkephalin analogues with potent anti-nociceptive activity. *Sci. Rep.* **2019**, *9*, 5771. [CrossRef]
53. Corbett, A.D.; Paterson, S.J.; Kosterlitz, H.W. Selectivity of ligands for opioid receptors. In *Handbook of Experimental Pharmacology Opioids I*; Herz, A., Ed.; Springer-Verlag: New York, NY, USA, 1993; Volume 104/1, pp. 645–679.
54. Zhang, S.; Tong, Y.; Tian, M.; Dehaven, R.N.; Cortesburgos, L.; Mansson, E.; Simonin, F.; Kieffer, B.; Yu, L. Dynorphin A as a potential endogenous ligand for four members of the opioid receptor gene family. *J. Pharmacol. Exp. Ther.* **1998**, *286*, 136–141. [PubMed]
55. Valenzano, K.J.; Miller, W.; Chen, Z.; Shan, S.; Crumley, G.; Victory, S.F.; Davies, E.; Huang, J.-C.; Allie, N.; Nolan, S.J.; et al. DiPOA ([8-(3,3-Diphenyl-propyl)-4-oxo-1-phenyl-1,3,8-triazaspiro[4.5]dec-3-yl]-acetic acid), a novel, systemically available, and peripherally restricted mu-opioid agonist with antihyperalgesic activity: I. In vitro pharmacological characterization and pharmacokinetic properties. *J. Pharmacol. Exp. Ther.* **2004**, *310*, 783–792. [PubMed]
56. Miyazaki, T.; Choi, I.Y.; Rubas, W.; Anand, N.K.; Ali, C.; Evans, J.; Gursahani, H.; Hennessy, M.; Kim, G.; McWeeney, D.; et al. NKTR-181: A novel mu-opioid analgesic with inherently low abuse potential. *J. Pharmacol. Exp. Ther.* **2017**, *363*, 104–113. [CrossRef] [PubMed]
57. Zheng, M.Q.; Nabulsi, N.; Kim, S.J.; Tomasi, G.; Lin, S.F.; Mitch, C.; Quimby, S.; Barth, V.; Rash, K.; Masters, J.; et al. Synthesis and evaluation of ^{11}C-LY2795050 as a kappa-opioid receptor antagonist radiotracer for PET imaging. *J. Nucl. Med.* **2013**, *54*, 455–463. [CrossRef]
58. Henriksen, G.; Platzer, S.; Marton, J.; Hauser, A.; Berthele, A.; Schwaiger, M.; Marinelli, L.; Lavecchia, A.; Novellino, E.; Wester, H.-J. Syntheses, biological evaluation, and molecular modeling of ^{18}F-labeled 4-anilidopiperidines as µ-opioid receptor imaging agents. *J. Med. Chem.* **2005**, *48*, 7720–7732. [CrossRef]
59. Frost, J.J.; Wagner, H.N.J.; Dannals, R.F.; Ravert, H.T.; Links, J.M.; Wilson, A.A.; Burns, H.D.; Wong, D.F.; McPherson, R.W.; Rosenbaum, A.E.; et al. Imaging opiate receptors in the human brain by positron tomography. *J. Comput. Assist. Tomogr.* **1985**, *9*, 231–236. [CrossRef]
60. Cometta-Morini, C.; Maguire, P.A.; Loew, G.H. Molecular determinants of mu receptor recognition for the fentanyl class of compounds. *Mol. Pharmacol.* **1992**, *41*, 185–196.
61. Raynor, K.; Kong, H.; Chen, Y.; Yasuda, K.; Yu, L.; Bell, G.I.; Reisine, T. Pharmacological characterization of the cloned kappa-, delta-, and mu-opioid receptors. *Mol. Pharmacol.* **1994**, *45*, 330–334.
62. Marton, J.; Schoultz, B.W.; Hjørnevik, T.; Drzezga, A.; Yousefi, B.H.; Wester, H.-J.; Willoch, F.; Henriksen, G. Synthesis and evaluation of a full-agonist orvinol for PET-Imaging of opioid receptors: [^{11}C]PEO. *J. Med. Chem.* **2009**, *52*, 5586–5589. [CrossRef]

63. Schoultz, B.W.; Hjørnevik, T.; Reed, B.J.; Marton, J.; Coello, C.S.; Willoch, F.; Henriksen, G. Synthesis and evaluation of three structurally related ^{18}F-labeled orvinols of different intrinsic activities: 6-O-[^{18}F]Fluoroethyl-diprenorphine ([^{18}F]FDPN), 6-O-[^{18}F]fluoroethyl-buprenorphine ([^{18}F]FBPN), and 6-O-[^{18}F]fluoroethyl-phenethyl-orvinol ([^{18}F]FPEO). *J. Med. Chem.* **2014**, *57*, 5464–5469.
64. Portoghese, P.S.; Sultana, M.; Takemori, A.E. Design of peptidomimetic delta opioid receptor antagonists using the message-address concept. *J. Med. Chem.* **1990**, *33*, 1714–1720. [CrossRef] [PubMed]
65. Schoultz, B.W.; Hjornevik, T.; Willoch, F.; Marton, J.; Noda, A.; Murakami, Y.; Miyoshi, S.; Nishimura, S.; Arstad, E.; Drzezga, A.; et al. Evaluation of the kappa-opioid receptor-selective tracer [^{11}C]GR103545 in awake rhesus macaques. *Eur. J. Nucl. Med. Mol. Imaging* **2010**, *37*, 1174–1180. [CrossRef] [PubMed]
66. Zheng, M.-Q.; Kim, S.J.; Holden, D.; Lin, S.-F.; Need, A.; Rash, K.; Barth, V.; Mitch, C.; Navarro, A.; Kapinos, M.; et al. An improved antagonist radiotracer for the kappa-opioid receptor: Synthesis and characterization of ^{11}C-LY2459989. *J. Nucl. Med.* **2014**, *55*, 1185–1191. [CrossRef] [PubMed]
67. Li, S.; Zheng, M.-Q.; Naganawa, M.; Gao, H.; Pracitto, R.; Shirali, A.; Lin, S.-F.; Teng, J.-K.; Ropchan, J.; Huang, Y. Novel kappa opioid receptor agonist as improved PET radiotracer: Development and in vivo evaluation. *Mol. Pharm.* **2019**, *16*, 1523–1531. [CrossRef] [PubMed]
68. Li, S.; Zheng, M.-Q.; Naganawa, M.; Kim, S.J.; Gao, H.; Kapinos, M.; Labaree, D.; Huang, Y. Development and in vivo evaluation of a kappa-opioid receptor agonist as a PET radiotracer with superior imaging characteristics. *J. Nucl. Med.* **2019**, *60*, 1023–1030. [CrossRef]
69. Poisnel, G.; Oueslati, F.; Dhilly, M.; Delamare, J.; Perrio, C.; Debruyne, D.; Barré, L. [^{11}C]-MeJDTic: A novel radioligand for kappa-opioid receptor positron emission tomography imaging. *Nucl. Med. Biol.* **2008**, *35*, 561–569. [CrossRef]
70. Harding, W.W.; Tidgewell, K.; Byrd, N.; Cobb, H.; Dersch, C.M.; Butelman, E.R.; Rothman, R.B.; Prisinzano, T.E. Neoclerodane diterpenes as a novel scaffold for mu opioid receptor ligands. *J. Med. Chem.* **2005**, *48*, 4765–4771. [CrossRef]
71. Pike, V.W.; Rash, K.S.; Chen, Z.; Pedregal, C.; Statnick, M.A.; Kimura, Y.; Hong, J.S.; Zoghbi, S.S.; Fujita, M.; Toledo, M.A.; et al. Synthesis and evaluation of radioligands for imaging brain nociceptin/orphanin FQ peptide (NOP) receptors with positron emission tomography. *J. Med. Chem.* **2011**, *54*, 2687–2700. [CrossRef]
72. Hostetler, E.D.; Sanabria-Bohórquez, S.; Eng, W.; Joshi, A.D.; Patel, S.; Gibson, R.E.; O'Malley, S.; Krause, S.M.; Ryan, C.; Riffel, K.; et al. Evaluation of [^{18}F]MK-0911, a positron emission tomography (PET) tracer for opioid receptor-like 1 (ORL1), in rhesus monkey and human. *NeuroImage* **2013**, *68*, 1–10. [CrossRef]
73. Luthra, S.K.; Pike, V.W.; Brady, F.; Horlock, P.L.; Prenant, C.; Crouzel, C. Preparation of [^{11}C]buprenorphine—A potential radioligand for the study of the opiate receptor system in vivo. *Int. J. Radiat. Appl. Instrum. Part A Appl. Radiat. Isot.* **1987**, *38*, 65–66. [CrossRef]
74. Luthra, S.K.; Pike, V.W.; Brady, F. The preparation of carbon-11 labelled diprenorphine: A new radioligand for the study of the opiate receptor system in vivo. *J. Chem. Soc. Chem. Commun.* **1985**, *20*, 1423–1425. [CrossRef]
75. Lever, J.R.; Mazza, S.M.; Dannals, R.F.; Ravert, H.T.; Wilson, A.A.; Wagner, H.N. Facile synthesis of [^{11}C]buprenorphine for positron emission tomographic studies of opioid receptors. *Int. J. Radiat. Appl. Instrum. Part A Appl. Radiat. Isot.* **1990**, *41*, 745–752. [CrossRef]
76. Luthra, S.K.; Brady, F.; Turton, D.R.; Brown, D.J.; Dowsett, K.; Waters, S.L.; Jones, A.K.P.; Matthews, R.W.; Crowder, J.C. Automated radiosyntheses of [6-O-methyl-^{11}C]diprenorphine and [6-O-methyl-^{11}C]buprenorphine from 3-O-trityl protected precursors. *Appl. Radiat. Isot.* **1994**, *45*, 857–873. [CrossRef]
77. Burns, H.D.; Lever, J.R.; Dannals, R.F.; Frost, J.J.; Wilson, A.A.; Ravert, H.T.; Subramanian, B.; Zeyman, S.E.; Langstrom, B.; Wagner, H.N., Jr. Synthesis of ligands for imaging opiate receptors by positron emission tomography: Carbon-11 labelled diprenorphine. *J. Label. Compd. Radiopharm.* **1984**, *22*, 1167–1169.
78. Lever, J.R.; Dannals, R.F.; Wilson, A.A.; Ravert, H.T.; Wagner, H.N. Synthesis of carbon-11 labeled diprenorphine: A radioligand for positron emission tomographic studies of opiate receptors. *Tetrahedron Lett.* **1987**, *28*, 4015–4018. [CrossRef]
79. Fairclough, M.; Prenant, C.; Brown, G.; McMahon, A.; Lowe, J.; Jones, A. The automated radiosynthesis and purification of the opioid receptor antagonist, [6-O-methyl-^{11}C]diprenorphine on the GE TRACERlab FXFE radiochemistry module. *J. Label. Compd. Radiopharm.* **2014**, *57*, 388–396. [CrossRef]

80. Galynker, I.; Schlyer, D.J.; Dewey, S.L.; Fowler, J.S.; Logan, J.; Gatley, S.J.; MacGregor, R.R.; Ferrieri, R.A.; Holland, M.J.; Brodie, J.; et al. Opioid receptor imaging and displacement studies with [6-O-[^{11}C]methyl]buprenorphine in baboon brain. *Nucl. Med. Biol.* **1996**, *23*, 325–331. [CrossRef]
81. Zubieta, J.; Greenwald, M.K.; Lombardi, U.; Woods, J.H.; Kilbourn, M.R.; Jewett, D.M.; Koeppe, R.A.; Schuster, C.R.; Johanson, C.E. Buprenorphine-induced changes in mu-opioid receptor availability in male heroin-dependent volunteers: A preliminary study. *Neuropsychopharmacology* **2000**, *23*, 326–334. [CrossRef]
82. Melichar, J.K.; Hume, S.P.; Williams, T.M.; Daglish, M.R.; Taylor, L.G.; Ahmad, R.; Malizia, A.L.; Brooks, D.J.; Myles, J.S.; Lingford-Hughes, A.; et al. Using [^{11}C]diprenorphine to image opioid receptor occupancy by methadone in opioid addiction: Clinical and preclinical studies. *J. Pharmacol. Exp. Ther.* **2005**, *312*, 309–315. [CrossRef]
83. Frost, J.J.; Mayberg, H.S.; Sadzot, B.; Dannals, R.F.; Lever, J.R.; Ravert, H.T.; Wilson, A.A.; Wagner, H.N.J.; Links, J.M. Comparison of [^{11}C]diprenorphine and [^{11}C]carfentanil binding to opiate receptors in humans by positron emission tomography. *J. Cereb. Blood Flow. Metab.* **1990**, *10*, 484–492. [CrossRef]
84. Schadrack, J.; Willoch, F.; Platzer, S.; Bartenstein, P.; Mahal, B.; Dworzak, D.; Wester, H.J.; Zieglgänsberger, W.; Tölle, T.R. Opioid receptors in the human cerebellum: Evidence from [^{11}C]diprenorphine PET, mRNA expression and autoradiography. *Neuroreport* **1999**, *10*, 619–624. [CrossRef] [PubMed]
85. Wester, H.-J.; Willoch, F.; Tölle, T.R.; Munz, F.; Herz, M.; Øye, I.; Schadrack, J.; Schwaiger, M.; Bartenstein, P. 6-O-(2-[^{18}F]Fluoroethyl-6-O-desmethyldiprenorphine ([^{18}F]DPN): Synthesis, biologic evaluation, and comparison with [^{11}C]DPN in humans. *J. Nucl. Med.* **2000**, *41*, 1279–1286. [PubMed]
86. Schoultz, B.W.; Reed, B.J.; Marton, J.; Willoch, F.; Henriksen, G. A fully automated radiosynthesis of [^{18}F]fluoroethyl-diprenorphine on a single module by use of SPE cartridges for preparation of high quality 2-[^{18}F]fluoroethyl tosylate. *Molecules* **2013**, *18*, 7271–7278. [CrossRef] [PubMed]
87. Baumgärtner, U.; Buchholz, H.G.; Bellosevich, A.; Magerl, W.; Siessmeier, T.; Rolke, R.; Höhnemann, S.; Piel, M.; Rösch, F.; Wester, H.J.; et al. High opiate receptor binding potential in the human lateral pain system. *NeuroImage* **2006**, *30*, 692–699. [CrossRef]
88. Henriksen, G.; Spilker, M.E.; Sprenger, T.; Hauser, A.; Platzer, S.; Boecker, H.; Toelle, T.R.; Schwaiger, M.; Wester, H.J. Gender dependent rate of metabolism of the opioid receptor-PET ligand [^{18}F]fluoroethyldiprenorphine. *Nuklearmedizin* **2006**, *45*, 197–200.
89. Vučković, S.; Prostran, M.; Ivanović, M.; Došen-Mićović, L.; Todorović, Z.; Nešić, Z.; Stojanović, R.; Divac, N.; Miković, Ž. Fentanyl analogs: Stuctrure-activity-relationship study. *Curr. Med. Chem.* **2009**, *16*, 2468–2474. [CrossRef]
90. Vardanyan, R.S.; Hruby, V.J. Fentanyl-related compounds and derivatives: Current status and future prospects for pharmaceutical applications. *Future Med. Chem.* **2014**, *6*, 385–412. [CrossRef]
91. Van Daele, P.G.; De Bruyn, M.F.; Boey, J.M.; Sanczuk, S.; Agten, J.T.; Janssen, P.A. Synthetic analgesics: *N*-(1-[2-arylethyl]-4-substituted 4-piperidinyl) *N*-arylalkanamides. *Arzneim. Forsch. Drug Res.* **1976**, *26*, 1521–1531.
92. Dannals, R.F.; Ravert, H.T.; Frost, J.J.; Wilson, A.A.; Burns, H.D.; Wagner, H.N.J. Radiosynthesis of an opiate receptor binding radiotracer: [^{11}C]carfentanil. *Int. J. Appl. Isot.* **1985**, *36*, 303–306. [CrossRef]
93. Jewett, D.M.; Kilbourn, M.R. In vivo evaluation of new carfentanil-based radioligands for the mu opiate receptor. *Nucl. Med. Biol.* **2004**, *31*, 321–325. [CrossRef]
94. Shafer, S.L. Carfentanil: A weapon of mass destruction. *Can. J. Anesth.* **2019**, *66*, 351–355. [CrossRef] [PubMed]
95. Frost, J.J.; Douglass, K.H.; Mayberg, H.S.; Dannals, R.F.; Links, J.M.; Wilson, A.A.; Ravert, H.T.; Crozier, W.C.; Wagner, H.N.J. Multicompartmental analysis of [^{11}C]-carfentanil binding to opiate receptors in humans measured by positron emission tomography. *J. Cereb. Blood Flow. Metab.* **1989**, *9*, 398–409. [CrossRef] [PubMed]
96. Hirvonen, J.; Aalto, S.; Hagelberg, N.; Maksimow, A.; Ingman, K.; Oikonen, V.; Virkkala, J.; Någren, K.; Scheinin, H. Measurement of central mu-opioid receptor binding in vivo with PET and [^{11}C]carfentanil: A test-retest study in healthy subjects. *Eur. J. Nucl. Med. Mol. Imaging* **2009**, *36*, 275–286. [CrossRef] [PubMed]
97. Bentley, K.W.; Hardy, D.G. Novel analgesics and molecular rearrangements in the morphine-thebaine group. III. Alcohols of the 6,14-*endo*-ethenotetrahydrooripavine series and derived analogs of *N*-allylnormorphine and -norcodeine. *J. Am. Chem. Soc.* **1967**, *89*, 3281–3292. [CrossRef]

98. Marton, J.; Henriksen, G. Design and synthesis of an ^{18}F-labeled version of phenylethyl orvinol ([^{18}F]FE-PEO) for PET-imaging of opioid receptors. *Molecules* **2012**, *17*, 11554–11569. [CrossRef]
99. Riss, P.J.; Hong, Y.T.; Marton, J.; Caprioli, D.; Williamson, D.J.; Ferrari, V.; Saigal, N.; Roth, B.L.; Henriksen, G.; Fryer, T.D.; et al. Synthesis and evaluation of ^{18}F-FE-PEO in rodents: An ^{18}F-labeled full agonist for opioid receptor imaging. *J. Nucl. Med.* **2013**, *54*, 299–305. [CrossRef]
100. Lever, J.R.; Kinter, C.M.; Ravert, H.T.; Musachio, J.L.; Mathews, W.B.; Dannals, R.F. Synthesis of N1'-([^{11}C]methyl)naltrindole ([^{11}C]MeNTI): A radioligand for positron emission tomographic studies of delta opioid receptors. *J. Label. Compd. Radiopharm.* **1995**, *36*, 137–145. [CrossRef]
101. Madar, I.; Lever, J.R.; Kinter, C.M.; Scheffel, U.; Ravert, H.T.; Musachio, J.L.; Mathews, W.B.; Dannals, R.F.; Frost, J.J. Imaging of delta opioid receptors in human brain by N1'-([^{11}C]methyl)naltrindole and PET. *Synapse* **1996**, *24*, 19–28. [CrossRef]
102. Smith, J.S.; Zubieta, J.K.; Price, J.C.; Flesher, J.E.; Madar, I.; Lever, J.R.; Kinter, C.M.; Dannals, R.F.; Frost, J.J. Quantification of delta-opioid receptors in human brain with N1'-([^{11}C]methyl) naltrindole and positron emission tomography. *J. Cereb. Blood Flow. Metab.* **1999**, *19*, 956–966. [CrossRef]
103. Mathews, W.B.; Kinter, C.M.; Palma, J.; Daniels, R.V.; Ravert, H.T.; Dannals, R.F.; Lever, J.R. Synthesis of N1'-([^{18}F]fluoroethyl)naltrindole ([^{18}F]FEtNTI): A radioligand for positron emission tomographic studies of delta opioid receptors. *J. Label. Compd. Radiopharm.* **1999**, *42*, 43–54. [CrossRef]
104. Tyacke, R.J.; Robinson, E.S.; Schnabel, R.; Lewis, J.W.; Husbands, S.M.; Nutt, D.J.; Hudson, A.L. N1'-fluoroethyl-naltrindole (BU97001) and N1'-fluoroethyl-(14-formylamino)-naltrindole (BU97018) potential delta-opioid receptor PET ligands. *Nucl. Med. Biol.* **2002**, *29*, 455–462. [CrossRef]
105. Bourdier, T.; Poisnel, G.; Dhilly, M.; Delamare, J.; Henry, J.; Debruyne, D.; Barré, L. Synthesis and biological evaluation of N-substituted quinolinimides, as potential ligands for in vivo imaging studies of delta-opioid receptors. *Bioconj. Chem.* **2007**, *18*, 538–548. [CrossRef] [PubMed]
106. Hayes, A.G.; Birch, P.J.; Hayward, N.J.; Sheehan, M.J.; Rogers, H.; Tyers, M.B.; Judd, D.B.; Scopes, D.I.C.; Naylor, A. A series of novel, highly potent and selective agonists for the kappa-opioid receptor. *Br. J. Pharmacol.* **1990**, *101*, 944–948. [CrossRef]
107. Birch, P.J.; Rogers, H.; Hayes, A.G.; Hayward, N.J.; Tyers, M.B.; Scopes, D.I.C.; Naylor, A.; Judd, D.B. Neuroprotective actions of GR89696, a highly potent and selective κ-opioid receptor agonist. *Br. J. Pharmacol.* **1991**, *103*, 1819–1823. [CrossRef] [PubMed]
108. Naylor, A.; Judd, D.B.; Lloyd, J.E.; Scopes, D.I.C.; Hayes, A.G.; Birch, P.J. A potent new class of kappa-receptor agonist: 4-substituted 1-(arylacetyl)-2-[(dialkylamino)methyl]piperazines. *J. Med. Chem.* **1993**, *36*, 2075–2083. [CrossRef] [PubMed]
109. Ravert, H.T.; Mathews, W.B.; Musachio, J.L.; Scheffel, U.; Finley, P.; Dannals, R.F. [^{11}C]-methyl 4-[(3,4-dichlorophenyl)acetyl]-3-[(1-pyrrolidinyl)methyl]-1-piperazinecarboxylate ([^{11}C]GR89696): Synthesis and in vivo binding to kappa opiate receptors. *Nucl. Med. Biol.* **1999**, *26*, 737–741. [CrossRef]
110. Ravert, H.T.; Scheffel, U.; Mathews, W.B.; Musachio, J.L.; Dannals, R.F. [^{11}C]-GR89696, a potent kappa opiate receptor radioligand; in vivo binding of the R and S enantiomers. *Nucl. Med. Biol.* **2002**, *29*, 47–53. [CrossRef]
111. Talbot, P.S.; Narendran, R.; Butelman, E.R.; Huang, Y.; Ngo, K.; Slifstein, M.; Martinez, D.; Laruelle, M.; Hwang, D.R. ^{11}C-GR103545, a radiotracer for imaging kappa-opioid receptors in vivo with PET: Synthesis and evaluation in baboons. *J. Nucl. Med.* **2005**, *46*, 484–494.
112. Schoultz, B.W.; Arstad, E.; Marton, J.; Willoch, F.; Drzezga, A.; Wester, H.J.; Henriksen, G. A new method for radiosynthesis of ^{11}C-labeled carbamate groups and its application for a highly efficient synthesis of the kappa-opioid receptor tracer [^{11}C]GR103545. *Open Med. Chem. J.* **2008**, *2*, 72–74. [CrossRef]
113. Wilson, A.A.; Gracia, A.; Houle, S.; Vasdev, N. Direct fixation of [^{11}C]-CO_2 by amines: Formation of [^{11}C-carbonyl]-methylcarbamates. *Org. Biomol. Chem.* **2010**, *8*, 428–432. [CrossRef]
114. Nabulsi, N.B.; Zheng, M.-Q.; Ropchan, J.; Labaree, D.; Ding, Y.-S.; Blumberg, L.; Huang, Y. [^{11}C]GR103545: Novel one-pot radiosynthesis with high specific activity. *Nucl. Med. Biol.* **2011**, *38*, 215–221. [CrossRef] [PubMed]
115. Tomasi, G.; Nabulsi, N.; Zheng, M.Q.; Weinzimmer, D.; Ropchan, J.; Blumberg, L.; Brown-Proctor, C.; Ding, Y.S.; Carson, R.E.; Huang, Y. Determination of in vivo Bmax and Kd for ^{11}C-GR103545, an agonist PET tracer for kappa-opioid receptors: A study in nonhuman primates. *J. Nucl. Med.* **2013**, *54*, 600–608. [CrossRef] [PubMed]

116. Naganawa, M.; Jacobsen, L.K.; Zheng, M.-Q.; Lin, S.-F.; Banerjee, A.; Byon, W.; Weinzimmer, D.; Tomasi, G.; Nabulsi, N.; Grimwood, S.; et al. Evaluation of the agonist PET radioligand [^{11}C]GR103545 to image kappa opioid receptor in humans: Kinetic model selection, test–retest reproducibility and receptor occupancy by the antagonist PF-04455242. *NeuroImage* **2014**, *99*, 69–79. [CrossRef] [PubMed]
117. Li, S.; Cai, Z.; Zheng, M.-Q.; Holden, D.; Naganawa, M.; Lin, S.-F.; Ropchan, J.; Labaree, D.; Kapinos, M.; Lara-Jaime, T.; et al. Novel ^{18}F-labeled kappa-opioid receptor antagonist as PET radiotracer: Synthesis and in vivo evaluation of ^{18}F-LY2459989 in nonhuman primates. *J. Nucl. Med.* **2018**, *59*, 140–146. [CrossRef]
118. Fichna, J.; Schicho, R.; Janecka, A.; Zjawiony, J.K.; Storr, M. Selective natural kappa opioid and cannabinoid receptor agonists with a potential role in the treatment of gastrointestinal dysfunction. *Drug News Perspect.* **2009**, *22*, 383–392. [CrossRef]
119. Butelman, E.R.; Kreek, M.J. Salvinorin A, a kappa-opioid receptor agonist hallucinogen: Pharmacology and potential template for novel pharmacotherapeutic agents in neuropsychiatric disorders. *Front. Pharmacol.* **2015**, *6*, 190.
120. Zjawiony, J.K.; Machado, A.S.; Menegatti, R.; Ghedini, P.C.; Costa, E.A.; Pedrino, G.R.; Lukas, S.E.; Franco, O.L.; Silva, O.N.; Fajemiroye, J.O. Cutting-edge search for safer opioid pain relief: Retrospective review of Salvinorin A and its analogs. *Front. Psychiatry* **2019**, *10*(157), 1–11. [CrossRef]
121. Hooker, J.M.; Xu, Y.; Schiffer, W.; Shea, C.; Carter, P.; Fowler, J.S. Pharmacokinetics of the potent hallucinogen, salvinorin A in primates parallels the rapid onset and short duration of effects in humans. *NeuroImage* **2008**, *41*, 1044–1050. [CrossRef]
122. Placzek, M.S.; Van de Bittner, G.C.; Wey, H.Y.; Lukas, S.E.; Hooker, J.M. Immediate and persistent effects of Salvinorin A on the kappa opioid receptor in rodents, monitored in vivo with PET. *Neuropsychopharmacology* **2015**, *40*, 2865–2872. [CrossRef]
123. Thomas, J.B.; Atkinson, R.N.; Rothman, R.B.; Fix, S.E.; Mascarella, S.W.; Vinson, N.A.; Xu, H.; Dersch, C.M.; Lu, Y.F.; Cantrell, B.E.; et al. Identification of the first trans-(3R,4R)-dimethyl-4-(3-hydroxyphenyl)piperidine derivative to possess highly potent and selective opioid kappa receptor antagonist activity. *J. Med. Chem.* **2001**, *44*, 2687–2690. [CrossRef]
124. Schmitt, S.; Delamare, J.; Tirel, O.; Fillesoye, F.; Dhilly, M.; Perrio, C. N-[^{18}F]-FluoropropylJDTic for kappa-opioid receptor PET imaging: Radiosynthesis, pre-clinical evaluation, and metabolic investigation in comparison with parent JDTic. *Nucl. Med. Biol.* **2017**, *44*, 50–61. [CrossRef] [PubMed]
125. Kim, S.J.; Zheng, M.Q.; Nabulsi, N.; Labaree, D.; Ropchan, J.; Najafzadeh, S.; Carson, R.E.; Huang, Y.; Morris, E.D. Determination of the in vivo selectivity of a new kappa-opioid receptor antagonist PET tracer ^{11}C-LY2795050 in the rhesus monkey. *J. Nucl. Med.* **2013**, *54*, 1668–1674. [CrossRef] [PubMed]
126. Naganawa, M.; Zheng, M.Q.; Nabulsi, N.; Tomasi, G.; Henry, S.; Lin, S.F.; Ropchan, J.; Labaree, D.; Tauscher, J.; Neumeister, A.; et al. Kinetic modeling of ^{11}C-LY2795050, a novel antagonist radiotracer for PET imaging of the kappa opioid receptor in humans. *J. Cereb. Blood Flow. Metab.* **2014**, *34*, 1818–1825. [CrossRef]
127. Naganawa, M.; Zheng, M.-Q.; Henry, S.; Nabulsi, N.; Lin, S.-F.; Ropchan, J.; Labaree, D.; Najafzadeh, S.; Kapinos, M.; Tauscher, J.; et al. Test–Retest Reproducibility of Binding Parameters in Humans with ^{11}C-LY2795050, an Antagonist PET Radiotracer for the kappa Opioid Receptor. *J. Nucl. Med.* **2015**, *56*, 243–248. [CrossRef] [PubMed]
128. Naganawa, M.; Dickinson, G.L.; Zheng, M.-Q.; Henry, S.; Vandenhende, F.; Witcher, J.; Bell, R.; Nabulsi, N.; Lin, S.-F.; Ropchan, J.; et al. Receptor occupancy of the kappa-opioid antagonist LY2456302 measured with positron emission tomography and the novel radiotracer ^{11}C-LY2795050. *J. Pharmacol. Exp. Ther.* **2016**, *356*, 260–266. [CrossRef]
129. Placzek, M.S.; Schroeder, F.A.; Che, T.; Wey, H.-Y.; Neelamegam, R.; Wang, C.; Roth, B.L.; Hooker, J.M. Discrepancies in kappa opioid agonist binding revealed through PET Imaging. *ACS Chem. Neurosci.* **2019**, *10*, 384–395. [CrossRef]
130. Chesis, P.L.; Welch, M.J. Synthesis and in vitro characterization of fluorinated U-50488 analogs for PET studies of kappa opioid receptors. *Int. J. Radiat. Applicat. Instrum. Part A Appl. Radiat. Isot.* **1990**, *41*, 267–273. [CrossRef]
131. Zaveri, N.T. Nociceptin opioid receptor (NOP) as a therapeutic target: Progress in translation from preclinical research to clinical utility. *J. Med. Chem.* **2016**, *59*, 7011–7028. [CrossRef]

132. Kimura, Y.; Fujita, M.; Hong, J.S.; Lohith, T.G.; Gladding, R.L.; Zoghbi, S.S.; Tauscher, J.A.; Goebl, N.; Rash, K.S.; Chen, Z.; et al. Brain and whole-body imaging in rhesus monkeys of ^{11}C-NOP-1A, a promising PET radioligand for nociceptin/orphanin FQ peptide receptors. *J. Nucl. Med.* **2011**, *52*, 1638–1645. [CrossRef]
133. Lohith, T.G.; Zoghbi, S.S.; Morse, C.L.; Araneta, M.F.; Barth, V.N.; Goebl, N.A.; Tauscher, J.T.; Pike, V.W.; Innis, R.B.; Fujita, M. Brain and whole-body imaging of nociceptin/orphanin FQ peptide receptor in humans using the PET Ligand ^{11}C-NOP-1A. *J. Nucl. Med.* **2012**, *53*, 385–392. [CrossRef]
134. Lohith, T.G.; Zoghbi, S.S.; Morse, C.L.; Araneta, M.D.F.; Barth, V.N.; Goebl, N.A.; Tauscher, J.T.; Pike, V.W.; Innis, R.B.; Fujita, M. Retest imaging of [^{11}C]NOP-1A binding to nociceptin/orphanin FQ peptide (NOP) receptors in the brain of healthy humans. *NeuroImage* **2014**, *87*, 89–95. [CrossRef] [PubMed]
135. Raddad, E.; Chappell, A.; Meyer, J.; Wilson, A.A.; Ruegg, C.E.; Tauscher, J.; Statnick, M.A.; Barth, V.; Zhang, X.; Verfaille, S.J. Occupancy of nociceptin/orphanin FQ peptide receptors by the antagonist LY2940094 in rats and healthy human subjects. *Drug Metab. Dispos.* **2016**, *44*, 1536–1542. [CrossRef] [PubMed]
136. Zhang, L.; Drummond, E.; Brodney, M.A.; Cianfrogna, J.; Drozda, S.E.; Grimwood, S.; Vanase-Frawley, M.A.; Villalobos, A. Design, synthesis and evaluation of [^{3}H]PF-7191, a highly specific nociceptin opioid peptide (NOP) receptor radiotracer for in vivo receptor occupancy (RO) studies. *Bioorg. Med. Chem. Lett.* **2014**, *24*, 5219–5223. [CrossRef] [PubMed]
137. Zubieta, J.K.; Dannals, R.F.; Frost, J.J. Gender and age influences on human brain mu-opioid receptor binding measured by PET. *Am. J. Psychiarty* **1999**, *156*, 842–848. [CrossRef] [PubMed]
138. Vijay, A.; Cavallo, D.; Goldberg, A.; de Laat, B.; Nabulsi, N.; Huang, Y.; Krishnan-Sarin, S.; Morris, E.D. PET imaging reveals lower kappa opioid receptor availability in alcoholics but no effect of age. *Neuropsychopharmacology* **2018**, *43*, 2539–2547. [CrossRef]
139. Vijay, A.; Wang, S.; Worhunsky, P.; Zheng, M.-Q.; Nabulsi, N.; Ropchan, J.; Krishnan-Sarin, S.; Huang, Y.; Morris, E.D. PET imaging reveals sex differences in kappa opioid receptor availability in humans, in vivo. *Am. J. Nucl. Med. Mol. Imaging* **2016**, *6*, 205–214. [PubMed]
140. Mayberg, H.S.; Sadzot, B.; Meltzer, C.C.; Fisher, R.S.; Lesser, R.P.; Dannals, R.F.; Lever, J.R.; Wilson, A.A.; Ravert, H.T.; Wagner, H.N.J.; et al. Quantification of mu and non–mu opiate receptors in temporal lobe epilepsy using positron emission tomography. *Ann. Neurol.* **1991**, *30*, 3–11. [CrossRef]
141. Madar, I.; Lesser, R.P.; Krauss, G.; Zubieta, J.K.; Lever, J.R.; Kinter, C.M.; Ravert, H.T.; Musachio, J.L.; Mathews, W.B.; Dannals, R.F.; et al. Imaging of δ- and μ-opioid receptors in temporal lobe epilepsy by positron emission tomography. *Ann. Neurol.* **1997**, *41*, 358–367. [CrossRef]
142. Hammers, A.; Asselin, M.-C.; Hinz, R.; Kitchen, I.; Brooks, D.J.; Duncan, J.S.; Koepp, M.J. Upregulation of opioid receptor binding following spontaneous epileptic seizures. *Brain* **2007**, *130*, 1009–1016. [CrossRef]
143. McGinnity, C.J.; Shidahara, M.; Feldmann, M.; Keihaninejad, S.; Riaño Barros, D.A.; Gousias, I.S.; Duncan, J.S.; Brooks, D.J.; Heckemann, R.A.; Turkheimer, F.E.; et al. Quantification of opioid receptor availability following spontaneous epileptic seizures: Correction of [^{11}C]diprenorphine PET data for the partial-volume effect. *NeuroImage* **2013**, *79*, 72–80. [CrossRef]
144. Koepp, M.J.; Richardson, M.P.; Brooks, D.J.; Duncan, J.S. Focal cortical release of endogenous opioids during reading induced seizures. *Lancet* **1998**, *352*, 952–955. [CrossRef]
145. Cohen, R.M.; Carson, R.E.; Aigner, T.G.; Doudet, D.J. Opiate receptor avidity is reduced in non-motor impaired MPTP-lesioned rhesus monkeys. *Brain Res.* **1998**, *806*, 292–296. [CrossRef]
146. Cohen, R.M.; Carson, R.E.; Wyatt, R.J.; Doudet, D.J. Opiate receptor avidity is reduced bilaterally in rhesus monkeys unilaterally lesioned with MPTP. *Synapse* **1999**, *33*, 282–288. [CrossRef]
147. Piccini, P.; Weeks, R.A.; Brooks, D.J. Alterations in opioid receptor binding in Parkinson's disease patients with levodopa-induced dyskinesias. *Ann. Neurol.* **1997**, *42*, 720–726. [CrossRef] [PubMed]
148. Whone, A.L.; Von Spiczak, S.; Edwards, M.; Valente, E.-M.; Hammers, A.; Bhatia, K.P.; Brooks, D.J. Opioid binding in DYT1 primary torsion dystonia: An ^{11}C-diprenorphine PET study. *Mov. Dis.* **2004**, *19*, 1498–1503. [CrossRef] [PubMed]
149. Von Spiczak, S.; Whone, A.L.; Hammers, A.; Asselin, M.-C.; Turkheimer, F.; Tings, T.; Happe, S.; Paulus, W.; Trenkwalder, C.; Brooks, D.J. The role of opioids in restless legs syndrome: An [^{11}C]diprenorphine PET study. *Brain* **2005**, *128*, 906–917. [CrossRef] [PubMed]
150. Weeks, R.A.; Cunningham, V.J.; Piccini, P.; Waters, S.; Harding, A.E.; Brooks, D.J. ^{11}C-Diprenorphine binding in Huntington's disease: A comparison of region of interest analysis with statistical parametric mapping. *J. Cereb. Blood Flow. Metab.* **1997**, *17*, 943–949. [CrossRef]

151. Willoch, F.; Tölle, T.R.; Wester, H.-J.; Munz, F.; Petzold, A.; Schwaiger, M.; Conrad, B.; Bartenstein, P. Central pain after pontine infarction is associated with changes in opioid receptor binding: A PET study with ^{11}C-Diprenorphine. *AJNR Am. J. Neuroradiol.* **1999**, *20*, 686–690.
152. Willoch, F.; Schindler, F.; Wester, H.-J.; Empl, M.; Straube, A.; Schwaiger, M.; Conrad, B.; Tölle, T.R. Central poststroke pain and reduced opioid receptor binding within pain processing circuitries: A [^{11}C]diprenorphine PET study. *Pain* **2004**, *108*, 213–220. [CrossRef]
153. Maarrawi, J.; Peyron, R.; Mertens, P.; Costes, N.; Magnin, M.; Sindou, M.; Laurent, B.; Garcia-Larrea, L. Differential brain opioid receptor availability in central and peripheral neuropathic pain. *Pain* **2007**, *127*, 183–194. [CrossRef]
154. Campbell, C.M.; Bounds, S.C.; Kuwabara, H.; Edwards, R.R.; Campbell, J.N.; Haythornthwaite, J.A.; Smith, M.T. Individual variation in sleep quality and duration is related to cerebral mu opioid receptor binding potential during tonic laboratory pain in healthy subjects. *Pain Med.* **2013**, *14*, 1882–1892. [CrossRef] [PubMed]
155. Bencherif, B.; Fuchs, P.N.; Sheth, R.; Dannals, R.F.; Campbell, J.N.; Frost, J.J. Pain activation of human supraspinal opioid pathways as demonstrated by [^{11}C]-carfentanil and positron emission tomography (PET). *Pain* **2002**, *99*, 589–598. [CrossRef]
156. Sprenger, T.; Valet, M.; Boecker, H.; Henriksen, G.; Spilker, M.E.; Willoch, F.; Wagner, K.J.; Wester, H.J.; Tölle, T.R. Opioidergic activation in the medial pain system after heat pain. *Pain* **2006**, *122*, 63–67. [CrossRef] [PubMed]
157. Zubieta, J.K.; Smith, Y.R.; Bueller, J.A.; Xu, Y.; Kilbourn, M.R.; Jewett, D.M.; Meyer, C.R.; Koeppe, R.A.; Stohler, C.S. Regional mu opioid receptor regulation of sensory and affective dimensions of pain. *Science* **2001**, *293*, 311–315. [CrossRef] [PubMed]
158. Zubieta, J.K.; Smith, Y.R.; Bueller, J.A.; Xu, Y.; Kilbourn, M.R.; Jewett, D.M.; Meyer, C.R.; Koeppe, R.A.; Stohler, C.S. Mu-opioid receptor-mediated antinociceptive responses differ in men and women. *J. Neurosci.* **2002**, *22*, 5100–5107. [CrossRef] [PubMed]
159. DosSantos, M.F.; Martikainen, I.K.; Nascimento, T.D.; Love, T.M.; Deboer, M.D.; Maslowski, E.C.; Monteiro, A.A.; Vincent, M.B.; Zubieta, J.K.; DaSilva, A.F. Reduced basal ganglia mu-opioid receptor availability in trigeminal neuropathic pain: A pilot study. *Mol. Pain* **2012**, *8*, 74. [CrossRef]
160. Scott, D.J.; Stohler, C.S.; Koeppe, R.A.; Zubieta, J.K. Time-course of change in [^{11}C]carfentanil and [^{11}C]raclopride binding potential after a nonpharmacological challenge. *Synapse* **2007**, *61*, 707–714. [CrossRef]
161. Hjornevik, T.; Schoultz, B.W.; Marton, J.; Gjerstad, J.; Drzezga, A.; Henriksen, G.; Willoch, F. Spinal long-term potentiation is associated with reduced opioid neurotransmission in the rat brain. *Clin. Physiol. Funct. Imaging* **2010**, *30*, 285–293. [CrossRef]
162. Boecker, H.; Sprenger, T.; Spilker, M.E.; Henriksen, G.; Koppenhoefer, M.; Wagner, K.J.; Valet, M.; Berthele, A.; Tolle, T.R. The runner's high: Opioidergic mechanisms in the human brain. *Cereb. Cortex* **2008**, *18*, 2523–2531. [CrossRef]
163. Saanijoki, T.; Tuominen, L.; Tuulari, J.J.; Nummenmaa, L.; Arponen, E.; Kalliokoski, K.; Hirvonen, J. Opioid release after high-intensity interval training in healthy human subjects. *Neuropsychopharmacol* **2018**, *43*, 246–254. [CrossRef]
164. Maarrawi, J.; Peyron, R.; Mertens, P.; Costes, N.; Magnin, M.; Sindou, M.; Laurent, B.; Garcia-Larrea, L. Motor cortex stimulation for pain control induces changes in the endogenous opioid system. *Neurology* **2007**, *69*, 827–834. [CrossRef] [PubMed]
165. DosSantos, M.F.; Love, T.M.; Martikainen, I.K.; Nascimento, T.D.; Fregni, F.; Cummiford, C.; Deboer, M.D.; Zubieta, J.K.; DaSilva, A.F. Immediate Effects of tDCS on the mu-opioid system of a chronic pain patient. *Front. Psychiatry* **2012**, *3*, 93. [PubMed]
166. Maarrawi, J.; Peyron, R.; Mertens, P.; Costes, N.; Magnin, M.; Sindou, M.; Laurent, B.; Garcia-Larrea, L. Brain opioid receptor density predicts motor cortex stimulation efficacy for chronic pain. *Pain* **2013**, *154*, 2563–2568. [CrossRef] [PubMed]
167. Sims-Williams, H.; Matthews, J.C.; Talbot, P.S.; Love-Jones, S.; Brooks, J.C.; Patel, N.K.; Pickering, A.E. Deep brain stimulation of the periaqueductal gray releases endogenous opioids in humans. *NeuroImage* **2017**, *146*, 833–842. [CrossRef]

168. Ly, H.G.; Dupont, P.; Geeraerts, B.; Bormans, G.; Van Laere, K.; Tack, J.; Van Oudenhove, L. Lack of endogenous opioid release during sustained visceral pain: A [^{11}C]carfentanil PET study. *Pain* **2013**, *154*, 2072–2077. [CrossRef]
169. Baier, B.; Bense, S.; Birklein, F.; Buchholz, H.-G.; Mischke, A.; Schreckenberger, M.; Dieterich, M. Evidence for modulation of opioidergic activity in central vestibular processing: A [^{18}F] diprenorphine PET study. *Hum. Brain Mapp.* **2010**, *31*, 550–555. [CrossRef]
170. Wager, T.D.; Scott, D.J.; Zubieta, J.K. Placebo effects on human mu-opioid activity during pain. *Proc. Natl. Acad. Sci. USA* **2007**, *104*, 11056–11061. [CrossRef]
171. DosSantos, M.F.; Martikainen, I.K.; Nascimento, T.D.; Love, T.M.; DeBoer, M.D.; Schambra, H.M.; Bikson, M.; Zubieta, J.K.; DaSilva, A.F. Building up analgesia in humans via the endogenous mu-opioid system by combining placebo and active tDCS: A preliminary report. *PLoS ONE* **2014**, *9*, e102350. [CrossRef]
172. Dougherty, D.D.; Kong, J.; Webb, M.; Bonab, A.A.; Fischman, A.J.; Gollub, R.L. A combined [^{11}C]diprenorphine PET study and fMRI study of acupuncture analgesia. *Behav. Brain Res.* **2008**, *193*, 63–68. [CrossRef]
173. Harris, R.E.; Zubieta, J.K.; Scott, D.J.; Napadow, V.; Gracely, R.H.; Clauw, D.J. Traditional Chinese acupuncture and placebo (sham) acupuncture are differentiated by their effects on mu-opioid receptors (MORs). *NeuroImage* **2009**, *47*, 1077–1085. [CrossRef]
174. Xiang, X.-H.; Chen, Y.-M.; Zhang, J.-M.; Tian, J.-H.; Han, J.-S.; Cui, C.-L. Low- and high-frequency transcutaneous electrical acupoint stimulation induces different effects on cerebral mu-opioid receptor availability in rhesus monkeys. *J. Neurosci. Res.* **2014**, *92*, 555–563. [CrossRef] [PubMed]
175. Mueller, C.; Klega, A.; Buchholz, H.-G.; Rolke, R.; Magerl, W.; Schirrmacher, R.; Schirrmacher, E.; Birklein, F.; Treede, R.-D.; Schreckenberger, M. Basal opioid receptor binding is associated with differences in sensory perception in healthy human subjects: A [^{18}F]diprenorphine PET study. *NeuroImage* **2010**, *49*, 731–737. [CrossRef] [PubMed]
176. Hagelberg, N.; Aalto, S.; Tuominen, L.; Pesonen, U.; Någren, K.; Hietala, J.; Scheinin, H.; Pertovaara, A.; Martikainen, I.K. Striatal mu-opioid receptor availability predicts cold pressor pain threshold in healthy human subjects. *Neurosci. Lett.* **2012**, *521*, 11–14. [CrossRef] [PubMed]
177. Thompson, S.J.; Pitcher, M.H.; Stone, L.S.; Tarum, F.; Niu, G.; Chen, X.; Kiesewetter, D.O.; Schweinhardt, P.; Bushnell, M.C. Chronic neuropathic pain reduces opioid receptor availability with associated anhedonia in rat. *Pain* **2018**, *159*, 1856–1866. [CrossRef] [PubMed]
178. Sprenger, T.; Willoch, F.; Miederer, M.; Schindler, F.; Valet, M.; Berthele, A.; Spilker, M.E.; Förderreuther, S.; Straube, A.; Stangier, I.; et al. Opioidergic changes in the pineal gland and hypothalamus in cluster headache: A ligand PET study. *Neurology* **2006**, *66*, 1108–1110. [CrossRef]
179. DaSilva, A.F.; Nascimento, T.D.; DosSantos, M.F.; Lucas, S.; van Holsbeeck, H.; DeBoer, M.; Maslowski, E.; Love, T.; Martikainen, I.K.; Koeppe, R.A.; et al. Mu-Opioid activation in the prefrontal cortex in migraine attacks—Brief report I. *Ann. Clin. Transl. Neurol.* **2014**, *1*, 439–444. [CrossRef]
180. Linnman, C.; Catana, C.; Petkov, M.P.; Chonde, D.B.; Becerra, L.; Hooker, J.; Borsook, D. Molecular and functional PET-fMRI measures of placebo analgesia in episodic migraine: Preliminary findings. *NeuroImage Clin.* **2018**, *17*, 680–690. [CrossRef]
181. Schreckenberger, M.; Klega, A.; Gründer, G.; Buchholz, H.-G.; Scheurich, A.; Schirrmacher, R.; Schirrmacher, E.; Müller, C.; Henriksen, G.; Bartenstein, P. Opioid receptor PET reveals the psychobiologic correlates of reward processing. *J. Nucl. Med.* **2008**, *49*, 1257–1261. [CrossRef]
182. Tuominen, L.; Salo, J.; Hirvonen, J.; Någren, K.; Laine, P.; Melartin, T.; Isometsä, E.; Viikari, J.; Raitakari, O.; Keltikangas-Järvinen, L.; et al. Temperament trait harm avoidance associates with mu-opioid receptor availability in frontal cortex: A PET study using [^{11}C]carfentanil. *NeuroImage* **2012**, *61*, 670–676. [CrossRef]
183. Karjalainen, T.; Tuominen, L.; Manninen, S.; Kalliokoski, K.K.; Nuutila, P.; Jääskeläinen, I.P.; Hari, R.; Sams, M.; Nummenmaa, L. Behavioural activation system sensitivity is associated with cerebral mu-opioid receptor availability. *Soc. Cogn. Affect. Neurosci.* **2016**, *11*, 1310–1316. [CrossRef]
184. Nummenmaa, L.; Manninen, S.; Tuominen, L.; Hirvonen, J.; Kalliokoski, K.K.; Nuutila, P.; Jääskeläinen, I.P.; Hari, R.; Dunbar, R.I.; Sams, M. Adult attachment style is associated with cerebral mu-opioid receptor availability in humans. *Hum. Brain Mapp.* **2015**, *36*, 3621–3628. [CrossRef]
185. Karjalainen, T.; Seppälä, K.; Glerean, E.; Karlsson, H.K.; Lahnakoski, J.M.; Nuutila, P.; Jääskeläinen, I.P.; Hari, R.; Sams, M.; Nummenmaa, L. Opioidergic Regulation of Emotional Arousal: A Combined PET–fMRI Study. *Cereb. Cortex* **2018**, *29*, 4006–4016. [CrossRef]

186. Nummenmaa, L.; Tuominen, L.; Dunbar, R.; Hirvonen, J.; Manninen, S.; Arponen, E.; Machin, A.; Hari, R.; Jääskeläinen, I.P.; Sams, M. Social touch modulates endogenous mu-opioid system activity in humans. *NeuroImage* **2016**, *138*, 242–247. [CrossRef]
187. Manninen, S.; Tuominen, L.; Dunbar, R.I.; Karjalainen, T.; Hirvonen, J.; Arponen, E.; Hari, R.; Jääskeläinen, I.P.; Sams, M.; Nummenmaa, L. Social laughter triggers endogenous opioid release in humans. *J. Neurosci.* **2017**, *37*, 6125–6131. [CrossRef]
188. Prossin, A.R.; Koch, A.E.; Campbell, P.L.; Barichello, T.; Zalcman, S.S.; Zubieta, J.-K. Acute experimental changes in mood state regulate immune function in relation to central opioid neurotransmission: A model of human CNS-peripheral inflammatory interaction. *Mol. Psychiatry* **2016**, *21*, 243–251. [CrossRef]
189. Kennedy, S.E.; Koeppe, R.A.; Young, E.A.; Zubieta, J.K. Dysregulation of Endogenous Opioid Emotion Regulation Circuitry in Major Depression in Women. *Arch. Gen. Psychiatry.* **2006**, *63*, 1199–1208. [CrossRef]
190. Miller, J.M.; Zanderigo, F.; Purushothaman, P.D.; DeLorenzo, C.; Rubin-Falcone, H.; Ogden, R.T.; Keilp, J.; Oquendo, M.A.; Nabulsi, N.; Huang, Y.H.; et al. Kappa opioid receptor binding in major depression: A pilot study. *Synapse* **2018**, *72*, e22042. [CrossRef]
191. Matuskey, D.; Dias, M.; Naganawa, M.; Pittman, B.; Henry, S.; Li, S.; Gao, H.; Ropchan, J.; Nabulsi, N.; Carson, R.E.; et al. Social status and demographic effects of the kappa opioid receptor: A PET imaging study with a novel agonist radiotracer in healthy volunteers. *Neuropsychopharmacology* **2019**, *44*, 1714–1719. [CrossRef]
192. Ashok, A.H.; Myers, J.; Marques, T.R.; Rabiner, E.A.; Howes, O.D. Reduced mu opioid receptor availability in schizophrenia revealed with [^{11}C]carfentanil positron emission tomographic Imaging. *Nat. Commun.* **2019**, *10*, 4493–4502. [CrossRef]
193. Prossin, A.R.; Zalcman, S.S.; Heitzeg, M.M.; Koch, A.E.; Campbell, P.L.; Phan, K.L.; Stohler, C.S.; Zubieta, J.K. Dynamic interactions between plasma IL-1 family cytokines and central endogenous opioid neurotransmitter function in humans. *Neuropsychopharmacology* **2015**, *40*, 554–565. [CrossRef]
194. Schrepf, A.; Harper, D.E.; Harte, S.E.; Wang, H.; Ichesco, E.; Hampson, J.P.; Zubieta, J.K.; Clauw, D.J.; Harris, R.E. Endogenous opioidergic dysregulation of pain in fibromyalgia: A PET and fMRI study. *Pain* **2016**, *157*, 2217–2225. [CrossRef] [PubMed]
195. Gorelick, D.A.; Kim, Y.K.; Bencherif, B.; Boyd, S.J.; Nelson, R.; Copersino, M.; Endres, C.J.; Dannals, R.F.; Frost, J.J. Imaging brain mu-opioid receptors in abstinent cocaine users: Time course and relation to cocaine craving. *Biol. Psychiatry* **2005**, *57*, 1573–1582. [CrossRef] [PubMed]
196. Gorelick, D.A.; Kim, Y.K.; Bencherif, B.; Boyd, S.J.; Nelson, R.; Copersino, M.L.; Dannals, R.F.; Frost, J.J. Brain mu-opioid receptor binding: Relationship to relapse to cocaine use after monitored abstinence. *Psychopharmacology* **2008**, *200*, 475–486. [CrossRef] [PubMed]
197. Martinez, D.; Slifstein, M.; Matuskey, D.; Nabulsi, N.; Zheng, M.-Q.; Lin, S.-F.; Ropchan, J.; Urban, N.; Grassetti, A.; Chang, D.; et al. Kappa-opioid receptors, dynorphin, and cocaine addiction: A positron emission tomography study. *Neuropsychopharmacology* **2019**, *44*, 1720–1727. [CrossRef]
198. Narendran, R.; Tollefson, S.; Himes, M.L.; Paris, J.; Lopresti, B.; Ciccocioppo, R.; Scott Mason, N. Nociceptin receptors upregulated in cocaine use disorder: A positron emission tomography imaging study using [^{11}C]NOP-1A. *Am. J. Psychiarty* **2019**, *176*, 468–476. [CrossRef]
199. Heinz, A.; Reimold, M.; Wrase, J.; Hermann, D.; Croissant, B.; Mundle, G.; Dohmen, B.M.; Braus, D.F.; Schumann, G.; Machulla, H.J.; et al. Correlation of stable elevations in striatal mu-opioid receptor availability in detoxified alcoholic patients with alcohol craving: A positron emission tomography study using carbon ^{11}C-labeled carfentanil. *Arch. Gen. Psychiatry* **2005**, *62*, 57–64. [CrossRef]
200. Turton, S.; Myers, J.F.M.; Mick, I.; Colasanti, A.; Venkataraman, A.; Durant, C.; Waldman, A.; Brailsford, A.; Parkin, M.C.; Dawe, G.; et al. Blunted endogenous opioid release following an oral dexamphetamine challenge in abstinent alcohol-dependent individuals. *Mol. Psychiatry* **2018**. [CrossRef]
201. Williams, T.M.; Davies, S.J.C.; Taylor, L.G.; Daglish, M.R.C.; Hammers, A.; Brooks, D.J.; Nutt, D.J.; Lingford-Hughes, A. Brain opioid receptor binding in early abstinence from alcohol dependence and relationship to craving: An [^{11}C]diprenorphine PET study. *Eur. Neuropsychopharmacol.* **2009**, *19*, 740–748. [CrossRef]

202. Hermann, D.; Hirth, N.; Reimold, M.; Batra, A.; Smolka, M.N.; Hoffmann, S.; Kiefer, F.; Noori, H.R.; Sommer, W.H.; Reischl, G.; et al. Low mu-opioid receptor status in alcohol dependence identified by combined positron emission tomography and post-mortem brain analysis. *Neuropsychopharmacology* **2016**, *42*, 606–614. [CrossRef]
203. Weerts, E.M.; Wand, G.S.; Kuwabara, H.; Munro, C.A.; Dannals, R.F.; Hilton, J.; Frost, J.J.; McCaul, M.E. Positron emission tomography imaging of Mu- and Delta-opioid receptor binding in alcohol-dependent and healthy control subjects. *Alcohol. Clin. Exp. Res.* **2011**, *35*, 2162–2173. [CrossRef]
204. Wand, G.S.; Weerts, E.M.; Kuwabara, H.; Frost, J.J.; Xu, X.; McCaul, M.E. Naloxone-induced cortisol predicts mu opioid receptor binding potential in specific brain regions of healthy subjects. *Psychoneuroendocrinology* **2011**, *36*, 1453–1459. [CrossRef] [PubMed]
205. Wand, G.S.; Weerts, E.M.; Kuwabara, H.; Wong, D.F.; Xu, X.; McCaul, M.E. The relationship between naloxone-induced cortisol and mu opioid receptor availability in mesolimbic structures is disrupted in alcohol dependent subjects. *Alcohol* **2012**, *46*, 511–517. [CrossRef] [PubMed]
206. Ray, R.; Ruparel, K.; Newberg, A.; Wileyto, E.P.; Loughead, J.W.; Divgi, C.; Blendy, J.A.; Logan, J.; Zubieta, J.-K.; Lerman, C. Human Mu Opioid Receptor (OPRM1 A118G) polymorphism is associated with brain mu-opioid receptor binding potential in smokers. *Proc. Natl. Acad. Sci. USA* **2011**, *108*, 9268–9273. [CrossRef] [PubMed]
207. Domino, E.F.; Hirasawa-Fujita, M.; Ni, L.; Guthrie, S.K.; Zubieta, J.K. Regional brain [^{11}C]carfentanil binding following tobacco smoking. *Progr. Neuro-Psychopharmacol. Biol. Psychiatry* **2015**, *59*, 100–104. [CrossRef]
208. Nuechterlein, E.B.; Ni, L.; Domino, E.F.; Zubieta, J.K. Nicotine-specific and non-specific effects of cigarette smoking on endogenous opioid mechanisms. *Progr. Neuro-Psychopharmacol. Biol. Psychiatry* **2016**, *69*, 69–77. [CrossRef]
209. Weerts, E.M.; Wand, G.S.; Kuwabara, H.; Xu, X.; Frost, J.J.; Wong, D.F.; McCaul, M.E. Association of smoking with mu-opioid receptor availability before and during naltrexone blockade in alcohol-dependent subjects. *Addict. Biol.* **2014**, *19*, 733–742. [CrossRef]
210. Kuwabara, H.; Heishman, S.J.; Brasic, J.R.; Contoreggi, C.; Cascella, N.; Mackowick, K.M.; Taylor, R.; Rousset, O.; Willis, W.; Huestis, M.A.; et al. Mu opioid receptor binding correlates with nicotine dependence and reward in smokers. *PLoS ONE* **2014**, *9*, e113694. [CrossRef]
211. Guterstam, J.; Jayaram-Lindström, N.; Cervenka, S.; Frost, J.J.; Farde, L.; Halldin, C.; Franck, J. Effects of amphetamine on the human brain opioid system—A positron emission tomography study. *Int. J. Neuropsychopharmacol.* **2013**, *16*, 763–769. [CrossRef]
212. Colasanti, A.; Searle, G.E.; Long, C.J.; Hill, S.P.; Reiley, R.R.; Quelch, D.; Erritzoe, D.; Tziortzi, A.C.; Reed, L.J.; Lingford-Hughes, A.; et al. Endogenous opioid release in the human brain reward system induced by acute amphetamine administration. *Biol. Psychiatry* **2012**, *72*, 371–377. [CrossRef]
213. Mick, I.; Myers, J.; Stokes, P.R.A.; Erritzoe, D.; Colasanti, A.; Bowden-Jones, H.; Clark, L.; Gunn, R.N.; Rabiner, E.A.; Searle, G.E.; et al. Amphetamine induced endogenous opioid release in the human brain detected with [^{11}C]carfentanil PET: Replication in an independent cohort. *Int. J. Neuropsychopharmacol.* **2014**, *17*, 2069–2074. [CrossRef]
214. Quelch, D.R.; Katsouri, L.; Nutt, D.J.; Parker, C.A.; Tyacke, R.J. Imaging endogenous opioid peptide release with [^{11}C]carfentanil and [^{3}H]diprenorphine: Influence of agonist-induced internalization. *J. Cereb. Blood Flow. Metab.* **2014**, *34*, 1604–1612. [CrossRef] [PubMed]
215. Mick, I.; Myers, J.; Ramos, A.C.; Stokes, P.R.A.; Erritzoe, D.; Colasanti, A.; Gunn, R.N.; Rabiner, E.A.; Searle, G.E.; Waldman, A.D.; et al. Blunted endogenous opioid release following an oral amphetamine challenge in pathological gamblers. *Neuropsychopharmacology* **2016**, *41*, 1742–1750. [CrossRef] [PubMed]
216. Majuri, J.; Joutsa, J.; Arponen, E.; Forsback, S.; Kaasinen, V. Dopamine synthesis capacity correlates with mu-opioid receptor availability in the human basal ganglia: A triple-tracer PET study. *NeuroImage* **2018**, *183*, 1–6. [CrossRef] [PubMed]
217. Bencherif, B.; Guarda, A.S.; Colantuoni, C.; Ravert, H.T.; Dannals, R.F.; Frost, J.J. Regional mu-opioid receptor binding in insular cortex is decreased in bulimia nervosa and correlates inversely with fasting behavior. *J. Nucl. Med.* **2005**, *46*, 1349–1351. [PubMed]
218. Karlsson, H.K.; Tuominen, L.; Tuulari, J.J.; Hirvonen, J.; Parkkola, R.; Helin, S.; Salminen, P.; Nuutila, P.; Nummenmaa, L. Obesity is associated with decreased mu-opioid but unaltered dopamine D_2 receptor availability in the brain. *J. Neurosci.* **2015**, *35*, 3959–3965. [CrossRef] [PubMed]

219. Joutsa, J.; Karlsson, H.K.; Majuri, J.; Nuutila, P.; Helin, S.; Kaasinen, V.; Nummenmaa, L. Binge eating disorder and morbid obesity are associated with lowered mu-opioid receptor availability in the brain. *Psychiatry Res. Neuroimaging* **2018**, *276*, 41–45. [CrossRef] [PubMed]
220. Karlsson, H.K.; Tuulari, J.J.; Tuominen, L.; Hirvonen, J.; Honka, H.; Parkkola, R.; Helin, S.; Salminen, P.; Nuutila, P.; Nummenmaa, L. Weight loss after bariatric surgery normalizes brain opioid receptors in morbid obesity. *Mol. Psychiatry* **2016**, *21*, 1057–1062. [CrossRef]
221. Tuominen, L.; Tuulari, J.; Karlsson, H.; Hirvonen, J.; Helin, S.; Salminen, P.; Parkkola, R.; Hietala, J.; Nuutila, P.; Nummenmaa, L. Aberrant mesolimbic dopamine–opiate interaction in obesity. *NeuroImage* **2015**, *122*, 80–86. [CrossRef]
222. Tuulari, J.J.; Tuominen, L.; de Boer, F.E.; Hirvonen, J.; Helin, S.; Nuutila, P.; Nummenmaa, L. Feeding releases endogenous opioids in humans. *J. Neurosci.* **2017**, *37*, 8284–8291. [CrossRef]
223. Nummenmaa, L.; Saanijoki, T.; Tuominen, L.; Hirvonen, J.; Tuulari, J.J.; Nuutila, P.; Kalliokoski, K. Mu-opioid receptor system mediates reward processing in humans. *Nat. Commun.* **2018**, *9*, 1500. [CrossRef]
224. Saanijoki, T.; Nummenmaa, L.; Tuulari, J.J.; Tuominen, L.; Arponen, E.; Kalliokoski, K.K.; Hirvonen, J. Aerobic exercise modulates anticipatory reward processing via the mu-opioid receptor system. *Hum. Brain Mapp.* **2018**, *39*, 3972–3983. [CrossRef] [PubMed]
225. Hiura, M.; Sakata, M.; Ishii, K.; Toyohara, J.; Oda, K.; Nariai, T.; Ishiwata, K. Central mu-opioidergic system activation evoked by heavy and severe-intensity cycling exercise in humans: A pilot study using positron emission tomography with ^{11}C-carfentanil. *Int. J. Sports Med.* **2017**, *38*, 19–26. [PubMed]
226. Pert, C.B.; Snyder, S.H. Opiate receptor binding—Enhancement by opiate administration in vivo. *Biochem. Pharmacol.* **1976**, *25*, 847–853. [CrossRef]
227. Burns, J.A.; Kroll, D.S.; Feldman, D.E.; Liu, C.K.; Manza, P.; Wiers, C.E.; Volkow, N.D.; Wang, G.-J. Molecular imaging of opioid and dopamine systems: Insights into the pharmacogenetics of opioid use disorders. *Front. Psychiatry* **2019**, *10*, 626. [CrossRef] [PubMed]
228. Gu, Z.-H.; Wang, B.; Kou, Z.-Z.; Bai, Y.; Chen, T.; Dong, Y.-L.; Li, H.; Li, Y.-Q. Endomorphins: Promising endogenous opioid peptides for the development of novel analgesics. *Neurosignals* **2017**, *25*, 98–116. [CrossRef] [PubMed]
229. Redila, V.A.; Chavkin, C. Stress-induced reinstatement of cocaine seeking is mediated by the kappa opioid system. *Psychopharmacology* **2008**, *200*, 59–70. [CrossRef] [PubMed]

 © 2019 by the authors. Licensee MDPI, Basel, Switzerland. This article is an open access article distributed under the terms and conditions of the Creative Commons Attribution (CC BY) license (http://creativecommons.org/licenses/by/4.0/).

Review

Current Landscape and Emerging Fields of PET Imaging in Patients with Brain Tumors

Jan-Michael Werner [1], Philipp Lohmann [2], Gereon R. Fink [1,2], Karl-Josef Langen [2,3] and Norbert Galldiks [1,2,*,†]

1. Department of Neurology, Faculty of Medicine and University Hospital Cologne, University of Cologne, Kerpener St. 62, 50937 Cologne, Germany; jan-michael.werner@uk-koeln.de (J.-M.W.); gereon.fink@uk-koeln.de (G.R.F.)
2. Institute of Neuroscience and Medicine (INM-3, -4), Research Center Juelich, Leo-Brandt-St., 52425 Juelich, Germany; p.lohmann@fz-juelich.de (P.L.); k.j.langen@fz-juelich.de (K.-J.L.)
3. Department of Nuclear Medicine, University Hospital Aachen, 52074 Aachen, Germany
* Correspondence: n.galldiks@fz-juelich.de or norbert.galldiks@uk-koeln.de; Tel.: +49-2461-61-9324 or +49-221-478-86124; Fax: +49-2461-61-1518 or +49-221-478-5669
† Center of Integrated Oncology (CIO), Universities of Aachen, Bonn, Cologne, and Duesseldorf, Germany.

Academic Editor: Peter Brust
Received: 28 February 2020; Accepted: 20 March 2020; Published: 24 March 2020

Abstract: The number of positron-emission tomography (PET) tracers used to evaluate patients with brain tumors has increased substantially over the last years. For the management of patients with brain tumors, the most important indications are the delineation of tumor extent (e.g., for planning of resection or radiotherapy), the assessment of treatment response to systemic treatment options such as alkylating chemotherapy, and the differentiation of treatment-related changes (e.g., pseudoprogression or radiation necrosis) from tumor progression. Furthermore, newer PET imaging approaches aim to address the need for noninvasive assessment of tumoral immune cell infiltration and response to immunotherapies (e.g., T-cell imaging). This review summarizes the clinical value of the landscape of tracers that have been used in recent years for the above-mentioned indications and also provides an overview of promising newer tracers for this group of patients.

Keywords: amino acid; FET; FACBC; FDOPA; immunoPET; molecular imaging; glioma; brain metastases

1. Introduction

For the management of patients with brain tumors, clinicians frequently need to rely on imaging information obtained from anatomical magnetic resonance imaging (MRI) before, during, and after the treatment. While contrast-enhanced MRI is of paramount value in neuro-oncology, its specificity for neoplastic tissue is low, and changes of the blood-brain barrier permeability as indicated by contrast enhancement are not limited to tumor tissue [1–7]. Nevertheless, precise delineation of tumor extent, including non-enhancing tumor subregions, is decisive for several diagnostic and therapeutic steps (e.g., planning of biopsy, surgery, or radiotherapy) [7,8]. Following radio- and/or chemotherapy, neurooncologists often encounter treatment-related changes. Some of these, e.g., pseudoprogression, are difficult to differentiate from actual tumor progression with conventional MRI alone [6–12]. Pseudoprogression describes a phenomenon characterized by an increase of contrast enhancement without clinical deterioration, and which disappears again over time without any treatment change [3,9,13–16]. Such treatment-related changes may occur early (in the case of pseudoprogression typically within the first 12 weeks after chemoradiation completion) or late (several months or even years after radiotherapy in the case 31 of radiation necrosis) [7,8,17].

Furthermore, surrogates of treatment response or progression obtained from MRI (e.g., a decrease of contrast enhancement or the fluid-attenuated inversion recovery (FLAIR) signal hyperintensity) may be unspecific. They can be influenced by inflammation, infarction, and reactive changes after surgery [2,6,12,18]. If treatment-related changes remain unidentified, an effective treatment may be erroneously terminated prematurely. The latter may also harm survival and mislead study results evaluating novel treatment approaches for tumor relapse [19].

To overcome these diagnostic challenges, imaging techniques with higher diagnostic accuracy than conventional MRI offering more than just anatomical information are needed. Apart from advanced MRI techniques, positron-emission tomography (PET) imaging has been evaluated over the past decades. It has been shown that PET imaging offers additional value in neuro-oncology since it enables the non-invasive evaluation of molecular and metabolic features of brain tumors. PET, therefore, is of great value for the indications mentioned above, which are of particular clinical interest [7,8,10,20]. Consequently, the PET task force of the Response Assessment in Neuro-Oncology (RANO) working group highlighted the additional clinical value of PET imaging using amino acid tracers compared to anatomical MRI. Accordingly, its widespread clinical use was recommended in patients with glioma and brain metastases [17,21].

The continuously growing landscape of PET tracers enables the evaluation of many biochemical processes in patients with brain tumors. With the advent of newer treatment options in neuro-oncology, in particular, targeted therapy and various immunotherapy options, the needs for additional information derived from neuroimaging in terms of characterization of the tumor environment, the evaluation of tumoral drug accumulation, immune cell infiltration, and the diagnosis of treatment-related changes following these newer treatment options are steadily increasing. Some of these requirements may be met by the existing landscape of well-established PET tracers, while others can be addressed by newer ones [22,23].

This review summarizes the value of PET tracers that have been used in brain tumors in recent years for the most relevant clinical indications. Furthermore, more unique but promising PET tracers are summarized and discussed.

2. Methods

A PubMed search using the terms "PET", "positron", "tracer", "glioma", "brain metastases", "FDG", "amino acid", "methionine", "FET", "FDOPA", "FACBC", "AMT", "TSPO", "GE-180", "FLT", "FAZA", "EGFR", "VEGF", "immunoPET", "isocitrate dehydrogenase", "radiotherapy", "T-cell imaging", "reporter gene", "radiation necrosis", "pseudoprogression", "tumor extent", "response assessment", "treatment-related changes", and combinations thereof was performed until January 2020. The PET tracers were evaluated regarding their clinical value for the delineation of tumor extent, diagnosis of treatment-related changes, the assessment of treatment response (Table 1), and according to the information provided by newer PET probes (Table 2).

Table 1. Frequently used PET tracers for the delineation of tumor extent, diagnosis of treatment-related changes, and the assessment of treatment response in brain tumor imaging.

Imaging Target and Corresponding Tracers	Delineation of Tumor Extent	Diagnosis of Treatment-Related Changes	Assessment of Treatment Response
Glucose metabolism			
[18F]FDG	-	+	-
Amino acid transport			
[18F]FET	++	++ [1]	++ [2]
[11C]MET	++	+	++
[18F]FDOPA	++	++	++
[11C]AMT	(++)	(++)	n.a.
[18F]FACBC	(++)	n.a.	n.a.
Mitochondrial translocator protein (TSPO)			
[18F]GE-180	unclear	n.a.	n.a.
Cellular proliferation			
[18F]FLT	-	+	++ [3]
Hypoxia			
[18F]FMISO	n.a.	n.a.	(++) [3]
[18F]FAZA	n.a.	n.a.	(++) [3]
Perfusion			
[15O]H$_2$O	n.a.	n.a.	n.a.
Angiogenesis			
[89Zr]bevacizumab	n.a.	n.a.	n.a.

++ = high diagnostic accuracy; (++) = high diagnostic accuracy, but limited data available; + = limited diagnostic accuracy; - = not helpful; [1] = increased accuracy when using dynamic [18F]FET PET; [2] = in enhancing and non-enhancing tumors; [3] = in patients undergoing antiangiogenic treatment with bevacizumab; [11C]AMT = α-[11C]-methyl-L-tryptophan; [18F]FACB = anti-1-amino-3-[18F]fluorocyclobutane-1-carboxylic acid; [18F]FAZA = [18F]flouroazomycin arabinoside; [18F]FDG = [18F]-2-fluoro-2-deoxy-D-glucose; [18F]FDOPA = 3,4-dihydroxy-6-[18F]fluoro-L-phenylalanine; [18F]FET = O-(2-[18F]fluoroethyl)-L-tyrosine; [18F]FLT = 3′-deoxy-3′-[18F]fluorothymidine; [18F]FMISO = [18F]fluoromisonidazole; [15O]H$_2$O = radiolabeled water; [11C]MET = [11C]methyl-L-methionine; n.a. = only preliminary or no data available.

Table 2. Promising PET tracers for the evaluation of newer treatment options.

Tracer	Target	Mechanism
Imaging of the EGFR family		
[^{11}C]erlotinib	EGFR	TKI-mediated imaging
[^{89}Zr]Zr-DFO-nimotuzumab	EGFR	Antibody-mediated imaging
[^{11}C]PD153035	EGFR	TKI-mediated imaging
[^{89}Zr]pertuzumab	HER2	Antibody-mediated imaging
[^{64}Cu]-DOTA-trastuzumab	HER2	Antibody-mediated imaging
Immuno-Imaging		
[^{89}Zr]nivolumab	PD-1	Antibody-mediated imaging
[^{89}Zr]atezolizumab	PD-L1	Antibody-mediated imaging
[^{18}F]BMS-986192	PD-L1	PET imaging using an engineered target-binding protein (adnectin)
[^{89}Zr]IAB22M2C	CD8+ T-cells	Antibody fragment-mediated imaging
[^{18}F]CFA	DCK	Targeting of the deoxy-cytidine kinase
[^{18}F]FHBG	HSV1-tk	Imaging of reporter gene expression
Imaging of IDH mutations		
[^{18}F]AGI-5198	IDH-mutant cells	Imaging of the mutant IDH enzyme using a radiolabeled IDH1 inhibitor
[^{18}F]-labeled triazinediamine analogue	IDH-mutant cells	Imaging of the mutant IDH enzyme
Radiolabeled butyl-phenyl sulfonamide	IDH-mutant cells	Imaging of the mutant IDH enzyme
[^{11}C]acetate	IDH-mutant cells	Metabolic trapping of the tracer in IDH-mutant cells

[^{18}F]CFA = 2-chloro-2′-deoxy-2′-[^{18}F]fluoro-9-b-D-arabinofuranosyl-adenine; DCK = deoxy-cytidine kinase EGFR = epidermal growth factor receptor; [^{18}F]FHBG = 9-[4-[^{18}F]fluoro-3-(hydroxymethyl)butyl]guanine; HER2 = human epidermal growth factor receptor 2; HSV1-tk = herpes simplex virus type 1 thymidine kinase; IDH = isocitrate dehydrogenase-1 or -2; PD-1 = programmed cell death receptor-1; PD-L1 = programmed cell death protein ligand 1; TKI = tyrosine kinase inhibitor.

3. Current Landscape of PET Imaging

PET allows targeting metabolic and molecular processes in patients with brain tumors relevant to diagnosis, treatment, and prognosis that cannot be assessed with anatomic computed tomography (CT) or MR imaging. A variety of PET tracers have been evaluated predominantly in glioma patients or patients with brain metastases with the main focus on glucose metabolism, amino acid transport, proliferation, hypoxia, blood flow, or angiogenesis. This section will provide an overview of PET tracers for brain tumors that have been evaluated in human subjects in the last years, especially for those above mentioned highly relevant indications in clinical routine. An overview is presented in Table 1.

3.1. PET Imaging of Glucose Metabolism

[^{18}F]-2-Fluoro-2-deoxy-D-glucose ([^{18}F]FDG) is the most widespread PET tracer in nuclear medicine. In neoplastic tissue, the uptake of [^{18}F]FDG reflects the increased expression of glucose transporters and hexokinase. The latter enzyme phosphorylates glucose and [^{18}F]FDG. In the central nervous system, the physiologically high and varying uptake of [^{18}F]FDG in healthy brain parenchyma hampers the accurate delineation of the brain tumor. This limits the diagnostic accuracy for the correct identification of treatment-related changes and assessment of treatment response in gliomas and brain metastases [17,21]. It has repeatedly been shown that the diagnostic accuracy of [^{18}F]FDG regarding the differentiation of radiation-induced changes from glioma and brain metastases recurrence is inferior to other imaging modalities, including advanced MRI and amino acid PET [17,24–26]. However, [^{18}F]FDG PET seems to be of value for the delineation of tumor extent and assessment of treatment response in patients with primary central nervous system (CNS) lymphoma [27–30].

3.2. PET Using Amino Acid PET Tracers

Radiolabeled amino acid tracers (Figure 1) are of great interest in brain tumor imaging because of the high tumor-to-brain contrast based on the relatively high specificity for neoplastic tissue and the low uptake in healthy brain tissue [7,8,21,31–33].

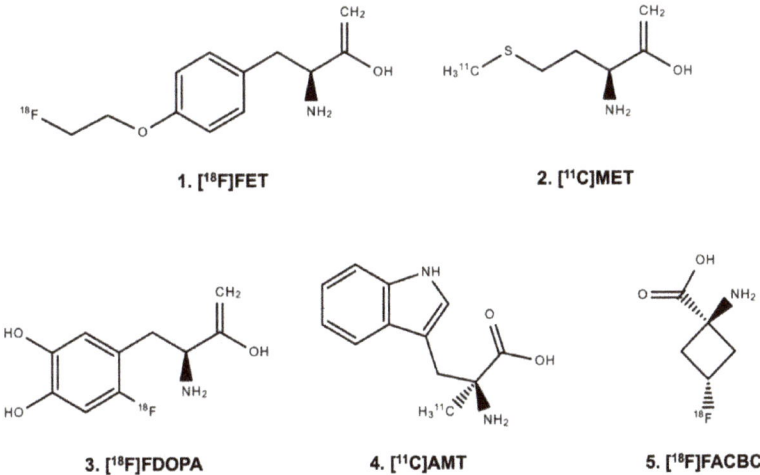

Figure 1. Chemical structure of radiolabeled amino acids.

3.2.1. Uptake Mechanisms of Amino Acid PET Tracers

The uptake of the amino acid tracers O-(2-[^{18}F]fluoroethyl)-L-tyrosine ([^{18}F]FET), [^{11}C]methyl-L-methionine ([^{11}C]MET), and 3,4-dihydroxy-6-[^{18}F]fluoro-L-phenylalanine ([^{18}F]FDOPA) is mainly based on the increased expression of large neutral amino acid transporters of the L-type (LAT) in gliomas and brain metastases (i.e., subtypes LAT1 and LAT2) [7,34–37]. Moreover, LAT1 overexpression correlates with malignant phenotypes and proliferation of gliomas. It is associated with glioma angiogenesis [38,39]. A critical consideration for the practical application of [^{11}C]MET compared to [^{18}F]FET or [^{18}F]FDOPA is the half-life of the [^{11}C]-isotope ([^{18}F] 110 vs. [^{11}C] 20 min) [40,41], which allows the transport of [^{18}F]FET and [^{18}F]FDOPA to PET facilities. In contrast, the use of [^{11}C]MET necessitates an on-site cyclotron. In many European centers, this logistical disadvantage has led to the replacement of [^{11}C]MET predominantly by [^{18}F]FET [7]. When using [^{18}F]FDOPA, the physiological uptake in the striatum may hamper the evaluation of tumor extent [7,42].

The L-tryptophan analogue α-[^{11}C]-methyl-L-tryptophan ([^{11}C]AMT) is another radiolabeled amino acid with uptake via the LAT system. Additionally, [^{11}C]AMT uptake is mediated via the kynurenine pathway and has a rate-limiting enzyme indoleamine 2,3-dioxygenase [43]. Indoleamine 2,3-dioxygenase is upregulated in various cancers including gliomas [44], which prompted the use of [^{11}C]AMT PET in patients with brain tumors [45].

Other tracers such as the synthetic amino acid analog anti-1-amino-3-[^{18}F]fluorocyclobutane-1-carboxylic acid ([^{18}F]FACBC or [^{18}F]fluciclovine) are also LAT-mediated but use additionally the alanine, serine, and cysteine transporter 2, which is upregulated in many human cancers [46–49].

3.2.2. Value of Amino Acid PET Tracers for Brain Tumor Patients

For the planning of diagnostic and therapeutic procedures, the precise delineation of tumor spread is essential. For example, the tumor extent as assessed by amino acid PET provides valuable information for planning stereotactic biopsies, resection, and radiotherapy [1,50–52]. For [^{11}C]MET and [^{18}F]FET, it has been shown that the delineation of tumor extent, particularly in non-enhancing gliomas, can be assessed with high accuracy using amino acid PET [51,53]. Preliminary data suggest that newer tracers such as [^{18}F]FACBC PET are also helpful in identifying metabolically active and non-contrast enhancing tumor regions in glioma patients [54–56]. Moreover, it has been shown that in the majority of cases, the metabolically active tumor burden as assessed by amino acid PET extends considerably beyond the volume of MRI contrast enhancement, which is of significant relevance for subsequent treatment planning [1,7,53,57]. Regarding the comparability of PET tracers for this indication, [^{11}C]MET, [^{18}F]FET, and [^{18}F]FDOPA seem to be equally informative [58–61]. Nevertheless, it has to be pointed out that 20–30% of grade II gliomas, according to the World Health Organization (WHO) classification of tumors of the central nervous system [62,63], show no amino acid uptake [64–66]. A negative amino acid PET, therefore, does not exclude glioma [8].

For the differentiation of treatment-related changes from tumor relapse, amino acid PET also provides valuable diagnostic information. Using [^{18}F]FET or [^{18}F]FDOPA, especially the differentiation of radiation injury from tumor relapse in glioma patients, as well as in patients with brain metastases, can be obtained with a relatively high diagnostic accuracy between 80–90% (Figure 2) [11,26,67–80]. Importantly, in glioma patients, parameters derived from dynamic [^{18}F]FET PET acquisition may further increase the diagnostic accuracy [67,74–76,78]. This has also been demonstrated in patients with brain metastases who underwent radiosurgery for brain metastases treatment [80,81]. The diagnostic accuracy of [^{11}C]MET PET regarding this clinical question is slightly lower (approximately 75%) [8,82,83], which is most probably related to the higher affinity of [^{11}C]MET for inflammation [84]. First PET studies using [^{11}C]AMT or [^{18}F]FACBC suggest that these tracers may also be of value for the differentiation of radiation injury from glioma progression [85,86].

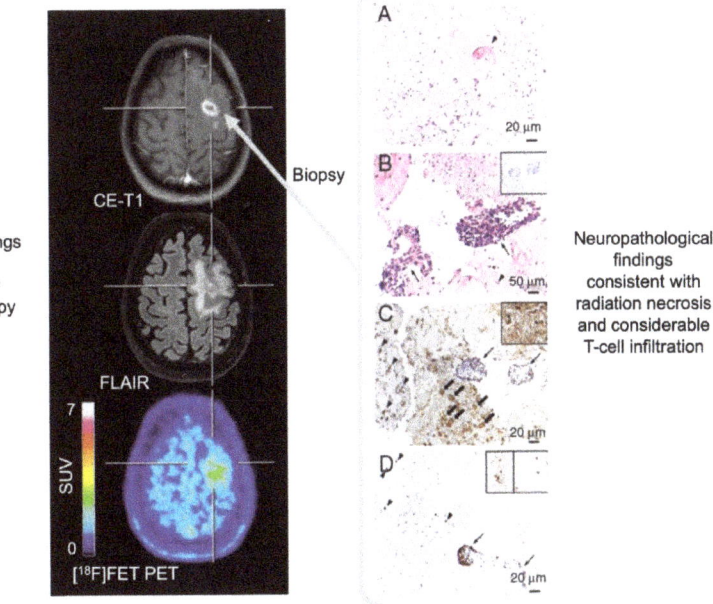

Figure 2. Radiation necrosis and chronic inflammation in a patient with brain metastases of a B-Raf proto-oncogene (BRAF)-mutated malignant melanoma who had been treated with whole-brain radiation therapy combined with concurrent dabrafenib plus trametinib. Twenty-four months later, the contrast-enhanced magnetic resonance imaging (MRI) indicates a recurrence of the brain metastases (left panel), whereas the O-(2-[^{18}F]fluoroethyl)-L-tyrosine ([^{18}F]FET) positron-emission tomography (PET) shows only insignificant metabolic activity and is consistent with the findings of treatment-related MRI changes. Neuropathological findings (right panel) after stereotactic biopsy show signs of radiation necrosis as well as considerable T-cell infiltration. (**A**) Hyaline, eosinophilic necrosis with evidence of a necrotic vessel wall (arrowhead). (**B**) Vital brain parenchyma besides necrosis with activated microglia cells (arrowhead), and blood vessels with lymphocyte infiltrates (arrows) without evidence of tumor cells (inserted box). (**C**) Adjacent to inflamed blood vessels (arrows), a resorption of necroses by macrophages (block arrows) as well as activated microglia cells (arrowheads) and astrocytes in the brain parenchyma (inserted box). (**D**) The main population of intra- and perivascular T-cell infiltrates are CD3$^+$ (arrow), but also CD4$^+$ (inserted box left) and CD8$^+$ (inserted box right) T-cells contribute to the infiltrates (modified from Galldiks et al. [10], with permission from Oxford University Press).

The advent of immunotherapy using immune checkpoint inhibitors and targeted therapy has improved the survival of cancer patients, particularly in melanoma and lung cancer. Recent trials suggest that patients with brain metastases from these tumor entities may also benefit from these agents alone or in combination [87]. Regarding patients with brain metastases treated with checkpoint inhibitors or targeted therapy (frequently combined with radiotherapy), initial data indicate that amino acid PET may provide valuable information for differentiating relapse from equivocal MRI findings related to immunotherapy-induced inflammation [10,88].

Recently, a variety of experimental treatment options has been introduced for treating patients with high-grade glioma. [^{18}F]FET PET was shown to differentiate benign MRI findings related to these experimental therapies, e.g., immunotherapy with dendritic cell vaccination or targeted therapy with regorafenib, from tumor relapse [89,90]. However, the number of patients treated with these therapies and monitored with [^{18}F]FET PET is still small, and the results should be interpreted with caution.

The assessment of the response to a particular neurooncological treatment is of clinical relevance since treatment decisions can be negatively affected by treatment-related changes. The accurate assessment of response helps both to discontinue an ineffective treatment option as early as possible and to prevent an effective treatment from being erroneously terminated prematurely with a potentially harmful influence on survival. Furthermore, the evaluation of response also helps to avoid possible treatment side effects, e.g., bone marrow depression or fatigue, and, therefore, to maintain or even improve life-quality. It has been shown in glioma patients that the assessment of response to alkylating chemotherapy (i.e., temozolomide or lomustine) using [^{11}C]MET or [^{18}F]FET PET provides valuable additional information compared to contrast-enhanced MRI. Importantly, metabolic PET responders (i.e., patients with a decrease of tumor-to-brain ratios or metabolically active tumor volumes at follow-up relative to baseline imaging) had a significantly longer survival than metabolic non-responders [91–98].

Following antiangiogenic therapy using bevacizumab, the use of reduced contrast enhancement as a surrogate marker for treatment response is not optimal due to a phenomenon called pseudoresponse [9]. Pseudoresponse describes a decrease of contrast enhancement related to a rapid restoration of the blood-brain barrier by antiangiogenic drugs [9]. However, a clinical benefit is not infrequently lacking in patients with an impressive radiological response (pseudoresponse). [^{18}F]FET and [^{18}F]FDOPA PET may provide valuable information regarding the identification of pseudoresponse [99–102]. Moreover, [^{18}F]FDOPA and [^{18}F]FET PET were also able to predict a favorable clinical outcome in bevacizumab responders [101–104].

After chemoradiation completion, newly diagnosed glioblastoma patients can be treated with tumor-treating fields therapy in addition to adjuvant temozolomide chemotherapy [105]. Initial studies suggest that amino acid PET might identify responding patients undergoing tumor-treating fields therapy, but it has to be considered that the response can also be related to the concurrently applied chemotherapy or delayed chemoradiation effects [106,107].

3.3. PET Imaging of the Mitochondrial Translocator Protein

PET ligands targeting the 18 kDa mitochondrial translocator protein (TSPO), located at the outer mitochondrial membrane and formerly known as the peripheral benzodiazepine receptor, are also of interest in neuro-oncology [108,109]. TSPO is associated with neuroinflammation due to its expression in activated microglia, endothelial cells, and infiltrating macrophages [109]. The PET ligand [^{11}C]PK11195 was one of the first ligands evaluated for TSPO expression in glioma patients [110–112].

The recently introduced TSPO ligand GE-180 labeled with [^{18}F] offers an increased binding specificity and was tested in patients with gliomas [113] and neuroinflammatory diseases such as multiple sclerosis [114–116]. Regarding the delineation of glioma extent, it has been demonstrated that the [^{18}F]GE-180 uptake volume is significantly larger than the volume of contrast enhancement [113,117]. However, when comparing [^{18}F]FET with [^{18}F]GE-180 uptake volumes intraindividually in terms of spatial distribution, the overlap is only moderate (Dice similarity coefficient, 0.55) despite comparable tumor volumes [117]. These differences might help to characterize glioma heterogeneity and warrant further studies with spatial correlation of imaging findings of [^{18}F]FET uptake to [^{18}F]GE-180 uptake with neuropathology.

3.4. PET Imaging of Cellular Proliferation

The radiolabeled nucleoside 3'-deoxy-3'-[^{18}F]fluorothymidine ([^{18}F]FLT) is a pyrimidine analogue. It is used to evaluate cellular proliferation because of its rapid incorporation into newly synthesized DNA [118]. [^{18}F]FLT is trapped intracellularly after phosphorylation by the thymidine kinase-1, a cytoplasmatic enzyme expressed during cell proliferation [46,119]. However, the requirement of a disrupted blood-brain barrier for [^{18}F]FLT uptake may limit its diagnostic use [2,120]. For example, in terms of tumor detection and delineation, [^{18}F]FLT PET was less sensitive than [^{11}C]MET PET to detect WHO grade II gliomas, which usually show no contrast enhancement [121]. Furthermore, a meta-analysis evaluating the value of [^{18}F]FLT PET for the diagnosis of glioma recurrence based on approximately 800 patients showed no superiority of [^{18}F]FLT (pooled sensitivity, 82%; pooled specificity, 76%) compared to [^{18}F]FDG (pooled sensitivity, 78%; pooled specificity, 77%) [122].

On the other hand, [^{18}F]FLT PET seems to be useful for the assessment of response to antiangiogenic therapy with bevacizumab in patients with recurrent malignant glioma. [^{18}F]FLT PET was able to identify a reduction of proliferative activity in responding patients with favorable outcome as an indicator for response compared to metabolic non-responders [123–125]. Furthermore, [^{18}F]FLT PET was used in patients with malignant melanoma brain metastases treated with targeted therapy or immunotherapy using checkpoint inhibitors [126]. In that study, responding patients showed a clearer reduction of proliferative activity as assessed by [^{18}F]FLT PET than the decrease of contrast enhancement on standard MRI.

3.5. PET Imaging of Tumor Hypoxia

Hypoxia is a key factor in treatment outcome in various cancers, including glioma. It has been shown that hypoxia is associated with tumor persistence and resistance to cancer treatment [127]. To further evaluate this phenomenon using PET, the tracer [^{18}F]fluoromisonidazole ([^{18}F]FMISO) has been developed, which is trapped in hypoxic but viable cells [46,128,129]. It has been demonstrated in glioblastoma patients that [^{18}F]FMISO PET delineates additional hypoxic tumor subregions which exceed the contrast-enhancing tumor parts, indicating that hypoxia may induce peripheral tumor growth [130]. A subsequent study showed that the metabolically active tumor volume in [^{11}C]MET PET strongly correlated with the hypoxic volume defined by [^{18}F]FMISO [131]. Importantly, the tumor area on [^{11}C]MET PET exceeded the area of the contrast enhancement on MRI in the range of 20–30%.

More recently, [^{18}F]FMISO PET has been used for monitoring the effects of antiangiogenic therapy with bevacizumab in patients with recurrent malignant glioma [132,133]. It was shown that patients who had a response in both contrast-enhanced MRI and [^{18}F]FMISO PET had significantly longer survival than patients who responded on MRI only [133].

[^{18}F]-labeled flouroazomycin arabinoside ([^{18}F]FAZA) may be a promising alternative to [^{18}F]FMISO offering an improved tumor-to-background ratio due to faster blood clearance [134,135]. Preliminary data in glioblastoma patients suggest that [^{18}F]FAZA PET might be of value for radiotherapy response assessment [136].

3.6. PET Imaging of Tumor Perfusion

The evaluation of regional cerebral blood flow (rCBF) allows identifying brain tumors with high vascularization. rCBF can be measured by PET using [^{15}O]-labeled water. However, [^{15}O]H$_2$O requires an on-site cyclotron because of its very short half-life (2 min) [137]. The development and easy accessibility of perfusion-weighted CT and MRI has led to various studies evaluating brain tumors outnumbering [^{15}O]H$_2$O PET studies by far. In direct comparisons, it has been shown that rCBF values differ considerably between CT- and MRI-based perfusion measurements and [^{15}O]H$_2$O PET [138,139].

In patients with malignant gliomas undergoing chemoradiation with nitrosoureas, a reduction of rCBF was evaluated using [^{15}O]H$_2$O PET, but data on subsequent survival as an indicator for treatment response are lacking [140,141].

3.7. PET Imaging of Angiogenesis

The vascular endothelial growth factor (VEGF) is overexpressed by most tumors, including brain tumors, and an important trigger for neovascularization [142]. Bevacizumab is a recombinant humanized monoclonal antibody against VEGF and appears to prolong progression-free survival and decrease steroid usage in patients with malignant glioma [142,143]. Based on these properties, [^{89}Zr]-labeled bevacizumab PET imaging has been evaluated for tumoral drug accumulation in pediatric patients with diffuse intrinsic pontine glioma [144]. That study demonstrated intertumoral heterogeneity of drug accumulation and may aid in selecting those patients with the greatest chance of benefit from bevacizumab [144,145]. Other studies have used [^{64}Cu]-labeled conjugates of VEGF with DOTA (1,4,7,10-tetra-azacylododecane N,N',N'',N'''-tetraacetic acid) for PET imaging in animal models [146–148]. Similarly, these approaches suggest the clinical value for the identification of patients who will benefit from anti-VEGF therapy.

In adult glioblastoma patients, it has been demonstrated that accumulation of [^{123}I]VEGF in the tumor region of glioblastomas can be assessed using single photon emission computed tomography (SPECT) imaging [149]. Importantly, high uptake of [^{123}I]VEGF was able to identify glioblastoma patients with a poor clinical outcome.

4. Emerging Fields of PET Imaging

Currently, blockade of immune checkpoints and other immunotherapy options (i.e., vaccination strategies, oncolytic virus approaches, cell-based immunotherapy such as chimeric antigen receptor T-cells (CAR-T cells) are under evaluation in patients with brain cancer including glioma and brain metastases. Furthermore, recent study results suggest that newer-generation targeted therapies are a promising treatment option, especially in a subset of patients with brain metastases [150]. As a new treatment option, mutations of the isocitrate dehydrogenase (*IDH*) gene, which frequently occur in WHO grade II and III gliomas, have also gained interest as a potential treatment target [151,152]. All these promising treatment options impose new demands on brain imaging (e.g., imaging of immune reactions in the brain). The current body of literature suggests that PET has the potential to adapt to these needs. An overview is presented in Table 2.

4.1. PET Imaging of the Epidermal Growth Factor Receptor Family

Both the epidermal growth factor receptor (EGFR) and the human epidermal growth factor receptor 2 (HER2) are transmembrane protein receptors and belong to the EGFR family. These receptors are targets for various growth factors that mediate various cellular processes such as differentiation or proliferation. In clinical oncology, various gene mutations may lead to overexpression of these proteins and are associated with the development of a variety of cancers. Importantly, these mutations also play a significant role in various treatment options, including tyrosine kinase inhibitors and monoclonal antibodies targeting EGFR, HER2, or both [153], as well as for imaging. Especially the evaluation of a response to these targeted therapy options in patients with brain metastases (from melanoma, lung, or breast cancer) as well as in glioma patients is the indication with the highest clinical potential for EGFR- and HER2-targeted PET [154,155]. In recent years, radiolabeled EGFR and HER2 antibodies, as well as tyrosine kinase inhibitors, have been used as PET imaging agents.

For identifying EGFR overexpression, PET ligands such as [^{11}C]erlotinib, [^{11}C]PD153035, and [^{89}Zr]Zr-DFO-nimotuzumab have been used [156–158]. The most relevant PET tracers for imaging of HER2 overexpression are [^{64}Cu]DOTA-trastuzumab and [^{89}Zr]pertuzumab (Figure 3) [159,160].

4.2. Immuno-Imaging: Immuno-PET and Imaging of T-Cells

In recent years, immunotherapy with antibodies directed against immune checkpoints such as the cytotoxic T-lymphocyte antigen-4 (CTLA-4; e.g., ipilimumab), the programmed cell death receptor-1 (PD-1; e.g., pembrolizumab, nivolumab), or the programmed cell death protein ligand 1 (PD-L1; e.g., atezolizumab) have gained paramount importance in clinical oncology and neuro-oncology. However, the efficacy and responsiveness of these agents may vary considerably among different cancer types and across individuals. Biomarkers obtained from tumor tissue, such as PD-1 and PD-L1 expression, can help to select patients. However, these tissue biomarkers are limited, and some patients show no response even if the target is present [161]. Therefore, the significance of PET for predicting response to immunotherapy and patient selection increases. Among other methods, immuno-PET combines antibodies or antibody fragments with a radionuclide and takes advantage of the specificity and affinity of antibodies and the sensitivity of PET [162]. Generally, targets for immuno-PET can be T-cell markers (e.g., CD4$^+$, CD8$^+$), immune checkpoints (e.g., CTLA-4, PD-1, PD-L1), or biomarkers of the immune response (e.g., interferon-γ, interleukin-2) [23].

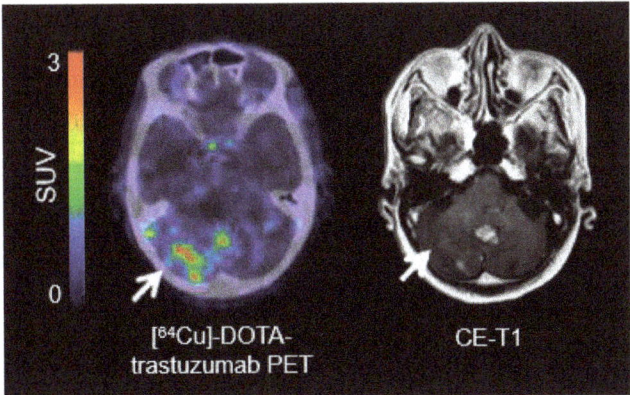

Figure 3. [^{64}Cu]-DOTA-trastuzumab positron-emission tomography (PET) and contrast-enhanced magnetic resonance imaging (MRI) performed one day after initiation of treatment with trastuzumab in a patient with a human epidermal growth factor receptor 2 (HER2)-positive breast cancer with brain metastases. In single brain metastases, [^{64}Cu]-DOTA-trastuzumab PET helps to improve lesion detection (arrow) (modified from Tamura et al. [160], with permission from the Society of Nuclear Medicine and Molecular Imaging).

First-in-human studies suggest that targeting PD-1 with [^{89}Zr]nivolumab [163] or PD-L1 with [^{89}Zr]atezolizumab [164] is useful as imaging biomarkers to non-invasively evaluate the expression of these immune checkpoints in patients with extra- and intracranial cancer [164,165]. Engineered target-binding proteins (adnectins) for PD-L1 ligands such as [^{18}F]BMS-986192 are currently under evaluation [166].

Tumor-infiltrating T-cells (such as CD8$^+$) play an essential role in the activation of immune cells in response to checkpoint inhibition [167]. Recently, it has been shown that a radiolabeled [^{89}Zr]IAB22M2C has the potential to visualize CD8$^+$ T-cell-enriched tumor tissue [168]. The assessment of immune cells infiltrating tumors has also been investigated with PET using radiolabeled clofarabine (2-chloro-2'-deoxy-2'-[^{18}F]fluoro-9-b-D-arabinofuranosyl-adenine; [^{18}F]CFA) [169]. [^{18}F]CFA is a substrate for the enzyme deoxy-cytidine kinase, which is overexpressed in immune cells such as CD8$^+$ T-cells [169]. Importantly, [^{18}F]CFA PET has shown a great clinical potential to localize and quantify immune responses in glioblastoma patients undergoing dendritic cell vaccination treatment combined with immune checkpoint blockade (Figure 4) [169].

Another interesting approach is the transfection of immune cells with a reporter gene that encodes a protein that can be specifically targeted by a radiolabeled reporter probe [170]. Imaging of reporter gene expression of cells transfected with the herpes simplex virus type 1 thymidine kinase reporter gene has been demonstrated using 9-[4-[^{18}F]fluoro-3-(hydroxymethyl)butyl]guanine ([^{18}F]FHBG) [171]. Moreover, this technique has the potential to image recently introduced cell-based therapies with CAR-T cells [23]. Accordingly, a study with recurrent high-grade glioma patients suggested that [^{18}F]FHBG PET can detect reporter gene expression in CAR-engineered cytotoxic T-lymphocytes [172].

4.3. PET Imaging of Isocitrate Dehydrogenase Mutations

In patients with malignant glioma, the modest outcome improvement following both standard therapy (i.e., chemoradiation with temozolomide) and newer treatment options (e.g., tumor-treating fields) has prompted various efforts to identify molecules that are fundamental to regulate tumor progression and provide additional options for personalized therapy in this group of patients.

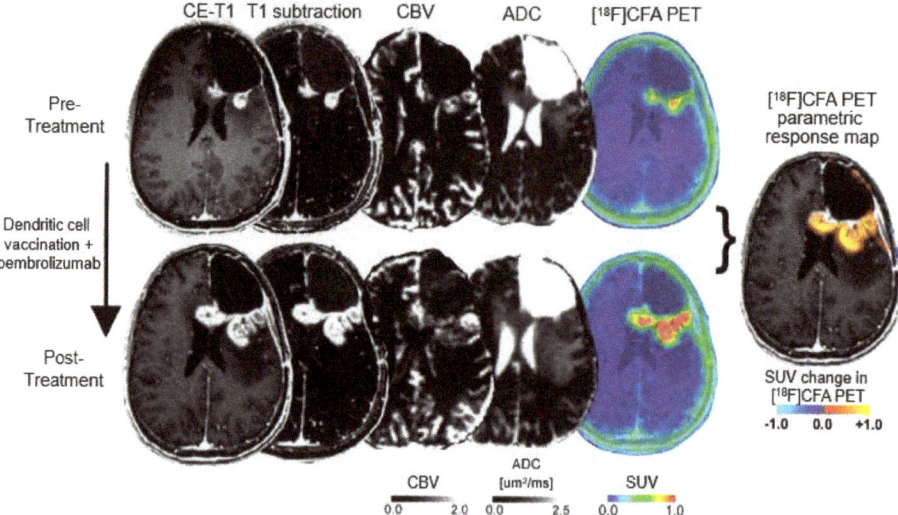

Figure 4. Detection of immune response in a patient with recurrent glioblastoma using 2-chloro-2′-deoxy-2′-[^{18}F]fluoro-9-b-D-arabinofuranosyl-adenine ([^{18}F]CFA) positron-emission tomography (PET) and advanced magnetic resonance imaging (MRI) before (upper panel) and after treatment with dendritic cell vaccination and programmed cell death receptor-1 (PD-1) blockade using pembrolizumab (lower panel). Following treatment, [^{18}F]CFA uptake is considerably increased, indicating an immune-cell infiltration, and helps distinguishing tumor progression from inflammation (modified from Antonios et al. [169], with permission from the National Academy of Sciences).

Accordingly, the enzyme isocitrate dehydrogenase (IDH) has gained interest as a potential target. IDH is an enzyme of the Krebs cycle, catalyzing the oxidative decarboxylation of isocitrate to alpha-ketoglutarate. Mutations in the *IDH1* and *IDH2* gene, frequently occurring in WHO grade II or III astrocytomas and oligodendrogliomas, result in a significant increase of the oncometabolite 2-hydroxyglutarate (2-HG) [173,174]. In cells with IDH mutant enzymes, the accumulation of 2-HG alters several downstream cellular activities, causing epigenetic dysregulation and, consequently, a block in cellular differentiation, leading to oncogenesis [175].

Therefore, mutant IDH proteins are highly attractive targets for inhibitory drugs. In glioma patients, selective oral IDH inhibitors of IDH1 (i.e., ivosidenib, BAY-1436032), pan-IDH1/2 (i.e., AG-881), and vaccination strategies targeting the IDH1R132H mutation are currently under clinical evaluation. Initial results predominantly from phase-1 studies are promising and suggest that these inhibitors are safe and have antitumoral activity [151,152,176]. Although immunohistochemistry and genomic sequencing are the methods of choice for the detection of an IDH mutation, these techniques are invasive and are not appropriate for treatment monitoring, which requires continual assessment. Furthermore, the use of magnetic resonance spectroscopy (MRS) to non-invasively evaluate 2-HG is technically challenging, and may be false-positive in 20% of cases [177,178].

Newer PET probes for imaging mutant IDH expression in gliomas may be an alternative imaging method. Recent radiochemical developments suggest that triazinediamine or butyl-phenyl sulfonamide analogs labeled with [^{18}F] are promising candidate radiotracers for noninvasive PET imaging of IDH mutations in gliomas [179,180]. Furthermore, a [^{18}F]-labeled IDH1 inhibitor (AGI-5198) has also been investigated [181]. Interestingly, Koyaso and colleagues demonstrated that [^{11}C]acetate uptake in IDH mutant cells is significantly higher than in IDH wild-type cells because of metabolic trapping [182]. Taken together, further efforts to translate these promising approaches for IDH imaging into clinical use are warranted.

4.4. PET-Based Theranostics

The combination of therapeutics and diagnostics, also termed as theranostics, supports the concept of precision oncology. One PET-based theranostic approach is the peptide receptor radionuclide therapy (PRRT), in which overexpressed tumor-specific receptors are used as a therapeutic target. By exchanging the radionuclide used for diagnostic PET such as [^{68}Ga] with a radiation source, typically ß-emitters like [^{177}Lu] or [^{90}Y], the same PET tracer can be used for therapy. Although there are currently no theranostic approaches clinically established for gliomas or brain metastases, there are promising concepts. For example in patients with glioblastoma, a potential target is the overexpressed chemokine receptor-4 (CXCR4) which is associated with a poor clinical outcome [183–185]. Visualization of CXCR4 expression using diagnostic PET with CXCR4-directed [^{68}Ga]Pentixafor® has been demonstrated in glioblastoma patients [186].

Another potential target is the prostate-specific membrane antigen (PSMA) which may be overexpressed in prostate cancer and also in predominantly malignant gliomas. Diagnostic PET imaging of PSMA expression in patients with malignant glioma can be obtained using [^{68}Ga]-labeled PSMA ligands [187–189]. The use of the theranostic agent [^{177}Lu]-PSMA-617 has demonstrated favorable safety and efficacy in patients with advanced prostate cancer, indicating the potential to be also of value for patients with malignant glioma [190]. Nevertheless, the potential of theranostics needs to be further evaluated in patients with malignant gliomas.

5. Discussion

Anatomical MRI is currently the method of choice for neuroimaging of brain tumors, but PET complements this technique and provides important biological information that cannot be obtained from anatomical MRI alone. Currently, best-established PET tracers in neuro-oncology are radiolabeled amino acids targeting L-system transporters. However, a considerable number of other PET tracers have been developed for brain tumor patients and allow the evaluation of a wide range of biochemical processes. A variety of PET biomarkers offers the potential to play a clinically significant role for the monitoring of newer treatment options such as targeted therapy and immunotherapy, e.g., by providing an early assessment of response to these options, and for a more accurate differentiation of viable tumor from treatment-related changes. A major current shortcoming is the lack of large, prospective clinical trials in patients with both glioma and brain metastases for many of these PET tracers. Despite encouraging early study results in the field, it has to be further demonstrated that these tracers improve considerably patient management and outcome.

Author Contributions: Study design, writing of manuscript drafts: J.W. and N.G. Revising manuscript, approving final content of the manuscript: All. All authors have read and agreed to the published version of the manuscript.

Funding: The Cologne Clinician Scientist-Program (CCSP) of the Deutsche Forschungsgemeinschaft (DFG, FI 773/15-1), Germany, supported this work.

Conflicts of Interest: Related to the present work, the authors disclosed no potential conflicts of interest.

References

1. Lohmann, P.; Stavrinou, P.; Lipke, K.; Bauer, E.K.; Ceccon, G.; Werner, J.M.; Neumaier, B.; Fink, G.R.; Shah, N.J.; Langen, K.J.; et al. FET PET reveals considerable spatial differences in tumour burden compared to conventional MRI in newly diagnosed glioblastoma. *Eur. J. Nucl. Med. Mol. Imaging* **2019**, *46*, 591–602. [CrossRef]
2. Dhermain, F.G.; Hau, P.; Lanfermann, H.; Jacobs, A.H.; van den Bent, M.J. Advanced MRI and PET imaging for assessment of treatment response in patients with gliomas. *Lancet Neurol.* **2010**, *9*, 906–920. [CrossRef]
3. Brandsma, D.; Stalpers, L.; Taal, W.; Sminia, P.; van den Bent, M.J. Clinical features, mechanisms, and management of pseudoprogression in malignant gliomas. *Lancet Oncol.* **2008**, *9*, 453–461. [CrossRef]

4. Hygino da Cruz, L.C., Jr.; Rodriguez, I.; Domingues, R.C.; Gasparetto, E.L.; Sorensen, A.G. Pseudoprogression and pseudoresponse: Imaging challenges in the assessment of posttreatment glioma. *AJNR Am. J. Neuroradiol.* **2011**, *32*, 1978–1985. [CrossRef]
5. Yang, I.; Aghi, M.K. New advances that enable identification of glioblastoma recurrence. *Nat. Rev. Clin. Oncol.* **2009**, *6*, 648–657. [CrossRef]
6. Kumar, A.J.; Leeds, N.E.; Fuller, G.N.; Van Tassel, P.; Maor, M.H.; Sawaya, R.E.; Levin, V.A. Malignant gliomas: MR imaging spectrum of radiation therapy- and chemotherapy-induced necrosis of the brain after treatment. *Radiology* **2000**, *217*, 377–384. [CrossRef] [PubMed]
7. Langen, K.J.; Galldiks, N.; Hattingen, E.; Shah, N.J. Advances in neuro-oncology imaging. *Nat. Rev. Neurol.* **2017**, *13*, 279–289. [CrossRef]
8. Galldiks, N.; Lohmann, P.; Albert, N.L.; Tonn, J.C.; Langen, K.-J. Current status of PET imaging in neuro-oncology. *Neurooncol. Adv.* **2019**, *1*. [CrossRef]
9. Brandsma, D.; van den Bent, M.J. Pseudoprogression and pseudoresponse in the treatment of gliomas. *Curr. Opin. Neurol.* **2009**, *22*, 633–638. [CrossRef]
10. Galldiks, N.; Kocher, M.; Ceccon, G.; Werner, J.M.; Brunn, A.; Deckert, M.; Pope, W.B.; Soffietti, R.; Le Rhun, E.; Weller, M.; et al. Imaging challenges of immunotherapy and targeted therapy in patients with brain metastases: Response, progression, and pseudoprogression. *Neuro Oncol.* **2020**, *22*, 17–30. [CrossRef]
11. Galldiks, N.; Dunkl, V.; Stoffels, G.; Hutterer, M.; Rapp, M.; Sabel, M.; Reifenberger, G.; Kebir, S.; Dorn, F.; Blau, T.; et al. Diagnosis of pseudoprogression in patients with glioblastoma using O-(2-[18F]fluoroethyl)-L-tyrosine PET. *Eur. J. Nucl. Med. Mol. Imaging* **2015**, *42*, 685–695. [CrossRef] [PubMed]
12. Ahluwalia, M.S.; Wen, P.Y. Antiangiogenic therapy for patients with glioblastoma: Current challenges in imaging and future directions. *Expert Rev. Anticancer Ther.* **2011**, *11*, 653–656. [CrossRef]
13. Taal, W.; Brandsma, D.; de Bruin, H.G.; Bromberg, J.E.; Swaak-Kragten, A.T.; Smitt, P.A.; van Es, C.A.; van den Bent, M.J. Incidence of early pseudo-progression in a cohort of malignant glioma patients treated with chemoirradiation with temozolomide. *Cancer* **2008**, *113*, 405–410. [CrossRef]
14. Radbruch, A.; Fladt, J.; Kickingereder, P.; Wiestler, B.; Nowosielski, M.; Baumer, P.; Schlemmer, H.P.; Wick, A.; Heiland, S.; Wick, W.; et al. Pseudoprogression in patients with glioblastoma: Clinical relevance despite low incidence. *Neuro Oncol.* **2015**, *17*, 151–159. [CrossRef]
15. Galldiks, N.; Kocher, M.; Langen, K.J. Pseudoprogression after glioma therapy: An update. *Expert Rev. Neurother.* **2017**, *17*, 1109–1115. [CrossRef] [PubMed]
16. Young, R.J.; Gupta, A.; Shah, A.D.; Graber, J.J.; Zhang, Z.; Shi, W.; Holodny, A.I.; Omuro, A.M. Potential utility of conventional MRI signs in diagnosing pseudoprogression in glioblastoma. *Neurology* **2011**, *76*, 1918–1924. [CrossRef] [PubMed]
17. Galldiks, N.; Langen, K.J.; Albert, N.L.; Chamberlain, M.; Soffietti, R.; Kim, M.M.; Law, I.; Le Rhun, E.; Chang, S.; Schwarting, J.; et al. PET imaging in patients with brain metastasis-report of the RANO/PET group. *Neuro Oncol.* **2019**, *21*, 585–595. [CrossRef] [PubMed]
18. Wen, P.Y.; Macdonald, D.R.; Reardon, D.A.; Cloughesy, T.F.; Sorensen, A.G.; Galanis, E.; Degroot, J.; Wick, W.; Gilbert, M.R.; Lassman, A.B.; et al. Updated response assessment criteria for high-grade gliomas: Response assessment in neuro-oncology working group. *J. Clin. Oncol.* **2010**, *28*, 1963–1972. [CrossRef]
19. Reardon, D.A.; Weller, M. Pseudoprogression: Fact or wishful thinking in neuro-oncology? *Lancet Oncol.* **2018**, *19*, 1561–1563. [CrossRef]
20. Nandu, H.; Wen, P.Y.; Huang, R.Y. Imaging in neuro-oncology. *Ther. Adv. Neurol. Disord.* **2018**, *11*. [CrossRef]
21. Albert, N.L.; Weller, M.; Suchorska, B.; Galldiks, N.; Soffietti, R.; Kim, M.M.; la Fougere, C.; Pope, W.; Law, I.; Arbizu, J.; et al. Response Assessment in Neuro-Oncology working group and European Association for Neuro-Oncology recommendations for the clinical use of PET imaging in gliomas. *Neuro Oncol.* **2016**, *18*, 1199–1208. [CrossRef] [PubMed]
22. Wei, W.; Ni, D.; Ehlerding, E.B.; Luo, Q.Y.; Cai, W. PET Imaging of Receptor Tyrosine Kinases in Cancer. *Mol. Cancer Ther.* **2018**, *17*, 1625–1636. [CrossRef] [PubMed]
23. Wei, W.; Jiang, D.; Ehlerding, E.B.; Luo, Q.; Cai, W. Noninvasive PET Imaging of T cells. *Trends Cancer* **2018**, *4*, 359–373. [CrossRef] [PubMed]

24. Chernov, M.; Hayashi, M.; Izawa, M.; Ochiai, T.; Usukura, M.; Abe, K.; Ono, Y.; Muragaki, Y.; Kubo, O.; Hori, T.; et al. Differentiation of the radiation-induced necrosis and tumor recurrence after gamma knife radiosurgery for brain metastases: Importance of multi-voxel proton MRS. *Min-Minim. Invasive Neurosurg.* **2005**, *48*, 228–234. [CrossRef]
25. Hatzoglou, V.; Yang, T.J.; Omuro, A.; Gavrilovic, I.; Ulaner, G.; Rubel, J.; Schneider, T.; Woo, K.M.; Zhang, Z.; Peck, K.K.; et al. A prospective trial of dynamic contrast-enhanced MRI perfusion and fluorine-18 FDG PET-CT in differentiating brain tumor progression from radiation injury after cranial irradiation. *Neuro Oncol.* **2016**, *18*, 873–880. [CrossRef]
26. Tomura, N.; Kokubun, M.; Saginoya, T.; Mizuno, Y.; Kikuchi, Y. Differentiation between Treatment-Induced Necrosis and Recurrent Tumors in Patients with Metastatic Brain Tumors: Comparison among (11)C-Methionine-PET, FDG-PET, MR Permeability Imaging, and MRI-ADC-Preliminary Results. *AJNR Am. J. Neuroradiol.* **2017**, *38*, 1520–1527. [CrossRef]
27. Palmedo, H.; Urbach, H.; Bender, H.; Schlegel, U.; Schmidt-Wolf, I.G.; Matthies, A.; Linnebank, M.; Joe, A.; Bucerius, J.; Biersack, H.J.; et al. FDG-PET in immunocompetent patients with primary central nervous system lymphoma: Correlation with MRI and clinical follow-up. *Eur. J. Nucl. Med. Mol. Imaging* **2006**, *33*, 164–168. [CrossRef]
28. Birsen, R.; Blanc, E.; Willems, L.; Burroni, B.; Legoff, M.; Le Ray, E.; Pilorge, S.; Salah, S.; Quentin, A.; Deau, B.; et al. Prognostic value of early 18F-FDG PET scanning evaluation in immunocompetent primary CNS lymphoma patients. *Oncotarget* **2018**, *9*, 16822–16831. [CrossRef]
29. Chiavazza, C.; Pellerino, A.; Ferrio, F.; Cistaro, A.; Soffietti, R.; Ruda, R. Primary CNS Lymphomas: Challenges in Diagnosis and Monitoring. *Biomed. Res. Int.* **2018**, *2018*, 3606970. [CrossRef]
30. Herholz, K.; Langen, K.J.; Schiepers, C.; Mountz, J.M. Brain tumors. *Semin. Nucl. Med.* **2012**, *42*, 356–370. [CrossRef]
31. Langen, K.J.; Watts, C. Neuro-oncology: Amino acid PET for brain tumours—Ready for the clinic? *Nat. Rev. Neurol.* **2016**, *12*, 375–376. [CrossRef] [PubMed]
32. Law, I.; Albert, N.L.; Arbizu, J.; Boellaard, R.; Drzezga, A.; Galldiks, N.; la Fougere, C.; Langen, K.J.; Lopci, E.; Lowe, V.; et al. Joint EANM/EANO/RANO practice guidelines/SNMMI procedure standards for imaging of gliomas using PET with radiolabelled amino acids and [(18)F]FDG: Version 1.0. *Eur. J. Nucl. Med. Mol. Imaging* **2019**, *46*, 540–557. [CrossRef] [PubMed]
33. Drake, L.R.; Hillmer, A.T.; Cai, Z. Approaches to PET Imaging of Glioblastoma. *Molecules* **2020**, *25*, 568. [CrossRef]
34. Youland, R.S.; Kitange, G.J.; Peterson, T.E.; Pafundi, D.H.; Ramiscal, J.A.; Pokorny, J.L.; Giannini, C.; Laack, N.N.; Parney, I.F.; Lowe, V.J.; et al. The role of LAT1 in (18)F-DOPA uptake in malignant gliomas. *J. Neurooncol.* **2013**, *111*, 11–18. [CrossRef] [PubMed]
35. Papin-Michault, C.; Bonnetaud, C.; Dufour, M.; Almairac, F.; Coutts, M.; Patouraux, S.; Virolle, T.; Darcourt, J.; Burel-Vandenbos, F. Study of LAT1 Expression in Brain Metastases: Towards a Better Understanding of the Results of Positron Emission Tomography Using Amino Acid Tracers. *PLoS ONE* **2016**, *11*, e0157139. [CrossRef] [PubMed]
36. Wiriyasermkul, P.; Nagamori, S.; Tominaga, H.; Oriuchi, N.; Kaira, K.; Nakao, H.; Kitashoji, T.; Ohgaki, R.; Tanaka, H.; Endou, H.; et al. Transport of 3-fluoro-L-alpha-methyl-tyrosine by tumor-upregulated L-type amino acid transporter 1: A cause of the tumor uptake in PET. *J. Nucl. Med.* **2012**, *53*, 1253–1261. [CrossRef]
37. Okubo, S.; Zhen, H.N.; Kawai, N.; Nishiyama, Y.; Haba, R.; Tamiya, T. Correlation of L-methyl-11C-methionine (MET) uptake with L-type amino acid transporter 1 in human gliomas. *J. Neurooncol.* **2010**, *99*, 217–225. [CrossRef] [PubMed]
38. Haining, Z.; Kawai, N.; Miyake, K.; Okada, M.; Okubo, S.; Zhang, X.; Fei, Z.; Tamiya, T. Relation of LAT1/4F2hc expression with pathological grade, proliferation and angiogenesis in human gliomas. *BMC Clin. Pathol.* **2012**, *12*, 4. [CrossRef]
39. Kracht, L.W.; Friese, M.; Herholz, K.; Schroeder, R.; Bauer, B.; Jacobs, A.; Heiss, W.D. Methyl-[11C]-l-methionine uptake as measured by positron emission tomography correlates to microvessel density in patients with glioma. *Eur. J. Nucl. Med. Mol. Imaging* **2003**, *30*, 868–873. [CrossRef]
40. Langen, K.J.; Hamacher, K.; Weckesser, M.; Floeth, F.; Stoffels, G.; Bauer, D.; Coenen, H.H.; Pauleit, D. O-(2-[18F]fluoroethyl)-L-tyrosine: Uptake mechanisms and clinical applications. *Nucl. Med. Biol.* **2006**, *33*, 287–294. [CrossRef]

41. Galldiks, N.; Law, I.; Pope, W.B.; Arbizu, J.; Langen, K.J. The use of amino acid PET and conventional MRI for monitoring of brain tumor therapy. *Neuroimage Clin.* **2017**, *13*, 386–394. [CrossRef] [PubMed]
42. Cicone, F.; Filss, C.P.; Minniti, G.; Rossi-Espagnet, C.; Papa, A.; Scaringi, C.; Galldiks, N.; Bozzao, A.; Shah, N.J.; Scopinaro, F.; et al. Volumetric assessment of recurrent or progressive gliomas: Comparison between F-DOPA PET and perfusion-weighted MRI. *Eur. J. Nucl. Med. Mol. Imaging* **2015**, *42*, 905–915. [CrossRef]
43. Juhasz, C.; Dwivedi, S.; Kamson, D.O.; Michelhaugh, S.K.; Mittal, S. Comparison of amino acid positron emission tomographic radiotracers for molecular imaging of primary and metastatic brain tumors. *Mol. Imaging* **2014**, *13*. [CrossRef]
44. Uyttenhove, C.; Pilotte, L.; Theate, I.; Stroobant, V.; Colau, D.; Parmentier, N.; Boon, T.; Van den Eynde, B.J. Evidence for a tumoral immune resistance mechanism based on tryptophan degradation by indoleamine 2,3-dioxygenase. *Nat. Med.* **2003**, *9*, 1269–1274. [CrossRef] [PubMed]
45. Batista, C.E.; Juhasz, C.; Muzik, O.; Kupsky, W.J.; Barger, G.; Chugani, H.T.; Mittal, S.; Sood, S.; Chakraborty, P.K.; Chugani, D.C. Imaging correlates of differential expression of indoleamine 2,3-dioxygenase in human brain tumors. *Mol. Imaging Biol.* **2009**, *11*, 460–466. [CrossRef] [PubMed]
46. Choudhary, G.; Langen, K.J.; Galldiks, N.; McConathy, J. Investigational PET tracers for high-grade gliomas. *Q. J. Nucl. Med. Mol. Imaging* **2018**, *62*, 281–294. [CrossRef] [PubMed]
47. Fuchs, B.C.; Bode, B.P. Amino acid transporters ASCT2 and LAT1 in cancer: Partners in crime? *Semin. Cancer Biol.* **2005**, *15*, 254–266. [CrossRef] [PubMed]
48. Scalise, M.; Pochini, L.; Console, L.; Losso, M.A.; Indiveri, C. The Human SLC1A5 (ASCT2) Amino Acid Transporter: From Function to Structure and Role in Cell Biology. *Front. Cell Dev. Biol.* **2018**, *6*, 96. [CrossRef]
49. Schuster, D.M.; Nanni, C.; Fanti, S.; Oka, S.; Okudaira, H.; Inoue, Y.; Sorensen, J.; Owenius, R.; Choyke, P.; Turkbey, B.; et al. Anti-1-amino-3-18F-fluorocyclobutane-1-carboxylic acid: Physiologic uptake patterns, incidental findings, and variants that may simulate disease. *J. Nucl. Med.* **2014**, *55*, 1986–1992. [CrossRef]
50. Hayes, A.R.; Jayamanne, D.; Hsiao, E.; Schembri, G.P.; Bailey, D.L.; Roach, P.J.; Khasraw, M.; Newey, A.; Wheeler, H.R.; Back, M. Utilizing 18F-fluoroethyltyrosine (FET) positron emission tomography (PET) to define suspected nonenhancing tumor for radiation therapy planning of glioblastoma. *Pract. Radiat. Oncol.* **2018**, *8*, 230–238. [CrossRef]
51. Kracht, L.W.; Miletic, H.; Busch, S.; Jacobs, A.H.; Voges, J.; Hoevels, M.; Klein, J.C.; Herholz, K.; Heiss, W.D. Delineation of brain tumor extent with [11C]L-methionine positron emission tomography: Local comparison with stereotactic histopathology. *Clin. Cancer Res.* **2004**, *10*, 7163–7170. [CrossRef] [PubMed]
52. Lopez, W.O.; Cordeiro, J.G.; Albicker, U.; Doostkam, S.; Nikkhah, G.; Kirch, R.D.; Trippel, M.; Reithmeier, T. Correlation of (18)F-fluoroethyl tyrosine positron-emission tomography uptake values and histomorphological findings by stereotactic serial biopsy in newly diagnosed brain tumors using a refined software tool. *Onco Targets Ther.* **2015**, *8*, 3803–3815. [CrossRef] [PubMed]
53. Pauleit, D.; Floeth, F.; Hamacher, K.; Riemenschneider, M.J.; Reifenberger, G.; Muller, H.W.; Zilles, K.; Coenen, H.H.; Langen, K.J. O-(2-[18F]fluoroethyl)-L-tyrosine PET combined with MRI improves the diagnostic assessment of cerebral gliomas. *Brain* **2005**, *128*, 678–687. [CrossRef] [PubMed]
54. Bogsrud, T.V.; Londalen, A.; Brandal, P.; Leske, H.; Panagopoulos, I.; Borghammer, P.; Bach-Gansmo, T. 18F-Fluciclovine PET/CT in Suspected Residual or Recurrent High-Grade Glioma. *Clin. Nucl. Med.* **2019**. [CrossRef] [PubMed]
55. Michaud, L.; Beattie, B.J.; Akhurst, T.; Dunphy, M.; Zanzonico, P.; Finn, R.; Mauguen, A.; Schoder, H.; Weber, W.A.; Lassman, A.B.; et al. (18)F-Fluciclovine ((18)F-FACBC) PET imaging of recurrent brain tumors. *Eur. J. Nucl. Med. Mol. Imaging* **2019**. [CrossRef] [PubMed]
56. Tsuyuguchi, N.; Terakawa, Y.; Uda, T.; Nakajo, K.; Kanemura, Y. Diagnosis of Brain Tumors Using Amino Acid Transport PET Imaging with (18)F-fluciclovine: A Comparative Study with L-methyl-(11)C-methionine PET Imaging. *Asia Ocean. J. Nucl. Med. Biol.* **2017**, *5*, 85–94. [CrossRef] [PubMed]
57. Galldiks, N.; Ullrich, R.; Schroeter, M.; Fink, G.R.; Jacobs, A.H.; Kracht, L.W. Volumetry of [(11)C]-methionine PET uptake and MRI contrast enhancement in patients with recurrent glioblastoma multiforme. *Eur. J. Nucl. Med. Mol. Imaging* **2010**, *37*, 84–92. [CrossRef]

58. Grosu, A.L.; Astner, S.T.; Riedel, E.; Nieder, C.; Wiedenmann, N.; Heinemann, F.; Schwaiger, M.; Molls, M.; Wester, H.J.; Weber, W.A. An interindividual comparison of O-(2-[18F]fluoroethyl)-L-tyrosine (FET)- and L-[methyl-11C]methionine (MET)-PET in patients with brain gliomas and metastases. *Int. J. Radiat. Oncol. Biol. Phys.* **2011**, *81*, 1049–1058. [CrossRef]
59. Becherer, A.; Karanikas, G.; Szabo, M.; Zettinig, G.; Asenbaum, S.; Marosi, C.; Henk, C.; Wunderbaldinger, P.; Czech, T.; Wadsak, W.; et al. Brain tumour imaging with PET: A comparison between [18F]fluorodopa and [11C]methionine. *Eur. J. Nucl. Med. Mol. Imaging* **2003**, *30*, 1561–1567. [CrossRef]
60. Kratochwil, C.; Combs, S.E.; Leotta, K.; Afshar-Oromieh, A.; Rieken, S.; Debus, J.; Haberkorn, U.; Giesel, F.L. Intra-individual comparison of (1)(8)F-FET and (1)(8)F-DOPA in PET imaging of recurrent brain tumors. *Neuro Oncol.* **2014**, *16*, 434–440. [CrossRef]
61. Lapa, C.; Linsenmann, T.; Monoranu, C.M.; Samnick, S.; Buck, A.K.; Bluemel, C.; Czernin, J.; Kessler, A.F.; Homola, G.A.; Ernestus, R.I.; et al. Comparison of the amino acid tracers 18F-FET and 18F-DOPA in high-grade glioma patients. *J. Nucl. Med.* **2014**, *55*, 1611–1616. [CrossRef] [PubMed]
62. Louis, D.N.; Perry, A.; Reifenberger, G.; von Deimling, A.; Figarella-Branger, D.; Cavenee, W.K.; Ohgaki, H.; Wiestler, O.D.; Kleihues, P.; Ellison, D.W. The 2016 World Health Organization Classification of Tumors of the Central Nervous System: A summary. *Acta Neuropathol.* **2016**, *131*, 803–820. [CrossRef] [PubMed]
63. Louis, D.N.; Ohgaki, H.; Wiestler, O.D.; Cavenee, W.K.; Burger, P.C.; Jouvet, A.; Scheithauer, B.W.; Kleihues, P. The 2007 WHO classification of tumours of the central nervous system. *Acta Neuropathol.* **2007**, *114*, 97–109. [CrossRef] [PubMed]
64. Pichler, R.; Dunzinger, A.; Wurm, G.; Pichler, J.; Weis, S.; Nussbaumer, K.; Topakian, R.; Aigner, R.M. Is there a place for FET PET in the initial evaluation of brain lesions with unknown significance? *Eur. J. Nucl. Med. Mol. Imaging* **2010**, *37*, 1521–1528. [CrossRef] [PubMed]
65. Hutterer, M.; Nowosielski, M.; Putzer, D.; Jansen, N.L.; Seiz, M.; Schocke, M.; McCoy, M.; Gobel, G.; la Fougere, C.; Virgolini, I.J.; et al. [18F]-fluoro-ethyl-L-tyrosine PET: A valuable diagnostic tool in neuro-oncology, but not all that glitters is glioma. *Neuro Oncol.* **2013**, *15*, 341–351. [CrossRef]
66. Jansen, N.L.; Graute, V.; Armbruster, L.; Suchorska, B.; Lutz, J.; Eigenbrod, S.; Cumming, P.; Bartenstein, P.; Tonn, J.C.; Kreth, F.W.; et al. MRI-suspected low-grade glioma: Is there a need to perform dynamic FET PET? *Eur. J. Nucl. Med. Mol. Imaging* **2012**, *39*, 1021–1029. [CrossRef]
67. Werner, J.M.; Stoffels, G.; Lichtenstein, T.; Borggrefe, J.; Lohmann, P.; Ceccon, G.; Shah, N.J.; Fink, G.R.; Langen, K.J.; Kabbasch, C.; et al. Differentiation of treatment-related changes from tumour progression: A direct comparison between dynamic FET PET and ADC values obtained from DWI MRI. *Eur. J. Nucl. Med. Mol. Imaging* **2019**, *46*, 1889–1901. [CrossRef]
68. Kebir, S.; Fimmers, R.; Galldiks, N.; Schafer, N.; Mack, F.; Schaub, C.; Stuplich, M.; Niessen, M.; Tzaridis, T.; Simon, M.; et al. Late Pseudoprogression in Glioblastoma: Diagnostic Value of Dynamic O-(2-[18F]fluoroethyl)-L-Tyrosine PET. *Clin. Cancer Res.* **2016**, *22*, 2190–2196. [CrossRef]
69. Popperl, G.; Gotz, C.; Rachinger, W.; Gildehaus, F.J.; Tonn, J.C.; Tatsch, K. Value of O-(2-[18F]fluoroethyl)-L-tyrosine PET for the diagnosis of recurrent glioma. *Eur. J. Nucl. Med. Mol. Imaging* **2004**, *31*, 1464–1470. [CrossRef]
70. Rachinger, W.; Goetz, C.; Popperl, G.; Gildehaus, F.J.; Kreth, F.W.; Holtmannspotter, M.; Herms, J.; Koch, W.; Tatsch, K.; Tonn, J.C. Positron emission tomography with O-(2-[18F]fluoroethyl)-l-tyrosine versus magnetic resonance imaging in the diagnosis of recurrent gliomas. *Neurosurgery* **2005**, *57*, 505–511. [CrossRef]
71. Mihovilovic, M.I.; Kertels, O.; Hanscheid, H.; Lohr, M.; Monoranu, C.M.; Kleinlein, I.; Samnick, S.; Kessler, A.F.; Linsenmann, T.; Ernestus, R.I.; et al. O-(2-((18)F)fluoroethyl)-L-tyrosine PET for the differentiation of tumour recurrence from late pseudoprogression in glioblastoma. *J. Neurol. Neurosurg. Psychiatry* **2019**, *90*, 238–239. [CrossRef] [PubMed]
72. Mehrkens, J.H.; Popperl, G.; Rachinger, W.; Herms, J.; Seelos, K.; Tatsch, K.; Tonn, J.C.; Kreth, F.W. The positive predictive value of O-(2-[18F]fluoroethyl)-L-tyrosine (FET) PET in the diagnosis of a glioma recurrence after multimodal treatment. *J. Neurooncol.* **2008**, *88*, 27–35. [CrossRef] [PubMed]
73. Jena, A.; Taneja, S.; Gambhir, A.; Mishra, A.K.; D'Souza, M.M.; Verma, S.M.; Hazari, P.P.; Negi, P.; Jhadav, G.K.; Sogani, S.K. Glioma Recurrence Versus Radiation Necrosis: Single-Session Multiparametric Approach Using Simultaneous O-(2-18F-Fluoroethyl)-L-Tyrosine PET/MRI. *Clin. Nucl. Med.* **2016**, *41*, e228–e236. [CrossRef] [PubMed]

74. Pyka, T.; Hiob, D.; Preibisch, C.; Gempt, J.; Wiestler, B.; Schlegel, J.; Straube, C.; Zimmer, C. Diagnosis of glioma recurrence using multiparametric dynamic 18F-fluoroethyl-tyrosine PET-MRI. *Eur. J. Radiol.* **2018**, *103*, 32–37. [CrossRef]
75. Galldiks, N.; Stoffels, G.; Filss, C.; Rapp, M.; Blau, T.; Tscherpel, C.; Ceccon, G.; Dunkl, V.; Weinzierl, M.; Stoffel, M.; et al. The use of dynamic O-(2-18F-fluoroethyl)-l-tyrosine PET in the diagnosis of patients with progressive and recurrent glioma. *Neuro Oncol.* **2015**, *17*, 1293–1300. [CrossRef]
76. Ceccon, G.; Lohmann, P.; Stoffels, G.; Judov, N.; Filss, C.P.; Rapp, M.; Bauer, E.; Hamisch, C.; Ruge, M.I.; Kocher, M.; et al. Dynamic O-(2-18F-fluoroethyl)-L-tyrosine positron emission tomography differentiates brain metastasis recurrence from radiation injury after radiotherapy. *Neuro Oncol.* **2017**, *19*, 281–288. [CrossRef]
77. Tsuyuguchi, N.; Sunada, I.; Iwai, Y.; Yamanaka, K.; Tanaka, K.; Takami, T.; Otsuka, Y.; Sakamoto, S.; Ohata, K.; Goto, T.; et al. Methionine positron emission tomography of recurrent metastatic brain tumor and radiation necrosis after stereotactic radiosurgery: Is a differential diagnosis possible? *J. Neurosurg.* **2003**, *98*, 1056–1064. [CrossRef]
78. Galldiks, N.; Stoffels, G.; Filss, C.P.; Piroth, M.D.; Sabel, M.; Ruge, M.I.; Herzog, H.; Shah, N.J.; Fink, G.R.; Coenen, H.H.; et al. Role of O-(2-(18)F-fluoroethyl)-L-tyrosine PET for differentiation of local recurrent brain metastasis from radiation necrosis. *J. Nucl. Med.* **2012**, *53*, 1367–1374. [CrossRef]
79. Lizarraga, K.J.; Allen-Auerbach, M.; Czernin, J.; DeSalles, A.A.; Yong, W.H.; Phelps, M.E.; Chen, W. (18)F-FDOPA PET for differentiating recurrent or progressive brain metastatic tumors from late or delayed radiation injury after radiation treatment. *J. Nucl. Med.* **2014**, *55*, 30–36. [CrossRef]
80. Cicone, F.; Minniti, G.; Romano, A.; Papa, A.; Scaringi, C.; Tavanti, F.; Bozzao, A.; Maurizi Enrici, R.; Scopinaro, F. Accuracy of F-DOPA PET and perfusion-MRI for differentiating radionecrotic from progressive brain metastases after radiosurgery. *Eur. J. Nucl. Med. Mol. Imaging* **2015**, *42*, 103–111. [CrossRef]
81. Terakawa, Y.; Tsuyuguchi, N.; Iwai, Y.; Yamanaka, K.; Higashiyama, S.; Takami, T.; Ohata, K. Diagnostic accuracy of 11C-methionine PET for differentiation of recurrent brain tumors from radiation necrosis after radiotherapy. *J. Nucl. Med.* **2008**, *49*, 694–699. [CrossRef] [PubMed]
82. Nihashi, T.; Dahabreh, I.J.; Terasawa, T. Diagnostic accuracy of PET for recurrent glioma diagnosis: A meta-analysis. *AJNR Am. J. Neuroradiol.* **2013**, *34*, 944–950. [CrossRef] [PubMed]
83. Minamimoto, R.; Saginoya, T.; Kondo, C.; Tomura, N.; Ito, K.; Matsuo, Y.; Matsunaga, S.; Shuto, T.; Akabane, A.; Miyata, Y.; et al. Differentiation of Brain Tumor Recurrence from Post-Radiotherapy Necrosis with 11C-Methionine PET: Visual Assessment versus Quantitative Assessment. *PLoS ONE* **2015**, *10*, e0132515. [CrossRef]
84. Salber, D.; Stoffels, G.; Pauleit, D.; Oros-Peusquens, A.M.; Shah, N.J.; Klauth, P.; Hamacher, K.; Coenen, H.H.; Langen, K.J. Differential uptake of O-(2-18F-fluoroethyl)-L-tyrosine, L-3H-methionine, and 3H-deoxyglucose in brain abscesses. *J. Nucl. Med.* **2007**, *48*, 2056–2062. [CrossRef] [PubMed]
85. Alkonyi, B.; Barger, G.R.; Mittal, S.; Muzik, O.; Chugani, D.C.; Bahl, G.; Robinette, N.L.; Kupsky, W.J.; Chakraborty, P.K.; Juhasz, C. Accurate differentiation of recurrent gliomas from radiation injury by kinetic analysis of alpha-11C-methyl-L-tryptophan PET. *J. Nucl. Med.* **2012**, *53*, 1058–1064. [CrossRef] [PubMed]
86. Henderson, F.; Brem, S.; O'Rourke, D.M.; Nasrallah, M.; Buch, V.P.; Young, A.J.; Doot, R.K.; Pantel, A.; Desai, A.; Bagley, S.J.; et al. 18F-Fluciclovine PET to distinguish treatment-related effects from disease progression in recurrent glioblastoma: PET fusion with MRI guides neurosurgical sampling. *Neurooncol. Pract.* **2019**. [CrossRef]
87. Long, G.V.; Atkinson, V.; Lo, S.; Sandhu, S.; Guminski, A.D.; Brown, M.P.; Wilmott, J.S.; Edwards, J.; Gonzalez, M.; Scolyer, R.A.; et al. Combination nivolumab and ipilimumab or nivolumab alone in melanoma brain metastases: A multicentre randomised phase 2 study. *Lancet Oncol.* **2018**, *19*, 672–681. [CrossRef]
88. Galldiks, N.; Abdulla, D.S.Y.; Scheffler, M.; Schweinsberg, V.; Schlaak, M.; Kreuzberg, N.; Landsberg, J.; Lohmann, P.; Ceccon, G.; Werner, J.M.; et al. Treatment monitoring of immunotherapy and targeted therapy using FET PET in patients with melanoma and lung cancer brain metastases: Initial experiences. *J. Clin. Oncol.* **2019**, *37*, e13525. [CrossRef]
89. Kristin Schmitz, A.; Sorg, R.V.; Stoffels, G.; Grauer, O.M.; Galldiks, N.; Steiger, H.J.; Kamp, M.A.; Langen, K.J.; Sabel, M.; Rapp, M. Diagnostic impact of additional O-(2-[18F]fluoroethyl)-L-tyrosine ((18)F-FET) PET following immunotherapy with dendritic cell vaccination in glioblastoma patients. *Br. J. Neurosurg.* **2019**. [CrossRef]

90. Galldiks, N.; Werner, J.M.; Tscherpel, C.; Fink, G.R.; Langen, K.J. Imaging findings following regorafenib in malignant gliomas: FET PET adds valuable information to anatomical MRI. *Neurooncol. Adv.* **2019**, 1. [CrossRef]
91. Galldiks, N.; Kracht, L.W.; Burghaus, L.; Ullrich, R.T.; Backes, H.; Brunn, A.; Heiss, W.D.; Jacobs, A.H. Patient-tailored, imaging-guided, long-term temozolomide chemotherapy in patients with glioblastoma. *Mol. Imaging* **2010**, *9*, 40–46. [CrossRef] [PubMed]
92. Galldiks, N.; Kracht, L.W.; Burghaus, L.; Thomas, A.; Jacobs, A.H.; Heiss, W.D.; Herholz, K. Use of 11C-methionine PET to monitor the effects of temozolomide chemotherapy in malignant gliomas. *Eur. J. Nucl. Med. Mol. Imaging* **2006**, *33*, 516–524. [CrossRef] [PubMed]
93. Herholz, K.; Kracht, L.W.; Heiss, W.D. Monitoring the effect of chemotherapy in a mixed glioma by C-11-methionine PET. *J. Neuroimaging* **2003**, *13*, 269–271. [CrossRef] [PubMed]
94. Galldiks, N.; Langen, K.J.; Holy, R.; Pinkawa, M.; Stoffels, G.; Nolte, K.W.; Kaiser, H.J.; Filss, C.P.; Fink, G.R.; Coenen, H.H.; et al. Assessment of treatment response in patients with glioblastoma using O-(2-18F-fluoroethyl)-L-tyrosine PET in comparison to MRI. *J. Nucl. Med.* **2012**, *53*, 1048–1057. [CrossRef] [PubMed]
95. Piroth, M.D.; Pinkawa, M.; Holy, R.; Klotz, J.; Nussen, S.; Stoffels, G.; Coenen, H.H.; Kaiser, H.J.; Langen, K.J.; Eble, M.J. Prognostic value of early [18F]fluoroethyltyrosine positron emission tomography after radiochemotherapy in glioblastoma multiforme. *Int. J. Radiat. Oncol. Biol. Phys.* **2011**, *80*, 176–184. [CrossRef] [PubMed]
96. Wyss, M.; Hofer, S.; Bruehlmeier, M.; Hefti, M.; Uhlmann, C.; Bartschi, E.; Buettner, U.W.; Roelcke, U. Early metabolic responses in temozolomide treated low-grade glioma patients. *J. Neurooncol.* **2009**, *95*, 87–93. [CrossRef]
97. Roelcke, U.; Wyss, M.T.; Nowosielski, M.; Ruda, R.; Roth, P.; Hofer, S.; Galldiks, N.; Crippa, F.; Weller, M.; Soffietti, R. Amino acid positron emission tomography to monitor chemotherapy response and predict seizure control and progression-free survival in WHO grade II gliomas. *Neuro Oncol.* **2016**, *18*, 744–751. [CrossRef]
98. Suchorska, B.; Unterrainer, M.; Biczok, A.; Sosnova, M.; Forbrig, R.; Bartenstein, P.; Tonn, J.C.; Albert, N.L.; Kreth, F.W. (18)F-FET-PET as a biomarker for therapy response in non-contrast enhancing glioma following chemotherapy. *J. Neurooncol.* **2018**, *139*, 721–730. [CrossRef]
99. Galldiks, N.; Rapp, M.; Stoffels, G.; Dunkl, V.; Sabel, M.; Langen, K.J. Earlier diagnosis of progressive disease during bevacizumab treatment using O-(2-18F-fluoroethyl)-L-tyrosine positron emission tomography in comparison with magnetic resonance imaging. *Mol. Imaging* **2013**, *12*, 273–276. [CrossRef]
100. Morana, G.; Piccardo, A.; Garre, M.L.; Nozza, P.; Consales, A.; Rossi, A. Multimodal magnetic resonance imaging and 18F-L-dihydroxyphenylalanine positron emission tomography in early characterization of pseudoresponse and nonenhancing tumor progression in a pediatric patient with malignant transformation of ganglioglioma treated with bevacizumab. *J. Clin. Oncol.* **2013**, *31*, e1. [CrossRef]
101. Hutterer, M.; Nowosielski, M.; Putzer, D.; Waitz, D.; Tinkhauser, G.; Kostron, H.; Muigg, A.; Virgolini, I.J.; Staffen, W.; Trinka, E.; et al. O-(2-18F-fluoroethyl)-L-tyrosine PET predicts failure of antiangiogenic treatment in patients with recurrent high-grade glioma. *J. Nucl. Med.* **2011**, *52*, 856–864. [CrossRef] [PubMed]
102. Galldiks, N.; Rapp, M.; Stoffels, G.; Fink, G.R.; Shah, N.J.; Coenen, H.H.; Sabel, M.; Langen, K.-J. Response assessment of bevacizumab in patients with recurrent malignant glioma using [18F]Fluoroethyl-l-tyrosine PET in comparison to MRI. *Eur. J. Nucl. Med. Mol. Imaging* **2012**, *40*, 22–33. [CrossRef] [PubMed]
103. Galldiks, N.; Dunkl, V.; Ceccon, G.; Tscherpel, C.; Stoffels, G.; Law, I.; Henriksen, O.M.; Muhic, A.; Poulsen, H.S.; Steger, J.; et al. Early treatment response evaluation using FET PET compared to MRI in glioblastoma patients at first progression treated with bevacizumab plus lomustine. *Eur. J. Nucl. Med. Mol. Imaging* **2018**, *45*, 2377–2386. [CrossRef] [PubMed]
104. Schwarzenberg, J.; Czernin, J.; Cloughesy, T.F.; Ellingson, B.M.; Pope, W.B.; Grogan, T.; Elashoff, D.; Geist, C.; Silverman, D.H.; Phelps, M.E.; et al. Treatment response evaluation using 18F-FDOPA PET in patients with recurrent malignant glioma on bevacizumab therapy. *Clin. Cancer Res.* **2014**, *20*, 3550–3559. [CrossRef]
105. Stupp, R.; Taillibert, S.; Kanner, A.; Read, W.; Steinberg, D.; Lhermitte, B.; Toms, S.; Idbaih, A.; Ahluwalia, M.S.; Fink, K.; et al. Effect of Tumor-Treating Fields Plus Maintenance Temozolomide vs Maintenance Temozolomide Alone on Survival in Patients With Glioblastoma: A Randomized Clinical Trial. *JAMA* **2017**, *318*, 2306–2316. [CrossRef]

106. Bosnyak, E.; Barger, G.R.; Michelhaugh, S.K.; Robinette, N.L.; Amit-Yousif, A.; Mittal, S.; Juhasz, C. Amino Acid PET Imaging of the Early Metabolic Response During Tumor-Treating Fields (TTFields) Therapy in Recurrent Glioblastoma. *Clin. Nucl. Med.* **2018**, *43*, 176–179. [CrossRef]
107. Ceccon, G.; Lazaridis, L.; Stoffels, G.; Rapp, M.; Weber, M.; Blau, T.; Lohmann, P.; Kebir, S.; Herrmann, K.; Fink, G.R.; et al. Use of FET PET in glioblastoma patients undergoing neurooncological treatment including tumour-treating fields: Initial experience. *Eur. J. Nucl. Med. Mol. Imaging* **2018**, *45*, 1626–1635. [CrossRef]
108. Chen, M.K.; Baidoo, K.; Verina, T.; Guilarte, T.R. Peripheral benzodiazepine receptor imaging in CNS demyelination: Functional implications of anatomical and cellular localization. *Brain* **2004**, *127*, 1379–1392. [CrossRef]
109. Papadopoulos, V.; Baraldi, M.; Guilarte, T.R.; Knudsen, T.B.; Lacapere, J.J.; Lindemann, P.; Norenberg, M.D.; Nutt, D.; Weizman, A.; Zhang, M.R.; et al. Translocator protein (18kDa): New nomenclature for the peripheral-type benzodiazepine receptor based on its structure and molecular function. *Trends Pharmacol. Sci.* **2006**, *27*, 402–409. [CrossRef]
110. Schweitzer, P.J.; Fallon, B.A.; Mann, J.J.; Kumar, J.S. PET tracers for the peripheral benzodiazepine receptor and uses thereof. *Drug Discov. Today* **2010**, *15*, 933–942. [CrossRef]
111. Su, Z.; Roncaroli, F.; Durrenberger, P.F.; Coope, D.J.; Karabatsou, K.; Hinz, R.; Thompson, G.; Turkheimer, F.E.; Janczar, K.; Du Plessis, D.; et al. The 18-kDa mitochondrial translocator protein in human gliomas: An 11C-(R)PK11195 PET imaging and neuropathology study. *J. Nucl. Med.* **2015**, *56*, 512–517. [CrossRef] [PubMed]
112. Su, Z.; Herholz, K.; Gerhard, A.; Roncaroli, F.; Du Plessis, D.; Jackson, A.; Turkheimer, F.; Hinz, R. [(1)(1)C]-(R)PK11195 tracer kinetics in the brain of glioma patients and a comparison of two referencing approaches. *Eur. J. Nucl. Med. Mol. Imaging* **2013**, *40*, 1406–1419. [CrossRef] [PubMed]
113. Albert, N.L.; Unterrainer, M.; Fleischmann, D.F.; Lindner, S.; Vettermann, F.; Brunegraf, A.; Vomacka, L.; Brendel, M.; Wenter, V.; Wetzel, C.; et al. TSPO PET for glioma imaging using the novel ligand (18)F-GE-180: First results in patients with glioblastoma. *Eur. J. Nucl. Med. Mol. Imaging* **2017**, *44*, 2230–2238. [CrossRef] [PubMed]
114. Vomacka, L.; Albert, N.L.; Lindner, S.; Unterrainer, M.; Mahler, C.; Brendel, M.; Ermoschkin, L.; Gosewisch, A.; Brunegraf, A.; Buckley, C.; et al. TSPO imaging using the novel PET ligand [(18)F]GE-180: Quantification approaches in patients with multiple sclerosis. *EJNMMI Res.* **2017**, *7*, 89. [CrossRef] [PubMed]
115. Unterrainer, M.; Mahler, C.; Vomacka, L.; Lindner, S.; Havla, J.; Brendel, M.; Boning, G.; Ertl-Wagner, B.; Kumpfel, T.; Milenkovic, V.M.; et al. TSPO PET with [(18)F]GE-180 sensitively detects focal neuroinflammation in patients with relapsing-remitting multiple sclerosis. *Eur. J. Nucl. Med. Mol. Imaging* **2018**, *45*, 1423–1431. [CrossRef] [PubMed]
116. Sridharan, S.; Raffel, J.; Nandoskar, A.; Record, C.; Brooks, D.J.; Owen, D.; Sharp, D.; Muraro, P.A.; Gunn, R.; Nicholas, R. Confirmation of Specific Binding of the 18-kDa Translocator Protein (TSPO) Radioligand [(18)F]GE-180: A Blocking Study Using XBD173 in Multiple Sclerosis Normal Appearing White and Grey Matter. *Mol. Imaging Biol.* **2019**, *21*, 935–944. [CrossRef]
117. Unterrainer, M.; Fleischmann, D.F.; Diekmann, C.; Vomacka, L.; Lindner, S.; Vettermann, F.; Brendel, M.; Wenter, V.; Ertl-Wagner, B.; Herms, J.; et al. Comparison of (18)F-GE-180 and dynamic (18)F-FET PET in high grade glioma: A double-tracer pilot study. *Eur. J. Nucl. Med. Mol. Imaging* **2019**, *46*, 580–590. [CrossRef]
118. Shields, A.F.; Grierson, J.R.; Dohmen, B.M.; Machulla, H.J.; Stayanoff, J.C.; Lawhorn-Crews, J.M.; Obradovich, J.E.; Muzik, O.; Mangner, T.J. Imaging proliferation in vivo with [F-18]FLT and positron emission tomography. *Nat. Med.* **1998**, *4*, 1334–1336. [CrossRef]
119. van Waarde, A.; Elsinga, P.H. Proliferation markers for the differential diagnosis of tumor and inflammation. *Curr. Pharm. Des.* **2008**, *14*, 3326–3339. [CrossRef]
120. Saga, T.; Kawashima, H.; Araki, N.; Takahashi, J.A.; Nakashima, Y.; Higashi, T.; Oya, N.; Mukai, T.; Hojo, M.; Hashimoto, N.; et al. Evaluation of primary brain tumors with FLT-PET: Usefulness and limitations. *Clin. Nucl. Med.* **2006**, *31*, 774–780. [CrossRef]
121. Jacobs, A.H.; Thomas, A.; Kracht, L.W.; Li, H.; Dittmar, C.; Garlip, G.; Galldiks, N.; Klein, J.C.; Sobesky, J.; Hilker, R.; et al. 18F-fluoro-L-thymidine and 11C-methylmethionine as markers of increased transport and proliferation in brain tumors. *J. Nucl. Med.* **2005**, *46*, 1948–1958. [PubMed]

122. Li, Z.; Yu, Y.; Zhang, H.; Xu, G.; Chen, L. A meta-analysis comparing 18F-FLT PET with 18F-FDG PET for assessment of brain tumor recurrence. *Nucl. Med. Commun.* **2015**, *36*, 695–701. [CrossRef] [PubMed]
123. Chen, W.; Delaloye, S.; Silverman, D.H.; Geist, C.; Czernin, J.; Sayre, J.; Satyamurthy, N.; Pope, W.; Lai, A.; Phelps, M.E.; et al. Predicting treatment response of malignant gliomas to bevacizumab and irinotecan by imaging proliferation with [18F] fluorothymidine positron emission tomography: A pilot study. *J. Clin. Oncol.* **2007**, *25*, 4714–4721. [CrossRef] [PubMed]
124. Schwarzenberg, J.; Czernin, J.; Cloughesy, T.F.; Ellingson, B.M.; Pope, W.B.; Geist, C.; Dahlbom, M.; Silverman, D.H.; Satyamurthy, N.; Phelps, M.E.; et al. 3′-deoxy-3′-18F-fluorothymidine PET and MRI for early survival predictions in patients with recurrent malignant glioma treated with bevacizumab. *J. Nucl. Med.* **2012**, *53*, 29–36. [CrossRef] [PubMed]
125. Harris, R.J.; Cloughesy, T.F.; Pope, W.B.; Nghiemphu, P.L.; Lai, A.; Zaw, T.; Czernin, J.; Phelps, M.E.; Chen, W.; Ellingson, B.M. 18F-FDOPA and 18F-FLT positron emission tomography parametric response maps predict response in recurrent malignant gliomas treated with bevacizumab. *Neuro Oncol.* **2012**, *14*, 1079–1089. [CrossRef]
126. Nguyen, N.C.; Yee, M.K.; Tuchayi, A.M.; Kirkwood, J.M.; Tawbi, H.; Mountz, J.M. Targeted Therapy and Immunotherapy Response Assessment with F-18 Fluorothymidine Positron-Emission Tomography/Magnetic Resonance Imaging in Melanoma Brain Metastasis: A Pilot Study. *Front. Oncol.* **2018**, *8*, 18. [CrossRef]
127. Bell, C.; Dowson, N.; Fay, M.; Thomas, P.; Puttick, S.; Gal, Y.; Rose, S. Hypoxia imaging in gliomas with 18F-fluoromisonidazole PET: Toward clinical translation. *Semin. Nucl. Med.* **2015**, *45*, 136–150. [CrossRef]
128. Rasey, J.S.; Koh, W.J.; Evans, M.L.; Peterson, L.M.; Lewellen, T.K.; Graham, M.M.; Krohn, K.A. Quantifying regional hypoxia in human tumors with positron emission tomography of [18F]fluoromisonidazole: A pretherapy study of 37 patients. *Int. J. Radiat. Oncol. Biol. Phys.* **1996**, *36*, 417–428. [CrossRef]
129. Rajendran, J.G.; Krohn, K.A. F-18 fluoromisonidazole for imaging tumor hypoxia: Imaging the microenvironment for personalized cancer therapy. *Semin. Nucl. Med.* **2015**, *45*, 151–162. [CrossRef]
130. Swanson, K.R.; Chakraborty, G.; Wang, C.H.; Rockne, R.; Harpold, H.L.; Muzi, M.; Adamsen, T.C.; Krohn, K.A.; Spence, A.M. Complementary but distinct roles for MRI and 18F-fluoromisonidazole PET in the assessment of human glioblastomas. *J. Nucl. Med.* **2009**, *50*, 36–44. [CrossRef]
131. Kawai, N.; Maeda, Y.; Kudomi, N.; Miyake, K.; Okada, M.; Yamamoto, Y.; Nishiyama, Y.; Tamiya, T. Correlation of biological aggressiveness assessed by 11C-methionine PET and hypoxic burden assessed by 18F-fluoromisonidazole PET in newly diagnosed glioblastoma. *Eur. J. Nucl. Med. Mol. Imaging* **2011**, *38*, 441–450. [CrossRef] [PubMed]
132. Barajas, R.F.; Krohn, K.A.; Link, J.M.; Hawkins, R.A.; Clarke, J.L.; Pampaloni, M.H.; Cha, S. Glioma FMISO PET/MR Imaging Concurrent with Antiangiogenic Therapy: Molecular Imaging as a Clinical Tool in the Burgeoning Era of Personalized Medicine. *Biomedicines* **2016**, *4*, 24. [CrossRef] [PubMed]
133. Yamaguchi, S.; Hirata, K.; Toyonaga, T.; Kobayashi, K.; Ishi, Y.; Motegi, H.; Kobayashi, H.; Shiga, T.; Tamaki, N.; Terasaka, S.; et al. Change in 18F-Fluoromisonidazole PET Is an Early Predictor of the Prognosis in the Patients with Recurrent High-Grade Glioma Receiving Bevacizumab Treatment. *PLoS ONE* **2016**, *11*, e0167917. [CrossRef] [PubMed]
134. Piert, M.; Machulla, H.J.; Picchio, M.; Reischl, G.; Ziegler, S.; Kumar, P.; Wester, H.J.; Beck, R.; McEwan, A.J.; Wiebe, L.I.; et al. Hypoxia-specific tumor imaging with 18F-fluoroazomycin arabinoside. *J. Nucl. Med.* **2005**, *46*, 106–113.
135. Postema, E.J.; McEwan, A.J.; Riauka, T.A.; Kumar, P.; Richmond, D.A.; Abrams, D.N.; Wiebe, L.I. Initial results of hypoxia imaging using 1-alpha-D: -(5-deoxy-5-[18F]-fluoroarabinofuranosyl)-2-nitroimidazole (18F-FAZA). *Eur. J. Nucl. Med. Mol. Imaging* **2009**, *36*, 1565–1573. [CrossRef]
136. Mapelli, P.; Zerbetto, F.; Incerti, E.; Conte, G.M.; Bettinardi, V.; Fallanca, F.; Anzalone, N.; Di Muzio, N.; Gianolli, L.; Picchio, M. 18F-FAZA PET/CT Hypoxia Imaging of High-Grade Glioma Before and After Radiotherapy. *Clin. Nucl. Med.* **2017**, *42*, e525–e526. [CrossRef]
137. Ter-Pogossian, M.M.; Eichling, J.O.; Davis, D.O.; Welch, M.J. The measure in vivo of regional cerebral oxygen utilization by means of oxyhemoglobin labeled with radioactive oxygen-15. *J. Clin. Investig.* **1970**, *49*, 381–391. [CrossRef]

138. Ludemann, L.; Warmuth, C.; Plotkin, M.; Forschler, A.; Gutberlet, M.; Wust, P.; Amthauer, H. Brain tumor perfusion: Comparison of dynamic contrast enhanced magnetic resonance imaging using T1, T2, and T2* contrast, pulsed arterial spin labeling, and H2(15)O positron emission tomography. *Eur. J. Radiol.* **2009**, *70*, 465–474. [CrossRef]
139. Gruner, J.M.; Paamand, R.; Kosteljanetz, M.; Broholm, H.; Hojgaard, L.; Law, I. Brain perfusion CT compared with (1)(5)O-H(2)O PET in patients with primary brain tumours. *Eur. J. Nucl. Med. Mol. Imaging* **2012**, *39*, 1691–1701. [CrossRef]
140. Ogawa, T.; Uemura, K.; Shishido, F.; Yamaguchi, T.; Murakami, M.; Inugami, A.; Kanno, I.; Sasaki, H.; Kato, T.; Hirata, K.; et al. Changes of cerebral blood flow, and oxygen and glucose metabolism following radiochemotherapy of gliomas: A PET study. *J. Comput. Assist. Tomogr.* **1988**, *12*, 290–297. [CrossRef]
141. Mineura, K.; Yasuda, T.; Kowada, M.; Ogawa, T.; Shishido, F.; Uemura, K. Positron emission tomographic evaluation of radiochemotherapeutic effect on regional cerebral hemocirculation and metabolism in patients with gliomas. *J. Neurooncol.* **1987**, *5*, 277–285. [CrossRef]
142. Jain, R.K.; di Tomaso, E.; Duda, D.G.; Loeffler, J.S.; Sorensen, A.G.; Batchelor, T.T. Angiogenesis in brain tumours. *Nat. Rev. Neurosci.* **2007**, *8*, 610–622. [CrossRef] [PubMed]
143. Levin, V.A.; Bidaut, L.; Hou, P.; Kumar, A.J.; Wefel, J.S.; Bekele, B.N.; Grewal, J.; Prabhu, S.; Loghin, M.; Gilbert, M.R.; et al. Randomized double-blind placebo-controlled trial of bevacizumab therapy for radiation necrosis of the central nervous system. *Int. J. Radiat. Oncol. Biol. Phys.* **2011**, *79*, 1487–1495. [CrossRef] [PubMed]
144. Jansen, M.H.; Veldhuijzen van Zanten, S.E.M.; van Vuurden, D.G.; Huisman, M.C.; Vugts, D.J.; Hoekstra, O.S.; van Dongen, G.A.; Kaspers, G.L. Molecular Drug Imaging: (89)Zr-Bevacizumab PET in Children with Diffuse Intrinsic Pontine Glioma. *J. Nucl. Med.* **2017**, *58*, 711–716. [CrossRef] [PubMed]
145. Veldhuijzen van Zanten, S.E.M.; Sewing, A.C.P.; van Lingen, A.; Hoekstra, O.S.; Wesseling, P.; Meel, M.H.; van Vuurden, D.G.; Kaspers, G.J.L.; Hulleman, E.; Bugiani, M. Multiregional Tumor Drug-Uptake Imaging by PET and Microvascular Morphology in End-Stage Diffuse Intrinsic Pontine Glioma. *J. Nucl. Med.* **2018**, *59*, 612–615. [CrossRef]
146. Hsu, A.R.; Cai, W.; Veeravagu, A.; Mohamedali, K.A.; Chen, K.; Kim, S.; Vogel, H.; Hou, L.C.; Tse, V.; Rosenblum, M.G.; et al. Multimodality molecular imaging of glioblastoma growth inhibition with vasculature-targeting fusion toxin VEGF121/rGel. *J. Nucl. Med.* **2007**, *48*, 445–454.
147. Chen, K.; Cai, W.; Li, Z.B.; Wang, H.; Chen, X. Quantitative PET imaging of VEGF receptor expression. *Mol. Imaging Biol.* **2009**, *11*, 15–22. [CrossRef]
148. Cai, W.; Chen, K.; Mohamedali, K.A.; Cao, Q.; Gambhir, S.S.; Rosenblum, M.G.; Chen, X. PET of vascular endothelial growth factor receptor expression. *J. Nucl. Med.* **2006**, *47*, 2048–2056.
149. Rainer, E.; Wang, H.; Traub-Weidinger, T.; Widhalm, G.; Fueger, B.; Chang, J.; Zhu, Z.; Marosi, C.; Haug, A.; Hacker, M.; et al. The prognostic value of [(123)I]-vascular endothelial growth factor ([(123)I]-VEGF) in glioma. *Eur. J. Nucl. Med. Mol. Imaging* **2018**, *45*, 2396–2403. [CrossRef]
150. Wu, Y.L.; Ahn, M.J.; Garassino, M.C.; Han, J.Y.; Katakami, N.; Kim, H.R.; Hodge, R.; Kaur, P.; Brown, A.P.; Ghiorghiu, D.; et al. CNS Efficacy of Osimertinib in Patients With T790M-Positive Advanced Non-Small-Cell Lung Cancer: Data From a Randomized Phase III Trial (AURA3). *J. Clin. Oncol.* **2018**, *36*, 2702–2709. [CrossRef]
151. Pusch, S.; Krausert, S.; Fischer, V.; Balss, J.; Ott, M.; Schrimpf, D.; Capper, D.; Sahm, F.; Eisel, J.; Beck, A.C.; et al. Pan-mutant IDH1 inhibitor BAY 1436032 for effective treatment of IDH1 mutant astrocytoma in vivo. *Acta Neuropathol.* **2017**, *133*, 629–644. [CrossRef] [PubMed]
152. Mellinghoff, I.K.; Penas-Prado, M.; Peters, K.B.; Cloughesy, T.F.; Burris, H.A.; Maher, E.A.; Janku, F.; Cote, G.M.; Fuente, M.I.D.L.; Clarke, J.; et al. Phase 1 study of AG-881, an inhibitor of mutant IDH1/IDH2, in patients with advanced IDH-mutant solid tumors, including glioma. *J. Clin. Oncol.* **2018**, *36*, 2002. [CrossRef]
153. Bublil, E.M.; Yarden, Y. The EGF receptor family: Spearheading a merger of signaling and therapeutics. *Curr. Opin. Cell Biol.* **2007**, *19*, 124–134. [CrossRef] [PubMed]
154. Hatanpaa, K.J.; Burma, S.; Zhao, D.; Habib, A.A. Epidermal growth factor receptor in glioma: Signal transduction, neuropathology, imaging, and radioresistance. *Neoplasia* **2010**, *12*, 675–684. [CrossRef]

155. Kelly, W.J.; Shah, N.J.; Subramaniam, D.S. Management of Brain Metastases in Epidermal Growth Factor Receptor Mutant Non-Small-Cell Lung Cancer. *Front. Oncol.* **2018**, *8*, 208. [CrossRef]
156. Weber, B.; Winterdahl, M.; Memon, A.; Sorensen, B.S.; Keiding, S.; Sorensen, L.; Nexo, E.; Meldgaard, P. Erlotinib accumulation in brain metastases from non-small cell lung cancer: Visualization by positron emission tomography in a patient harboring a mutation in the epidermal growth factor receptor. *J. Thorac. Oncol.* **2011**, *6*, 1287–1289. [CrossRef]
157. Tang, Y.; Hu, Y.; Liu, W.; Chen, L.; Zhao, Y.; Ma, H.; Yang, J.; Yang, Y.; Liao, J.; Cai, J.; et al. A radiopharmaceutical [(89)Zr]Zr-DFO-nimotuzumab for immunoPET with epidermal growth factor receptor expression in vivo. *Nucl. Med. Biol.* **2019**, *70*, 23–31. [CrossRef]
158. Sun, J.; Cai, L.; Zhang, K.; Zhang, A.; Pu, P.; Yang, W.; Gao, S. A pilot study on EGFR-targeted molecular imaging of PET/CT With 11C-PD153035 in human gliomas. *Clin. Nucl. Med.* **2014**, *39*, e20–e26. [CrossRef]
159. Ulaner, G.A.; Lyashchenko, S.K.; Riedl, C.; Ruan, S.; Zanzonico, P.B.; Lake, D.; Jhaveri, K.; Zeglis, B.; Lewis, J.S.; O'Donoghue, J.A. First-in-Human Human Epidermal Growth Factor Receptor 2-Targeted Imaging Using (89)Zr-Pertuzumab PET/CT: Dosimetry and Clinical Application in Patients with Breast Cancer. *J. Nucl. Med.* **2018**, *59*, 900–906. [CrossRef]
160. Tamura, K.; Kurihara, H.; Yonemori, K.; Tsuda, H.; Suzuki, J.; Kono, Y.; Honda, N.; Kodaira, M.; Yamamoto, H.; Yunokawa, M.; et al. 64Cu-DOTA-trastuzumab PET imaging in patients with HER2-positive breast cancer. *J. Nucl. Med.* **2013**, *54*, 1869–1875. [CrossRef]
161. van der Veen, E.L.; Bensch, F.; Glaudemans, A.; Lub-de Hooge, M.N.; de Vries, E.G.E. Molecular imaging to enlighten cancer immunotherapies and underlying involved processes. *Cancer Treat. Rev.* **2018**, *70*, 232–244. [CrossRef] [PubMed]
162. Decazes, P.; Bohn, P. Immunotherapy by Immune Checkpoint Inhibitors and Nuclear Medicine Imaging: Current and Future Applications. *Cancers* **2020**, *12*, 371. [CrossRef] [PubMed]
163. Cole, E.L.; Kim, J.; Donnelly, D.J.; Smith, R.A.; Cohen, D.; Lafont, V.; Morin, P.E.; Huang, R.Y.; Chow, P.L.; Hayes, W.; et al. Radiosynthesis and preclinical PET evaluation of (89)Zr-nivolumab (BMS-936558) in healthy non-human primates. *Bioorg. Med. Chem.* **2017**, *25*, 5407–5414. [CrossRef]
164. Bensch, F.; van der Veen, E.L.; Lub-de Hooge, M.N.; Jorritsma-Smit, A.; Boellaard, R.; Kok, I.C.; Oosting, S.F.; Schroder, C.P.; Hiltermann, T.J.N.; van der Wekken, A.J.; et al. (89)Zr-atezolizumab imaging as a non-invasive approach to assess clinical response to PD-L1 blockade in cancer. *Nat. Med.* **2018**, *24*, 1852–1858. [CrossRef] [PubMed]
165. Niemeijer, A.N.; Leung, D.; Huisman, M.C.; Bahce, I.; Hoekstra, O.S.; van Dongen, G.; Boellaard, R.; Du, S.; Hayes, W.; Smith, R.; et al. Whole body PD-1 and PD-L1 positron emission tomography in patients with non-small-cell lung cancer. *Nat. Commun.* **2018**, *9*, 4664. [CrossRef] [PubMed]
166. Donnelly, D.J.; Smith, R.A.; Morin, P.; Lipovsek, D.; Gokemeijer, J.; Cohen, D.; Lafont, V.; Tran, T.; Cole, E.L.; Wright, M.; et al. Synthesis and Biologic Evaluation of a Novel (18)F-Labeled Adnectin as a PET Radioligand for Imaging PD-L1 Expression. *J. Nucl. Med.* **2018**, *59*, 529–535. [CrossRef]
167. Tumeh, P.C.; Harview, C.L.; Yearley, J.H.; Shintaku, I.P.; Taylor, E.J.; Robert, L.; Chmielowski, B.; Spasic, M.; Henry, G.; Ciobanu, V.; et al. PD-1 blockade induces responses by inhibiting adaptive immune resistance. *Nature* **2014**, *515*, 568–571. [CrossRef]
168. Pandit-Taskar, N.; Postow, M.; Hellmann, M.; Harding, J.; Barker, C.; O'Donoghue, J.; Ziolkowska, M.; Ruan, S.; Lyashchenko, S.; Tsai, F.; et al. First-in-human imaging with (89)Zr-Df-IAB22M2C anti-CD8 minibody in patients with solid malignancies: Preliminary pharmacokinetics, biodistribution, and lesion targeting. *J. Nucl. Med.* **2019**. [CrossRef]
169. Antonios, J.P.; Soto, H.; Everson, R.G.; Moughon, D.L.; Wang, A.C.; Orpilla, J.; Radu, C.; Ellingson, B.M.; Lee, J.T.; Cloughesy, T.; et al. Detection of immune responses after immunotherapy in glioblastoma using PET and MRI. *Proc. Natl. Acad. Sci. USA* **2017**, *114*, 10220–10225. [CrossRef]
170. Nair-Gill, E.D.; Shu, C.J.; Radu, C.G.; Witte, O.N. Non-invasive imaging of adaptive immunity using positron emission tomography. *Immunol. Rev.* **2008**, *221*, 214–228. [CrossRef]
171. Min, J.J.; Iyer, M.; Gambhir, S.S. Comparison of [18F]FHBG and [14C]FIAU for imaging of HSV1-tk reporter gene expression: Adenoviral infection vs stable transfection. *Eur. J. Nucl. Med. Mol. Imaging* **2003**, *30*, 1547–1560. [CrossRef] [PubMed]

172. Keu, K.V.; Witney, T.H.; Yaghoubi, S.; Rosenberg, J.; Kurien, A.; Magnusson, R.; Williams, J.; Habte, F.; Wagner, J.R.; Forman, S.; et al. Reporter gene imaging of targeted T cell immunotherapy in recurrent glioma. *Sci. Transl. Med.* **2017**, *9*. [CrossRef] [PubMed]
173. Yan, H.; Parsons, D.W.; Jin, G.; McLendon, R.; Rasheed, B.A.; Yuan, W.; Kos, I.; Batinic-Haberle, I.; Jones, S.; Riggins, G.J.; et al. IDH1 and IDH2 mutations in gliomas. *N. Engl. J. Med.* **2009**, *360*, 765–773. [CrossRef] [PubMed]
174. Dang, L.; White, D.W.; Gross, S.; Bennett, B.D.; Bittinger, M.A.; Driggers, E.M.; Fantin, V.R.; Jang, H.G.; Jin, S.; Keenan, M.C.; et al. Cancer-associated IDH1 mutations produce 2-hydroxyglutarate. *Nature* **2009**, *462*, 739–744. [CrossRef]
175. Lu, C.; Ward, P.S.; Kapoor, G.S.; Rohle, D.; Turcan, S.; Abdel-Wahab, O.; Edwards, C.R.; Khanin, R.; Figueroa, M.E.; Melnick, A.; et al. IDH mutation impairs histone demethylation and results in a block to cell differentiation. *Nature* **2012**, *483*, 474–478. [CrossRef]
176. Platten, M.; Schilling, D.; Bunse, L.; Wick, A.; Bunse, T.; Riehl, D.; Karapanagiotou-Schenkel, I.; Harting, I.; Sahm, F.; Schmitt, A.; et al. A mutation-specific peptide vaccine targeting IDH1R132H in patients with newly diagnosed malignant astrocytomas: A first-in-man multicenter phase I clinical trial of the German Neurooncology Working Group (NOA-16). *J. Clin. Oncol.* **2018**, *36*, 2001. [CrossRef]
177. Suh, C.H.; Kim, H.S.; Paik, W.; Choi, C.; Ryu, K.H.; Kim, D.; Woo, D.C.; Park, J.E.; Jung, S.C.; Choi, C.G.; et al. False-Positive Measurement at 2-Hydroxyglutarate MR Spectroscopy in Isocitrate Dehydrogenase Wild-Type Glioblastoma: A Multifactorial Analysis. *Radiology* **2019**, *291*, 752–762. [CrossRef]
178. Choi, C.; Ganji, S.K.; DeBerardinis, R.J.; Hatanpaa, K.J.; Rakheja, D.; Kovacs, Z.; Yang, X.L.; Mashimo, T.; Raisanen, J.M.; Marin-Valencia, I.; et al. 2-hydroxyglutarate detection by magnetic resonance spectroscopy in IDH-mutated patients with gliomas. *Nat. Med.* **2012**, *18*, 624–629. [CrossRef]
179. Chitneni, S.K.; Reitman, Z.J.; Gooden, D.M.; Yan, H.; Zalutsky, M.R. Radiolabeled inhibitors as probes for imaging mutant IDH1 expression in gliomas: Synthesis and preliminary evaluation of labeled butyl-phenyl sulfonamide analogs. *Eur. J. Med. Chem.* **2016**, *119*, 218–230. [CrossRef]
180. Chitneni, S.K.; Yan, H.; Zalutsky, M.R. Synthesis and Evaluation of a (18)F-Labeled Triazinediamine Analogue for Imaging Mutant IDH1 Expression in Gliomas by PET. *ACS Med. Chem. Lett.* **2018**, *9*, 606–611. [CrossRef]
181. Chitneni, S.K.; Reitman, Z.J.; Spicehandler, R.; Gooden, D.M.; Yan, H.; Zalutsky, M.R. Synthesis and evaluation of radiolabeled AGI-5198 analogues as candidate radiotracers for imaging mutant IDH1 expression in tumors. *Bioorg. Med. Chem. Lett.* **2018**, *28*, 694–699. [CrossRef] [PubMed]
182. Koyasu, S.; Shimizu, Y.; Morinibu, A.; Saga, T.; Nakamoto, Y.; Togashi, K.; Harada, H. Increased (14)C-acetate accumulation in IDH-mutated human glioblastoma: Implications for detecting IDH-mutated glioblastoma with (11)C-acetate PET imaging. *J. Neurooncol.* **2019**, *145*, 441–447. [CrossRef] [PubMed]
183. Burger, J.A.; Kipps, T.J. CXCR4: A key receptor in the crosstalk between tumor cells and their microenvironment. *Blood* **2006**, *107*, 1761–1767. [CrossRef]
184. Bian, X.W.; Yang, S.X.; Chen, J.H.; Ping, Y.F.; Zhou, X.D.; Wang, Q.L.; Jiang, X.F.; Gong, W.; Xiao, H.L.; Du, L.L.; et al. Preferential expression of chemokine receptor CXCR4 by highly malignant human gliomas and its association with poor patient survival. *Neurosurgery* **2007**, *61*, 570–579. [CrossRef] [PubMed]
185. Bailly, C.; Vidal, A.; Bonnemaire, C.; Kraeber-Bodere, F.; Cherel, M.; Pallardy, A.; Rousseau, C.; Garcion, E.; Lacoeuille, F.; Hindre, F.; et al. Potential for Nuclear Medicine Therapy for Glioblastoma Treatment. *Front. Pharmacol.* **2019**, *10*, 772. [CrossRef]
186. Lapa, C.; Luckerath, K.; Kleinlein, I.; Monoranu, C.M.; Linsenmann, T.; Kessler, A.F.; Rudelius, M.; Kropf, S.; Buck, A.K.; Ernestus, R.I.; et al. (68)Ga-Pentixafor-PET/CT for Imaging of Chemokine Receptor 4 Expression in Glioblastoma. *Theranostics* **2016**, *6*, 428–434. [CrossRef]
187. Wernicke, A.G.; Edgar, M.A.; Lavi, E.; Liu, H.; Salerno, P.; Bander, N.H.; Gutin, P.H. Prostate-specific membrane antigen as a potential novel vascular target for treatment of glioblastoma multiforme. *Arch. Pathol. Lab. Med.* **2011**, *135*, 1486–1489. [CrossRef]
188. Schwenck, J.; Tabatabai, G.; Skardelly, M.; Reischl, G.; Beschorner, R.; Pichler, B.; la Fougere, C. In vivo visualization of prostate-specific membrane antigen in glioblastoma. *Eur. J. Nucl. Med. Mol. Imaging* **2015**, *42*, 170–171. [CrossRef]

189. Unterrainer, M.; Niyazi, M.; Ruf, V.; Bartenstein, P.; Albert, N.L. The endothelial prostate-specific membrane antigen is highly expressed in gliosarcoma and visualized by [68Ga]-PSMA-11 PET: A theranostic outlook for brain tumor patients? *Neuro Oncol.* **2017**, *19*, 1698–1699. [CrossRef]
190. Rahbar, K.; Ahmadzadehfar, H.; Kratochwil, C.; Haberkorn, U.; Schafers, M.; Essler, M.; Baum, R.P.; Kulkarni, H.R.; Schmidt, M.; Drzezga, A.; et al. German Multicenter Study Investigating 177Lu-PSMA-617 Radioligand Therapy in Advanced Prostate Cancer Patients. *J. Nucl. Med.* **2017**, *58*, 85–90. [CrossRef]

© 2020 by the authors. Licensee MDPI, Basel, Switzerland. This article is an open access article distributed under the terms and conditions of the Creative Commons Attribution (CC BY) license (http://creativecommons.org/licenses/by/4.0/).

Review

Approaches to PET Imaging of Glioblastoma

Lindsey R. Drake [1,2,*], Ansel T. Hillmer [1,2,3,4] and Zhengxin Cai [1,2]

1. Yale PET Center, Yale University School of Medicine, New Haven, CT 06511, USA; ansel.hillmer@yale.edu (A.T.H.); jason.cai@yale.edu (Z.C.)
2. Department of Radiology and Bioimaging Sciences, Yale University School of Medicine, New Haven, CT 06511, USA
3. Department of Psychiatry, Yale University School of Medicine, New Haven, CT 06511, USA
4. Department of Biomedical Engineering, Yale School of Engineering and Applied Science, New Haven, CT 06511, USA
* Correspondence: lindsey.drake@yale.edu; Tel.: +1-203-737-2978

Academic Editor: Peter Brust
Received: 11 December 2019; Accepted: 23 January 2020; Published: 28 January 2020

Abstract: Glioblastoma multiforme (GBM) is the deadliest type of brain tumor, affecting approximately three in 100,000 adults annually. Positron emission tomography (PET) imaging provides an important non-invasive method of measuring biochemically specific targets at GBM lesions. These powerful data can characterize tumors, predict treatment effectiveness, and monitor treatment. This review will discuss the PET imaging agents that have already been evaluated in GBM patients so far, and new imaging targets with promise for future use. Previously used PET imaging agents include the tracers for markers of proliferation ([^{11}C]methionine; [^{18}F]fluoro-ethyl-L-tyrosine, [^{18}F]Fluorodopa, [^{18}F]fluoro-thymidine, and [^{18}F]clofarabine), hypoxia sensing ([^{18}F]FMISO, [^{18}F]FET-NIM, [^{18}F]EF5, [^{18}F]HX4, and [^{64}Cu]ATSM), and ligands for inflammation. As cancer therapeutics evolve toward personalized medicine and therapies centered on tumor biomarkers, the development of complimentary selective PET agents can dramatically enhance these efforts. Newer biomarkers for GBM PET imaging are discussed, with some already in use for PET imaging other cancers and neurological disorders. These targets include Sigma 1, Sigma 2, programmed death ligand 1, poly-ADP-ribose polymerase, and isocitrate dehydrogenase. For GBM, these imaging agents come with additional considerations such as blood–brain barrier penetration, quantitative modeling approaches, and nonspecific binding.

Keywords: PET imaging; GBM; biomarkers; Sigma 1; Sigma 2; PD-L1; PARP; IDH

---

1. Introduction

Glioblastoma Multiforme (GBM) is a fast growing, invasive brain tumor that typically results in death in the first 15 months after diagnosis [1]. It develops from glial cells, astrocytes or oligodendrocytes, and can evolve from lower-grade tumors or de novo. Previously, GBM was characterized as 'grade IV' astrocytoma. Recently, the World Health Organization (WHO) updated the classification of brain tumors to include genotypic markers, building on the histological markers considered previously [2]. Glioblastoma can be classified by a single nucleotide polymorphism in the isocitrate dehydrogenase (IDH) gene as wild-type or mutant. Approximately 10% of glioblastomas are IDH-mutant [2]. IDH-mutant status weakly predicts long-term survival (over 3 years post diagnosis) [3]. GBM tumors are heterogenous in location (with 25%–43% incidence in frontal lobes), histopathology, and the tumor microenvironment [4]. The first line of treatment for GBM is surgery, followed by radiation and chemotherapy [1]. Temozolomide, a DNA alkylating agent is often used for chemotherapy. In 2015, the vascular endothelial growth factor inhibitor Bevacizumab was fast-tracked for use in GBM after demonstrating efficacy in shrinking or halting tumor growth. However, it has failed to show improvement in overall survival [5]. Patients with GBMs have a very low survival rate with very few treatment options, making this a particularly acute health challenge.

Medical imaging provides critical information for diagnosing, staging, and monitoring the treatment of GBM. While formal diagnosis relies on histopathology and genetic markers for grading, structural magnetic resonance images (MRIs) are routinely acquired and can be used in guiding surgery. Additional structural MRI methods can accurately classify and grade tumors with high accuracy, though it has not been adopted yet as common practice [6]. Positron emission tomography (PET) imaging provides important complementary information to anatomical MRI data. In this functional type of imaging, biochemical information about the tumor and the tissue surrounding it can be measured non-invasively. GBMs typically are fast growing, giving an important role for specific PET radioligands to quantify proliferation. PET imaging is also uniquely positioned to identify ideal cases for targeted treatments and evaluate treatment progression.

This article provides an overview of the novel imaging tracers used in PET imaging of brain tumors. Discussion includes the strengths, limitations, and pitfalls of individual imaging biomarker strategies, and general challenges associated with PET imaging of brain tumors. We first provide a brief overview of established PET imaging biomarkers (glycolysis, amino acid metabolism, DNA replication, hypoxia, and inflammation), followed by newer imaging targets (Sigma 1/ 2, programmed death ligand 1, poly-ADP-ribose polymerase, and isocitrate dehydrogenase) with promise to image glioblastoma lesions. None of these biomarkers are unique to glioblastoma, though their presence has been found in resected brain tumors. This work concludes with important quantitative considerations for use of these imaging biomarkers in the evaluation and treatment of GBM patients.

2. Overview of PET Imaging Agents for Brain Tumor

2.1. Sustained Proliferation Markers: Glycolysis, Amino Acid Transportation, and DNA Replication

The classic approach to imaging tumors in general, and in application to GBM, has been to probe the functional necessities of proliferation. These necessities include glucose metabolism, protein synthesis, and DNA replication. From a biochemical prospective, these functions highlight the 'building block' small molecules that compose macromolecules: sugars, nucleotide bases, and amino acids.

Radionuclide-labeled forms of these building blocks have been employed to study these functions with PET imaging. The gold standard of most cancer imaging is [^{18}F]FDG (**1**), a fluorine-18 glucose analogue. This radiotracer is actively taken up by the glucose transporter and participates in the first step of glucose metabolism (phosphorylation), then becomes trapped in the cell [7]. [^{18}F]FDG PET allows for the functional imaging of glucose metabolism, a relative gold mine of information in most cancers. However, the brain has naturally high uptake of [^{18}F]FDG, which complicates interpretation of GBM lesions near gray matter. Further efforts to image proliferation through the 'building block' strategy include neutral amino acid analogues ([^{11}C]methionine (**2**); [^{18}F]fluoro-ethyl-L-tyrosine (**3**); [^{18}F]Fluorodopa (**4**)) and deoxynucleoside bases ([^{18}F]fluoro-thymidine (**5**); and [^{18}F]clofarabine (**6**)) (Figure 1).

[^{11}C]Methionine ([^{11}C]MET) was developed shortly after [^{18}F]FDG [8,9]; it was considered a valuable tracer because methionine, an essential amino acid, can be used in protein synthesis. Both L-and D-isomers were synthesized, and no difference in accumulation was observed [10]. This lack of selectivity between stereoisomers indicated that [^{11}C]MET was not being incorporated for protein synthesis, as only L-amino acids are incorporated into proteins. Despite this, [^{11}C]MET was still used as an alternative to [^{18}F]FDG PET imaging because it is more sensitive (see review [11]). [^{18}F]fluoroethyl-tyrosine ([^{18}F]FET) was developed as another amino acid based alternative to [^{18}F]FDG [12]. Unlike [^{11}C]MET, [^{18}F]FET is a modified amino acid, which led to questions about the radiotracer's ability to be taken up into cells. In cellular experiments and tumor-bearing mice, it was found that [^{18}F]FET is actively transported into cells [13]. The transporter responsible was later identified as L-amino acid transporter (LAT-1) [14]. [^{18}F]FET PET combined with MRI can dramatically improve glioma identification and tumor diagnosis [15]. In terms of the diagnostic performance, MRI alone yielded a 96% sensitivity and 53% specificity; the combined technique achieved 93% sensitivity and 94% specificity [15]. While MRI is the gold standard for diagnosis and tumor staging, [^{18}F]FET PET imaging provides complementary information in the form of increased specificity.

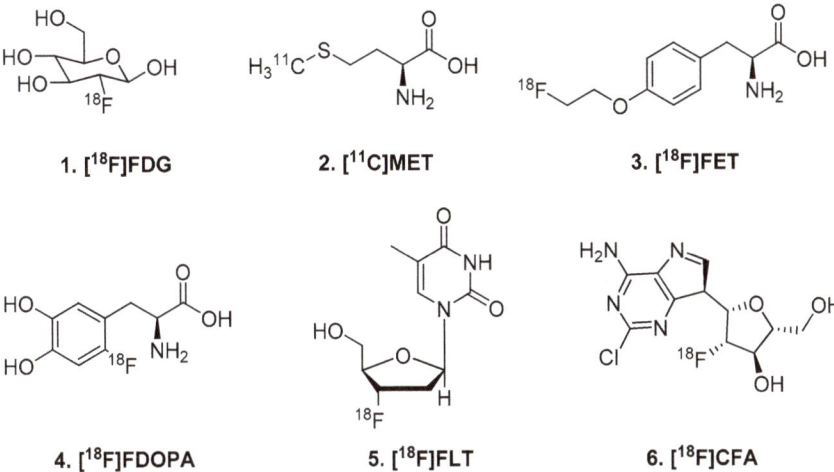

Figure 1. Radiotracers for markers of cellular proliferation.

The nonessential amino acid and neurotransmitter dopamine has also been used as a PET tracer in the form of [^{18}F]L-fluoro-dihydroxyphenylalanine ([^{18}F]FDOPA), which becomes a dopamine analogue in vivo after decarboxylation. [^{18}F]Fluoro-dopamine can then be further taken up into vesicles by VMAT, or metabolized by monoamine oxidases or catechol-O-methyltransferase. This gives [^{18}F]FDOPA a particular utility in neuroendocrine tumors [16] in addition to the obvious neurological application. [^{18}F]FDOPA imaging in pre-operative glioma patients has shown a significant correlation between WHO grade and the volume of MRI contrast enhancement, volume of T2 hyperintensity, and [^{18}F]FDOPA uptake (as SUV$_{max}$ ratio of tumor to normal tissue) [17]. From this 45 patient cohort, a multivariate Cox regression suggested that [^{18}F]FDOPA PET and age were significant prognostic factors for overall survival [17]. A major limitation of [^{18}F]FDOPA use is the radiosynthesis. One of the first described high yielding radiosynthesis involved electrophilic fluorination [18]. Electrophilic fluorination is not a desirable method for routine clinical production because of the hazardous nature of F_2 gas, however, improved syntheses were recently published involving nucleophilic fluorination [19].

Concurrent with the use of amino acid analogues, the deoxynucleoside base derivative [^{18}F]fluorothymidine ([^{18}F]FLT) was developed. [^{18}F]FLT is trapped in tissues after phosphorylation by thymidine kinase [20]. This trapping was thought to be due to accumulation into growing DNA chains, as fluorothymidine was originally developed to terminate DNA, and used in HIV therapy. In reality, less than 1% of [^{18}F]FLT was found to be incorporated in DNA in cellular studies, though it correlated highly with [^{3}H]thymidine uptake [21]. Despite this lack of accumulation in DNA, [^{18}F]FLT imaging confirms cellular uptake and correlates to Ki-67 expression on the corresponding resected tumor tissue [22]. Thymidine kinase is a principal enzyme in the DNA salvage pathway, and is most active in G1 and S phases of the cell cycle [23]. Because of its lower uptake in normal brain than [^{18}F]FDG, glioma patients underwent [^{18}F]FLT imaging and a similar correlation was observed with proliferative tissue markers [24]. When [^{18}F]FLT PET was used to monitor treatment in malignant glioma trials and compared to MRI responses, [^{18}F]FLT-PET was more predictive of overall survival than MRI [25]. Despite this utility, [^{18}F]FLT is less widely used because of nonspecific binding in individuals [26], and because blood–brain barrier (BBB) penetration is limited, which limits its use specifically for glioma imaging.

An alternative method to [^{18}F]FLT, which is phosphorylated by thymidine kinase, is to target deoxycytidine kinase (dCK). The activity of dCK, TK, and the other deoxyribonucleoside kinases are to provide an alternative pathway to de novo synthesis of DNA precursors [27]. Additionally, the activity of these kinases is critical for the activation of nucleoside analogues that are used for chemotherapy, like clofarabine [28]. Clofarabine is a chemotherapy used in pediatric patients with relapsed or refectory acute lymphoblastic leukemia. The radiolabeled [^{18}F]clofarabine ([^{18}F]CFA) has been evaluated in healthy humans and found to be BBB permeable [29]. Biodistribution of [^{18}F]CFA showed uptake in lymph nodes, consistent with known dCK activity being required for T- and B-cell development [30]. In two patients with recurrent GBM, PET imaging with MRI was able to delineate specific regions of immune activity [31]. In one individual, the post-treatment PET-MRI scan demonstrated a 300% increase in immune cells in the tumor microenvironment, with the tumor volume remaining consistent. This combined technique has been shown in this preliminary data set to be useful in differentiating tumor progression from immune cell infiltration in a treatment monitoring scenario.

PET imaging of these classic markers of proliferation is useful to image GBM due to their simplistic design. However, changes in glucose metabolism, DNA replication, protein synthesis, and neurotransmitter homeostasis are not unique to cancers; they can describe a number of disease states. This generality makes these imaging approaches most appropriate for disease monitoring in cancer treatments like chemotherapy and targeted radiation. For further discussion of these tracers and their prognostic value in brain tumors, see a recent systematic review [32].

2.2. Hypoxia-Sensing Tracers

The first-line treatment strategy for gliomas is surgical resection if possible, followed by chemotherapy and targeted radiation [33]. Chemotherapy and radiation are therapies that hinge on cell death by damaging DNA and initiating apoptosis. However, some tumors are resistant to these strategies. For example, it has been long known that poorly oxygenated tissue, or hypoxic tissue, is less sensitive to radiation. Gray et al. confirmed in 1953 that X ray therapy was more effective in mice that were breathing oxygen at a higher pressures than normal atmosphere [34]. Hypoxia has become a recognized key feature of most solid tumors [35]. In a hypoxic tumor microenvironment, radiation therapy could be more effective at a higher dose; however, this requires an accurate identification of that cell population.

PET radiotracers that sense oxygen levels in cells can be used to visualize hypoxia (Figure 2). [^{18}F]Fluoromisoinodazole ([^{18}F]FMISO; **7**) contains a nitroimidazole which is reduced to RNO_2 radical after entering a viable cell. In the presence of oxygen, it will be re-oxidized and diffuse from the cell. If the cell is hypoxic, however, the radiotracer will be trapped. This tracer was first evaluated in V-79 cells using low O_2 levels in the incubation to mimic hypoxia [36] and followed up with cancer models in animals [37]. Glioma patients have undergone [^{18}F]FMISO PET imaging, though with limited success [38]. Chakhoyan et al. were able to build ptO_2 maps from [^{18}F]FMISO PET images and compare them to perfusion weighted imaging (MRI) and 1H-MR mono-voxel magnetic resonance spectroscopy (MRS) [39]. The correspondence with MRI and MRS imaging confirms the direct relationship between [^{18}F]FMISO and oxygen levels in tissue. However, Valk et al. observed retention of [^{18}F]FMISO in a GBM subject and anaplastic astrocytoma subject, but no retention of [^{18}F]FMISO in another GBM subject [38]. [^{18}F]FMISO PET imaging has a limited range in sensitivity between normoxic and hypoxic tissue [40]. Additionally, BBB penetration is low for [^{18}F]FMISO which does not make it attractive for glioma imaging.

There are multiple hypoxia radiotracers developed around the same time as [^{18}F]FMISO including [^{18}F]FET-NIM (**8**), [^{18}F]EF5 (**9**), and [^{18}F]HX4 (**10**) (see review [35]). All of these ligands use the nitroimidazole moiety to sense oxygen level in vivo, and the differences in the chemical structures are primarily focused on the linker groups and location of fluorine-18 label (Figure 2). [^{18}F]EF5 exhibits greater cell membrane permeability, slower clearance, and improved tumor uptake [41–44]. Its structurally very similar to [^{18}F]FMISO and [^{18}F]FETNIM, but incorporates a pentafluoro ethyl group via an amide linker. Labeling at this position requires electrophilic fluorination in the radiosynthesis, limiting regular production of this radiotracer. [^{18}F]FETNIM was developed concurrently with [^{18}F]FMISO, and is the most structurally similar with the addition of a hydroxyl group alpha to the nitroimidazole ring [45–47]. [^{18}F]HX4 still contains the nitroimidazole, though added a triazole linker between this and the fluorine-18 label [48–51]. When compared to [^{18}F]FMISO, [^{18}F]HX4 provides the higher image contrast 4 h post-injection, but has high variability [52,53]. A second generation version of [^{18}F]FMISO, [^{18}F]DiFA (12) is slightly less lipophilic. This radiotracer aims to have a faster clearance rate, and thus improved signal-to-noise ratio over [^{18}F]FMISO [54].

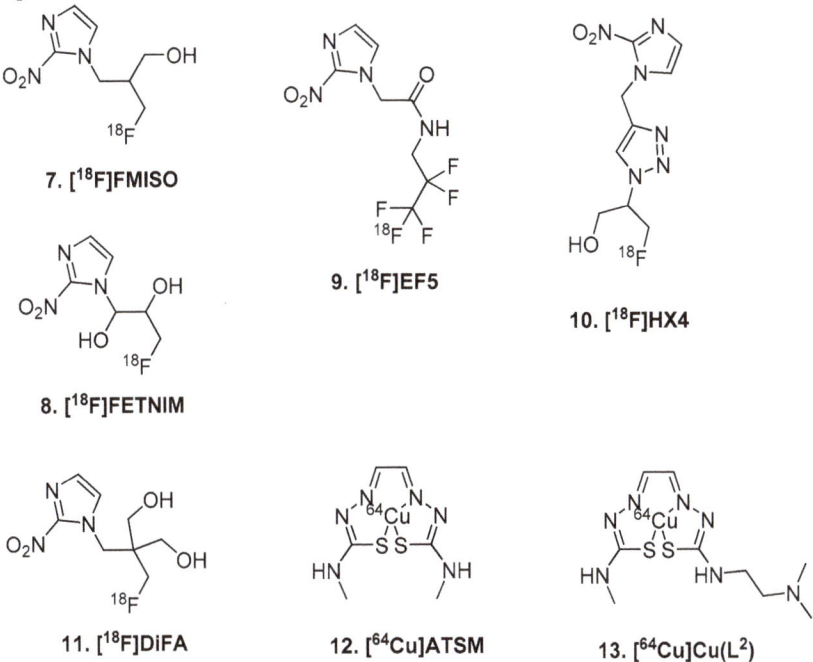

Figure 2. Hypoxia-sensing tracers.

An alternative to [^{18}F]FMISO, and the other organic oxygen sensing compounds, is [^{64}Cu]ATSM (12), which uses a metal to sense oxidation changes (Cu(II) to Cu(I)) [55]. In comparison to [^{18}F]FMISO, it has faster tracer kinetics and can reveal 'hypoxic' tissues in 15 min post injection. [^{64}Cu]ATSM uptake is influenced not only by hypoxia, but also cellular concentrations of reducing species such as NADH [56]. Using [^{64}Cu]ATSM imaging in glioma patients, SUV was found to be an independent predictor of both progression free survival and overall survival in this study [57]. In the GBM subgroup analysis, however, max SUV only showed significant prediction of progression free survival [57], using copper-64 allowed for synergy between PET and MRI, in this proof of concept using a closely related [^{64}Cu]ATSM compound ([^{64}Cu]Cu(L$_2$); 13), which has only been evaluated so far in vitro [58,59]. Though, it is likely to have similar BBB penetration issues as [^{64}Cu]ATSM.

Hypoxia imaging techniques are important for the treatment planning of solid tumors to ensure effective regression. While current therapies do not target the hypoxia machinery in cells, this biochemical process remains a critical consideration for radiotherapy. Many tumor types, including gliomas, are described as being hypoxic and utilize PET or MR imaging methods for treatment planning [60]. This method also serves to monitor treatment after targeted radiotherapies.

2.3. Inflammation

An important part of the immune response is inflammation and, like peripheral cancers, brain cancer cells will trigger this response. A histological analysis of lesions from 1265 patients with glioblastoma multiforme identified the presence of lymphocytes and reactive astrocytes [61]. Inflammation has been explored as a therapeutic target for such cases. Nonsteroidal anti-inflammatory drugs (NSAIDs) have been evaluated to suppress the growth of tumor cells in vitro [62]. While NSAIDs are an inappropriate therapy for GBM because of the lack of targeting, inflammation is a characteristic of GBM pathology and could be useful for imaging.

Translocator protein (18 kDa), TSPO, is an integral mitochondrial membrane protein responsible for cholesterol transport and responds to cell stress. Broadly, TSPO is treated as a biomarker sensitive to pro-inflammatory stimuli, and small molecule inhibitors of TSPO have been utilized in PET imaging [63]. [^{11}C]PK11195 (**14**, Figure 3) was developed for general use of inflammation imaging, and even in glioma patients [64]. Immunohistochemical experiments with patient tissue further confirmed increase TSPO mRNA and protein levels [65]. Furthermore TSPO PET imaging was correlated to outcome [66]. In a larger prospective trial which included tumor biopsy, imaging and histology were found to correlate, and BP_{ND} in high-grade gliomas was significantly higher than in low-grade astrocytomas and low-grade oligodendrogliomas [67]. Although second and third generation TSPO radioligands with higher specific binding have since been developed, TSPO imaging in GBM has major limitations including tumor heterogeneity and inability to distinguish signal caused by radiation therapy from signal due to the tumor microenvironment.

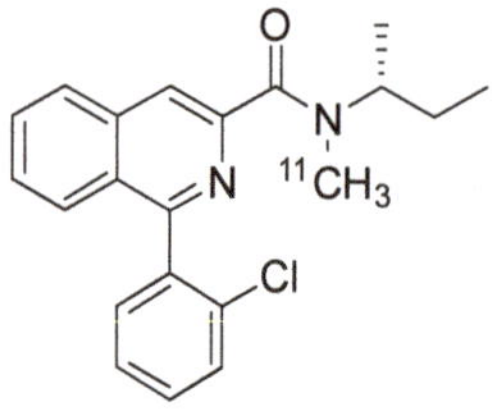

14. [^{11}C]PK11195

Figure 3. TSPO ligand.

3. New Biomarkers for GBM PET Imaging

New strategies for cancer therapeutics target proliferation (sigma 2), immunity (sigma 1, PD-L1), and genetic modification (PARP, IDH). PET imaging agents have developed in tandem with advancement of these therapies. These therapies were initially designed to treat more prevalent, peripheral tumors. The nature of heterogeneity in brain tumors inspires the use of biomarker specific imaging agents, as opposed to the more general ligands for proliferation, hypoxia, and inflammation. Applications to brain cancers comes with the significant considerations of blood–brain barrier penetration and effectiveness. The PET imaging agents discussed in the following section are not yet utilized in human brain imaging but represent promising candidates for the new therapeutic and/or biomarker strategies.

3.1. Sigma 1

A key characteristic of GBMs is their invasiveness, which leads to the very low survival rate. Sigma 1 and sigma 2 receptors are expressed in the human tumor cell lines: C6 glioma, NIE-115 neuroblastoma, and NG108-15 neuroblastoma- glioma hybrid [68]. While sigma 1 is associated with invasiveness, sigma 2 receptors have been associated with proliferation. Both receptors are interesting PET imaging targets for glioblastoma.

Sigma 1 receptor (S1R) was identified as the site of action of the antipsychotic haloperidol [69]. Its role in the CNS has been investigated for neurodegenerative disorders [70] in addition to GBM. The first PET ligand for S1R was [^{11}C]1-(3,4-dimethoxyphenethyl)-4-(3-phenylpropyl)piperazine ([^{11}C]SA4503: **15**, Figure 4) [71]. While it showed nanomolar affinity to S1R, it unfortunately has high affinities for other receptors, ion channels, and second messenger systems [72]. Both S1R and S2R have structural similarity to opiate receptors, which necessitates specificity and selectivity when designing ligands. Additionally, in the development of fluorine-18 sigma-1 receptor ligands, very high binding affinity has corresponded to very slow clearance rate. For example, [^{18}F]1-3-fluoropropyl-4-((4-cyanophenoxy)-methyl)piperidine ([^{18}F]FPS; **16**) [73] did not reach pseudo-equilibrium by 4 h in human [74]. Another ligand, [^{18}F]6-(3-fluoropropyl)-3-(2-(azepan-1-yl)ethyl)benzo[d]thiazol-2(3H)-one ([^{18}F]FTC-146; **17**) was also found to be irreversible and not suitable for neuroimaging [75]. Recently, [^{18}F](S)-Fluspidine (**18**) [76] was developed and evaluated in human [77]. The pharmacokinetics are improved, and with xenograft mouse models of glioblastoma have visible increases in radioligand binding.

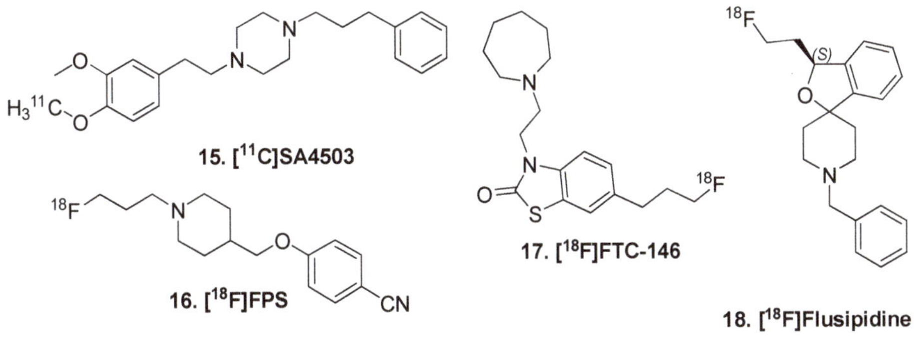

Figure 4. Sigma 1 receptor radioligands.

3.2. Sigma 2

The amino acid and nucleoside base PET imaging approach is a classic, straightforward way to assess functional proliferation. The sigma-2 receptor (S2R; TMEM97) has recently been implicated in cancer biology; S2R levels increased 5-fold in proliferating tumor cells compared to quiescent tumor cells [78]. This allows for radioligand design for a specific protein more akin to drug design. For an in-depth history of S2R ligand development, see references [79,80].

An early imaging effort for S2R was using 4-[^{125}I]BP (**19**, Figure 5), a small molecule with high affinity for both sigma 1 and 2 receptors [81]. Further efforts by these authors led to [^{125}I]PIMBA (**20**), though this radioligand suffered from high background binding [82]. [^{18}F]RHM-4 (**21**) was developed and demonstrated S2R overexpression in pancreatic cancer, though BBB penetration was not investigated [83]. The radioligand [^{18}F]ISO (**22**) was also developed, structurally similar to **21** except lacking the aryl methoxy group [84]. A positive correlation is observed between **22** binding and tumor Ki67 expression [85]. In treatment monitoring of CDK4/6 inhibition plus endocrine therapy in breast cancer xenograft animals, **22** was found to assess more delayed changes related to cell cycle arrest compared to [^{18}F]FLT [86]. However, **22** is not taken up into the brain, based on organ residence studies in rodent [85]. This has inspired alternative ligands with BBB penetration as the design goal.

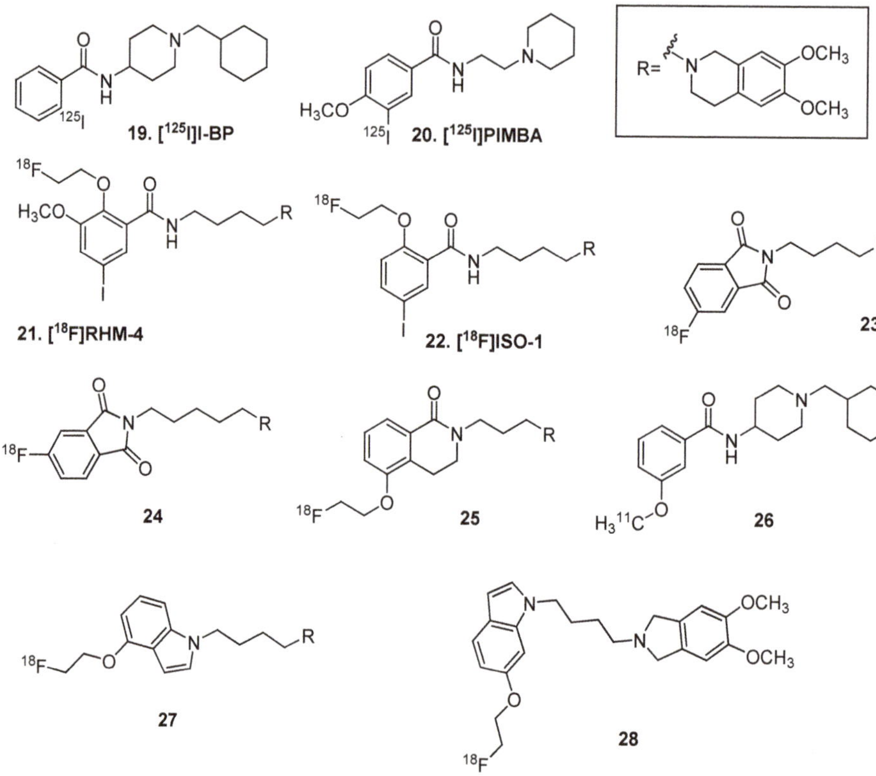

Figure 5. Sigma 2 receptor radioligands.

A scaffold incorporating pthalimides (**23** and **24**) showed elevated brain uptake and specific binding (displaceable with cold ligand and haloperidol), though the tumor to background ratio was low [87]. Abate et al. described **25**, which demonstrated good in vitro binding and specificity; however, as a P-glycoprotein (PGP) substrate it is not suitable for brain imaging [88]. The same group continued with a carbon-11 effort (**26**), based on the inhibitor PB28 and very similar to **19**; however, brain uptake was low and the compound did not display high enough specific binding [89]. Wang et al. reported two fluorine-18 inhibitors, **27** and **28**, with high brain uptake in mice in 2017, though it may be a PGP substrate as well [90]. These efforts have demonstrated some challenges in the radioligand design of sigma 2 receptor inhibitors for brain imaging, i.e., PGP efflux and limited specific binding.

3.3. PD-L1

Avoiding destruction by immune cells is a powerful strategy utilized by cancer cells. Until recently, imaging research in this space has focused on antibody-based strategies, which is challenging due to limited brain penetration. The molecules used in PET imaging for Programmed death ligand 1 (PD-L1) are large molecule therapeutics, including antibodies, antibody fragments, and peptides. An early imaging effort developed at Johns Hopkins University adapted the therapeutic antibody for PET imaging: [^{64}Cu]Azetozolizumab [91]. The same group later developed a peptide, [^{64}Cu]WL12, although brain was not listed in the biodistribution study [92]. Another protein effort from the Gambhir lab at Stanford, [^{64}Cu]NOTA-HACA-PD1 and a gallium 68 version, determined no brain uptake definitively [93]. Merck developed an affibody Z$_{PD-L1\_1}$ that was fluorine-18 labeled; however, brain penetration has not been demonstrated, though an affibody is more likely to be BBB penetrant than the preceding antibodies [94].

Although small molecule inhibitors have been in development, to date no small molecule radioligands have been described. Bristol Meyers Squib has the first small molecule inhibitor reported, BMS-202 (**29**, Figure 6) [95] and has characterized the binding in a crystal structure [96]. The mechanism of action is thought to be selectively induced dimerization of PD-L1, which inhibits binding to PD-1 (see reviews [97–99])

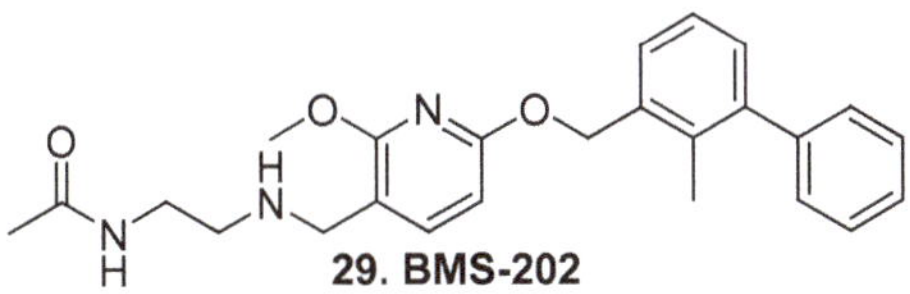

Figure 6. Small molecule PD-L1 inhibitor.

3.4. PARP

ADP-ribose polymerase (PARP) is the enzyme responsible for attaching linear or branched polymers of ADP onto broken DNA and other biomolecules. PARP-1 recognizes single and double-strand breaks, crossovers, cruciform, and supercoils. Additionally, PARP-1 maintains the stability of replication forks, and the basal activity is very low [100]. In cancer cells containing a BRCA2 deficiency, inhibiting PARP causes synthetic lethality [101]. This treatment strategy has been used to develop multiple small molecule therapeutics [102].

The first in-class inhibitor olaparib has been adapted for use in PET imaging: [^{18}F]BO (biorthogonal olaparib; **30**, Figure 7) [103]. [^{18}F]BO demonstrated ubiquitous distribution in cancer cells and localization within the nucleus in cancer cells [104]. Structurally similar to olaparib, the inhibitor radioligand [^{18}F]PARPi (**31**), and corresponding fluorescent version (**32**), has been used in GBM cell lines [104] and in rodent [105]. Although this shows specific binding in peripheral tumors, there was no significant blocking with brain uptake at 2 h biodistribution. Brain penetration was very low initially. In comparison to [^{11}C]choline and [^{18}F]FLT, PARPi demonstrated a lower mean uptake in tumor than the other two PET ligands; however, the lower background uptake enabled PARPi to delineate brain tumors in rodent models with more clear contrast [106]. The Gouverneur group recently radiolabeled olaparib itself with fluorine-18, though limited imaging studies have been done with this tracer [107]. As expected, [^{18}F]olaparib (**33**) was taken up selectively in PARP-1 expressing cells and in mouse tumors. Additionally, radioligand uptake was increased by 70% after tumor irradiation, indicated a great potential for monitoring radiation damage. However, brain was not included in biodistribution calculations and there does not appear to be brain uptake in the dynamic PET images. Unlike the preceding PARP radioligands, the inhibitor [^{18}F]fluorthanatrace ([^{18}F]FTT; **34**) is not based on olaparib, but rather rucaparib. This benzimidazole carboxamide derivative is highly potent with an IC$_{50}$ of 6.3 nM [108]. It has been evaluated in humans, though with low brain penetration [109]. There remains a need for BBB penetrating small molecule inhibitors of PARP in order to be useful specifically in glioblastoma imaging.

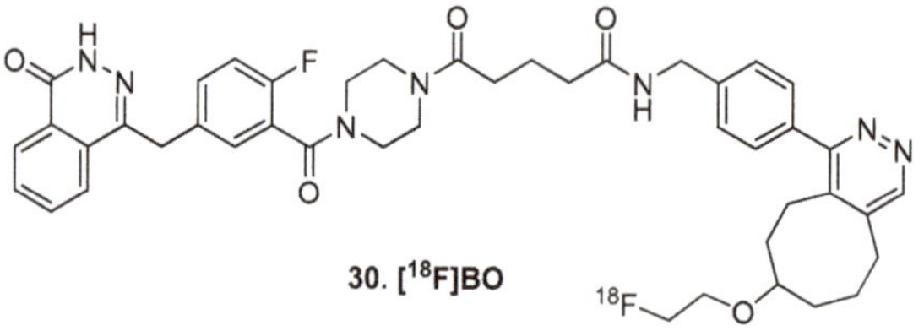

30. [^{18}F]BO

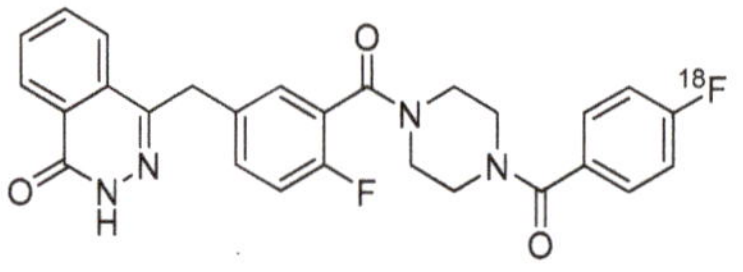

31. [^{18}F]PARPi

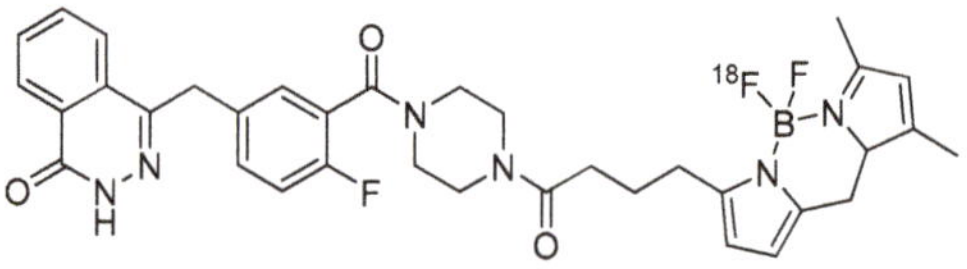

32. [^{18}F]PARPi-FL

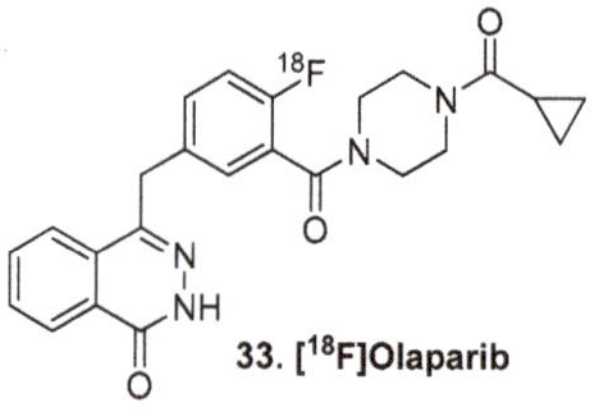

33. [^{18}F]Olaparib

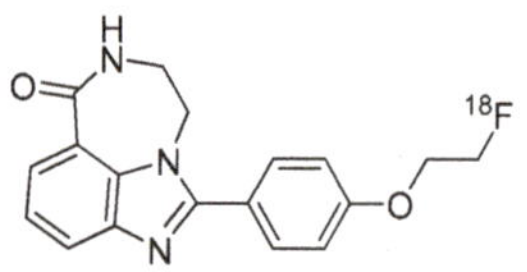

34. [^{18}F]Fluorthanatrace (FTT)

Figure 7. PARP ligands.

3.5. Isocitrate Dehydrogenase (IDH)

Through genome wide association studies (GWASs), the common mutation R132H, located on the isocitrate dehydrogenase (IDH) 1 gene, was found in more than 70% of grade II and III astrocytomas, oligodendrogliomas, and glioblastomas that developed from these lower-grade lesions [110]. However, only 10% of glioblastomas are IDH mutant [1], and this is weakly associated with tumor aggressiveness [3]. IDH- mutant GBM has a significantly longer survival rate compared to IDH wild type, 31 months compared to 15 months with standard treatment.

The observed IDH mutation essentially eliminates all enzymatic activity [111]. In normal cells, IDH1 converts isocitrate to α-ketoglutarate and form NADPH, which maintains a pool of reduced glutathione and peroxiredoxin. When isocitrate cannot be converted to α-ketoglutarate, it instead is converted to R(-)-2-hydroxyglutarate [112]. 2-Hyrdoxyglutarate and α-ketoglutarate are cofactors for many enzymes and their availability influences DNA methylation status. 2-HG inhibits 5mC hydroxylase (TET2) and lysine demethylases (KDM) leading to demethylation of DNA and histone, respectively. This changes gene expression and thus tumorigenesis. R-2HG stimulates EgIN1, which promotes HIF1a degradation by hydroxylation [113]. Therapeutic inhibitors are being developed for IDH1 to influence this pathway and similarly slow tumorigenesis and promote survival.

Ivosidenib (AG-120) is the first in-class IDH1 reversible inhibitor [114]. In rats, some brain penetration has been demonstrated, 4.1% after 50 mg/kg dose, which could indicate some effectiveness in glioblastoma. The phase I trial in low grade glioma subjects is ongoing (NCT03343197). Another inhibitor (IDH305) for mutant IDH1 is in development by Novartis, with strong data for binding in brain homogenate [115]. In a phase I clinical trial, safety was evaluated in glioma, AML/MDS, and other/ non-CNS solid tumors with IDH mutation [116]. However, phase II trials of IDH305 in glioma has been withdrawn, possibly due to liver toxicity (NCT02977689 [117]).

Before small molecule inhibitors of IDH were available, PET radioligands were evaluated for their relationship to IDH mutation status. While [^{18}F]FDOPA does not correlate with IDH mutation status [118], it appears that [^{18}F]FET imaging does significantly associate with IDH mutation status [119–121], although there does not appear to be a biochemical relationship between [^{18}F]FET uptake and IDH mutation status. With the disclosure of small molecule IDH inhibitors came the preliminary development of IDH-selective PET radioligands. Chitneni et al. utilized an iodine-131 and fluorine-18 version (**35**, Figure 8) of AGI-5198; however, these compounds lacked selectivity for mutant IDH over wild type [122]. In a follow up from the same group, [^{18}F]triazinediamine (**36**) analogues were radiosynthesized, based on Enasidenib (AG-221) [123], The K_d (**36**) was calculated to be 40 nM with a B_{max} of 4426 gmol/mg in a mutant anaplastic astrocytoma cell line, which is promising. However, biodistribution studies showed bone uptake from radiolytic defluorination, so further design is required [123].

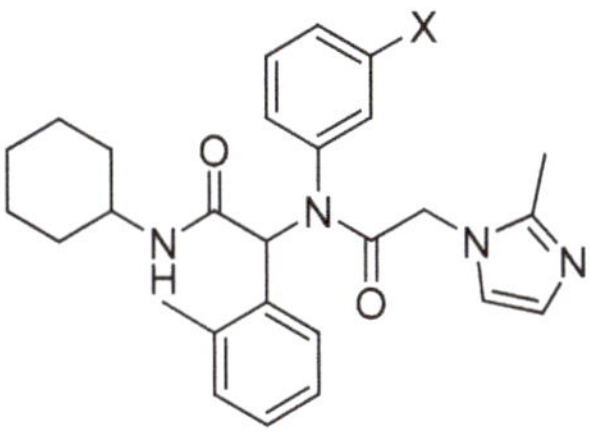

Figure 8. Isocitrate dehydrogenase (IDH) radioligands.

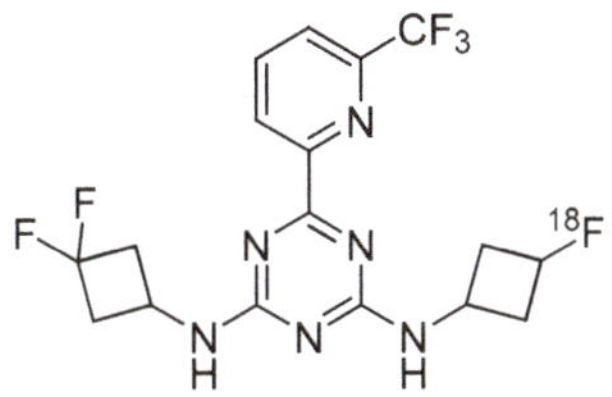

4. General Imaging Considerations

Brain PET imaging for any disease state is challenging. The blood–brain barrier (BBB) poses a major obstacle for any successful radiotracer targeting glioblastoma. Even if the radiotracer enters the brain, many compounds exhibit slow brain entry (i.e., small K_1 values). Since glioblastoma lesions often compromise BBB, increased radioactivity concentrations at the lesion site may reflect increased nonspecific radiotracer instead of, or in addition to, increased target signal. For example, a combined [^{18}F]FMISO PET and MRI study noted high uptake of [^{18}F]FMISO in areas of BBB disruption as well as in necrotic tissue [124]. This raises a challenge in quantification at a suspected glioblastoma lesion. Kinetic modeling approaches incorporating dynamic data, in some cases, may help distinguish nonspecific signal from specific signal in some cases. Although these scanning protocols can require longer scanning time and possibly arterial blood sampling, the potential for improved quantification of specific radiotracer uptake offers important benefits to evaluating diagnosis, staging, or treatment efficacy that must be considered when using PET to image GBM.

Reference region approaches offer alternatives for quantitative analyses that may shorten the length of scanning time and do not require arterial blood sampling. While such approaches have major limitations in cases of ubiquitously expressed targets, such as TSPO, in the case of brain tumor imaging the reference region may be drawn as a larger region that is removed from the lesion. For example, in a case of [^{18}F]FET imaging a fixed-size reference ROI was placed in hemispheres contralateral to tumorous tissue, which yielded fully image-derived measures that correlated with disease-free survival [125]. Image-derived input function allow for alternative non-invasive modeling approaches. For example, [^{18}F]FMISO has no reference region, but image-derived tissue-to-blood ratios provide reasonable proxies for measured parent radioactivity in venous blood [126]. Such approaches can maintain quantitative accuracy while reducing logistical complexities introduced by full dynamic scans with blood sampling.

A final major obstacle for PET radiotracers imaging GBM is high nonspecific binding. For example, [^{18}F]FMISO, [^{18}F]FLT, and [^{64}Cu]ATSM exhibit elevated nonspecific binding, which limits their usefulness. Such radiotracers exhibit low signal to noise ratio (SNR), which makes it more difficult to differentiate areas of low uptake from noise. In contrast, radiotracers with low nonspecific binding may be amiable to reference region analysis if non-tumor brain regions exhibit negligible specific binding.

Off-target binding poses another related challenge. For example, sigma 1 and sigma 2 receptors have significant structural similarities to opiate receptors. These challenges highlight the need for blocking studies with candidate radiotracers to confirm suitable sensitivity and specificity without high nonspecific binding.

5. Conclusions

PET imaging provides important functional information about GBM tumors and surrounding tissue environment. Markers of proliferation, hypoxia, and inflammation have been used to image the lesions of GBM patients. Exciting new frontiers for PET imaging targets for GBM include PD-L1 for immune status, PARP-1 for DNA damage, sigma 2 receptors as alternative markers of proliferation, and isocitrate dehydrogenase for tumorigenesis activity. These PET imaging targets have the potential to enhance diagnosis, staging, and treatment approaches for GBM. As GBM PET imaging techniques advance, it is critical to consider blood brain barrier penetration and nonspecific binding in the evaluation of new radioligands.

Author Contributions: L.R.D. performed literature search and drafted the initial manuscript. Z.C. and A.T.H. revised and edited subsequent manuscript iterations. All authors have read and agreed to the published version of the manuscript.

Funding: This research was supported by NIH grants K01AA024788, R21EB027872, R01AG058773. Z.C. is an Archer Foundation Research Scientist.

Conflicts of Interest: The authors declare no conflict of interest.

References

1. Glioblastoma Multiforme. Available online: https://www.aans.org/Patients/Neurosurgical-Conditions-and-Treatments/Glioblastoma-Multiforme (accessed on 15 November 2019).
2. Louis, D.N.; Perry, A.; Reifenberger, G.; Von Deimling, A.; Figarella-Branger, D.; Cavenee, W.K.; Ohgaki, H.; Wiestler, O.D.; Kleihues, P.; Ellison, D.W. The 2016 World Health Organization classification of tumors of the central nervous system: A summary. *Acta Neuropathol.* **2016**, *131*, 803–820. [CrossRef] [PubMed]
3. Amelot, A.; De Cremoux, P.; Quillien, V.; Polivka, M.; Adle-Biassette, H.; Lehmann-Che, J.; Francoise, L.; Carpentier, A.F.; George, B.; Mandonnet, E.; et al. IDH-mutation is a weak predictor of long-term survival in glioblastoma patients. *PLoS ONE* **2015**, *10*, e0130596. [CrossRef]
4. Perrin, S.L.; Samuel, M.S.; Koszyca, B.; Brown, M.P.; Ebert, L.M.; Oksdath, M.; Gomez, G.A. Glioblastoma heterogeneity and the tumour microenvironment: Implications for preclinical research and development of new treatments. *Biochem. Soc. Trans.* **2019**, *47*, 625–638. [CrossRef] [PubMed]
5. Ozdemir-Kaynak, E.; Qutub, A.A.; Yesil-Celiktas, O. Advances in Glioblastoma Multiforme Treatment: New Models for Nanoparticle Therapy. *Front. Physiol.* **2018**, *9*. [CrossRef] [PubMed]
6. Reza, S.M.S.; Samad, M.D.; Shboul, Z.A.; Jones, K.A.; Iftekharuddin, K.M. Glioma grading using structural magnetic resonance imaging and molecular data. *J. Med. Imaging* **2019**, *6*, 024501. [CrossRef]
7. Som, P.; Atkins, H.L.; Bandoypadhyay, D.; Fowler, J.S.; MacGregor, R.R.; Matsui, K.; Oster, Z.H.; Sacker, D.F.; Shiue, C.Y.; Turner, H.; et al. A fluorinated glucose analog, 2-fluoro-2-deoxy-D-glucose (F-18): Nontoxic tracer for rapid tumor detection. *J. Nucl Med.* **1980**, *21*, 670–675. [CrossRef]
8. Langstrom, B.; Lundqvist, H. The preparation of carbon-11-labeled methyl iodide and its use in the synthesis of carbon-11-labeled methyl-L-methionine. *Int. J. Appl. Radiat. Isot.* **1976**, *27*, 357–363. [CrossRef]
9. Comar, D.; Cartron, J.C.; Maziere, M.; Marazano, C. Labeling and metabolism of methionine-methyl-11C. *Eur. J. Nucl. Med.* **1976**, *1*, 11–14. [CrossRef]
10. Schober, O.; Duden, C.; Meyer, G.J.; Muller, J.A.; Hundeshagen, H. Non selective transport of [11C-methyl]-L-and D-methionine into a malignant glioma. *Eur J. Nucl. Med.* **1987**, *13*, 103–105. [CrossRef]
11. Singhal, T.; Narayanan, T.K.; Jain, V.; Mukherjee, J.; Mantil, J. 11C-L-methionine positron emission tomography in the clinical management of cerebral gliomas. *Mol. Imaging Biol.* **2008**, *10*, 1–18.
12. Wester, H.J.; Herz, M.; Weber, W.; Heiss, P.; Senekowitsch-Schmidtke, R.; Schwaiger, M.; Stocklin, G. Synthesis and radiopharmacology of O-(2-[18F]fluoroethyl)-L-tyrosine for tumor imaging. *J. Nucl. Med.* **1999**, *40*, 205–212. [PubMed]

13. Heiss, P.; Mayer, S.; Herz, M.; Wester, H.-J.; Schwaiger, M.; Senekowitsch-Schmidtke, R. Investigation of transport mechanism and uptake kinetics of O-(2-[18F]fluoroethyl)-L-tyrosine in vitro and in vivo. *J. Nucl. Med.* **1999**, *40*, 1367–1373. [PubMed]
14. Habermeier, A.; Graf, J.; Sandhoefer, B.F.; Boissel, J.P.; Roesch, F.; Closs, E.I. System L amino acid transporter LAT1 accumulates O-(2-fluoroethyl)-L-tyrosine (FET). *Amino Acids* **2015**, *47*, 335–344. [CrossRef] [PubMed]
15. Pauleit, D.; Floeth, F.; Hamacher, K.; Riemenschneider, M.J.; Reifenberger, G.; Muller, H.-W.; Zilles, K.; Coenen, H.H.; Langen, K.-J. O-(2-[18F]fluoroethyl)-L-tyrosine PET combined with MRI improves the diagnostic assessment of cerebral gliomas. *Brain* **2005**, *128*, 678–687. [CrossRef]
16. Minn, H.; Kauhanen, S.; Seppanen, M.; Nuutila, P. 18F-FDOPA: A multiple-target molecule. *J. Nucl. Med.* **2009**, *50*, 1915–1918. [CrossRef]
17. Patel, C.B.; Fazzari, E.; Chakhoyan, A.; Yao, J.; Raymond, C.; Nguyen, H.; Manoukian, J.; Nguyen, N.; Pope, W.; Cloughesy, T.F.; et al. 18F-FDOPA PET and MRI characteristics correlate with degree of malignancy and predict survival in treatment-naive gliomas: A cross-sectional study. *J. Neuro-Oncol.* **2018**, *139*, 399–409. [CrossRef]
18. Chirakal, R.; Firnau, G.; Garnett, E. High yield synthesis of 6-[18F] fluoro-L-dopa. *J. Nucl. Med.* **1986**, *27*, 417–421.
19. Mossine, A.V.; Tanzey, S.S.; Brooks, A.F.; Makaravage, K.J.; Ichiishi, N.; Miller, J.M.; Henderson, B.D.; Skaddan, M.B.; Sanford, M.S.; Scott, P.J.H. One-pot synthesis of high molar activity 6-[18F]fluoro-L-DOPA by Cu-mediated fluorination of a BPin precursor. *Org. Biomol. Chem.* **2019**. Ahead of Print. [CrossRef]
20. Rasey, J.S.; Grierson, J.R.; Wiens, L.W.; Kolb, P.D.; Schwartz, J.L. Validation of FLT uptake as a measure of thymidine kinase-1 activity in A549 carcinoma cells. *J. Nucl. Med.* **2002**, *43*, 1210–1217.
21. Toyohara, J.; Waki, A.; Takamatsu, S.; Yonekura, Y.; Magata, Y.; Fujibayashi, Y. Basis of FLT as a cell proliferation marker: Comparative uptake studies with [3H] thymidine and [3H] arabinothymidine, and cell-analysis in 22 asynchronously growing tumor cell lines. *Nucl. Med. Biol.* **2002**, *29*, 281–287. [CrossRef]
22. Vesselle, H.; Grierson, J.; Muzi, M.; Pugsley, J.M.; Schmidt, R.A.; Rabinowitz, P.; Peterson, L.M.; Vallieres, E.; Wood, D.E. In vivo validation of 3'deoxy-3'-[18F]fluorothymidine ([18F]FLT) as a proliferation imaging tracer in humans: Correlation of [18F]FLT uptake by positron emission tomography with Ki-67 immunohistochemistry and flow cytometry in human lung tumors. *Clin. Cancer Res.* **2002**, *8*, 3315–3323. [PubMed]
23. Been, L.B.; Suurmeijer, A.J.H.; Cobben, D.C.P.; Jager, P.L.; Hoekstra, H.J.; Elsinga, P.H. [18F]FLT-PET in oncology: Current status and opportunities. *Eur. J. Nucl. Med. Mol. Imaging* **2004**, *31*, 1659–1672. [CrossRef] [PubMed]
24. Yamamoto, Y.; Ono, Y.; Aga, F.; Kawai, N.; Kudomi, N.; Nishiyama, Y. Correlation of 18F-FLT uptake with tumor grade and Ki-67 immunohistochemistry in patients with newly diagnosed and recurrent gliomas. *J. Nucl. Med.* **2012**, *53*, 1911–1915. [CrossRef] [PubMed]
25. Chen, W.; Delaloye, S.; Silverman, D.H.S.; Geist, C.; Czernin, J.; Sayre, J.; Satyamurthy, N.; Pope, W.; Lai, A.; Phelps, M.E.; et al. Predicting treatment response of malignant gliomas to bevacizumab and irinotecan by imaging proliferation with [18F] fluorothymidine positron emission tomography: A pilot study. *J. Clin. Oncol.* **2007**, *25*, 4714–4721. [CrossRef] [PubMed]
26. Shields, A.F. PET imaging with 18F-FLT and thymidine analogs: Promise and pitfalls. *J. Nucl. Med.* **2003**, *44*, 1432–1434.
27. Arner, E.S.; Eriksson, S. Mammalian deoxyribonuclease kinases. *Pharmacol. Ther.* **1995**, *67*, 155–186. [CrossRef]
28. Ghanem, H.; Kantarjian, H.; Ohanian, M.; Jabbour, E. The role of clofarabine in acute myeloid leukemia. *Leuk. Lymphoma* **2013**, *54*, 688–698. [CrossRef]
29. Barrio, M.J.; Spick, C.; Radu, C.G.; Lassmann, M.; Eberlein, U.; Allen-Auerbach, M.; Schiepers, C.; Slavik, R.; Czernin, J.; Herrmann, K. Human biodistribution and radiation dosimetry of 18F-clofarabine, a PET probe targeting the deoxyribonucleoside salvage pathway. *J. Nucl. Med.* **2017**, *58*, 374–378. [CrossRef]
30. Toy, G.; Austin, W.R.; Liao, H.-I.; Cheng, D.; Singh, A.; Campbell, D.O.; Ishikawa, T.-o.; Lehmann, L.W.; Satyamurthy, N.; Phelps, M.E.; et al. Requirement for deoxycytidine kinase in T and B lymphocyte development. *Proc. Natl. Acad. Sci. USA* **2010**, *107*, 5551–5556. [CrossRef]

31. Antonios, J.P.; Soto, H.; Everson, R.G.; Moughon, D.L.; Wang, A.C.; Orpilla, J.; Radu, C.; Ellingson, B.M.; Lee, J.T.; Cloughesy, T.; et al. Detection of immune responses after immunotherapy in glioblastoma using PET and MRI. *Proc. Natl. Acad. Sci. USA.* **2017**, *114*, 10220–10225. [CrossRef]
32. Treglia, G.; Muoio, B.; Trevisi, G.; Mattoli, M.V.; Albano, D.; Bertagna, F.; Giovanella, L. Diagnostic Performance and Prognostic Value of PET/CT with Different Tracers for Brain Tumors: A Systematic Review of Published Meta-Analyses. *Int. J. Mol. Sci.* **2019**, *20*, 4669. [CrossRef] [PubMed]
33. Omuro, A.; DeAngelis, L.M. Glioblastoma and other malignant gliomas: A clinical review. *JAMA J. Am. Med. Assoc.* **2013**, *310*, 1842–1850. [CrossRef] [PubMed]
34. Gray, L.H.; Conger, A.D.; Ebert, M.; Hornsey, S.; Scott, O. The concentration of oxygen dissolved in tissues at the time of irradiation as a factor in radiotherapy. *Br. J. Radiol.* **1953**, *26*, 638–648. [CrossRef] [PubMed]
35. Horsman, M.R.; Mortensen, L.S.; Petersen, J.B.; Busk, M.; Overgaard, J. Imaging hypoxia to improve radiotherapy outcome. *Nat. Rev. Clin. Oncol.* **2012**, *9*, 674–687. [CrossRef]
36. Rasey, J.S.; Grunbaum, Z.; Magee, S.; Nelson, N.J.; Olive, P.L.; Durand, R.E.; Krohn, K.A. Characterization of radiolabeled fluoromisonidazole as a probe for hypoxic cells. *Radiat. Res.* **1987**, *111*, 292–304. [CrossRef]
37. Rasey, J.S.; Koh, W.-J.; Grierson, J.R.; Grunbaum, Z.; Krohn, K.A. Radiolabeled fluoromisonidazole as an imaging agent for tumor hypoxia. *Int. J. Radiat. Oncol. Biol. Phys.* **1989**, *17*, 985–991. [CrossRef]
38. Valk, P.E.; Mathis, C.A.; Prados, M.D.; Gilbert, J.C.; Budinger, T.F. Hypoxia in human gliomas: Demonstration by PET with fluorine-18-fluoromisonidazole. *J. Nucl Med.* **1992**, *33*, 2133–2137.
39. Chakhoyan, A.; Guillamo, J.-S.; Collet, S.; Kauffmann, F.; Delcroix, N.; Lechapt-Zalcman, E.; Constans, J.-M.; Petit, E.; MacKenzie, E.T.; Barre, L.; et al. FMISO-PET-derived brain oxygen tension maps: Application to glioblastoma and less aggressive gliomas. *Sci. Rep.* **2017**, *7*, 1–9. [CrossRef]
40. Bell, C.; Dowson, N.; Fay, M.; Thomas, P.; Puttick, S.; Gal, Y.; Rose, S. Hypoxia imaging in gliomas with 18F-fluoromisonidazole PET: Toward clinical translation. *Semin. Nucl. Med.* **2015**, *45*, 136–150. [CrossRef]
41. Koch, C.J.; Scheuermann, J.S.; Divgi, C.; Judy, K.D.; Kachur, A.V.; Freifelder, R.; Reddin, J.S.; Karp, J.; Stubbs, J.B.; Hahn, S.M.; et al. Biodistribution and dosimetry of 18F-EF5 in cancer patients with preliminary comparison of 18F-EF5 uptake versus EF5 binding in human glioblastoma. *Eur. J. Nucl. Med. Mol. Imaging* **2010**, *37*, 2048–2059. [CrossRef]
42. Komar, G.; Seppanen, M.; Eskola, O.; Lindholm, P.; Gronroos, T.J.; Forsback, S.; Sipila, H.; Evans, S.M.; Solin, O.; Minn, H. 18F-EF5: A new PET tracer for imaging hypoxia in head and neck cancer. *J. Nucl. Med.* **2008**, *49*, 1944–1951. [CrossRef]
43. Ziemer, L.S.; Evans, S.M.; Kachur, A.V.; Shuman, A.L.; Cardi, C.A.; Jenkins, W.T.; Karp, J.S.; Alavi, A.; Dolbier, W.R.; Koch, C.J. Noninvasive imaging of tumor hypoxia in rats using the 2-nitroimidazole 18F-EF5. *Eur. J. Nucl. Med. Mol. Imaging* **2003**, *30*, 259–266. [CrossRef]
44. Dolbier, W.R., Jr.; Li, A.-R.; Koch, C.J.; Shiue, C.-Y.; Kachur, A.V. [18F]-EF5, a marker for PET detection of hypoxia: Synthesis of precursor and a new fluorination procedure. *Appl. Radiat. Isot.* **2001**, *54*, 73–80. [CrossRef]
45. Lehtioe, K.; Oikonen, V.; Nyman, S.; Groenroos, T.; Roivainen, A.; Eskola, O.; Minn, H. Quantifying tumour hypoxia with fluorine-18 fluoroerythronitroimidazole ([18F]FETNIM) and PET using the tumour to plasma ratio. *Eur. J. Nucl. Med. Mol. Imaging* **2003**, *30*, 101–108. [CrossRef]
46. Gronroos, T.; Eskola, O.; Lehtio, K.; Minn, H.; Marjamaki, P.; Bergman, J.; Haaparanta, M.; Forsback, S.; Solin, O. Pharmacokinetics of [18F]FETNIM: A potential hypoxia marker for PET. *J. Nucl. Med.* **2001**, *42*, 1397–1404. [PubMed]
47. Yang, D.J.; Wallace, S.; Cherif, A.; Li, C.; Gretzer, M.B.; Kim, E.E.; Podoloff, D.A. Development of F-18-labeled fluoroerythronitroimidazole as a PET agent for imaging tumor hypoxia. *Radiology* **1995**, *194*, 795–800. [CrossRef]
48. Verwer, E.E.; Windhorst, A.D.; Boellaard, R.; Verwer, E.E.; Zegers, C.M.L.; van, E.W.; Lambin, P.; Wierts, R.; Mottaghy, F.M.; Mottaghy, F.M.; et al. Pharmacokinetic modeling of a novel hypoxia PET tracer [(18)F]HX4 in patients with non-small cell lung cancer. *EJNMMI Phys.* **2016**, *3*, 30. [CrossRef]
49. Dubois, L.J.; Lieuwes, N.G.; Janssen, M.H.M.; Peeters, W.J.M.; Windhorst, A.D.; Walsh, J.C.; Kolb, H.C.; Oellers, M.C.; Bussink, J.; van Dongen, G.A.M.S.; et al. Preclinical evaluation and validation of [18F]HX4, a promising hypoxia marker for PET imaging. *Proc. Natl. Acad. Sci. USA.* **2011**, *108*, 14620–14625. [CrossRef]

50. Van Loon, J.; Janssen, M.H.M.; Oellers, M.; Aerts, H.J.W.L.; Dubois, L.; Hochstenbag, M.; Dingemans, A.-M.C.; Lalisang, R.; Brans, B.; Windhorst, B.; et al. PET imaging of hypoxia using [18F]HX4: A phase I trial. *Eur. J. Nucl. Med. Mol. Imaging* **2010**, *37*, 1663–1668. [CrossRef]
51. Doss, M.; Zhang, J.J.; Belanger, M.-J.; Stubbs, J.B.; Hostetler, E.D.; Alpaugh, K.; Kolb, H.C.; Yu, J.Q. Biodistribution and radiation dosimetry of the hypoxia marker 18F-HX4 in monkeys and human determined by using whole-body PET/CT. *Nucl. Med. Commun.* **2010**, *31*, 1016–1024. [CrossRef]
52. Wack, L.J.; Moennich, D.; van Elmpt, W.; Zegers, C.M.L.; Troost, E.G.C.; Zips, D.; Thorwarth, D. Comparison of [18F]-FMISO, [18F]-FAZA and [18F]-HX4 for PET imaging of hypoxia - a simulation study. *Acta Oncol.* **2015**, *54*, 1370–1377. [CrossRef] [PubMed]
53. Peeters, S.G.J.A.; Zegers, C.M.L.; Lieuwes, N.G.; van Elmpt, W.; Eriksson, J.; van Dongen, G.A.M.S.; Dubois, L.; Lambin, P. A Comparative Study of the Hypoxia PET Tracers [18F]HX4, [18F]FAZA, and [18F]FMISO in a Preclinical Tumor Model. *Int. J. Radiat. Oncol. Biol. Phys.* **2015**, *91*, 351–359. [CrossRef] [PubMed]
54. Watanabe, S.; Shiga, T.; Hirata, K.; Magota, K.; Okamoto, S.; Toyonaga, T.; Higashikawa, K.; Yasui, H.; Kobayashi, J.; Nishijima, K.-I.; et al. Biodistribution and radiation dosimetry of the novel hypoxia PET probe [18F]DiFA and comparison with [18F]FMISO. *EJNMMI Res.* **2019**. [CrossRef] [PubMed]
55. Dunphy, M.P.S.; Lewis, J.S. Radiopharmaceuticals in preclinical and clinical development for monitoring of therapy with PET. *J. Nucl. Med.* **2009**, *50*, 106S–121S. [CrossRef] [PubMed]
56. Floberg, J.M.; Wang, L.; Bandara, N.; Rashmi, R.; Mpoy, C.; Garbow, J.R.; Rogers, B.E.; Patti, G.J.; Schwarz, J.K. Altering cellular reducing potential changes 64Cu-ATSM signal with or without hypoxia. *J. Nucl. Med.* **2019**. [CrossRef] [PubMed]
57. Toriihara, A.; Yoneyama, T.; Kitazume, Y.; Tateishi, U.; Ohtake, M.; Tateishi, K.; Kawahara, N.; Hino-Shishikura, A.; Inoue, T. Prognostic implications of (62)Cu-diacetyl-bis (N(4)-methylthiosemicarbazone) PET/CT in patients with glioma. *Ann. Nucl Med.* **2018**, *32*, 264–271. [CrossRef]
58. Xie, D.; Kim, S.; Kohli, V.; Banerjee, A.; Yu, M.; Enriquez, J.S.; Luci, J.J.; Que, E.L. Hypoxia-Responsive 19F MRI Probes with Improved Redox Properties and Biocompatibility. *Inorg. Chem.* **2017**, *56*, 6429–6437. [CrossRef]
59. Xie, D.; King, T.L.; Banerjee, A.; Kohli, V.; Que, E.L. Exploiting Copper Redox for 19F Magnetic Resonance-Based Detection of Cellular Hypoxia. *J. Am. Chem. Soc.* **2016**, *138*, 2937–2940. [CrossRef]
60. Challapalli, A.; Carroll, L.; Aboagye, E.O. Molecular mechanisms of hypoxia in cancer. *Clin. Transl. Imaging* **2017**, *5*, 225–253. [CrossRef]
61. Burger, P.C.; Vogel, F.S.; Green, S.B.; Strike, T.A. Glioblastoma multiforme and anaplastic astrocytoma. Pathologic criteria and prognostic implications. *Cancer* **1985**, *56*, 1106–1111. [CrossRef]
62. King, J.G., Jr.; Khalili, K. Inhibition of human brain tumor cell growth by the anti-inflammatory drug, flurbiprofen. *Oncogene* **2001**, *20*, 6864–6870. [CrossRef] [PubMed]
63. Roncaroli, F.; Su, Z.; Herholz, K.; Gerhard, A.; Turkheimer, F.E. TSPO expression in brain tumours: Is TSPO a target for brain tumour imaging? *Clin. Transl. Imaging* **2016**, *4*, 145–156. [CrossRef] [PubMed]
64. Pappata, S.; Cornu, P.; Samson, Y.; Prenant, C.; Benavides, J.; Scatton, B.; Crouzel, C.; Hauw, J.; Syrota, A. PET study of carbon-11-PK 11195 binding to peripheral type benzodiazepine sites in glioblastoma: A case report. *J. Nucl. Med.* **1991**, *32*, 1608–1610. [PubMed]
65. Miettinen, H.; Kononen, J.; Haapasalo, H.; Helén, P.; Sallinen, P.; Harjuntausta, T.; Helin, H.; Alho, H. Expression of peripheral-type benzodiazepine receptor and diazepam binding inhibitor in human astrocytomas: Relationship to cell proliferation. *Cancer Res.* **1995**, *55*, 2691–2695. [PubMed]
66. Vlodavsky, E.; Soustiel, J.F. Immunohistochemical expression of peripheral benzodiazepine receptors in human astrocytomas and its correlation with grade of malignancy, proliferation, apoptosis and survival. *J. Neuro-Oncol.* **2007**, *81*, 1–7. [CrossRef]
67. Su, Z.; Roncaroli, F.; Durrenberger, P.F.; Coope, D.J.; Karabatsou, K.; Hinz, R.; Thompson, G.; Turkheimer, F.E.; Janczar, K.; Du Plessis, D. The 18-kDa mitochondrial translocator protein in human gliomas: An 11C-(R) PK11195 PET imaging and neuropathology study. *J. Nucl. Med.* **2015**, *56*, 512–517. [CrossRef]
68. Vilner, B.J.; John, C.S.; Bowen, W.D. Sigma-1 and sigma-2 receptors are expressed in a wide variety of human and rodent tumor cell lines. *Cancer Res.* **1995**, *55*, 408–413.
69. Megalizzi, V.; Decaestecker, C.; Debeir, O.; Spiegl-Kreinecker, S.; Berger, W.; Lefranc, F.; Kast, R.E.; Kiss, R. Screening of anti-glioma effects induced by sigma-1 receptor ligands: Potential new use for old anti-psychiatric medicines. *Eur. J. Cancer* **2009**, *45*, 2893–2905. [CrossRef]

70. Jia, H.; Zhang, Y.; Huang, Y. Imaging sigma receptors in the brain: New opportunities for diagnosis of Alzheimer's disease and therapeutic development. *Neurosci. Lett.* **2019**, *691*, 3–10. [CrossRef]
71. Toyohara, J.; Kobayashi, T.; Mita, S.; Ishiwata, K. Application of [11C]SA4503 to selection of novel σ1 selective agonists. *Nucl. Med. Biol.* **2012**, *39*, 1117–1121. [CrossRef]
72. Matsuno, K.; Nakazawa, M.; Okamoto, K.; Kawashima, Y.; Mita, S. Binding properties of SA4503, a novel and selective σ1 receptor agonist. *Eur. J. Pharmacol.* **1996**, *306*, 271–279. [CrossRef]
73. Lee Collier, T.; O'Brien, J.C.; Waterhouse, R.N. Synthesis of [18F]-1-(3-Fluoropropyl)-4-(4-cyanophenoxymethyl)-piperidine: A potential sigma-1 receptor radioligand for PET. *J. Label. Compd. Radiopharm.* **1996**, *38*, 785–794. [CrossRef]
74. Waterhouse, R.N.; Chang, R.C.; Atuehene, N.; Collier, T.L. In vitro and in vivo binding of neuroactive steroids to the sigma-1 receptor as measured with the positron emission tomography radioligand [18F]FPS. *SYNAPSE* **2007**, *61*, 540–546. [CrossRef] [PubMed]
75. James, M.L.; Shen, B.; Zavaleta, C.L.; Nielsen, C.H.; Mesangeau, C.; Vuppala, P.K.; Chan, C.; Avery, B.A.; Fishback, J.A.; Matsumoto, R.R.; et al. New Positron Emission Tomography (PET) Radioligand for Imaging σ-1 Receptors in Living Subjects. *J. Med. Chem.* **2012**, *55*, 8272–8282. [CrossRef] [PubMed]
76. Kranz, M.; Bergmann, R.; Kniess, T.; Belter, B.; Neuber, C.; Cai, Z.; Deng, G.; Fischer, S.; Zhou, J.; Huang, Y.; et al. Bridging from brain to tumor imaging: (S)-(-)- and (R)-(+)-[18F]fluspidine for investigation of sigma-1 receptors in tumor-bearing mice. *Molecules* **2018**, *23*, 702. [CrossRef]
77. Kranz, M.; Sattler, B.; Wüst, N.; Deuther-Conrad, W.; Patt, M.; Meyer, P.; Fischer, S.; Donat, C.; Wünsch, B.; Hesse, S. Evaluation of the enantiomer specific biokinetics and radiation doses of [18F] fluspidine—A new tracer in clinical translation for imaging of σ1 receptors. *Molecules* **2016**, *21*, 1164. [CrossRef]
78. Wheeler, K.T.; Wang, L.M.; Wallen, C.A.; Childers, S.R.; Cline, J.M.; Keng, P.C.; Mach, R.H. Sigma-2 receptors as a biomarker of proliferation in solid tumours. *Br. J. Cancer* **2000**, *82*, 1223–1232. [CrossRef]
79. Zeng, C.; McDonald, E.S.; Mach, R.H. Molecular probes for imaging the sigma-2 receptor: In vitro and in vivo imaging studies. *Handb. Exp. Pharmacol.* **2017**, *244*, 309–330. [CrossRef]
80. Mach, R.H.; Zeng, C.; Hawkins, W.G. The σ2 Receptor: A Novel Protein for the Imaging and Treatment of Cancer. *J. Med. Chem.* **2013**, *56*, 7137–7160. [CrossRef]
81. John, C.S.; Wilner, B.J.; Gulden, M.E.; Efange, S.M.N.; Langason, R.B.; Moody, T.W.; Bowen, W.D. Synthesis and pharmacological characterization of 4-[125I]-N-(N-benzylpiperidin-4-yl)-4-iodobenzamide: A high affinity or receptor ligand for potential imaging of breast cancer. *Cancer Res.* **1995**, *55*, 3022–3027.
82. John, C.S.; Bowen, W.D.; Fisher, S.J.; Lim, B.B.; Geyer, B.C.; Vilner, B.J.; Wahl, R.L. Synthesis, in vitro pharmacologic characterization, and preclinical evaluation of N-[2-(1'-piperidinyl)ethyl]-3-[125I]iodo-4-methoxybenzamide (P[125I]MBA) for imaging breast cancer. *Nucl. Med. Biol.* **1999**, *26*, 377–382. [CrossRef]
83. Kashiwagi, H.; McDunn, J.E.; Simon, P.O., Jr.; Goedegebuure, P.S.; Xu, J.; Jones, L.; Chang, K.; Johnston, F.; Trinkaus, K.; Hotchkiss, R.S.; et al. Selective sigma-2 ligands preferentially bind to pancreatic adenocarcinomas: Applications in diagnostic imaging and therapy. *Mol. Cancer* **2007**, *6*, 48. [CrossRef] [PubMed]
84. Dehdashti, F.; Laforest, R.; Gao, F.; Shoghi, K.I.; Aft, R.L.; Nussenbaum, B.; Kreisel, F.H.; Bartlett, N.L.; Cashen, A.; Wagner-Johnston, N.; et al. Assessment of cellular proliferation in tumors by PET using 18F-ISO-1. *J. Nucl. Med.* **2013**, *54*, 350–357. [CrossRef] [PubMed]
85. Shoghi, K.I.; Xu, J.; Su, Y.; He, J.; Rowland, D.; Yan, Y.; Garbow, J.R.; Tu, Z.; Jones, L.A.; Higashikubo, R.; et al. Quantitative receptor-based imaging of tumor proliferation with the sigma-2 ligand [18F]ISO-1. *PLoS ONE* **2013**, *8*, e74188. [CrossRef]
86. Elmi, A.; Makvandi, M.; Weng, C.-C.; Hou, C.; Mach, R.H.; Mankoff, D.A.; Clark, A.S. Cell-Proliferation Imaging for Monitoring Response to CDK4/6 Inhibition Combined with Endocrine-Therapy in Breast Cancer: Comparison of [(18)F]FLT and [(18)F]ISO-1 PET/CT. *Clin. Cancer Res.* **2019**, *25*, 3063–3073. [CrossRef]
87. Nguyen, V.H.; Pham, T.; Fookes, C.; Berghofer, P.; Greguric, I.; Arthur, A.; Mattner, F.; Rahardjo, G.; Davis, E.; Howell, N.; et al. Synthesis and biological characterization of 18F-SIG343 and 18F-SIG353, novel and high selectivity σ2 radiotracers, for tumor imaging properties. *EJNMMI Res.* **2013**, *3*, 80. [CrossRef]
88. Abate, C.; Selivanova, S.V.; Muller, A.; Kramer, S.D.; Schibli, R.; Marottoli, R.; Perrone, R.; Berardi, F.; Niso, M.; Ametamey, S.M. Development of 3,4-dihydroisoquinolin-1(2H)-one derivatives for the Positron Emission Tomography (PET) imaging of σ2 receptors. *Eur. J. Med. Chem.* **2013**, *69*, 920–930. [CrossRef]

89. Selivanova, S.V.; Toscano, A.; Abate, C.; Berardi, F.; Muller, A.; Kramer, S.D.; Schibli, R.; Ametamey, S.M. Synthesis and pharmacological evaluation of 11C-labeled piperazine derivative as a PET probe for sigma-2 receptor imaging. *Nucl. Med. Biol.* **2015**, *42*, 399–405. [CrossRef]
90. Wang, L.; Ye, J.; He, Y.; Deuther-Conrad, W.; Zhang, J.; Zhang, X.; Cui, M.; Steinbach, J.; Huang, Y.; Brust, P.; et al. 18F-Labeled indole-based analogs as highly selective radioligands for imaging sigma-2 receptors in the brain. *Bioorg. Med. Chem.* **2017**, *25*, 3792–3802. [CrossRef]
91. Lesniak, W.G.; Chatterjee, S.; Gabrielson, M.; Lisok, A.; Wharram, B.; Pomper, M.G.; Nimmagadda, S. PD-L1 Detection in Tumors Using [(64)Cu]Atezolizumab with PET. *Bioconjug Chem.* **2016**, *27*, 2103–2110. [CrossRef]
92. Chatterjee, S.; Lesniak, W.G.; Lisok, A.; Wharram, B.; Kumar, D.; Gabrielson, M.; Miller, M.S.; Sikorska, E.; Pomper, M.G.; Gabelli, S.B.; et al. Rapid PD-L1 detection in tumors with PET using a highly specific peptide. *Biochem. Biophys. Res. Commun.* **2017**, *483*, 258–263. [CrossRef]
93. Mayer, A.T.; Gambhir, S.S.; Mayer, A.T.; Natarajan, A.; Gambhir, S.S.; Gordon, S.R.; Maute, R.L.; McCracken, M.N.; Weissman, I.L.; Gordon, S.R.; et al. Practical Immuno-PET Radiotracer Design Considerations for Human Immune Checkpoint Imaging. *J. Nucl. Med.* **2017**, *58*, 538–546.
94. Gonzalez, T.D.E.; Meng, X.; McQuade, P.; Rubins, D.; Klimas, M.; Zeng, Z.; Connolly, B.M.; Miller, P.J.; O'Malley, S.S.; Lin, S.-A.; et al. In Vivo Imaging of the Programmed Death Ligand 1 by (18)F PET. *J. Nucl. Med.* **2017**, *58*, 1852–1857. [CrossRef]
95. Miller, M.M.; Mapelli, C.; Allen, M.P.; Bowsher, M.S.; Boy, K.M.; Gillis, E.P.; Langley, D.R.; Mull, E.; Poirier, M.A.; Sanghvi, N. Macrocyclic inhibitors of the PD-1/PD-L1 and CD80 (B7-1)/PD-L1 protein/protein interactions. US9308236B2, 7 July 2016.
96. Zak, K.M.; Grudnik, P.; Guzik, K.; Zieba, B.J.; Musielak, B.; Dömling, A.; Dubin, G.; Holak, T.A. Structural basis for small molecule targeting of the programmed death ligand 1 (PD-L1). *Oncotarget* **2016**, *7*, 30323–30335. [CrossRef]
97. Konstantinidou, M.; Zarganes-Tzitzikas, T.; Domling, A.; Magiera-Mularz, K.; Holak, T.A.; Holak, T.A. Immune Checkpoint PD-1/PD-L1: Is There Life Beyond Antibodies? *Angew. Chem. Int. Ed. Engl.* **2018**, *57*, 4840–4848. [CrossRef]
98. Chen, T.; Li, Q.; Liu, Z.; Chen, Y.; Feng, F.; Sun, H. Peptide-based and small synthetic molecule inhibitors on PD-1/PD-L1 pathway: A new choice for immunotherapy? *Eur. J. Med. Chem.* **2019**, *161*, 378–398. [CrossRef]
99. Lee, H.T.; Lee, S.H.; Heo, Y.-S. Molecular interactions of antibody drugs targeting PD-1, PD-L1, and CTLA-4 in immuno-oncology. *Molecules* **2019**, *24*, 1190. [CrossRef]
100. Kim, M.Y.; Zhang, T.; Kraus, W.L. Poly(ADP-ribosyl)ation by PARP-1: 'PAR-laying' NAD+ into a nuclear signal. *Genes Dev.* **2005**, *19*, 1951–1967. [CrossRef]
101. Ray Chaudhuri, A.; Nussenzweig, A. The multifaceted roles of PARP1 in DNA repair and chromatin remodelling. *Nat. Rev. Mol. Cell Biol.* **2017**, *18*, 610–621. [CrossRef]
102. Jain, P.G.; Patel, B.D. Medicinal chemistry approaches of poly ADP-Ribose polymerase 1 (PARP1) inhibitors as anticancer agents—A recent update. *Eur. J. Med. Chem.* **2019**, *165*, 198–215. [CrossRef]
103. Reiner, T.; Lacy, J.; Keliher, E.J.; Yang, K.S.; Ullal, A.; Kohler, R.H.; Vinegoni, C.; Weissleder, R. Imaging therapeutic PARP inhibition in vivo through bioorthogonally developed companion imaging agents. *Neoplasia* **2012**, *14*, 169–177. [CrossRef]
104. Irwin, C.P.; Portorreal, Y.; Brand, C.; Zhang, Y.; Desai, P.; Salinas, B.; Weber, W.A.; Reiner, T. PARPi-FL-a fluorescent PARP1 inhibitor for glioblastoma imaging. *Neoplasia* **2014**, *16*, 432–440. [CrossRef] [PubMed]
105. Carney, B.; Carlucci, G.; Salinas, B.; Di Gialleonardo, V.; Kossatz, S.; Vansteene, A.; Longo, V.A.; Bolaender, A.; Chiosis, G.; Keshari, K.R. Non-invasive PET imaging of PARP1 expression in glioblastoma models. *Mol. Imaging Biol.* **2016**, *18*, 386–392. [CrossRef] [PubMed]
106. Kossatz, S.; Carney, B.; Schweitzer, M.; Carlucci, G.; Miloushev, V.Z.; Maachani, U.B.; Rajappa, P.; Keshari, K.R.; Pisapia, D.; Weber, W.A.; et al. Biomarker-Based PET Imaging of Diffuse Intrinsic Pontine Glioma in Mouse Models. *Cancer Res.* **2017**, *77*, 2112. [CrossRef] [PubMed]
107. Wilson, T.C.; Xavier, M.-A.; Knight, J.; Verhoog, S.; Torres, J.B.; Mosley, M.; Hopkins, S.L.; Wallington, S.; Allen, P.D.; Kersemans, V.; et al. PET Imaging of PARP Expression Using 18F-Olaparib. *J. Nucl. Med.* **2019**, *60*, 504–510. [CrossRef]
108. Zhou, D.; Chu, W.; Xu, J.; Jones, L.A.; Peng, X.; Li, S.; Chen, D.L.; Mach, R.H. Synthesis, [18F] radiolabeling, and evaluation of poly (ADP-ribose) polymerase-1 (PARP-1) inhibitors for in vivo imaging of PARP-1 using positron emission tomography. *Bioorg. Med. Chem.* **2014**, *22*, 1700–1707. [CrossRef]

109. Michel, L.S.; Dyroff, S.; Brooks, F.J.; Spayd, K.J.; Lim, S.; Engle, J.T.; Phillips, S.; Tan, B.; Wang-Gillam, A.; Bognar, C.; et al. PET of Poly (ADP-Ribose) Polymerase Activity in Cancer: Preclinical Assessment and First In-Human Studies. *Radiology* **2016**, *282*, 453–463. [CrossRef]
110. Yan, H.; Parsons, D.W.; Jin, G.; McLendon, R.; Rasheed, B.A.; Yuan, W.; Kos, I.; Batinic-Haberle, I.; Jones, S.; Riggins, G.J.; et al. IDH1 and IDH2 Mutations in Gliomas. *New Engl. J. Med.* **2009**, *360*, 765–773. [CrossRef]
111. Sanson, M.; Marie, Y.; Paris, S.; Idbaih, A.; Laffaire, J.; Ducray, F.; El Hallani, S.; Boisselier, B.; Mokhtari, K.; Hoang-Xuan, K. Isocitrate dehydrogenase 1 codon 132 mutation is an important prognostic biomarker in gliomas. *J. Clin. Oncol.* **2009**, *27*, 4150–4154. [CrossRef]
112. Ye, D.; Ma, S.; Xiong, Y.; Guan, K.-L. R-2-hydroxyglutarate as the key effector of IDH mutations promoting oncogenesis. *Cancer Cell* **2013**, *23*, 274–276. [CrossRef]
113. Kaminska, B.; Czapski, B.; Guzik, R.; Krol, S.K.; Gielniewski, B. Consequences of IDH1/2 mutations in gliomas and an assessment of inhibitors targeting mutated IDH proteins. *Molecules* **2019**, *24*, 968. [CrossRef]
114. Popovici-Muller, J.; Lemieux, R.M.; Artin, E.; Saunders, J.O.; Salituro, F.G.; Travins, J.; Cianchetta, G.; Cai, Z.; Zhou, D.; Cui, D.; et al. Discovery of AG-120 (Ivosidenib): A First-in-Class Mutant IDH1 Inhibitor for the Treatment of IDH1 Mutant Cancers. *ACS Med. Chem. Lett.* **2018**, *9*, 300–305. [CrossRef] [PubMed]
115. Cho, Y.S.; Levell, J.R.; Liu, G.; Caferro, T.; Sutton, J.; Shafer, C.M.; Costales, A.; Manning, J.R.; Zhao, Q.; Sendzik, M.; et al. Discovery and Evaluation of Clinical Candidate IDH305, a Brain Penetrant Mutant IDH1 Inhibitor. *ACS Med. Chem. Lett.* **2017**, *8*, 1116–1121. [CrossRef] [PubMed]
116. DiNardo, C.D.; Schimmer, A.D.; Yee, K.W.L.; Hochhaus, A.; Kraemer, A.; Carvajal, R.D.; Janku, F.; Bedard, P.; Carpio, C.; Wick, A.; et al. A Phase I Study of IDH305 in Patients with Advanced Malignancies Including Relapsed/Refractory AML and MDS That Harbor $IDH1^{R132}$ Mutations. *Blood* **2016**, *128*, 1073. [CrossRef]
117. Chi, A.S. Trial of IDH305 in IDH1 Mutant Grade II or III Glioma. Available online: https://clinicaltrials.gov/ct2/show/NCT02977689 (accessed on 30 September 2019).
118. Cicone, F.; Carideo, L.; Scaringi, C.; Arcella, A.; Giangaspero, F.; Scopinaro, F.; Minniti, G. 18F-DOPA uptake does not correlate with IDH mutation status and 1p/19q co-deletion in glioma. *Ann. Nucl. Med.* **2019**, *33*, 295–302. [CrossRef]
119. Blanc-Durand, P.; Van Der Gucht, A.; Verger, A.; Langen, K.-J.; Dunet, V.; Bloch, J.; Brouland, J.-P.; Nicod-Lalonde, M.; Schaefer, N.; Prior, J.O. Voxel-based 18F-FET PET segmentation and automatic clustering of tumor voxels: A significant association with IDH1 mutation status and survival in patients with gliomas. *PLoS ONE* **2018**, *13*, e0199379. [CrossRef]
120. Unterrrainer, M.; Winkelmann, I.; Suchorska, B.; Giese, A.; Wenter, V.; Kreth, F.W.; Herms, J.; Bartenstein, P.; Tonn, J.C.; Albert, N.L. Biological tumour volumes of gliomas in early and standard 20-40 min 18F-FET PET images differ according to IDH mutation status. *Eur. J. Nucl. Med. Mol. Imaging* **2018**, *45*, 1242–1249. [CrossRef]
121. Verger, A.; Stoffels, G.; Bauer, E.K.; Lohmann, P.; Blau, T.; Fink, G.R.; Neumaier, B.; Shah, N.J.; Langen, K.-J.; Galldiks, N. Static and dynamic 18F-FET PET for the characterization of gliomas defined by IDH and 1p/19q status. *Eur. J. Nucl. Med. Mol. Imaging* **2018**, *45*, 443–451. [CrossRef]
122. Chitneni, S.K.; Reitman, Z.J.; Spicehandler, R.; Gooden, D.M.; Yan, H.; Zalutsky, M.R. Synthesis and evaluation of radiolabeled AGI-5198 analogues as candidate radiotracers for imaging mutant IDH1 expression in tumors. *Bioorg. Med. Chem. Lett.* **2018**, *28*, 694–699. [CrossRef]
123. Chitneni, S.K.; Yan, H.; Zalutsky, M.R. Synthesis and Evaluation of a 18F-Labeled Triazinediamine Analogue for Imaging Mutant IDH1 Expression in Gliomas by PET. *ACS Med. Chem. Lett.* **2018**, *9*, 606–611. [CrossRef]
124. Preibisch, C.; Shi, K.; Kluge, A.; Lukas, M.; Wiestler, B.; Goettler, J.; Gempt, J.; Ringel, F.; Al Jaberi, M.; Schlegel, J.; et al. Characterizing hypoxia in human glioma: A simultaneous multimodal MRI and PET study. *NMR Biomed.* **2017**, *30*, e3775. [CrossRef]

125. Thiele, F.; Ehmer, J.; Piroth, M.D.; Eble, M.J.; Coenen, H.H.; Kaiser, H.-J.; Schaefer, W.M.; Buell, U.; Boy, C. The quantification of dynamic FET PET imaging and correlation with the clinical outcome in patients with glioblastoma. *Phys. Med. Biol.* **2009**, *54*, 5525–5539. [CrossRef] [PubMed]
126. Muzi, M.; Peterson, L.M.; O'Sullivan, J.N.; Fink, J.R.; Rajendran, J.G.; McLaughlin, L.J.; Muzi, J.P.; Mankoff, D.A.; Krohn, K.A. 18F-fluoromisonidazole quantification of hypoxia in human cancer patients using image-derived blood surrogate tissue reference regions. *J. Nucl. Med.* **2015**, *56*, 1223–1228. [CrossRef] [PubMed]

© 2020 by the authors. Licensee MDPI, Basel, Switzerland. This article is an open access article distributed under the terms and conditions of the Creative Commons Attribution (CC BY) license (http://creativecommons.org/licenses/by/4.0/).

MDPI\
St. Alban-Anlage 66\
4052 Basel\
Switzerland\
Tel. +41 61 683 77 34\
Fax +41 61 302 89 18\
www.mdpi.com

Molecules Editorial Office\
E-mail: molecules@mdpi.com\
www.mdpi.com/journal/molecules

www.ingramcontent.com/pod-product-compliance
Lightning Source LLC
LaVergne TN
LVHW070048120526
838202LV00101B/1594